Sensorless Control of Permanent Magnet Synchronous Machine Drives

IEEE Press
445 Hoes Lane
Piscataway, NJ 08854

Sensorless Control of Permanent Magnet Synchronous Machine Drives

Zi Qiang Zhu

and

Xi Meng Wu

University of Sheffield, UK

IEEE Press Series on Control Systems Theory and Applications

Maria Domenica Di Benedetto, Series Editor

IEEE PRESS

WILEY

Published by John Wiley & Sons, Inc., Hoboken, New Jersey.
Published simultaneously in Canada.

Library of Congress Cataloging-in-Publication Data Applied for:

Hardback ISBN: 9781394194353

Cover Design: Wiley
Cover Image: © Bosca78/Getty Images

Set in 9.5/12.5pt STIXTwoText by Straive, Pondicherry, India

Contents

About the Authors

Professor Zi Qiang Zhu received the BEng and MSc degrees from Zhejiang University, Hangzhou, China, in 1982 and 1984, respectively, and the Ph.D. degree from The University of Sheffield, Sheffield, U.K., in 1991, all in electrical engineering.

After working as Lecturer/Assistant Lecturer at Zhejiang University from 1984 to 1988, he has been with The University of Sheffield since 1988, initially as Visiting Research Fellow sponsored by the British Council (1988–1989), then Research Associate working with Philips (1989-1992), Senior Research Scientist/Officer––an established university post (1992–2000), Professor (2000–present), Royal Academy of Engineering/Siemens Research Chair (2014–present), and Head of the Electrical Machines and Drives Research Group (2008–present). As the Founding Director, he has helped several industries in establishing their research centers, most notably, Siemens Gamesa Renewable Energy Research Center at Sheffield (2009–present) and Midea Electrical Machines and Control Research Centers at Shanghai and Sheffield (2010–present).

His research interests include design and control of permanent magnet brushless machines and drives for applications ranging from electrified transportation (electric vehicles, fast trains, and more electric aircraft) to domestic appliances to wind power generation. He has published more than 1400 papers, including more than 550 for IEEE Transactions and IET Proceedings.

He is a Fellow of the Royal Academy of Engineering, UK; Institute of Electrical and Electronics Engineers (IEEE), USA; Institution of Engineering and Technology (IET), UK; and Chinese Society for Electrical Engineering (CSEE), China. He is the recipient of the 2019 IEEE Industry Applications Society Outstanding Achievement Award and the 2021 IEEE Nikola Tesla Award.

Dr Xi Meng Wu received the BEng degree in electrical engineering and automation from Hefei University of Technology, Hefei, China, in 2011, and MSc and PhD degrees in electrical engineering from The University of Sheffield,

Sheffield, UK in 2016 and 2020, respectively. Since 2020, he has been a Postgraduate Research Associate at the University of Sheffield associated with the Sheffield Siemens Gamesa Renewable Energy Research Centre. His research interests include control of permanent magnet synchronous machines.

Preface

Permanent magnet (PM) synchronous machine (PMSM) drives, including PM brushless ac and dc drives, exhibit many advantages such as high efficiency and high torque density. A high-performance PM brushless ac or dc drive needs accurate rotor position information. This is usually obtained by using a hardware rotor position sensor, such as a resolver, encoder, or Hall sensor. However, these sensors increase drive size and cost and reduce reliability, particularly in harsh environments. Therefore, it is desirable to replace hardware rotor position sensors with software-based rotor position sensorless techniques.

This book aims to comprehensively describe sensorless control techniques of PMSM drives. We have strived to highlight the global research achievements and also many new techniques developed at the University of Sheffield. The basic principles and state-of-the-art rotor position sensorless control techniques are explained, together with their challenges and practical solutions. The scope is very broad, and readers may find the summary diagram in section 1.7 useful.

Thirty years ago, sensorless control of PMSM drives had limited application and was mostly used for driving ventilation fans with brushless dc drives. However, over the last 10 years, the field has rapidly expanded. Today, numerous commercial products are using sensorless control techniques for a wide variety of applications, for example wind power generators, automotive compressors, water and oil cooling pumps, electric bicycles, drones, general purpose variable frequency drives, and household appliances (e.g. air-conditioning and refrigerator compressors and fans, washing machines, dishwasher pumps, heat circulating pumps, and vacuum cleaners), as well as fault-tolerance drives in electrified transportation and aerospace applications. Despite these successes, the use of sensorless control for applications that require high torque for rapid starting remains challenging. As the technology continues to evolve and improve, it is likely that sensorless control will find even broader applications, offering a reliable, cost-effective, and efficient solution for a wide range of industrial and commercial needs.

Beginning in 1998, I have been working with my PhD students on various topics of sensorless control of PMSM drives. I would like to take this opportunity to thank them for their contributions, including J. Ede, J. X. Shen, Y. F. Shi, Y. Liu, Y. Li, L. M. Gong, J. M. Liu, T. C. Lin, A. H. Almarhoon, P. L. Xu, H. L. Zhan, X. M. Wu, L. Yang, B. Shuang, and T. Y. Liu. In particular, Dr. Xi Meng Wu, currently a post-doctoral researcher at the University of Sheffield and the co-author of this book, has made significant contributions to sensorless control, especially the novel techniques for initial rotor position detection.

I would like to acknowledge the pioneering work on sensorless control of PMSM drives by Professor Robert D. Lorenz at University of Wisconsin-Madison, USA, Professor Seung-Ki Sul at Seoul National University, South Korea, and Professor Fernando Briz and Professor David Reigosa at University of Oviedo, Spain, as well as Professor Nobuyuki Matsui, Dr. Zhiqian Chen, Professor Shigeo Morimoto, Professor Ion Boldea, and Professor Kaushik Rajashekara, and others.

I dedicate my thanks to our sponsors, including the UK government, particularly the UK EPSRC Prosperity Partnership ("A New Partnership in Offshore Wind" under Grant No. EP/R004900/1), and our partners in industry such as Siemens Gamesa Renewable Energy Ltd. on wind power, Midea group on household appliances, Nissan and Toyota group on electric vehicles.

My co-author and I sincerely hope this book will be useful to industrial engineers and researchers, university professionals, post-doctoral researchers, and students alike.

Professor Zi Qiang Zhu
FREng, FIEEE, FIET, FCSEE
University of Sheffield, UK

List of Abbreviations

ac	Alternating current
AD	Analog to digital
ADC	Analog-to-digital converter
AF	Active flux
ANF	Adaptive notch filter
BDS	Boundary selection strategy
BLAC	Brushless ac
BLDC	Brushless dc
BPF	Band-pass filter
CCS-MPC	Continuous-control-set model predictive control
CM	Current model
CMV	Common mode voltage
CPU	Central processing unit
dc	Direct current
DEA	Differential evolution algorithm
DFT	Discrete Fourier transform
DQZ	Direct-quadrature-zero
DSP	Digital signal processor
DTC	Direct torque control
DTP-PMSM	Dual three-phase permanent magnet synchronous machine
EEMF	Extended electromotive force
EKF	Extended Kalman filter
EMF	Electromotive force
FCS-MPC	Finite-control-set model predictive control
FE	Finite element
FEA	Finite element analysis
FO	Flux-linkage observer
FOC	Field oriented control
HF	High frequency
HPF	High-pass filter

INFORM	Indirect flux detection by online reactance measurement
IPM	Interior PM
IPMSM	Interior permanent magnet synchronous machine
LDF	Lower-diode freewheeling
LMS	Least-mean-squares
LPF	Low-pass filter
LUT	Look-up table
MMF	Magneto-motive force
MPC	Model predictive control
MRAS	Model reference adaptive system
MTPA	Maximum torque per ampere
OW-PMSM	Open-winding permanent magnet synchronous machine
PI	Proportional integral
PLL	Phase-locked loop
PM	Permanent magnet
PMSM	Permanent magnet synchronous machine
PO	Position observer
PWM	Pulse width modulation
QSG	Quadrature signal generator
RSA	Reliable selection area
RVD	Resistance voltage divider
SMO	Sliding mode observer
SNR	Signal-to-noise ratio
SOA	Safe operating area
SOGI	Second order generalized integrator
SPM	Surface-mounted PM
SPMSM	Surface-mounted permanent magnet synchronous machine
SSOA	Sensorless safe operation area
STP-PMSM	Single three-phase permanent magnet synchronous machine
SVPWM	Space vector pulse width modulation
THD	Total harmonic distortion
UDF	Upper-diode freewheeling
VC	Vector control
VM	Voltage model
VSD	Vector space decomposition
VSI	Voltage source inverter
ZCD	Zero-crossing detection
ZCP	Zero-crossing point
ZSC	Zero sequence current
ZSV	Zero sequence voltage
ZVC	Zero vector current
ZVCD	Zero vector current derivative

List of Symbols

Symbol	Description	Unit
A_{ac}	Amplitude of injected ac current reference	A
$A_{d,2f}$, $A_{q,2f}$	Amplitudes of second order harmonic of d- and q-axis estimated back-EMFs	V
A_{mp}	Amplitude of third harmonic flux-linkage	Wb
Amp_i_{Ah}, Amp_i_{Bh}, Amp_i_{Ch}	Amplitudes of three-phase HF currents	A
B	Viscous damping factor	Ns/m
B_3	Amplitude of third harmonic component of excitation flux density	T
D	Duty ratio	
$E_A(s)$, $E_B(s)$, $E_C(s)$	Three-phase back-EMFs after Laplace transform	V
e_A, e_B, e_C	Three-phase back-EMFs	V
e_{A5}, e_{B5}, e_{C5}	Three-phase fifth order harmonic back-EMFs	V
e_{A7}, e_{B7}, e_{C7}	Three-phase seventh order harmonic back-EMFs	V
$E_{a,d}$, $E_{a,q}$	dq-axis back-EMFs of active flux model	V
e_c	Back-EMF of current model	V
$\hat{E}_d$, $\hat{E}_q$	Estimated d- and q-axis back-EMFs	V
$\hat{E}_{d0}$, $\hat{E}_{q0}$	dc components of d- and q-axis estimated back-EMFs	V
$\hat{E}_{d,2f}$, $\hat{E}_{q,2f}$	Second order harmonic of d- and q-axis estimated back-EMFs	V
E_{ex}	Extended back-EMF	V

(Continued)

(Continued)

Symbol	Description	Unit
$\hat{E}_{ex,d}$, $\hat{E}_{ex,q}$	Estimated d- and q-axis extended back-EMFs	V
E_{ex_imp}	Improved extended back-EMF	V
E_{ex1}	Equivalent extended back-EMF of DTP-PMSM	V
$e_{ex,\alpha}$, $e_{ex,\beta}$	α- and β-axis extended back-EMFs	V
e_{H}, e_{L}, e_{O}	High-level, low-level, and floating phase back-EMFs	V
E_m	Peak value of phase back-EMF	V
E_{m3}	Amplitude of third harmonic back-EMF	V
e_α, e_β	α- and β-axis back-EMFs	V
e_0	Zero sequence back-EMF	V
$\hat{e}_0$	Estimated zero sequence back-EMF	V
E_3	Amplitude of third harmonic back-EMF	V
e_{3_set1}, e_{3_set2}	Third harmonic back-EMFs in two sets	V
e_{3_set1}, e_{9_set1}, e_{15_set1}	Third, ninth, and fifteenth harmonic back-EMFs of first set of DTP-PMSM	V
e_{3_set2}, e_{9_set2}, e_{15_set2}	Third, ninth, and fifteenth harmonic back-EMFs of second set of DTP-PMSM	V
e_9	Ninth order harmonic back-EMF	V
e_{15}	Fifteenth order harmonic back-EMF	V
f_e	Electrical rotor frequency	Hz
f_h	Frequency of injected high-frequency voltage signal	Hz
$f(\Delta\theta)$	Position error signal	
I^*	Amplitude of extra injected current signal	A
$I_{\mathrm{A}}(s)$, $I_{\mathrm{B}}(s)$, $I_{\mathrm{C}}(s)$	Three-phase currents after Laplace transform	A
i_{A}, i_{B}, i_{C}	Three-phase stator currents	A
I_A^P, I_B^P, I_C^P	Three-phase primary current response peak values	A
I_A^S, I_B^S, I_C^S	Three-phase secondary current response peak values	A
$i_{\mathrm{ABC}h}$	Three-phase high-frequency current responses	A
I_{AD_error}	Disturbance current vector due to current measurement error	A
I_d, I_q	Amplitudes of dq-axis currents	A
i_d, i_q	d- and q-axis currents	A
$\hat{i}_d$, $\hat{i}_q$	Estimated d- and q-axis currents	A
i_{dc}	dc-link current	A

(Continued)

Symbol	Description	Unit
i_{dc}^*	dc-link current reference	A
$\hat{i}_{d,ac}$, $\hat{i}_{q,ac}$	Estimated d- and q-axis ac current components	A
$i_{d,CM}$, $i_{q,CM}$	d- and q-axis currents of current model	A
$\hat{i}_{d,dc}$, $\hat{i}_{q,dc}$	Estimated d- and q-axis dc current components	A
i_{df}, i_{qf}	Fundamental d- and q-axis currents	A
$\hat{I}_{dh}$, $\hat{I}_{qh}$	Amplitudes of estimated dq-axis high-frequency currents	A
i_{dh}, i_{qh}	d-and q-axis high-frequency currents	A
$\hat{i}_{dh}$, $\hat{i}_{qh}$	Estimated d- and q-axis high-frequency currents	A
i_{dh}^v, i_{qh}^v	Virtual d- and q-axis high-frequency currents	A
i_{dq}^*	References of d- and q-axis currents	A
i_{dq}^p	Predicted d- and q-axis currents	A
ΔI_{error}	Error between real and recorded currents	A
$i_{(Extra)}^*$	Extra injected current signal	A
i_{H}, i_{L}, i_{O}	High-level, low-level, and floating phase currents	A
I_m	Peak value of phase current	A
$I_{\max}$	Maximum current response peak value	A
I_{mean}	Average current response peak value	A
I_n	Amplitude of negative sequence current response	A
i_n	Negative sequence current response	A
$\hat{i}_n$	Estimated negative sequence current response	A
I_n^{SQ}	Amplitude of negative sequence current response of square-wave injection	A
$\hat{i}_{nd}$, $\hat{i}_{nq}$	Estimated d- and q-axis negative sequence current responses	A
I_{neg_α}, I_{neg_β}	α- and β-axis components of negative sequence HF current	A
I_p	Amplitude of positive sequence current response	A
i_p	Positive sequence current response	A
I_p^{SQ}	Amplitude of positive sequence current response of square-wave injection	A

(Continued)

(Continued)

Symbol	Description	Unit
I_{pos_α}, I_{pos_β}	α- and β-axis components of positive sequence HF current	A
I_{q_MAX}	Maximum q-axis current	A
I_{qu}	Quantum current of analog-to-digital converter	A
I_{real}	Real current	A
I_{record}	Recorded current	A
I_s	Amplitude of stator current	A
i_s	Stator currents	A
ΔI_{th}	Threshold current value	A
i_X, i_Y, i_Z	Phase currents of second winding set of DTP-PMSM	A
i_{Xf}	Fundamental current in arbitrary phase	A
i_{Xh}	High-frequency current in arbitrary phase	A
i_{z1z2}	Stator current in z_1z_2 subspaces	A
i_α, i_β	α- and β-axis currents	A
$\hat{i}_\alpha$, $\hat{i}_\beta$	α- and β-axis estimated currents	A
Δi_α, Δi_β	α- and β-axis current estimation errors	A
$I_{\alpha\beta h}$	Amplitudes of α-and β-axis high-frequency currents	A
$i_{\alpha\beta h}$	α- and β-axis high-frequency currents	A
$i^*_{\alpha\beta h}$	α- and β-axis high-frequency currents before compensating positive current	A
$i^{**}_{\alpha\beta h}$	α- and β-axis high-frequency currents after compensating for positive current	A
I_0	Amplitude of dc component of three-phase current responses	A
i_0	Zero sequence current	A
I_2	Amplitude of second order harmonic component of three-phase current responses	A
i_{2nd}	Secondary positive sequence harmonics in HF current response	A
J	Inertia	$kg.m^2$
k_c	Compensation factor for cross-coupling inductances	mH/A
K_i	Integration gain of PI controller	
$K_{\Delta L_q}$	Deviation factor of q-axis inductance	rad/A
$K_{\Delta L_{q1}}$	Deviation factor of q-axis self-inductance in Set 1	rad/A

(Continued)

Symbol	Description	Unit
$K_{\Delta M_{q21}}$	Deviation factor of q-axis mutual-inductance between two sets	rad/A
K_p	Proportional gain of PI controller	
k_{p3}, k_{d3}, k_{s3}	Coil pitch factor, distribution factor, and skew factor for third harmonic	
K_{R}	Equivalent gain of resistance voltage divider	
k_r	Compensation factor for cross-coupling error angle	rad/A
$K_{\Delta R_s}$, $K_{\Delta L_Q}$	Deviation factors of resistance and q-axis equivalent inductance	rad/A
k_{w3}	Winding factor for third harmonic	
K_{ω}	Slope of back-EMF envelope around ZCP	V
L	Phase self-inductance of BLDC	mH
ΔL	Asymmetric inductance	mH
L_{AA}, L_{BB}, L_{CC}	Three-phase self-inductances	mH
ΔL_{AB}, ΔL_{BC}, ΔL_{CA}	Three-phase inductance asymmetric errors of BLDC	mH
L_c	Second order harmonic amplitude of sine inductance term in self-inductance	mH
L_D, L_Q	Equivalent d- and q-axis inductances of DTP-PMSM	mH
L_d, L_q	d- and q-axis self-inductances	mH
$\tilde{L}_d$, $\tilde{L}_q$	Nominal values of d- and q-axis inductances	mH
ΔL_d, ΔL_q	Mismatch values of d- and q-axis inductances	mH
L_{dh}, L_{qh}	dq-axis incremental self-inductances	mH
L_{dq}, L_{qd}	Cross-coupling dq-axis inductances	mH
L_{dqh}, L_{qdh}	Cross-coupling dq-axis incremental inductances	mH
L_{d1}, L_{d2}, L_{q1}, L_{q2}	d- and q-axis self-inductances of two winding sets	mH
L_{eq}	Equivalent inductance	mH
L_{H}, L_{L}	High- and low-level inductances	mH
$L_{\Delta h}$	Amplitude of hth spatial inductance	mH
L_{lm}	Boundary inductance	mH
L_{ls}	Leakage self-inductance	mH
L_{MAX}, L_{MIN}	Maximum and minimum inductances	mH

(Continued)

(Continued)

Symbol	Description	Unit
L_n	Negative sequence inductance	mH
L_p	Positive sequence inductance	mH
L_{qds}	Approximated cross-coupling inductances	mH
L_{sa}	Average between *d*- and *q*-axis incremental inductances	mH
L_{sd}	Difference of *d*- and *q*-axis incremental inductances	mH
L_{sj}, L_{sk}	*j*th and *k*th order self-inductances	mH
L_{s0}	Average value of self-inductance	Wb
L_{s2}	Amplitude of second order harmonic component of self-inductance	Wb
L_{XX}, L_{YY}, L_{ZZ}	Three-phase self-inductances of first winding set in DTP-PMSM	mH
$L_{\alpha\alpha}$, $L_{\beta\beta}$	α- and β-axis self-inductances	mH
L_0	Zero sequence inductance	mH
M	Phase mutual-inductance of BLDC	mH
M_{AB}, M_{BA}, M_{AC}, M_{CA}, M_{BC}, M_{CB}	Three-phase mutual-inductances	mH
M_c	Second order harmonic amplitudes of sine inductance terms in mutual-inductances	mH
M_{d12}, M_{d21}, M_{q12}, M_{q21}	*d*- and *q*-axis mutual-inductances between two winding sets	mH
ΔM_{q21}, ΔM_{q12}	Deviation values of *q*-axis inductances between two sets	mH
M_{sj}, M_{sk}	*j*th and *k*th order mutual-inductances	mH
M_{s0}	Average value of mutual-inductance	Wb
M_{s2}	Second order harmonic of mutual-inductance	Wb
M_{XY}, M_{YX}, M_{YZ}, M_{ZY}, M_{ZX}, M_{XZ}	Three phase mutual-inductances of second winding set in DTP-PMSM	mH
$M_{\alpha\beta}$, $M_{\beta\alpha}$	α- and β-axis mutual-inductances	mH
N_s	Number of sample points	
$P(k)$	Covariance matrix of EKF	
P	Number of pole pairs	
p	Derivative operator	
$Q(k)$, $R(k)$	Covariances of process noise and measurement noise of EKF	
R	Phase resistance of BLDC	Ω
ΔR_A, ΔR_B, ΔR_C	Asymmetric resistances components	Ω

(Continued)

Symbol	Description	Unit
ΔR_{ave}	dc offset due to resistance asymmetry	Ω
R_d, R_q	d- and q-axis resistances	Ω
R_{dc}	dc-link resistance	Ω
R_{dq}	dq-axis mutual-resistance	Ω
R_{eq}	Equivalent resistance	Ω
R_N	Resistance of auxiliary resistor network	Ω
R_s	Phase resistance	Ω
$\tilde{R}_s$	Nominal value of phase resistance	Ω
$\overline{R}_s$	Balanced part of three-phase resistances	Ω
ΔR_s	Mismatch value of phase resistance	Ω
R_{Xh}	Equivalent HF resistance of inverter in arbitrary phase	Ω
R_1, R_2	Nominal values of low and high side resistances of resistance voltage divider	Ω
R_1', R_2'	Actual values of low and high side resistances of resistance voltage divider	Ω
$\boldsymbol{S}$	Sliding mode surface	
$\mathrm{S_A}$, $\mathrm{S_B}$, $\mathrm{S_C}$	Switching states of three legs of VSI	
t	Time	s
ΔT	Period of injected square-wave voltage signal	s
Δt	Time step	s
T_c	Time constant of LPF	s
t_d	Time interval of half cycle between two zero-crossing points	s
t_{dd}	Turn-off delay of power device	s
t_{dt}	Deadtime	s
t_{du}	Turn-on delay of power device	s
T_{inj}	Period of extra injected current signal	s
T_{i1}, T_{i2}	Periods of first and second injected HF voltage signals	s
T_L	Load torque	Nm
T_{m_BLAC}	Electromagnetic torque of a BLAC machine	Nm
T_{m_BLDC}	Electromagnetic torque of a BLDC machine	Nm
T_{opt}	Optimal duration of voltage pulse	s
T_P	Duration of voltage pulse	s
t_{period}	Fundamental period	s

(Continued)

(Continued)

Symbol	Description	Unit
T_{P_MAX}, T_{P_MIN}	Maximum and minimum durations of voltage pulse	s
t_r	Remainder of division of time by injection period	s
T_s	Sampling period	s
$t_{\varphi}[n]$	nth time delay for commutation instant	s
$t_{\theta}[n]$	nth ZCP time interval	s
T_0	Duration of zero voltage vector	s
T_1	Duration of voltage vector 1	s
T_2	Duration of voltage vector 2	s
t_{23}, t_{34}, t_{45}, t_{25}, t_{52}, t_{56}, t_{61}, t_{12}	Period between sectors	s
u_{VA}, u_{VB}, u_{VC}	Three-phase vertical error correction common-mode bias	V
u_1, u_2, u_3, u_4, u_5, u_6	Zero-crossing thresholds	V
ΔV	Average terminal voltage error	V
v_{A}, v_{B}, v_{C}	Three-phase stator voltages	V
Δv_{AB}, Δv_{BC}, Δv_{CA}	Three-phase horizontal voltage shifts	V
V_{AD}	Maximum sampling voltage	V
$V_{\mathrm{AG}}(s)$, $V_{\mathrm{BG}}(s)$, $V_{\mathrm{CG}}(s)$	Three-phase terminal voltages after Laplace transform	V
v_{AG}, v_{BG}, v_{CG}	Voltage between phase terminal and ground	V
$v_{\mathrm{A}h}$, $v_{\mathrm{B}h}$, $v_{\mathrm{C}h}$	Injected three-phase HF voltages	V
v_{AN}, v_{BN}, v_{BN}	Phase voltages of a PM machine in ABC reference frame	V
v'_{BG}	Acquired phase B terminal voltage	V
V_c	Amplitude of equivalent voltage source	V
v_d, v_q	d- and q-axis voltages	V
$\hat{v}_d$, $\hat{v}_q$	Estimated d- and q-axis voltages	V
V_{dc}	dc-link voltage	V
ΔV_{dc}	dc-link voltage variation	V
$\hat{v}_{d\_ff}$	Estimated d-axis feed-forward voltage	V
v_{dh}, v_{qh}	d- and q-axis high-frequency voltages	V
$\hat{v}_{dh}$, $\hat{v}_{qh}$	Estimated d- and q-axis high-frequency voltages	V
v^v_{dh}, v^v_{qh}	Virtual d- and q-axis high-frequency voltages	V
v_{dh1}, v_{dh2}	First and second injected HF voltage signals	V
$\hat{v}_{d,VM}$, $\hat{v}_{q,VM}$	Voltages of voltage model	V

(Continued)

Symbol	Description	Unit
Δv_f	Fundamental disturbance voltage	V
$\mathbf{v}_f^*$	Fundamental voltage reference solution	V
V_h	Amplitude of injected high-frequency voltage	V
$\Delta V_{\mathrm{H}}(s)$	Voltage shift caused by asymmetric parameters after Laplace transform	V
Δv_{H}	Horizontal voltage shift	V
Δv_h	High-frequency disturbance voltage	V
$\mathbf{v}_{hf}^*$	High-frequency voltage signal reference	V
v_{HG}, v_{LG}	High-level and low-level terminal voltages	V
Δv_{HL}	Voltage error caused by asymmetric parameters	V
V_{h1}, V_{h2}	Amplitudes of first and second injected HF voltage signals	V
$v(k)$	Measurement noise	V
v_{N}	Neutral voltage	V
$V_{\mathrm{NG}}(s)$	Zero sequence voltage after Laplace transform	V
V_{N1N2}	Amplitude of zero sequence HF voltage of DTP-PMSM	V
v_{N1N2}	Zero sequence HF voltage of DTP-PMSM	V
V_P	Magnitude of voltage pulse	V
V_{P_MAX}, V_{P_MIN}	Maximum and minimum magnitudes of voltage pulse	V
V_{RL}	Voltage drop on winding resistance and inductance	V
V_s^*	Amplitude of voltage reference	V
$\mathbf{v}_s^*$	Voltage reference vector	V
v_{SM}	Voltage between network central point and capacitor mid-point	V
$\hat{v}_{\mathrm{SM}}$	Upper envelope of measured virtual third harmonic voltage	V
v_{SN}	Measured third harmonic voltage between network central point and machine neutral point	V
$v_{\mathrm{SN\_cos}}$	Cosine component of zero sequence HF voltage	V
$v_{\mathrm{SN\_sin}}$	Sine component of zero sequence HF voltage	V
$v_{\mathrm{SN}\_unified}$	Unified third harmonic voltage	V

(Continued)

(Continued)

Symbol	Description	Unit
v_{S1N1}, v_{S2N2}	Measured third harmonic voltages of two sets of three-phase windings	V
Δv_V	Vertical voltage shift	V
Δv_{VA}, Δv_{VB}, Δv_{VC}	Three-phase vertical voltage shifts	V
v_{Xh}, v_{Yh}, v_{Zh}	Injected three-phase HF voltages of second winding set in DTP-PMSM	V
v_{XN}, v_{YN}, v_{ZN}	Phase voltages of second winding set	V
Δv_{XO}	Terminal voltage error in arbitrary phase	V
Δv_{XY}	Voltage shift caused by asymmetric parameters	V
v_{z1z2}	Stator voltage in z_1z_2 subspaces	V
v_α, v_β	α- and β-axis voltages	V
v_{ah}, $v_{\beta h}$	α- and β-axis high-frequency voltages	V
v_0	Zero sequence voltage	V
$\mathbf{v}_{0,\ 7}$	Zero voltage vectors	V
$\mathbf{v}_{1,\ \ldots 6}$	Active voltage vectors	V
$\hat{v}_{2d}$, $\hat{v}_{2q}$	d_2- and q_2-axis voltages	V
v_{2nd}	Secondary positive sequence harmonic HF zero sequence voltage responses	V
$w(k)$	Process noise	V
W_s	Number of sampling points in a DFT window	
$x(k)$	State variables	
Z_H, Z_L	High- and low-level impedances	Ω
z_α, z_β	α- and β-axis switching functions	V
$\mathcal{L}^{-1}$	Inverse Laplace transform	
Λ	PWM switching function	
$\mathbf{\Phi}$	Exponential matrix of linearized system	
δ_L	Phase angle under inductance asymmetry	rad
$\delta_{\Delta L_Q}$	Position difference due to q-axis inductance mismatch after injection	rad
$\delta_{\Delta R_s}$	Position difference due to resistance mismatch after injection	rad
ε	Input error of controller	
$\varepsilon_{d\Delta L}$	d-axis voltage error due to inductance asymmetry	V
$\varepsilon_{d\Delta R}$	d-axis voltage error due to resistance asymmetry	V
$\varepsilon_{d2,\ \Delta\psi}$	d-axis voltage error due to flux-linkage asymmetry	V

(Continued)

Symbol	Description	Unit
$\varepsilon_{q\Delta L}$	q-axis voltage error due to inductance asymmetry	V
$\varepsilon_{q\Delta R}$	q-axis voltage error due to resistance asymmetry	V
$\varepsilon_{q2,\ \Delta\psi}$	q-axis voltage error due to flux-linkage asymmetry	V
ε_s	Quantization error	A
ε_T	Torque controller error	Nm
$\varepsilon_{\mathrm{VA}}$, $\varepsilon_{\mathrm{VB}}$, $\varepsilon_{\mathrm{VC}}$	Three-phase vertical error correction angles	rad
ε_ψ	Flux controller error	Wb
η	Current descent rate	A/s
η_{m1}^{on}, η_{m1}^{off}	Current descent rates of chop-on and chop-off periods in mode 1	A/s
η_{m2}^{on}, η_{m2}^{off}	Current descent rates of chop-on and chop-off periods in mode 2	A/s
θ	Angle of voltage reference	rad
$\Delta\theta$	Position error	rad
$\Delta\hat{\theta}$	Estimated position error	rad
$\Delta\theta_{AD}$	Position estimation error of analogue to digital conversion	rad
θ_{bound}	Boundary angle	rad
θ_c	Compensation angle for cross-coupling inductances	rad
$\Delta\theta_{\mathrm{HA}}$, $\Delta\theta_{\mathrm{HB}}$, $\Delta\theta_{\mathrm{HC}}$	Three-phase position shifts due to horizontal voltage shifts	rad
$\Delta\theta_{\Delta L_d}$	Estimated position error due to d-axis inductance mismatch	rad
$\Delta\theta_{\Delta L_q}$	Estimated position error due to q-axis inductance mismatch	rad
θ_m	Cross-saturation angle	rad
θ_{middle}	Middle angle	rad
$\Delta\theta_{par}$	Estimated position error due to parameter mismatches	rad
$\Delta\theta_{par1}$, $\Delta\theta_{par2}$	Position error caused by parameter deviation of Set 1 and Set 2	rad
$\Delta\theta_{par1}^{inj}$, $\Delta\theta_{par2}^{inj}$	Position estimated error after additional current signal injection in two sets	rad

(Continued)

(Continued)

Symbol	Description	Unit
$\Delta\theta_{p\_L}$	Position estimation error of HFSI due to inductance asymmetry	rad
θ_r	Electrical rotor position	rad
$\hat{\theta}_r$	Estimated electrical rotor position	rad
θ_r^v	Virtual rotor position	rad
$\hat{\theta}_{r\_fo}$	Estimated rotor position of fundamental model-based method	rad
$\hat{\theta}_{r\_hf}$	Estimated rotor position of HFSI method	rad
$\Delta\theta_{\Delta R_s}$	Estimated position error due to phase resistance mismatch	rad
$\hat{\theta}_{r0}$	Original estimated position from a position observer	rad
$\hat{\theta}_{r1}$, $\hat{\theta}_{r2}$	Estimated electrical rotor position of two sets of three-phase windings	rad
$\hat{\theta}_{r1(new)}$, $\hat{\theta}_{r2(new)}$	Estimated position after additional current signal injection in two sets	rad
θ_{s1}	Angle of cross-coupling component in phase self-inductance	rad
θ_{s2}	Angle of cross-coupling component in phase mutual-inductance	rad
θ_{s3}	Phase shift of zero sequence HF voltage due to cross-saturation effect	rad
$\Delta\theta_{\mathrm{VA}}$, $\Delta\theta_{\mathrm{VB}}$, $\Delta\theta_{\mathrm{VC}}$	Three-phase position shifts due to vertical voltage shifts	rad
θ_{xy}	ZCP interval from sector x to sector y	rad
$\hat{\theta}_{xy}$	Measured ZCP interval from sector x to sector y	rad
θ_0	Estimated rotor position of last zero-crossing point	rad
θ_{12}, θ_{23}, θ_{34}, θ_{45}, θ_{56}, θ_{61}, θ_{25}, θ_{52}	Nominal sector-ZCP-interval angle between sectors	rad
$\Delta\theta_{12}$	Shifted angle between two sets of three-phase windings	rad
$\hat{\theta}_{12}$, $\hat{\theta}_{23}$, $\hat{\theta}_{34}$, $\hat{\theta}_{45}$, $\hat{\theta}_{56}$, $\hat{\theta}_{61}$, $\hat{\theta}_{25}$, $\hat{\theta}_{52}$	Sector-ZCP-interval angle between sectors	rad
λ	Cross-coupling compensation factor for q-axis current	
λ_{comp}	Compensation factor of inverter nonlinearity	rad
μ	Bandwidth of peaking filter and ANF	rad/s

(Continued)

Symbol	Description	Unit
ξ	Damping ratio of PLL	
σ	Tolerance ratio	
τ_e	Electrical time constant	s
τ_{e_MAX}, τ_{e_MIN}	Maximum and minimum electrical time constants	s
τ_{eN}	Nominal electrical time constant	s
φ_{Ah}, φ_{Bh}, φ_{Ch}	Phases of three-phase HF currents	rad
φ_d	Phase shift due to signal-processing delays	rad
$\varphi_{d,\ 2f}$, $\varphi_{q,\ 2f}$	Phase shifts of second order harmonic of d- and q-axis estimated back-EMFs	rad
φ_{m1}	Freewheeling angles	rad
φ_{m1}^{sup}	Maximum freewheeling angle	rad
φ_{m1}, φ_{m2}	Measured freewheeling angles	rad
φ_{12}	Phase shift between two injected high-frequency signals	rad
ψ_A, ψ_B, ψ_C	Three-phase stator flux-linkages	Wb
ψ_a	Amplitude of active flux	Wb
$\vec{\psi}_a$	Active flux vector	Wb
$\Delta\psi_A$, $\Delta\psi_B$, $\Delta\psi_C$	Asymmetric flux-linkage components	Wb
$\Delta\psi_{ave}$	DC offset due to flux-linkage asymmetry	Wb
ψ_d, ψ_q	d- and q-axis flux-linkages	Wb
$\hat{\psi}_d$, $\hat{\psi}_q$	Estimated d- and q-axis flux-linkages	Wb
ψ_m	PM flux-linkage	Wb
$\psi_{m\alpha}$, $\psi_{m\beta}$	α- and β-axis PM flux-linkages	Wb
ψ_{m3}	Amplitude of third harmonic flux-linkage	Wb
ψ_{SN}	Third harmonic flux-linkage	Wb
ψ_{z1z2}	Stator flux in z_1z_2 subspaces	Wb
ψ_α, ψ_β	α- and β-axis flux-linkages	Wb
$\psi_\alpha(0)$, $\psi_\beta(0)$	Initial values of α- and β-axis flux-linkages	Wb
ψ_0	Zero sequence flux-linkage	Wb
ψ_{3_set1}, ψ_{3_set2}	Third harmonic flux-linkages in two sets of three-phase windings	Wb
$\Delta\omega$	Speed estimation error	rad/s
ω_{ac}	Frequency of injected ac current reference	rad/s
ω_{base}	Base rotor speed	rad/s
ω_c	Cut-off frequency	rad/s

(Continued)

(Continued)

Symbol	Description	Unit
ω_{comp}	Speed compensation	rad/s
ω_{cor}	Speed correction	rad/s
$\hat{\omega}_d$	LPF filtered speed	rad/s
ω_f	Cut-off frequency of low-pass filter	rad/s
ω_h	Frequency of injected high-frequency voltage	rad/s
$\omega_{h,\ ANF}$	Center frequency of ANF	rad/s
ω_{high}	High-speed border	rad/s
$\hat{\omega}_{hybrid}$	Hybrid speed estimation	rad/s
ω_{low}	Low-speed border	rad/s
ω_m	Mechanical rotor speed	rad/s
ω_n	Natural frequency of PLL	rad/s
ω_{pu}	Per-unit rotor speed	rad/s
ω_r	Electrical rotor speed	rad/s
ω_r^*	Electrical rotor speed reference	rad/s
$\hat{\omega}_r$	Estimated electrical rotor speed	rad/s
$\hat{\omega}_{ra}$	Average estimated rotor speed	rad/s
$\hat{\omega}_r^{ac}$	ac component of estimated rotor speed	rad/s
$\hat{\omega}_{r1}$, $\hat{\omega}_{r2}$	Estimated electrical rotor speed of two sets of three-phase windings	rad/s
$\hat{\omega}_v$	Speed estimation based on machine model	rad/s

1

General Introduction

1.1 Introduction

This chapter introduces permanent magnet (PM) synchronous machines (PMSMs), i.e. brushless ac (BLAC) machines and PM brushless dc (BLDC) machines. It also describes the basic principles of BLAC and BLDC drives, including modeling and control techniques. In the literature, PMSMs are often referred to as PM BLAC machines only. In this book, the terminologies of both PM BLAC machines and PMSMs will be used for convenience, but PM BLAC drives will only be used together with PM BLDC drives. Moreover, the term "PM machines" used in this book specifically refers to PM brushless machines. This chapter introduces various rotor position sensorless control techniques, as well as a comprehensive classification and their applications. Finally, the scope of this book will be illustrated.

1.2 Permanent Magnet Machines

Over the last several decades, significant advancements in PM materials, power electronics devices, and microprocessors have played a crucial role in the development of PM brushless machines PM brushless machines exhibit a range of advantages, including high efficiency, high torque density, high power density, easy maintenance, and excellent control performance. Hence, they have been used for a wide range of applications, including electric vehicles, renewable energy systems, robotics, industrial automation, domestic appliances, aerospace, etc. With continued research and development, it is likely that the performance and

Sensorless Control of Permanent Magnet Synchronous Machine Drives, First Edition.
Zi Qiang Zhu and Xi Meng Wu.

capabilities of PM brushless machines will continue to improve, making them an even more compelling choice for a range of industrial and commercial applications.

1.2.1 Topologies

Within the last few decades, there has been much research into various machine topologies of PMSMs, which can be rotary or linear. Rotary PMSMs can be classified as in Figure 1.1 [1].

As shown in Figure 1.1, from the type of air-gap fields, PM machines can be divided into radial-field, axial-field, transverse-flux, and hybrid-fields machines, respectively. Compared to radial-field PM machines, axial-field and transverse-flux PM machines can have higher torque and power density [2]–[5]. However, complex manufacturing processes restrict their applications. In terms of stator winding configurations, PM machines can be grouped mainly into two categories:

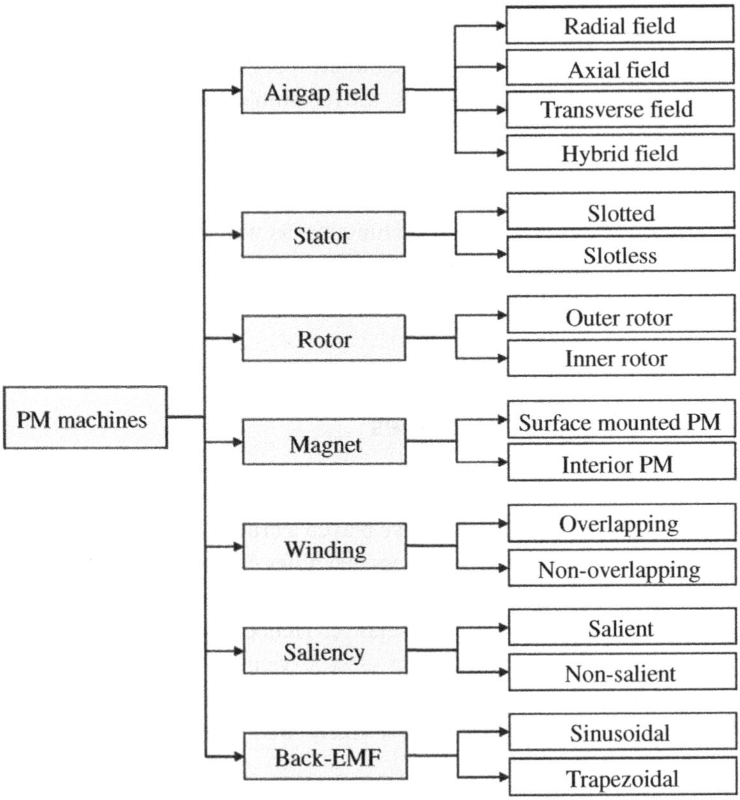

Figure 1.1 Classification of PM machine topologies.

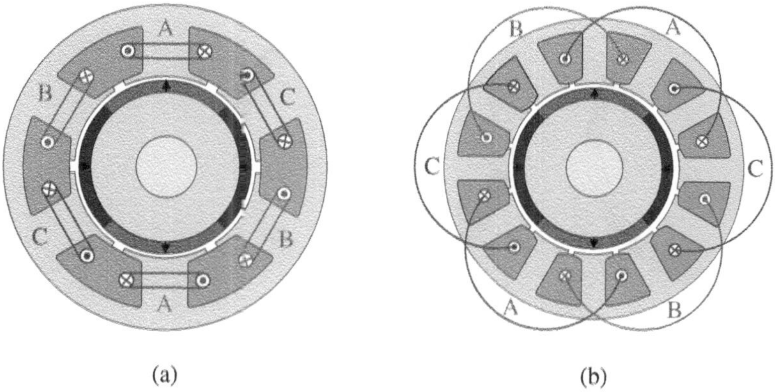

Figure 1.2 Alternate winding configurations for PM machines. (a) Non-overlapping windings. (b) Overlapping windings.

fractional-slot machines with non-overlapping concentrated windings as shown in Figure 1.2a, and integral-slot machines with overlapping distributed windings in Figure 1.2b. Compared to distributed windings, concentrated windings have the advantages of shorter end-windings and a higher slot-packing factor, leading to the reduction of total machine volume, mass, and copper losses [1], [6], [7]. Consequently, the overall efficiency and torque density are improved. However, compared to integral-slot machines, fractional-slot PM machines suffer from high contents of stator magneto-motive force (MMF) harmonics, which could cause increased PM eddy current losses, localized magnetic saturation, acoustic noises and vibrations, etc. [8].

On the other hand, the rotor structures of PM machines can be classified mainly into surface-mounted PM machines (SPM) and interior PM machines (IPM).

As shown in Figure 1.3a, the SPM machine is widely used. It has the advantages of a simple structure and low manufacturing cost. Moreover, the magnet shape can be easily optimized to produce more sinusoidal-distributed flux density in the airgap. However, due to direct exposure to the armature reaction field, the magnets may suffer from irreversible demagnetization risk [1]. In general, the magnets should be retained by a stainless steel sleeve or glass/carbon fiber. Another special SPM machine topology, namely the inset PM machine, is shown in Figure 1.3b, where the magnets are inserted in the rotor with the iron between them. Compared to the SPM machine, the inset PM machine has considerable rotor saliency due to high permeability iron steels between the PMs. Accordingly, it can utilize the reluctance torque to increase the torque density and/or reduce the amount of PM.

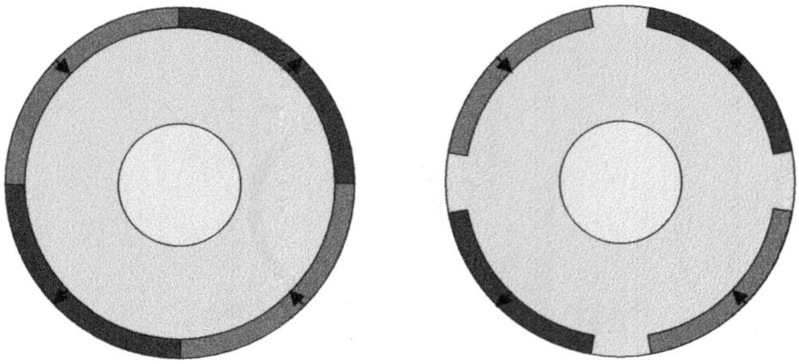

(a) (b)

Figure 1.3 Surface-mounted PM rotors. (a) Surface-mounted PM. (b) Surface-inset PM.

In contrast, the IPM structures in which the magnets are placed inside the rotor iron body, shown in Figure 1.4, have the merits of high rotor saliency, good flux-weakening capability, and a low risk of PM demagnetization. According to the rotor PM configurations, the IPM machines can be divided into single-layer IPM rotors, e.g. I-shape and V-shape as shown in Figure 1.4a–b; and double-layer

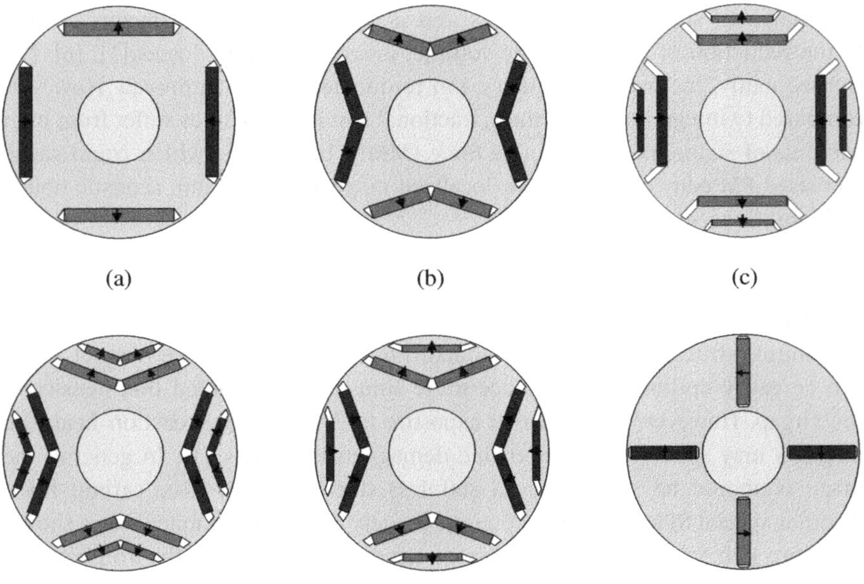

(a) (b) (c)

(d) (e) (f)

Figure 1.4 Interior PM rotors. (a) Single I-shape PM. (b) Single V-shape PM. (c) Double I-shape PM. (d) Double V-shape PM. (e) Delta-shape PM. (f) Spoke-shape PM.

IPM rotors, e.g. double I-shape, delta-shape, and double V-shape as shown in Figure 1.4c–e; as well as spoke-shape rotors as illustrated in Figure 1.4f, etc.

The I-shape IPM rotor, Figure 1.4a, has a simple construction with rotor saliency. Furthermore, the single-layer V-shape IPM rotor, Figure 1.4b, increases the cross-section areas for PMs and enhances the flux focusing effect, resulting in a better torque capability. For double-layer variants shown in Figure 1.4c–e, i.e. the double I-shape, delta-shape, and double V-shape IPM rotors, the rotor saliency and reluctance torque are further enhanced, but the manufacturing process becomes more complex. In addition, in the spoke-type IPM rotor, Figure 1.4f, the flux focusing effect can be significant if a higher pole number is employed, but a higher leakage flux exists due to additional magnetic bridges near the shaft, and a non-magnetic shaft should be employed and airspace flux barriers are often required in the laminations near the shaft.

1.2.2 Drives

A PM brushless drive consists of a controller, an inverter, and a PM machine, as shown in Figure 1.5. The currents can be measured through three-phase currents, two-phase currents, and dc-link current. The rotor speed and position are usually measured by a position sensor, unless one of the rotor position sensorless techniques is employed, as will be introduced in this book. A voltage sensor is used to measure the dc-link voltage. These feedback signals are sent to the controller, which generates the switching signals for the inverter to achieve the control objective.

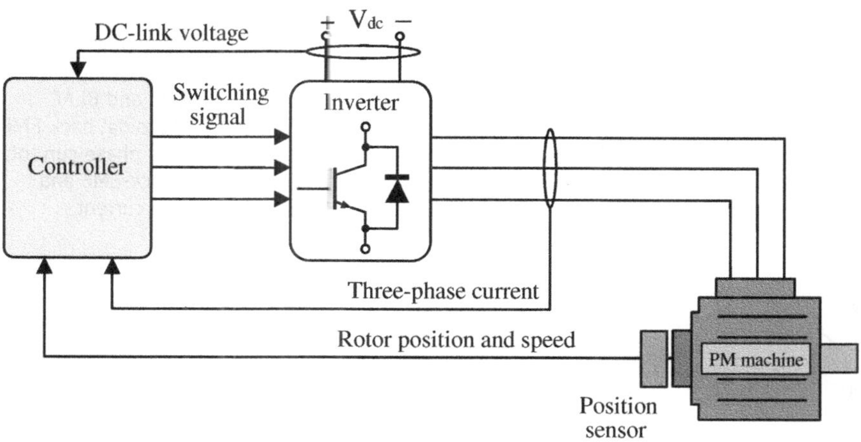

Figure 1.5 Block diagram of a three-phase PMSM drive.

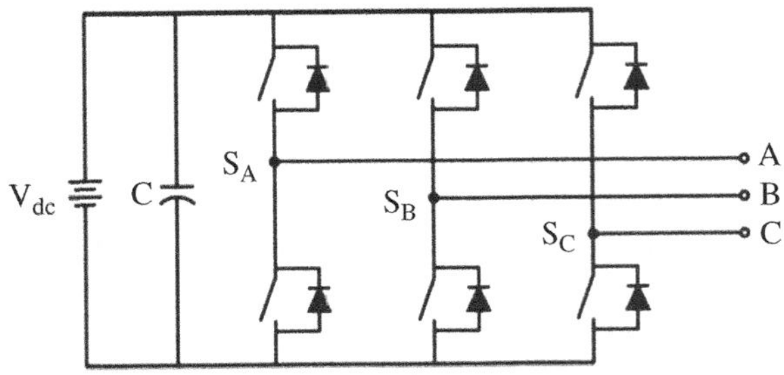

Figure 1.6 Block diagram of a three-phase VSI.

A three-phase voltage source inverter (VSI), Figure 1.6, is commonly employed in a three-phase PM brushless drive system. As shown in Figure 1.6, a VSI consists of six power semiconductor switches, usually IGBTs (insulated gate bipolar transistors) or MOSFETs (metal-oxide-semiconductor field-effect transistors). These six switches form a three-phase bridge, with each leg of the bridge consisting of two switches. The middle point of each leg is connected to the one phase terminal of the PM machine stator winding. The three legs are connected to a dc voltage source with a dc-link capacitor, which is used to smooth the dc-link voltage.

With the rotation of the PM rotor, the back electromotive force (back-EMF) will be generated in the stator windings. The PM machines can be classified according to the back-EMF waveform as either sine-wave or trapezoidal-wave (ideally square-wave), as illustrated in Figure 1.7. The control strategies, which determine the inverter switching status, can be classified as either BLDC or BLAC.

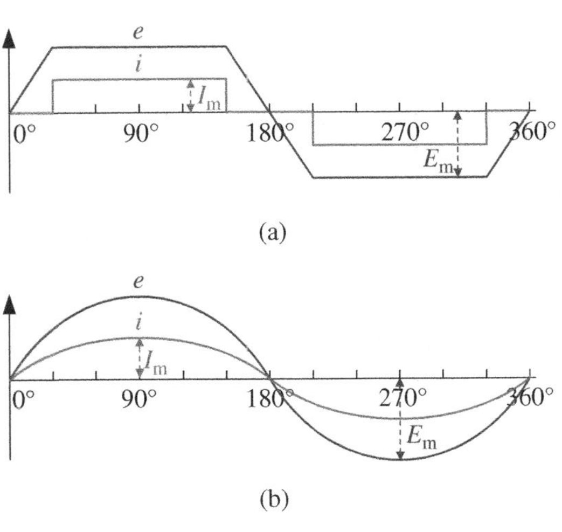

Figure 1.7 Back-EMF waveforms and excitation currents of BLDC and BLAC drives. (a) Trapezoidal back-EMF and square-wave phase current. (b) Sinusoidal back-EMF and sine-wave phase current.

In BLDC operation the inverter ideally supplies square-wave currents, while in BLAC operation the inverter supplies sine-wave currents, as shown in Figure 1.7.

Although both BLAC and BLDC drives use the same VSI, their switching schemes are different. Specifically, BLAC drives employ the three-phase conduction with space vector pulse-width modulation (SVPWM) control to produce the sine-wave current, whereas BLDC drives employ the two-phase conduction six-step control to produce the square-wave current. The switching schemes of BLAC and BLDC drives are described in the following sections.

1.3 Basic Principle of PM BLAC (PMSM) Drives

In this section, the mathematical models and control strategies of BLAC drives [9] are described, providing the reader with a basic understanding of the working principle of PM machine drives. The mathematical models and control strategies of BLDC drives [10], as well as the comparison between BLAC and BLDC drives, will be given in Sections 1.4 and 1.5, respectively.

1.3.1 Modeling

Three-phase PMSMs can be modeled in different reference frames, including the ABC reference frame, stationary reference frame, and synchronous reference frame (which will be shown later in this section). To simplify the modeling, the following assumptions are made:

- The stator winding is distributed symmetrically in space, resulting in a sinusoidal armature reactive MMF.
- The EMF induced in the stator windings by the PM rotor is sinusoidal.
- The three-phase windings are star connected without neutral connection.
- The iron loss and magnetic saturation effect are ignored.
- Parameter changes due to temperature and load are ignored.

1.3.1.1 ABC Reference Frame

In an ABC reference frame, the voltage equation is given by:

$$\begin{bmatrix} v_\mathrm{A} \\ v_\mathrm{B} \\ v_\mathrm{C} \end{bmatrix} = \begin{bmatrix} R_s & 0 & 0 \\ 0 & R_s & 0 \\ 0 & 0 & R_s \end{bmatrix} \cdot \begin{bmatrix} i_\mathrm{A} \\ i_\mathrm{B} \\ i_\mathrm{C} \end{bmatrix} + p \begin{bmatrix} \psi_\mathrm{A} \\ \psi_\mathrm{B} \\ \psi_\mathrm{C} \end{bmatrix} \tag{1.1}$$

where $v_\mathrm{A}, v_\mathrm{B}, v_\mathrm{C}$ are the three-phase voltages, $i_\mathrm{A}, i_\mathrm{B}, i_\mathrm{C}$ are the three-phase currents, $e_\mathrm{A}, e_\mathrm{B}, e_\mathrm{C}$ are the three-phase back-EMFs, $\psi_\mathrm{A}, \psi_\mathrm{B}, \psi_\mathrm{C}$ are the three-phase flux-linkages. R_s is the phase resistance, and p is the derivative operator, i.e. $p = d/dt$.

The three-phase flux-linkages can be written as:

$$\begin{bmatrix} \psi_A \\ \psi_B \\ \psi_C \end{bmatrix} = \begin{bmatrix} L_{AA} & M_{AB} & M_{AC} \\ M_{BA} & L_{BB} & M_{BC} \\ M_{CA} & M_{CB} & L_{CC} \end{bmatrix} \cdot \begin{bmatrix} i_A \\ i_B \\ i_C \end{bmatrix} + \psi_m \begin{bmatrix} \cos\theta_r \\ \cos(\theta_r - 2\pi/3) \\ \cos(\theta_r + 2\pi/3) \end{bmatrix} \tag{1.2}$$

where L_{AA}, L_{BB}, L_{CC} are the three-phase self-inductances, M_{AB}, M_{BA}, M_{AC}, M_{CA}, M_{BC}, M_{CB} are the three-phase mutual-inductances, ψ_m is the PM flux-linkage, and θ_r is the electrical rotor position. For the inductance matrix, the elements are given by:

$$\begin{aligned} L_{AA} &= L_{s0} - L_{s2}\cos 2\theta_r \\ L_{BB} &= L_{s0} - L_{s2}\cos 2(\theta_r - 2\pi/3) \\ L_{CC} &= L_{s0} - L_{s2}\cos 2(\theta_r + 2\pi/3) \end{aligned} \tag{1.3}$$

where L_{s0} and L_{s2} are the amplitudes of dc and the second order harmonic component of self-inductances, respectively.

$$\begin{aligned} M_{AB} &= M_{BA} = M_{s0} - M_{s2}\cos 2(\theta_r + 2\pi/3) \\ M_{BC} &= M_{CB} = M_{s0} - M_{s2}\cos 2\theta_r \\ M_{CA} &= M_{AC} = M_{s0} - M_{s2}\cos 2(\theta_r - 2\pi/3) \end{aligned} \tag{1.4}$$

where M_{s0} and M_{s2} are the amplitudes of dc and the second order harmonic component of mutual-inductances, respectively. Furthermore, there is no second order harmonic component in the non-salient machines, i.e. $L_{s2} = M_{s2} = 0$.

From (1.1) to (1.4), it can be seen that the machine model in the ABC reference frame is coupled with the rotor position, which makes the control very complex. Therefore, the machine model can be transformed into the stationary reference frame and the synchronous reference frame.

1.3.1.2 Stationary Reference Frame

The machine model can be transformed from the ABC reference frame to the stationary reference frame by using the Clarke transformation, which is given in (1.5).

$$T_{ABC-\alpha\beta0} = \frac{2}{3}\begin{bmatrix} 1 & -\frac{1}{2} & -\frac{1}{2} \\ 0 & \frac{\sqrt{3}}{2} & -\frac{\sqrt{3}}{2} \\ \frac{1}{2} & \frac{1}{2} & \frac{1}{2} \end{bmatrix}, \quad T_{\alpha\beta0-ABC} = \begin{bmatrix} 1 & 0 & 1 \\ -\frac{1}{2} & \frac{\sqrt{3}}{2} & 1 \\ -\frac{1}{2} & -\frac{\sqrt{3}}{2} & 1 \end{bmatrix} \tag{1.5}$$

The relationship between the ABC reference frame and the stationary reference frame is shown in Figure 1.8.

Assuming that the three-phase windings are star connected without neutral connection, the sum of the three-phase currents is zero, and the corresponding zero sequence component is zero. After transformation, the voltage and flux-linkage equations in the stationary reference frame are given by:

$$\begin{bmatrix} v_\alpha \\ v_\beta \end{bmatrix} = \begin{bmatrix} R_s & 0 \\ 0 & R_s \end{bmatrix} \cdot \begin{bmatrix} i_\alpha \\ i_\beta \end{bmatrix} + p\begin{bmatrix} \psi_\alpha \\ \psi_\beta \end{bmatrix} \tag{1.6}$$

$$\begin{bmatrix} \psi_\alpha \\ \psi_\beta \end{bmatrix} = \begin{bmatrix} L_{\alpha\alpha} & M_{\alpha\beta} \\ M_{\beta\alpha} & L_{\beta\beta} \end{bmatrix} \cdot \begin{bmatrix} i_\alpha \\ i_\beta \end{bmatrix} + \psi_m \begin{bmatrix} \cos\theta_r \\ \sin\theta_r \end{bmatrix} \tag{1.7}$$

where v_α, v_β, i_α, i_β, ψ_α, and ψ_β are the $\alpha\beta$-axis voltages, currents, and flux-linkages in the stationary reference frame, respectively. $L_{\alpha\alpha}$, $L_{\beta\beta}$, $M_{\alpha\beta}$, and $M_{\beta\alpha}$ are the self- and mutual- inductances in the stationary reference frame, respectively. Furthermore, the inductance elements can be represented as:

$$\begin{aligned} L_{\alpha\alpha} &= \frac{L_d + L_q}{2} + \frac{L_d - L_q}{2}\cos 2\theta_r \\ L_{\beta\beta} &= \frac{L_d + L_q}{2} - \frac{L_d - L_q}{2}\cos 2\theta_r \\ M_{\alpha\beta} = M_{\beta\alpha} &= \frac{L_d - L_q}{2}\cos 2\theta_r \end{aligned} \tag{1.8}$$

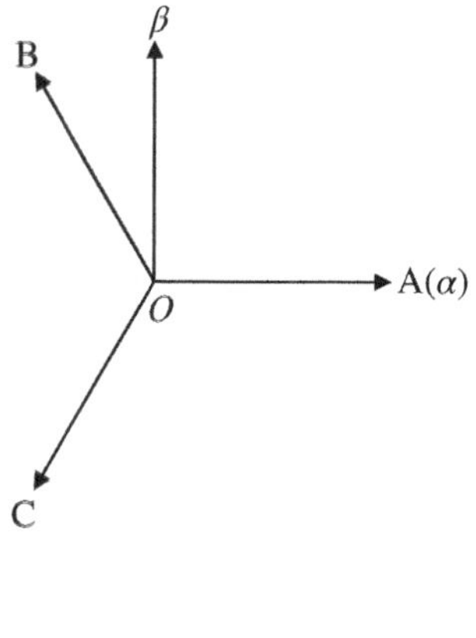

Figure 1.8 Relationship between ABC reference frame and stationary reference frame.

where L_d and L_q are the d- and q-axis inductances. If the machine is non-salient, i.e. SPM, the relationship of the inductances is then given by:

$$L_d = L_q = L_s \tag{1.9}$$

1.3.1.3 Synchronous Reference Frame

The machine model can be transformed from the stationary reference to the synchronous reference frame by using a Park transformation, which is given by:

$$T_{\alpha\beta 0-dq0} = \begin{bmatrix} \cos\theta_r & \sin\theta_r & 0 \\ -\sin\theta_r & \cos\theta_r & 0 \\ 0 & 0 & 1 \end{bmatrix} \tag{1.10}$$

$$T_{dq0-\alpha\beta 0} = \begin{bmatrix} \cos\theta_r & -\sin\theta_r & 0 \\ \sin\theta_r & \cos\theta_r & 0 \\ 0 & 0 & 1 \end{bmatrix} \tag{1.11}$$

Alternatively, the machine model can be transformed directly from the ABC reference frame to the synchronous reference frame by using a direct-quadrature-zero (DQZ) transformation, which is the product of Clarke transformation and Park transformation. Then, the DQZ transformation matrices are derived as:

$$\begin{aligned} T_{\mathrm{ABC}-dq0} &= T_{\mathrm{abc}-\alpha\beta 0} \cdot T_{\alpha\beta 0-dq0} \\ &= \frac{2}{3}\begin{bmatrix} \cos\theta_r & \cos(\theta_r - 2\pi/3) & \cos(\theta_r + 2\pi/3) \\ -\sin\theta_r & -\sin(\theta_r - 2\pi/3) & -\sin(\theta_r + 2\pi/3) \\ \frac{1}{2} & \frac{1}{2} & \frac{1}{2} \end{bmatrix} \end{aligned} \tag{1.12}$$

$$T_{dq0-\mathrm{ABC}} = T_{dq0-\alpha\beta 0} \cdot T_{\alpha\beta 0-\mathrm{ABC}} = \begin{bmatrix} \cos\theta_r & -\sin\theta_r & 1 \\ \cos(\theta_r - 2\pi/3) & -\sin(\theta_r - 2\pi/3) & 1 \\ \cos(\theta_r + 2\pi/3) & -\sin(\theta_r + 2\pi/3) & 1 \end{bmatrix} \tag{1.13}$$

In addition, the relationship between the ABC reference frame, stationary reference frame, and synchronous reference frame is shown in Figure 1.9.

After transformation, by neglecting the zero-sequence component, the voltage and flux-linkage equations in the synchronous reference frame are given by:

$$\begin{bmatrix} v_d \\ v_q \end{bmatrix} = \begin{bmatrix} R_s & 0 \\ 0 & R_s \end{bmatrix} \cdot \begin{bmatrix} i_d \\ i_q \end{bmatrix} + p\begin{bmatrix} \psi_d \\ \psi_q \end{bmatrix} + \omega_r \begin{bmatrix} -\psi_q \\ \psi_d \end{bmatrix} \tag{1.14}$$

and

$$\begin{bmatrix} \psi_d \\ \psi_q \end{bmatrix} = \begin{bmatrix} L_d & 0 \\ 0 & L_q \end{bmatrix} \cdot \begin{bmatrix} i_d \\ i_q \end{bmatrix} + \psi_m \begin{bmatrix} 1 \\ 0 \end{bmatrix} \tag{1.15}$$

where v_d, v_q, i_d, i_q, ψ_d, and ψ_q are the dq-axis voltages, currents, and flux-linkages in the synchronous reference frame, respectively, and ω_r is the electrical rotor speed.

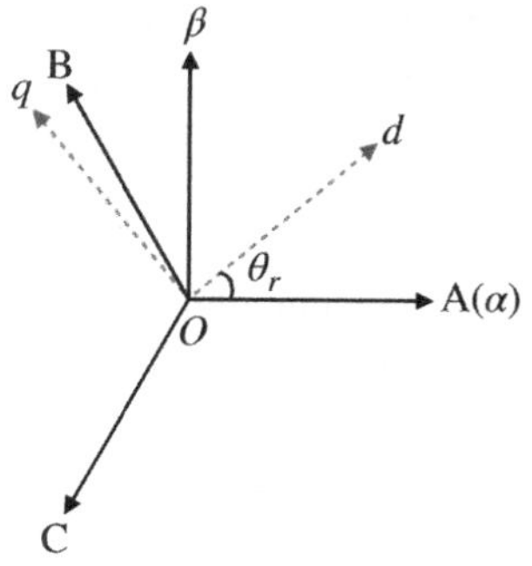

Figure 1.9 Relationship between ABC reference frame and synchronous reference frame.

1.3.2 Control Strategies

As shown in Figure 1.10, it is desirable for the three-phase back-EMFs of a PM BLAC machine to be sinusoidal. Hence, three-phase currents should be controlled to be sine-wave and aligned with or in advance of the back-EMF to maximize the output torque. In order to realize the three-phase sine-wave excitation currents, SVPWM is commonly used to control the VSI to generate the desired stator voltage to the machine. The reference of the stator voltage can be obtained by control strategies, including field-oriented control (FOC), direct torque control (DTC), and model predictive control (MPC).

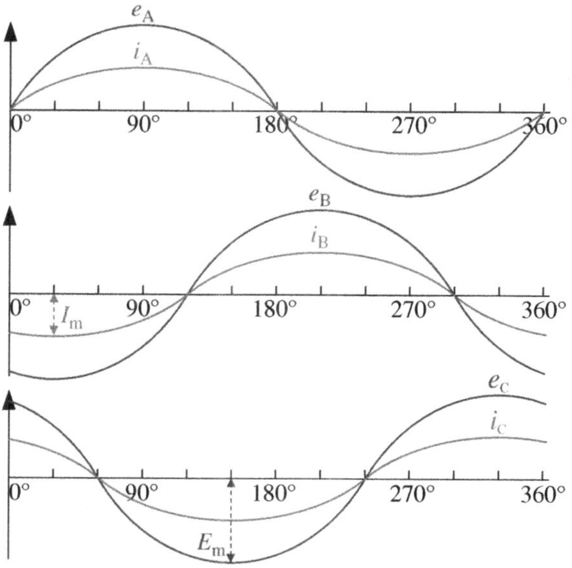

Figure 1.10 Sinusoidal back-EMF waveforms and sine-wave phase currents of BLAC drive.

1.3.2.1 Space Vector PWM

For BLAC drives, SVPWM is a popular technique used to control the output voltage of the VSI to drive the machine at the desired speed and torque [11]–[13].

As shown in Figure 1.6, S_A, S_B, S_C are the switching states of the three legs of the VSI. To avoid the short circuit, the upper and lower switches in one leg work in the complementary mode with a deadtime. Therefore, if the upper switch is on and the lower switch is off, the value of the switching state is "1", and if the upper switch is off and the lower switcher is on, the value of the switching state is "0". In total, there are eight switching states of the VSI as depicted in Figure 1.11. For all

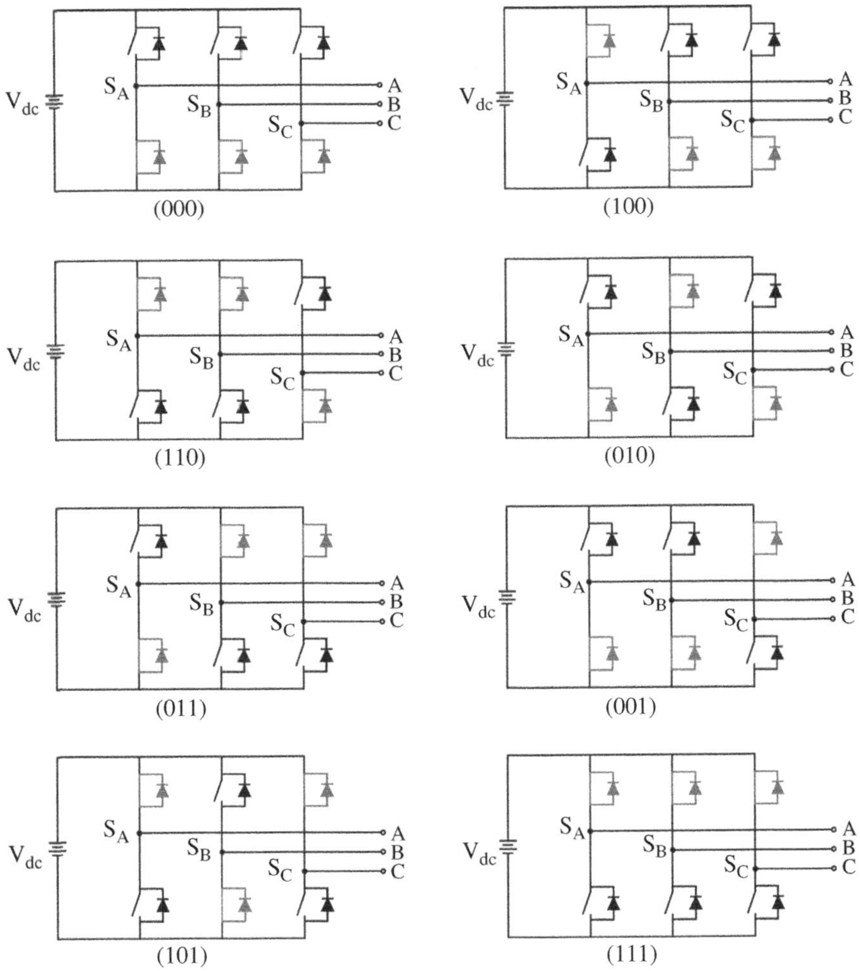

Figure 1.11 Block diagram of VSI at different switching states.

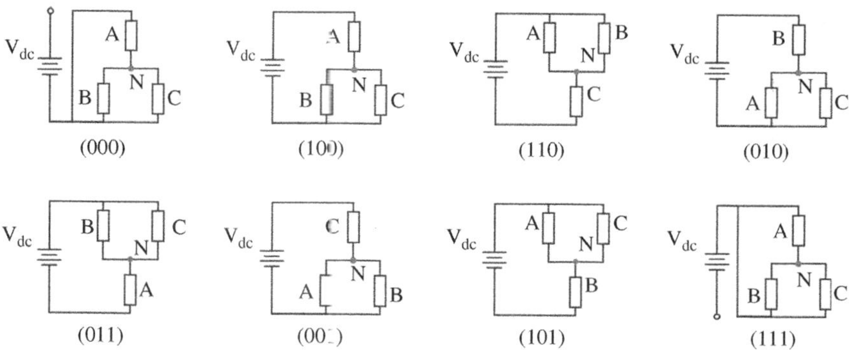

Figure 1.12 Equivalent circuits of switching states.

switching states, the corresponding equivalent circuits including the PM machine are shown in Figure 1.12. Based on the equivalent circuits, the three-phase voltages for all switching states can be derived as shown in Table 1.1.

SVPWM is based on the concept of a rotating space vector [14]. For VSI, the three-phase voltages can form an equivalent voltage vector v_s as:

$$\mathbf{v}_s = v_{AN} + v_{BN}e^{j\frac{2}{3}\pi} + v_{CN}e^{j\frac{4}{3}\pi} \tag{1.16}$$

where v_{AN}, v_{BN}, v_{CN} are the phase voltages of a PM machine in the ABC reference frame. The voltage vector in (1.16) can also be represented in the stationary reference frame as:

$$\mathbf{v}_s = v_{AN} + v_{BN}e^{j\frac{2}{3}\pi} + v_{CN}e^{j\frac{4}{3}\pi} = v_\alpha + jv_\beta \tag{1.17}$$

Table 1.1 Switching states and corresponding phase voltages of VSI.

Switching state (S_A S_B S_C)	**000**	**100**	**110**	**010**	**011**	**001**	**101**	**111**
v_{AN}	0	$\frac{2}{3}V_{dc}$	$\frac{1}{3}V_{dc}$	$-\frac{1}{3}V_{dc}$	$-\frac{2}{3}V_{dc}$	$-\frac{1}{3}V_{dc}$	$\frac{1}{3}V_{dc}$	0
v_{BN}	0	$-\frac{1}{3}V_{dc}$	$\frac{1}{3}V_{dc}$	$\frac{2}{3}V_{dc}$	$\frac{1}{3}V_{dc}$	$-\frac{1}{3}V_{dc}$	$-\frac{2}{3}V_{dc}$	0
v_{CN}	0	$-\frac{1}{3}V_{dc}$	$-\frac{2}{3}V_{dc}$	$-\frac{1}{3}V_{dc}$	$\frac{1}{3}V_{dc}$	$\frac{2}{3}V_{dc}$	$\frac{1}{3}V_{dc}$	0

Table 1.2 Switching states and voltage vectors of VSI.

Switching state (S_A S_B S_C)	**000**	**100**	**110**	**010**	**011**	**001**	**101**	**111**
Voltage vector	$\mathbf{v}_0$	$\mathbf{v}_1$	$\mathbf{v}_2$	$\mathbf{v}_3$	$\mathbf{v}_4$	$\mathbf{v}_5$	$\mathbf{v}_6$	$\mathbf{v}_7$
Value	0	$\frac{2}{3}V_{dc}e^{j0}$	$\frac{2}{3}V_{dc}e^{j\frac{1}{3}\pi}$	$\frac{2}{3}V_{dc}e^{j\frac{2}{3}\pi}$	$\frac{2}{3}V_{dc}e^{j\pi}$	$\frac{2}{3}V_{dc}e^{j\frac{4}{3}\pi}$	$\frac{2}{3}V_{dc}e^{j\frac{5}{3}\pi}$	0

Considering the eight switching states of the VSI, there are eight basic voltage vectors in total including six active vectors, i.e. $\mathbf{v}_{1,\ldots 6}$, and two zero vectors, i.e. $\mathbf{v}_{0,7}$. The eight basic voltage vectors for different switching states are summarized in Table 1.2.

As shown in Figure 1.13, in the stationary reference frame, the active voltage vectors form a hexagon with an amplitude of $2V_{dc}/3$, and the whole area is divided into six sectors by the voltage vectors.

Based on the eight basic voltage vectors $\mathbf{v}_{0,\ldots 7}$, the VSI can generate any rotating voltage vector by synthesizing two adjacent active vectors and a zero vector within a sampling period T_s. For example, as shown in Figure 1.13, a reference voltage vector $\mathbf{v}_s^*$ in Sector I is given, with two active vectors $\mathbf{v}_1$, $\mathbf{v}_2$ and the zero vector $\mathbf{v}_0$ are used for the synthesis. The synthesis is based on the volt-second balancing, which can be expressed as:

$$\mathbf{v}_s^* T_s = \mathbf{v}_1 T_1 + \mathbf{v}_2 T_2 + \mathbf{v}_0 T_0 \tag{1.18}$$

$$T_s = T_1 + T_2 + T_0 \tag{1.19}$$

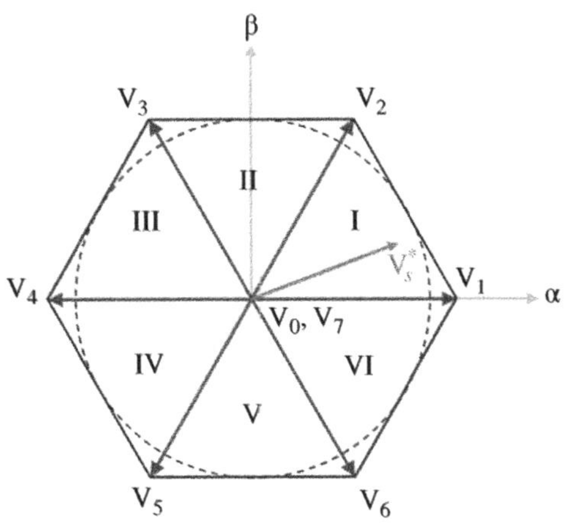

Figure 1.13 Voltage vectors.

where T_1, T_2, and T_0 are the durations of the voltage vectors $\mathbf{v}_1$, $\mathbf{v}_2$, and $\mathbf{v}_0$, respectively. Furthermore, the synthesis is illustrated in Figure 1.14.

Accordingly, the duration time for each vector can be derived as:

$$\begin{cases} T_1 = \dfrac{\sqrt{3}V_s^*}{V_{dc}} T_s \sin\left(\dfrac{\pi}{3} - \theta\right) \\ T_2 = \dfrac{\sqrt{3}V_s^*}{V_{dc}} T_s \sin(\theta), \quad 0 \le \theta \le \dfrac{\pi}{3} \\ T_0 = T_s - T_1 - T_2 \end{cases} \tag{1.20}$$

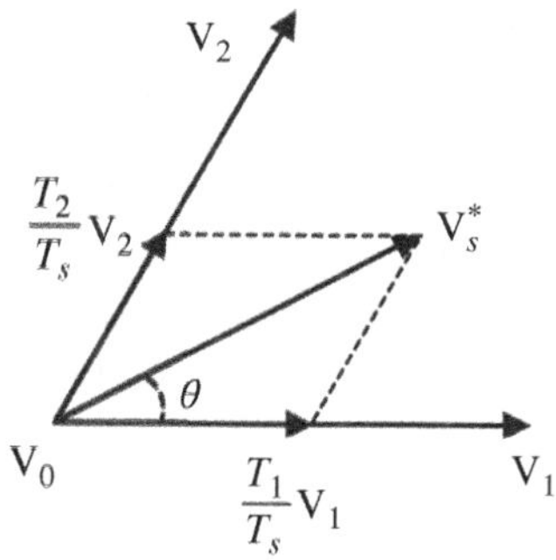

Figure 1.14 Stator voltage reference vector in Sector I.

where V_s^* is the magnitude of the stator voltage reference, and θ is the angle of the stator voltage reference. For the cases in other sectors, the duration time can be derived similarly.

After obtaining the durations of voltage vectors, the next step is to determine the switching sequence. The optimum switching sequence of SVPWM [11] is commonly used. An example of the switching sequence in Sector I is depicted in Figure 1.15. To reduce the switching losses, the transition from one switching state to the next is performed by switching only one leg. Therefore, the zero vector $\mathbf{v}_7$ is applied in the middle instead of the zero vector $\mathbf{v}_0$. Moreover, as shown in Figure 1.15, zero vectors are allocated to produce a symmetrical PWM so that the harmonic distortion in the output voltage can be minimized.

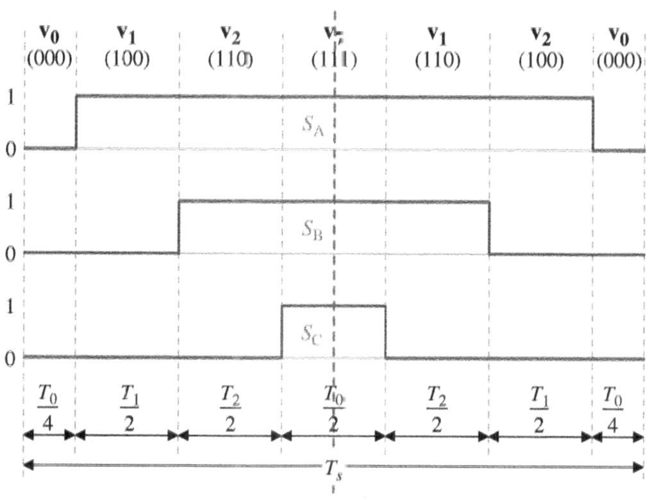

Figure 1.15 Switching pattern for SVPWM in Sector I.

When the reference voltage vector is within the inscribed circle of the hexagon, the trajectory of the rotating voltage vector generated by the SVPWM algorithm can be a circle within the hexagon, as in Figure 1.13. This ensures that the fundamental components of the output three-phase voltages are sine waves with a constant amplitude and frequency. However, if the reference voltage vector is outside the inscribed circle of the hexagon, over-modulation schemes [11], [13] need to be used to generate the desired output voltage.

1.3.2.2 Field-Oriented Control

FOC, or vector control (VC), was initially developed in the early 1970s [15], [16]. By adopting FOC, the three-phase alternative current (AC) machine can be controlled as a separately excited direct current (dc) machine. As described in Section 1.3.1.3, based on the reference frame transformation, the three-phase sinusoidal stator currents are transformed into two orthogonal dc components: one defines the magnetic flux and the other defines the torque. With FOC, an excellent performance of the PMSM (BLAC) drive is achieved. The PMSMs can operate over the full speed range and provide desired torque control performance even at zero speed. In addition, FOC shows good dynamic performance and control accuracy.

The block diagram of a typical FOC used for PMSM speed control is shown in Figure 1.16. The q-axis current is related to the torque and the d-axis current is related to the flux, and they can be regulated separately [17]. As shown in Figure 1.16, a SVPWM module is employed in the PMSM drives. As introduced in Section 1.3.2.1, the stator voltage reference generated by the current controller can be applied to the machine by SVPWM.

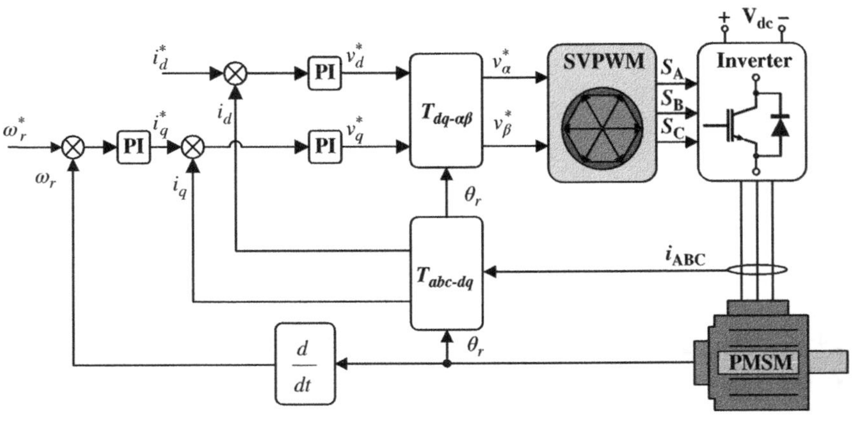

Figure 1.16 Block diagram of field-oriented control.

1.3.2.3 Direct Torque Control

DTC has been developed due to its simple structure, excellent transient response, and good robustness against rotor parameters. DTC was first proposed in [18]–[21] for induction machines and has been extended to other machines, such as PMSMs [22]. Based on the stator currents and voltages, the flux and electromagnetic torque are estimated at first. Then, hysteresis regulators are usually implemented in the conventional switching table-based DTC strategy for the regulation of both electromagnetic torque and stator flux as shown in Figure 1.17. The switching table for DTC is given in Table 1.3, in which ε_T and ε_ψ are the torque and flux error output by hysteresis controllers. The voltage vectors for each case of ε_T and ε_ψ at different rotor positions are listed together with corresponding switching states of the inverter in the parentheses.

Compared to the FOC, coordinate transformations and specific modulations are not required. Hence, the transient torque control performance can be significantly improved. However, in a digital controller, the responses of both torque and stator flux controllers can exceed the predefined hysteresis bands due to the time delay and the fixed sampling frequency, potentially leading to large torque and stator flux ripples. In order to reduce the torque and flux ripples, DTC with SVPWM [23]–[25] can be used with a fixed switching frequency.

1.3.2.4 Model Predictive Control

The advancements in fast and powerful microprocessors have led to increased interest in utilizing MPC for PMSM drive systems [26]–[29]. Research on PMSM drives has focused primarily on predictive current control, predictive torque

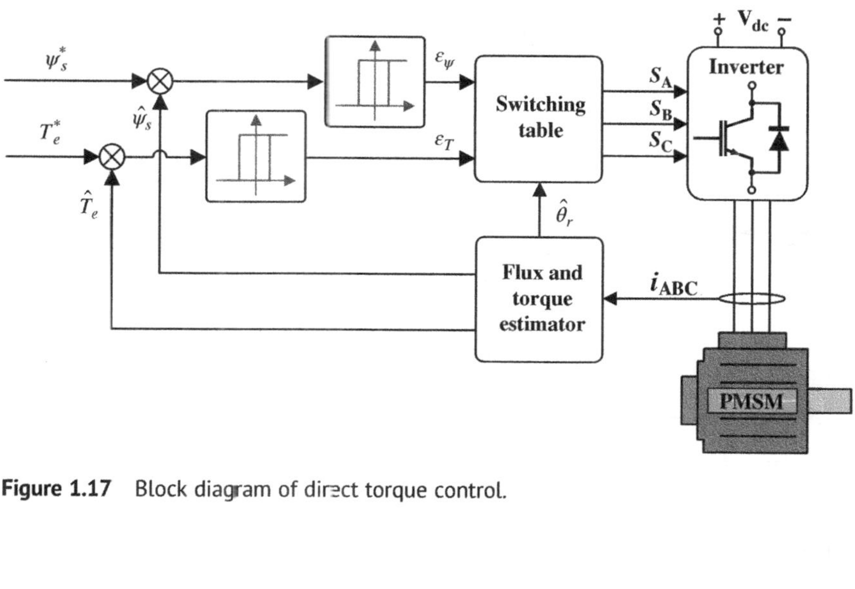

Figure 1.17 Block diagram of direct torque control.

Table 1.3 Switching table of DTC.

ε_ψ	ε_T	Rotor position					
		330–30°	30–90°	90–150°	150–210°	210–270°	270–330°
1	1	$\mathbf{v}_2$(110)	$\mathbf{v}_3$(010)	$\mathbf{v}_4$(011)	$\mathbf{v}_5$(001)	$\mathbf{v}_6$(101)	$\mathbf{v}_1$(100)
	0	$\mathbf{v}_6$(101)	$\mathbf{v}_1$(100)	$\mathbf{v}_2$(110)	$\mathbf{v}_3$(010)	$\mathbf{v}_4$(011)	$\mathbf{v}_5$(001)
0	1	$\mathbf{v}_3$(010)	$\mathbf{v}_4$(011)	$\mathbf{v}_5$(001)	$\mathbf{v}_6$(101)	$\mathbf{v}_1$(100)	$\mathbf{v}_2$(110)
	0	$\mathbf{v}_5$(001)	$\mathbf{v}_6$(101)	$\mathbf{v}_1$(100)	$\mathbf{v}_2$(110)	$\mathbf{v}_3$(010)	$\mathbf{v}_4$(011)

control, and predictive speed control [29]. A typical model of predictive current control of PMSM is shown in Figure 1.18.

Based on the machine model and the current at the kth sampling instant, i.e. $i_{dq}(k)$, the current at the $(k+1)$th sampling instant, i.e. $i_{dq}(k+1)$, is predicted iteratively with different inverter switching states. Based on the machine model in Section 1.3.1.3, the prediction model can be expressed as:

$$\begin{bmatrix} i_d^p(k+1) \\ i_q^p(k+1) \end{bmatrix} = \begin{bmatrix} 1-\dfrac{T_s}{L_s}R_s & \omega_r T_s \\ -\omega_r T_s & 1-\dfrac{T_s}{L_s}R_s \end{bmatrix} \begin{bmatrix} i_d(k) \\ i_q(k) \end{bmatrix} + \begin{bmatrix} \dfrac{T_s}{L_s} & 0 \\ 0 & \dfrac{T_s}{L_s} \end{bmatrix} \begin{bmatrix} v_d(k) \\ v_q(k) \end{bmatrix} + \begin{bmatrix} 0 \\ -\dfrac{T_s}{L_s}\omega_r\psi_m \end{bmatrix} \tag{1.21}$$

where T_s is the sampling period.

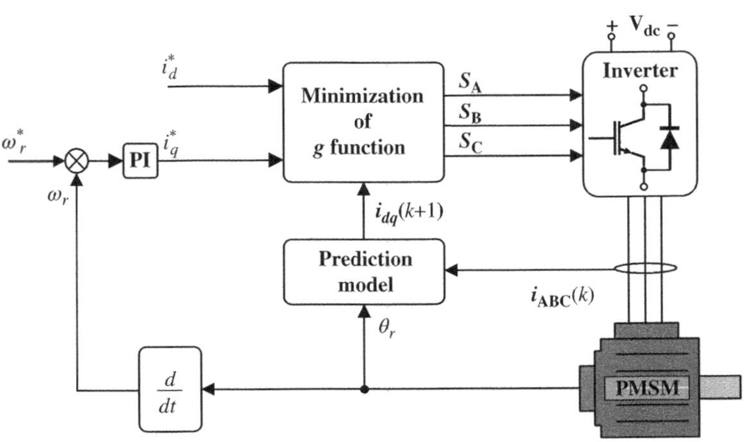

Figure 1.18 Block diagram of model predictive current control.

Then, a cost function in (1.22) is designed to select the voltage vector that minimizes the error between the reference current and the predicted current.

$$g = \left|i_d^* - i_d^p(k+1)\right| + \left|i_q^* - i_q^p(k+1)\right| \tag{1.22}$$

The dynamics of the MPC are desirable as it utilizes predictive variables to determine the switching state, avoiding the need for PWM. However, a major issue of MPC is the current tracking error caused by uncertainties in the model. To address this issue, additional compensations are required.

1.4 Basic Principle of PM BLDC Drives

1.4.1 Modeling

As with the BLAC drives, to simplify the modeling, the same assumptions of BLAC drives in Section 1.3 are made for BLDC drives. Then, the BLDC three-phase stator voltage equations can be expressed as:

$$\begin{bmatrix} v_A \\ v_B \\ v_C \end{bmatrix} = \begin{bmatrix} R_s & 0 & 0 \\ 0 & R_s & 0 \\ 0 & 0 & R_s \end{bmatrix} \cdot \begin{bmatrix} i_A \\ i_B \\ i_C \end{bmatrix} + p \begin{bmatrix} L_{AA} & M_{AB} & M_{AC} \\ M_{BA} & L_{BB} & M_{BC} \\ M_{CA} & M_{CB} & L_{CC} \end{bmatrix} \cdot \begin{bmatrix} i_A \\ i_B \\ i_C \end{bmatrix} + \begin{bmatrix} e_A \\ e_B \\ e_C \end{bmatrix} \tag{1.23}$$

where e_A, e_B, e_C are the three-phase back-EMFs.

For BLDC machines in which PMs are often surface-mounted, it can be assumed that there is no change in the inductance with rotor position and the three phases are symmetrical. Then, the self- and mutual-inductances can be represented as:

$$L_{AA} = L_{BB} = L_{CC} = L, \;\; M_{AB} = M_{BA} = M_{BC} = M_{CB} = M_{AC} = M_{CA} = M \tag{1.24}$$

Therefore, (1.23) can be re-written as:

$$\begin{bmatrix} v_A \\ v_B \\ v_C \end{bmatrix} = \begin{bmatrix} R_s & 0 & 0 \\ 0 & R_s & 0 \\ 0 & 0 & R_s \end{bmatrix} \begin{bmatrix} i_A \\ i_B \\ i_C \end{bmatrix} + \begin{bmatrix} L & M & M \\ M & L & M \\ M & M & L \end{bmatrix} \cdot p \begin{bmatrix} i_A \\ i_B \\ i_C \end{bmatrix} + \begin{bmatrix} e_A \\ e_B \\ e_C \end{bmatrix} \tag{1.25}$$

Since the three phase currents are balanced, i.e. $i_A + i_B + i_C = 0$, (1.25) can be further simplified as:

$$\begin{bmatrix} v_A \\ v_B \\ v_C \end{bmatrix} = \begin{bmatrix} R_s & 0 & 0 \\ 0 & R_s & 0 \\ 0 & 0 & R_s \end{bmatrix} \cdot \begin{bmatrix} i_A \\ i_B \\ i_C \end{bmatrix} + \begin{bmatrix} L-M & 0 & 0 \\ 0 & L-M & 0 \\ 0 & 0 & L-M \end{bmatrix} \cdot p \begin{bmatrix} i_A \\ i_B \\ i_C \end{bmatrix} + \begin{bmatrix} e_A \\ e_B \\ e_C \end{bmatrix} \tag{1.26}$$

Ideally, the three-phase back-EMFs are trapezoidal as shown in Figure 1.7.

1.4.2 Control Strategies

In the BLDC operation, the three-phase current should be aligned with the back-EMF so that the angle between the stator flux and the rotor flux is kept within the range from 60° to 120° to produce the maximum torque. Generally, two-phase 120° conduction six-step control is employed in BLDC operation, which is shown in Figure 1.19. As can be seen, the phase current is switched on 30° after the zero-crossing of its phase back-EMF waveform.

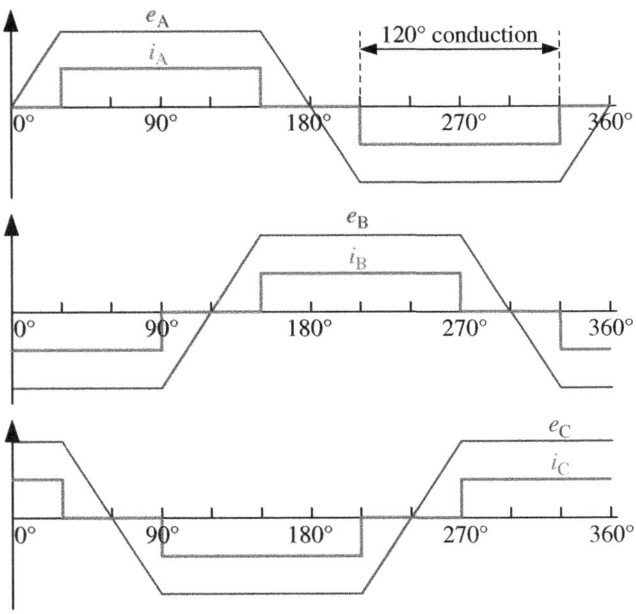

Figure 1.19 Illustration of 120° conduction six-step control strategies.

In the two-phase 120° conduction six-step control, a fundamental cycle is divided into six sectors as shown in Figure 1.2a. In each sector, two of the three phases are conducted and the third phase is floating. For each phase, it is switched on for 120° and is then switched off for 60°. In this 60° floating interval, the back-EMF can be measured, which provides a way to detect the rotor position.

Figure 1.20 shows a block diagram for the 120° conduction six-step control, where ω_r is the machine speed and i_{dc} is the dc-link current. The superscript * denotes the reference value.

As shown in Table 1.4, six sectors are defined according to the rotor position, which determines the conduction modes of the three phases. The conduction sequence across the six sectors is stored in the commutation logic block in Figure 1.20.

The standard control system of 120° conduction six-step control contains two control loops, i.e. the outer loop for speed control and the inner loop for current control, which are connected in cascade. The measured and estimated mechanical and electrical variables are fed back in each closed loop. The output of the speed

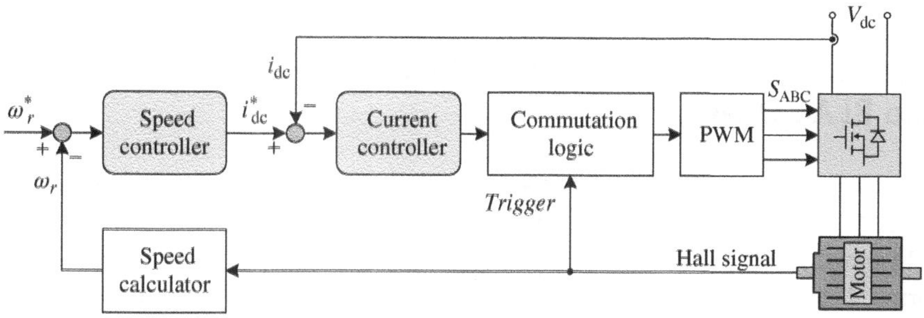

Figure 1.20 Block diagram of BLDC operation based on six-step control.

Table 1.4 Definition of sectors and conduction modes.

Sector	Rotor position (°)	Positive phase	Negative phase	Floating phase
I	30–90	A	B	C
II	90–150	A	C	B
III	150–210	B	C	A
IV	210–270	B	A	C
V	270–330	C	A	B
VI	330–30	C	B	A

controller is the reference input of the current controller i_{dc}^*. By adjusting the PWM duty ratio (output voltage), the feedback current i_{dc} can track the reference input i_{dc}^* and the feedback speed ω_r can track the reference input ω_r^*. That is the mechanism of the typical 120° conduction six-step control. Since the time constant of electrical variables (resistance and inductance) is far lower than the one of mechanical variables, the inner loop is much faster than the outer loop. In this way, there are no interferences between the current and speed controllers, and thus, they can be designed and tuned independently.

For the current control loop, since only two phases are conducted, the stator current can be directly reflected on the dc-link. For low-cost applications, the dc-link current signal is usually utilized in the current control loop. For the speed control loop, the machine speed and rotor position (commutation instants) information can be derived from position sensors. As shown in Figure 1.20, for a sensored BLDC drive, the commutation is often triggered by the Hall-effect sensors. The machine speed can be calculated from the time duration between two subsequent transitions of Hall-effect sensors. For sensorless operation of BLDC drives, refer to Chapters 10 and 11.

1.5 Comparison Between PM BLDC (PMSM) and BLAC Drives

A brief comparison between BLDC and BLAC operation modes is given in Table 1.5.

In practice, the back-EMF may be significantly different from the ideal waveforms. Besides, the selection of BLDC or BLAC operation mode also depends on the specific application. Various factors should be taken into consideration, e.g. the implementation complexity, the system cost, the performance demand, etc. Hence, irrespective of the back-EMF waveforms, the PM brushless machines may be operated in either BLDC or BLAC mode.

1.5.1 Square-Wave Back-EMF Machine

Ideally, a square-wave back-EMF machine should be operated in the BLDC mode. By way of example, if both the phase current and back-EMF are of ideal waveforms, as shown in Figure 1.7a, i.e. the back-EMF is trapezoidal with a flat top of 120° electrically, while the current waveform is square-wave, the electromagnetic torque of a BLDC machine will be free of ripple, and can be expressed as:

$$T_{m\_BLDC} = \frac{2E_m I_m}{\omega_m} \tag{1.27}$$

Table 1.5 Comparison between BLDC and BLAC operation [30].

	BLDC operation	BLAC operation
Control variable	Square-wave current	Sine-wave current
Position sensor resolution	Low	High
Ripple in electromagnetic torque	High or moderate[a]	Low
Speed and/or position control	Coarse	Precise
Inverter switching loss[b]	Low	High
Current harmonics	High	Low
Control simplicity	Simple	Complicated
Flux-weakening capability [31][c]	Low	High

[a] Because the ideal square-wave back-EMF waveform is difficult to achieve in practical machines.
[b] In BLDC operation, there is switching loss in one or two power electronic devices, depending on how PWM is applied. In BLAC operation there is switching loss in all six devices at any instant.
[c] Assuming the same machine running in BLDC and BLAC modes, respectively.

where ω_m is the mechanical angular speed. Both the phase current and back-EMF are rich in harmonics, as listed in Table 1.6. The interaction between current and back-EMF harmonics of the same order will result in an effective magnetic torque, as also listed in Table 1.6.

From Table 1.6, it can be seen that:

- There is no third harmonic in the phase current, but the higher order harmonics are significant.
- The most significant harmonic in the phase back-EMF is the third harmonic, while the higher harmonics are relatively low. However, the third-harmonic

Table 1.6 Harmonics of back-EMF and phase current, and electromagnetic torque due to the same-order harmonics in an ideal BLDC drive [30].

Order of harmonic (ν)	Amplitude of back-EMF harmonics ($E_{m\nu}/E_m$) (%)	Amplitude of phase current harmonics ($I_{m\nu}/I_m$) (%)	Torque due to same-order harmonics ($T_{m\nu}/T_m$) (%)
1	121.6	110.3	100.6
3	27.0	0	0
5	4.9	−22.1	−0.8
7	−2.5	−15.8	0.3
…	…	…	…

Table 1.7 Current and torque comparison for a square-wave back-EMF machine in BLDC and BLAC modes [30].

Amplitude of 120°-trapezoidal back-EMF			E_m	
BLDC operation	Amplitude of square-wave phase current		I_m	
	Electromagnetic torque		$2\,E_m\,I_m/\Omega$	
BLAC operation	Amplitude of sine-wave phase current	$1.096\,I_m$	I_m	$1.155\,I_m$
	Electromagnetic torque	$2\,E_m\,I_m/\omega_m$	$1.825\,E_m\,I_m/\omega_m$	$2.107\,E_m\,I_m/\omega_m$
	Condition	Same torque as BLDC operation	Same current peak as BLDC operation	Same current RMS as BLDC operation

back-EMF does not contribute to the electromagnetic torque. Therefore, only the fundamental component of the back-EMF is dominant.

- The effective electromagnetic torque is produced mainly by the interaction between the fundamental back-EMF and current. High-order harmonics produce little effective electromagnetic torque.

In practical applications, the square-wave machine can also operate in the BLAC mode. A performance comparison is listed in Table 1.7.

1.5.2 Sine-Wave Back-EMF Machine

A sine-wave machine should ideally operate in the BLAC mode. When both the phase current and back-EMF are ideal sinusoidal waveforms, as shown in Figure 1.7b, the electromagnetic torque will be ripple free, and can be expressed as:

$$T_{m\_\mathrm{BLAC}} = \frac{3E_m I_m}{2\omega_m} \tag{1.28}$$

However, in practical applications it can also be operated in the BLDC mode, since there is no third harmonic in the phase current while the higher harmonics are usually low. Hence, the fundamental component of current is dominant. In such a case, the performance comparison is listed in Table 1.8.

Table 1.8 Current and torque comparison for a sine-wave back-EMF machine in BLAC and BLDC modes [30].

Amplitude of sinusoidal back-EMF		E_m		
BLAC operation	Amplitude of sine-wave phase current	I_m		
	Electromagnetic torque	$1.5\, E_m\, I_m\, /\omega_m$		
BLDC operation	Amplitude of square-wave phase current	$0.907\, I_m$	I_m	$0.866\, I_m$
	Electromagnetic torque	$1.5\, E_m\, I_m\, /\omega_m$	$1.654\, E_m\, I_m\, /\omega_m$	$1.432\, E_m\, I_m\, /\omega_m$
	Condition	Same torque as BLAC operation	Same current peak as BLAC operation	Same current RMS as BLAC operation

Note that if a square-wave machine is operated in the BLAC mode, or a sine-wave machine is run in the BLDC mode, or if the machine back-EMF waveform is neither a pure sine wave nor a pure square wave or trapezoidal waveform with 120° flat top, torque ripples will arise. In some machines, the line-to-neutral back-EMF waveform is not sinusoidal since it contains zero sequence harmonics, such as the third harmonic. However, the line-to-line back-EMF waveform can be sinusoidal, since zero sequence harmonics do not exist. Such machines are still eminently suitable for BLAC operation.

1.6 Sensorless Control Techniques and Applications

High-performance PM machine drives need rotor position information. They are usually obtained by employing a hardware rotor position sensor, such as a resolver, encoder, Hall effect sensor, etc. However, doing so will increase the size and reduce the reliability particularly under harsh environments. It is always desirable to eliminate the rotor hardware rotor position sensor by using software-based sensorless techniques [32].

1.6.1 Classification

There are various sensorless control methods, and a brief classification of sensorless control techniques for BLAC and BLDC drives is shown in Figure 1.21.

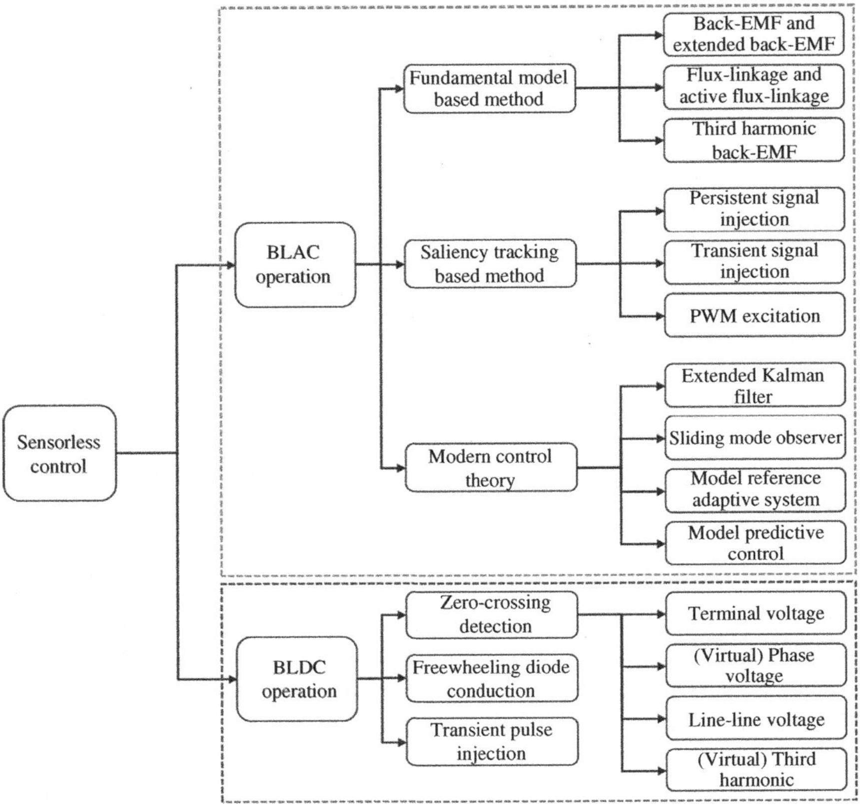

Figure 1.21 Classification of sensorless control techniques for PM machines.

For BLAC operation, fundamental model-based methods are based on the back-EMF or flux-linkage; accurate parameters and observable machine model are necessary. Therefore, they are preferred in middle and high-speed ranges. For SPMs, back-EMF and flux-linkage can be utilized, while for IPMs, the extended back-EMF and active flux-linkage are used for sensorless control. Apart from the fundamental component, the third harmonic of back-EMF and flux-linkage can also be used for position estimation. For saliency-based methods, the PM machine saliency is independent of the rotor speed. With different signal injection methods, this type of method can be used in zero and low-speed ranges. Moreover, modern control theories such as the extended Kalman filter (EKF), sliding mode observer (SMO), and model adaptive reference system (MRAS) can be applied to conventional sensorless techniques to enhance the estimation performance.

For BLDC operation, the zero-crossing points (ZCPs) detection-based method is widely adopted, which provides a good performance at middle and high speeds since it relies on the amplitude of the back-EMF. The back-EMF can be measured in various ways as shown in Figure 1.21. Moreover, the freewheeling diode conduction is another way to indirectly detect the ZCPs. To detect the rotor position at zero speed, the transient pulse injection-based method can be employed.

1.6.2 Applications

In general, sensorless techniques have proven to be highly effective for applications with low torque requirements at start-up. In recent years, these techniques have been successfully commercialized for a variety of applications, including ventilation fans, pumps, vacuum cleaners, hand dryers, hair dryers, and, more recently, wind power generators. Despite this success, the use of sensorless control for applications that require high torque for regular starting, such as general industrial drives and electric vehicle propulsion, remains challenging.

However, there have been successful developments in the use of sensorless control for commercial washing machines by utilizing high-frequency signal injection for starting since the starting period is not so critical, although it is desirable that it be as short as possible. Researchers are continuing to investigate the potential of this technology for electric vehicles, with extensive research activities currently underway. It is anticipated that over the next five to ten years, sensorless control will be increasingly used for fault-tolerance drives in electric vehicles and aerospace applications, as researchers work to improve the technology and address the challenges associated with high-torque applications.

Despite the current limitations of sensorless control techniques, they offer numerous advantages over traditional methods that rely on sensors. For example, sensorless control can reduce costs and complexity by eliminating the need for dedicated sensors, while also improving reliability by reducing the number of failure points in the system. As the technology continues to evolve and improve, it is likely that sensorless control will find even broader application, offering a reliable, cost-effective, and efficient solution for a wide range of industrial and commercial needs.

1.7 Scope of This Book

In this book, the basic principles and the state-of-the-art position sensorless control techniques for PM machine drives will be described, including both PM BLAC and BLDC drives, single three-phase PM machines, dual three-phase PM

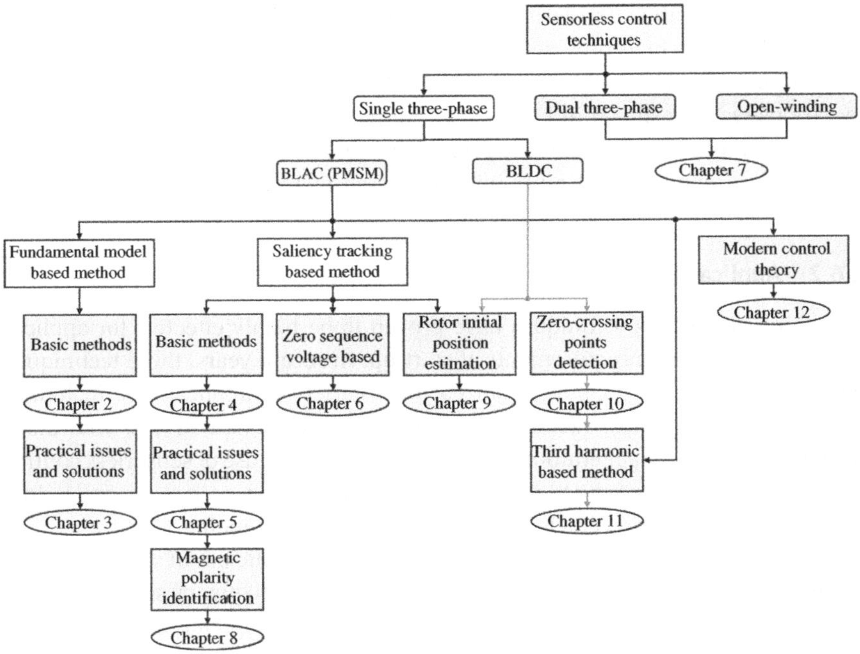

Figure 1.22 Scope of this book.

machines, and open-winding PM machines. The overall scope of this book is illustrated in Figure 1.22.

For PM BLAC drives, Chapter 2 covers the fundamental model-based methods, including flux observer, active flux observer, back-EMF, and extended back-EMF. Chapter 3 presents practical issues and corresponding solutions, e.g. integration drifting and filter phase delay, back-EMF and current harmonics, cross-coupling magnetic saturation, parameter mismatch, asymmetric parameters, etc. For the saliency tracking-based method, Chapter 4 introduces the basic methods using high-frequency signal injection in various reference frames according to injection signal and response voltage and current types, such as rotating, pulsating, and square wave. In Chapter 5, special issues of the saliency tracking-based method will be fully discussed, including various parasitic effect issues and solutions, e.g. cross-coupling magnetic saturation, multi-saliency, asymmetric parameters, inverter nonlinearity, signal delay, machine design, load effect, etc. For high-frequency signal injection, the methods for selecting the amplitudes and frequencies for injection signals will be systematically described. Moreover, for the saliency tracking-based method, the utilization of the zero-sequence voltage is described in Chapter 6. Apart from single three-phase PM machines, sensorless techniques for dual three-phase machines and open-winding machines are detailed in Chapter 7. Further, comprehensive techniques for detecting rotor magnetic polarity and initial rotor position will be introduced in Chapters 8 and 9, respectively.

For PM BLDC drives, sensorless techniques for detecting zero crossing of back-EMF waveforms under voltage and current PWM control will be described in Chapter 10, together with those based on third harmonic back-EMF in Chapter 11, which is also applicable to BLAC drives.

Finally, in Chapter 12, this book will also highlight some modern control theory-based sensorless techniques, including model reference adaptive system, sliding mode, EKF, and MPC.

References

1 Z. Q. Zhu and D. Howe, "Electrical machines and drives for electric, hybrid, and fuel cell vehicles," *Proc. IEEE*, vol. 95, no. 4, pp. 746–765, Apr. 2007.

2 J. R. Hendershot and T. J. E. Miller, *Design of Brushless Permanent-magnet Motors*. Oxford, UK: Magna Physics and Oxford Science Publications, 1994.

3 J. F. Gieras, R. Wang, and M. J. Kamper, *Axial Flux Permanent Magnet Brushless Machines*. Dordrecht, Netherlands: Springer, 2008.

4 H. Weh and H. May, "Achievable force densities for permanent magnet excited machines in new configurations," *Proc. Int. Conf. Elect. Mach. Drives*, pp. 1107–1111, 1986.

5 H. Weh and J. Jiang, "Calculation and design consideration for synchronous machines with transverse flux configuration," *Archiv fur Elektrotechnik*, vol. 71, pp. 187–198, 1988.

6 D. Ishak, Z. Q. Zhu, and D. Howe, "Comparison of PM brushless motors, having either all teeth or alternate teeth wound," *IEEE Trans. Energy Convers.*, vol. 21, no. 1, pp. 95–106, Mar. 2006.

7 A. M. EL-Refaie, "Fractional-slot concentrated-windings synchronous permanent magnet machines: opportunities and challenges," *IEEE Trans. Ind. Electron.*, vol. 57, no. 1, pp. 107–121, Jan. 2010.

8 G. Dajaku, W. Xie, and D. Gerling, "Reduction of low space harmonics for the fractional slot concentrated windings using a novel stator design," *IEEE Trans. Magn.*, vol. 50, no. 5, pp. 1–12, May 2014.

9 P. Pillay and R. Krishnan, "Modeling, simulation, and analysis of permanent-magnet motor drives. I. The permanent-magnet synchronous motor drive," *IEEE Trans. Ind. Appl.*, vol. 25, no. 2, pp. 265–273, Mar. 1989.

10 P. Pillay and R. Krishnan, “Modeling, simulation, and analysis of permanent-magnet motor drives. II. The brushless dc motor drive,” *IEEE Trans. Ind. Appl.*, vol. 25, no. 2, pp. 274–279, Mar. 1989.

11 H. W. van der Broeck, H.-C. Skudelny, and G. V. Stanke, “Analysis and realization of a pulse width modulator based on voltage space vectors,” *IEEE Trans. Ind. Appl.*, vol. 24, no. 1, pp. 142–150, Jan. 1988.

12 O. Ogasawara, H. Akagi, and A. Nabae, “A novel PWM scheme of voltage source inverters based on space vector theory,” *EPE Eur. Conf. Power Electron. Appl.*, Aachen, Germany, 1989, pp. 1197–1202.

13 J. Holtz, “Pulsewidth modulation––a survey,” *IEEE Trans. Ind. Elect.*, vol. 39, no. 5, pp. 410–420, Oct. 1992, https://doi.org/10.1109/41.161472.

14 K. P. Kovacs and I. Racz, *Transiente Vorgange in Wechselstrom-maschinen*, vol. I and II. Budapest, Hungary: Verlag d. Ungar. Akad. d. Wissensch, 1959.

15 F. Blaschke, “The principle of field orientation as applied to the new transvector closed loop control system for rotating field machines,” *Siemens Rev.*, vol. 34, pp. 217–220, 1972.

16 K. H. Bayer, H. Waldmann, and M. Weibelzahl, “Field-oriented closed-loop control of a synchronous machine with the new transvector control system,” *Siemens Rev.*, vol. 39, pp. 220–223, 1972.

17 P. Vas, *Sensorless Vector and Direct Torque Control.* Oxford, UK: Oxford Univ. Press, 2003.

18 I. Takahashi and T. Noguchi, “A new quick-response and high efficiency control strategy of an induction-motor,” *IEEE Trans. Ind. Appl.*, vol. 22, no. 5, pp. 820–827, 1986.

19 I. Takahashi and Y. Ohmori, “High-performance direct torque control of an induction motor,” *IEEE Trans. Ind. Appl.*, vol. 25, no. 2, pp. 257–264, Mar. 1989.

20 M. Depenbrock, “Direct self-control (DSC) of inverter-fed induction machine,” *IEEE Trans. Power Electron.*, vol. 3, no. 4, pp. 420–429, 1988.

21 U. Baader, M. Depenbrock, and G. Gierse, “Direct self control (DSC) of inverter-fed induction machine: a basis for speed control without speed measurement,” *IEEE Trans. Ind. Appl.*, vol. 28, no. 3, pp. 581–588, May 1992.

22 L. Zhong, M. F. Rahman, W. Y. Hu, and K. W. Lim, “Analysis of direct torque control in permanent magnet synchronous motor drives,” *IEEE Trans. Power Electron.*, vol. 12, no. 3, pp. 528–536, 1997.

23 T. G. Habetler, F. Profumo, M. Pastorelli, and L. M. Tolbert, “Direct torque control of induction machines using space vector modulation,” *IEEE Trans. Ind Appl.*, vol. 28, no. 5, pp. 1045–1053, Oct. 1992.

24 G. S. Buja and M. P. Kazmierkowski, “Direct torque control of PWM inverter-fed AC motors—a survey,” *IEEE Trans. Ind. Electron.*, vol. 51, no. 4, pp. 744–757, Aug. 2004.

25 L. Tang, L. Zhong, M. F. Rahman, and Y. Hu, "A novel direct torque controlled interior permanent magnet synchronous machine drive with low ripple in flux and torque and fixed switching frequency," *IEEE Trans. Power Electron.*, vol. 19, no. 2, pp. 346–354, Mar. 2004.

26 J. Rodriguez *et al.*, "Predictive current control of a voltage source inverter," *IEEE Trans. Ind. Electron.*, vol. 54, no. 1, pp. 495–503, Feb. 2007.

27 S. Kouro, P. Cortes, R. Vargas, U. Ammann, and J. Rodriguez, "Model predictive control—a simple and powerful method to control power converters," *IEEE Trans. Ind. Electron.*, vol. 56, no. 6, pp. 1826–1838, Jun. 2009.

28 J. Rodriguez *et al.*, "Latest advances of model predictive control in electrical drives—part I: basic concepts and advanced strategies," *IEEE Trans. Power Electron.*, vol. 37, no. 4, pp. 3927–3942, Apr. 2022.

29 J. Rodriguez *et al.*, "Latest advances of model predictive control in electrical drives—part II: applications and benchmarking with classical control methods," *IEEE Trans. Power Electron.*, vol. 37, no. 5, pp. 5047–5061, May 2022.

30 J. X. Shen, "Sensorless control of permanent magnet brushless drives," Ph.D. thesis, University of Sheffield, UK, 2003.

31 Y. F. Shi, Z. Q. Zhu, and D. Howe, "Torque-speed characteristics of interior-magnet machines in brushless ac and dc modes, with particular reference to their flux-weakening performance," *CES/IEEE 5th Int. Power Elect. Motion Control Conf.*, vol. 3, pp. 1–5, Aug. 2006.

32 K. Rajashekara, A. Kawamura, and K. Matsuse, *Sensorless Control of AC Motors.* New York, NY, USA: IEEE Press, 1996.

2

Fundamental Model-Based Sensorless Control

2.1 Introduction

For fundamental model-based sensorless control methods, the basic idea is to estimate the back-EMF or flux-linkage based on the fundamental model of PM machines. Most of them can fit into the frame shown in Figure 2.1. For each part of the frame, different types of sensorless control techniques are shown in Figure 2.1 and will be introduced in this chapter.

For different sensorless techniques, they can be characterized in terms of parts A, B, C, and D in Figure 2.1. In part A, the machine model coordinate, based on which of the sensorless control techniques is established, is classified. Then, the majority of the research focuses on part B. Different techniques are developed to estimate the back-EMF, flux-linkage, or position error as shown in Figure 2.1. For part B, the basic methods for both non-salient PMSMs (SPMSM) and salient PMSMs (IPMSM) are all included. For part C, after back-EMF, flux-linkage, or position error is estimated, a position observer is applied to generate the speed and position information. Part D includes some practical implementation issues, including cross-coupling inductance, back-EMF and current harmonics, parameter mismatches, and parameter asymmetries, which will be introduced in the next chapter.

2.2 Flux-Linkage-Based Method

Flux-linkage estimation-based sensorless control is widely used, and its basic idea is simple. Various flux estimation methods are applied to estimate the flux produced by the PM and consequently to estimate the rotor position for PMSMs. These methods are well developed in different reference frames for both non-salient [1], [2] and salient PMSMs [3], which will be introduced in this section.

Sensorless Control of Permanent Magnet Synchronous Machine Drives, First Edition.
Zi Qiang Zhu and Xi Meng Wu.

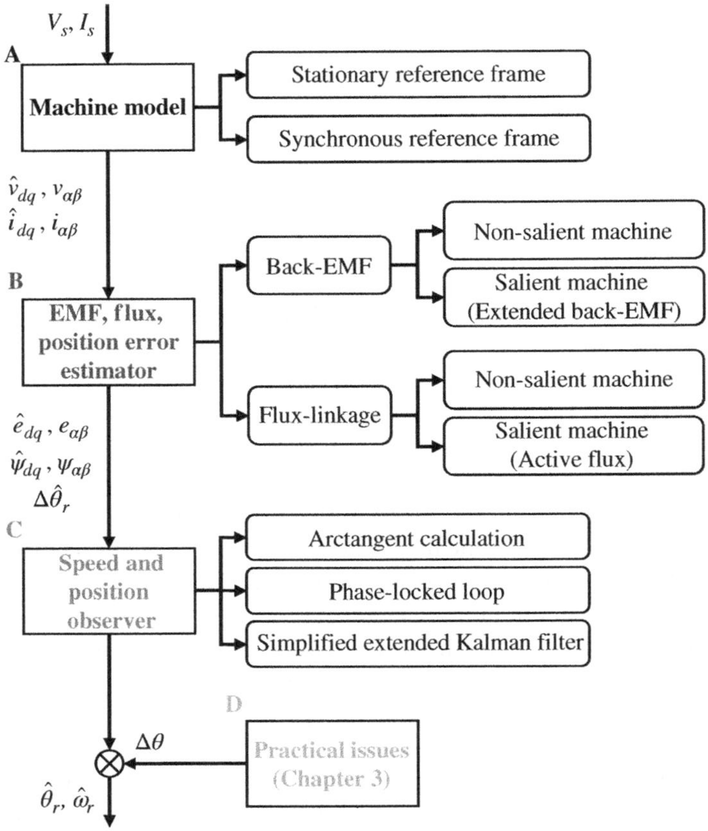

Figure 2.1 Classification of fundamental model-based sensorless control techniques.

2.2.1 Flux-Linkage Method for Non-salient PMSMs

The flux-linkages for non-salient PMSMs in the stationary reference frame are expressed as:

$$\begin{bmatrix} \psi_\alpha \\ \psi_\beta \end{bmatrix} = \begin{bmatrix} L_\alpha & 0 \\ 0 & L_\beta \end{bmatrix} \begin{bmatrix} i_\alpha \\ i_\beta \end{bmatrix} + \begin{bmatrix} \psi_{m\alpha} \\ \psi_{m\beta} \end{bmatrix}, \quad \begin{bmatrix} \psi_{m\alpha} \\ \psi_{m\beta} \end{bmatrix} = \psi_m \begin{bmatrix} \cos\theta_r \\ \sin\theta_r \end{bmatrix} \tag{2.1}$$

where ψ_α, ψ_β are the $\alpha\beta$-axis stator flux-linkages, $\psi_{m\alpha}$, $\psi_{m\beta}$ are the $\alpha\beta$-axis PM excitation flux-linkages, ψ_m is the magnitude of the PM flux-linkage, θ_r is the rotor position, i_α and i_β are the $\alpha\beta$-axis stator currents, and $L_\alpha = L_\beta = L_s$ is the phase inductance for non-salient PMSMs. The flux-linkage can be illustrated by the

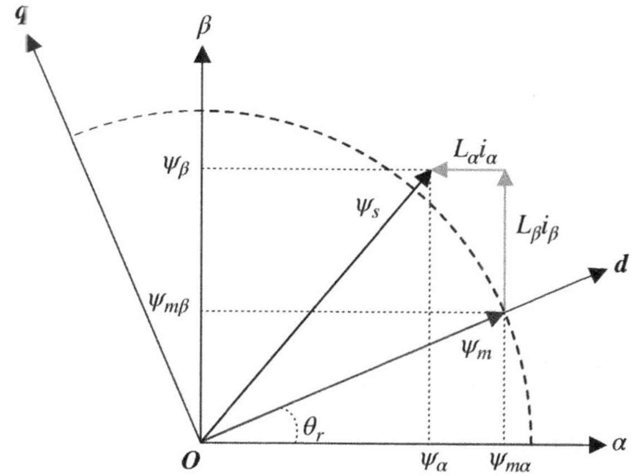

Figure 2.2 Phasor diagram of flux-linkage of a PMSM.

phasor diagram as shown in Figure 2.2, which corresponds to equation (2.1). Then, the rotor position can be obtained:

$$\theta_r = \tan^{-1}\left(\frac{\psi_\beta - L_s i_\beta}{\psi_\alpha - L_s i_\alpha}\right) = \tan^{-1}\left(\frac{\psi_{m\beta}}{\psi_{m\alpha}}\right) \tag{2.2}$$

The $\alpha\beta$-axis stator flux-linkages can be calculated by integrating the back-EMF voltages as

$$\begin{aligned} \psi_\alpha &= \int (v_\alpha - i_\alpha R_s)dt \\ \psi_\beta &= \int (v_\beta - i_\beta R_s)dt \end{aligned} \tag{2.3}$$

where R_s is the phase resistance.

Based on the prototype SPMSM-I (its parameters are given in Appendix B), some simulation results of position estimation performance with the flux-linkage method are shown in Figures 2.3–2.6. In Figures 2.3 and 2.4, the steady-state position estimation performance under no load and full load are shown. The rotor speed is controlled at 300 r/min. In Figures 2.5 and 2.6, the dynamic-state performance is shown including a step load test (from no load to full load) and a step speed test (300–400–500 r/min). Clearly, as can be seen from the results, a good position estimation performance can be obtained in both steady and dynamic states.

However, in practice, the implementation of a pure integration has several issues that need to be considered [4], such as initial value and drift issues, which will be discussed in detail in Chapter 3.

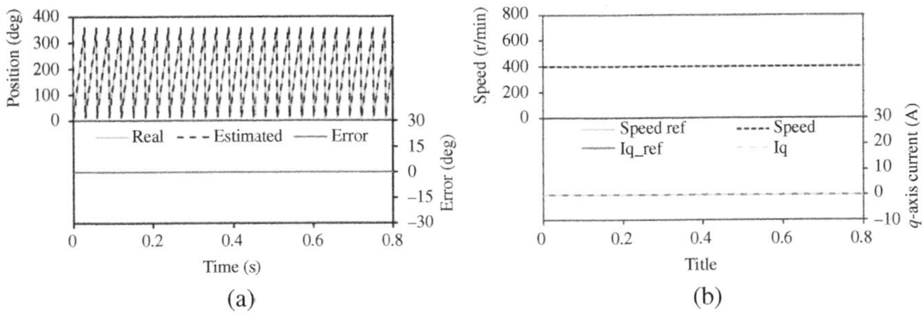

Figure 2.3 Steady-state position estimation with flux-linkage method under no load. (a) Position estimation. (b) Rotor speed and q-axis current.

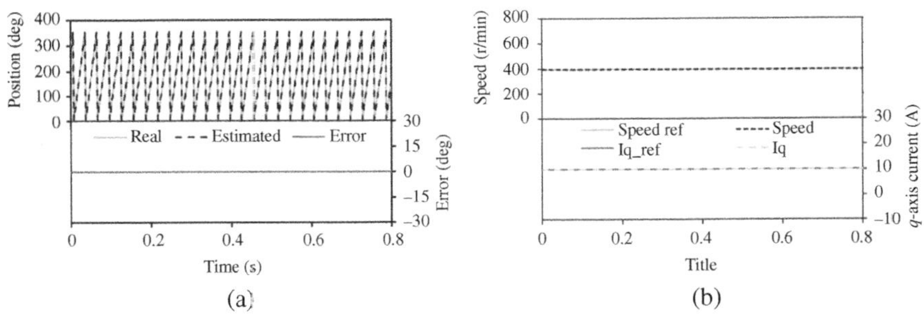

Figure 2.4 Steady-state position estimation with flux-linkage method under full load. (a) Position estimation. (b) Rotor speed and q-axis current.

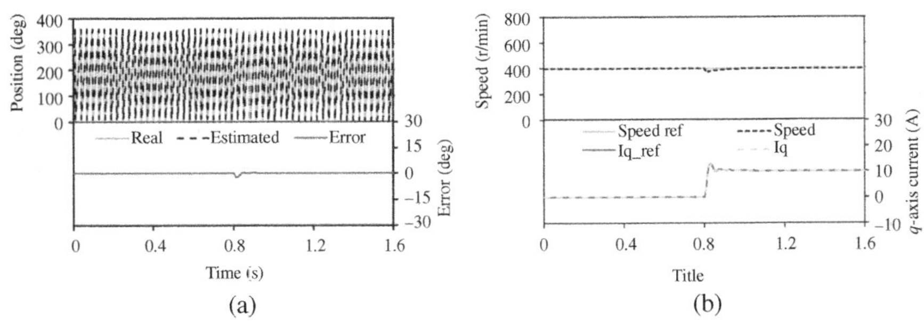

Figure 2.5 Dynamic-state position estimation with flux-linkage method under step load. (a) Position estimation. (b) Rotor speed and q-axis current.

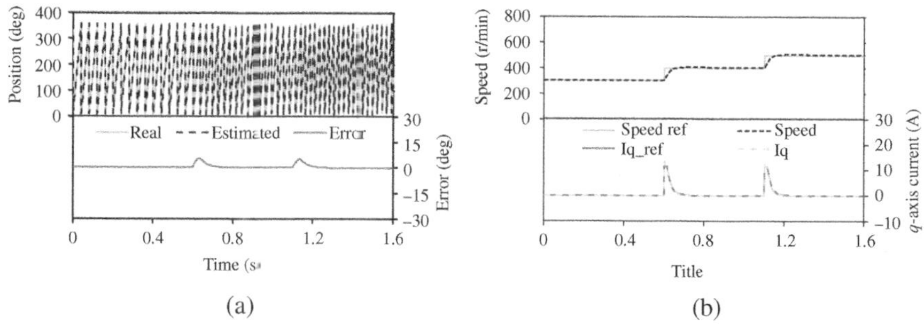

Figure 2.6 Dynamic-state position estimation with flux-linkage method under step speed. (a) Position estimation. (b) Rotor speed and q-axis current.

2.2.2 Active Flux-Linkage Method for Salient PMSMs

The flux-linkage observer introduced in Section 2.2.1 is for non-salient PMSMs. However, for a salient PMSM, the two phase inductances on the $\alpha\beta$-axis frame are coupled with the rotor position, which makes it difficult to separate the PM flux-linkages. The stator flux-linkages ψ_α, ψ_β for machines with salient pole are given by:

$$\begin{bmatrix} \psi_\alpha \\ \psi_\beta \end{bmatrix} = \begin{bmatrix} L_{sa} + L_{sd}\cos(2\theta_r) & L_{sa}\sin(2\theta_r) \\ L_{sa}\sin(2\theta_r) & L_{sa} - L_{sd}\cos(2\theta_r) \end{bmatrix} \begin{bmatrix} i_\alpha \\ i_\beta \end{bmatrix} + \psi_m \begin{bmatrix} \cos\theta_r \\ \sin\theta_r \end{bmatrix} \tag{2.4}$$

where $L_{sa} = (L_d + L_q)/2$ and $L_{sd} = (L_q - L_d)/2$. It is difficult to calculate the rotor position directly by (2.4). To simplify the position estimation process, the concept of active-flux $\vec{\psi}_a$, which is aligned with the PM flux-linkage as shown in Figure 2.7, is introduced in [3]. In Figure 2.7, $I_s = \sqrt{i_d^2 + i_q^2}$. The method in [3] unifies all the salient machines into fictitious non-salient machines and hence simplifies the position estimation.

The active flux can be expressed as:

$$\psi_a = \psi_m + (L_d - L_q)i_d \tag{2.5}$$

Then, the stator flux-linkages in stationary reference frame can be expressed as:

$$\begin{bmatrix} \psi_\alpha \\ \psi_\beta \end{bmatrix} = \begin{bmatrix} L_q & 0 \\ 0 & L_q \end{bmatrix} \begin{bmatrix} i_\alpha \\ i_\beta \end{bmatrix} + \psi_a \begin{bmatrix} \cos\theta_r \\ \sin\theta_r \end{bmatrix} \tag{2.6}$$

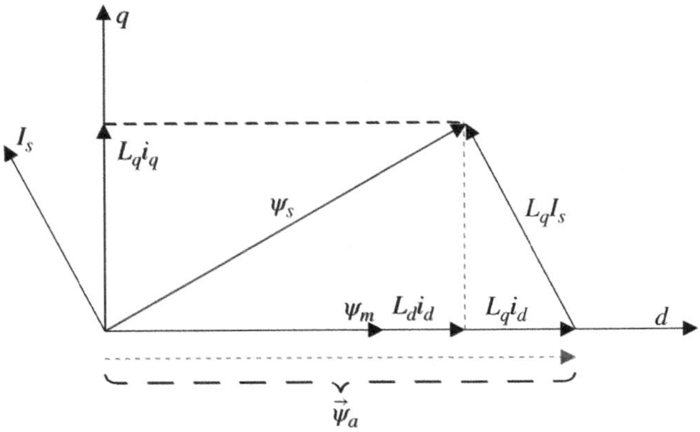

Figure 2.7 Active flux in vector diagram.

Thus, the rotor position can be calculated as:

$$\theta_r = \tan^{-1}\left(\frac{\psi_\beta - L_q i_\beta}{\psi_\alpha - L_q i_\alpha}\right) \tag{2.7}$$

The active flux-based method can also be implemented based on the estimated synchronous reference frame. With the active flux, the flux equation can be generalized as:

$$\begin{bmatrix} \psi_d \\ \psi_q \end{bmatrix} = \begin{bmatrix} L_q & 0 \\ 0 & L_q \end{bmatrix} \begin{bmatrix} i_d \\ i_q \end{bmatrix} + \begin{bmatrix} \psi_a \\ 0 \end{bmatrix} \tag{2.8}$$

Since the rotor position is unknown in the practical sensorless control system, the flux-linkage should be transformed into the estimated synchronous reference frame as:

$$\begin{bmatrix} \hat{\psi}_d \\ \hat{\psi}_q \end{bmatrix} = \begin{bmatrix} L_q & 0 \\ 0 & L_q \end{bmatrix} \begin{bmatrix} \hat{i}_d \\ \hat{i}_q \end{bmatrix} + \psi_a \begin{bmatrix} \cos\Delta\theta \\ \sin\Delta\theta \end{bmatrix} \tag{2.9}$$

Based on (2.9), the rotor position can be estimated by an observer by forcing the estimated q-axis flux-linkage $\hat{\psi}_q$ to zero.

Similarly, the active flux-linkage based method is simulated based on the prototype IPMSM-I (its parameters are given in Appendix B). The results are shown in Figures 2.8–2.11. The rotor speed is controlled at 300 r/min in the simulation. In Figures 2.8 and 2.9, the steady-state position estimation performance under both no load and full load are shown. Furthermore, Figures 2.10 and 2.11 demonstrate the dynamic-state position estimation including a step load test (from no load to

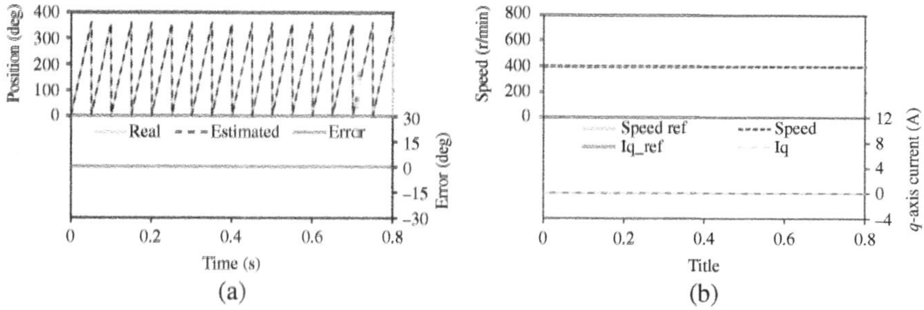

Figure 2.8 Steady-state position estimation with active flux-linkage method under no load. (a) Position estimation. (b) Rotor speed and q-axis current.

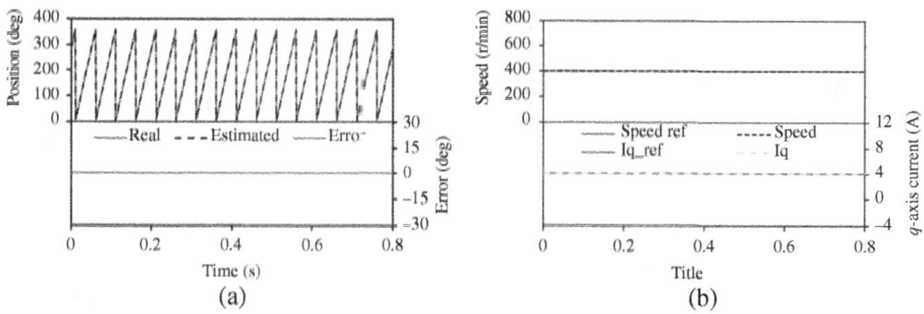

Figure 2.9 Steady-state position estimation with active flux-linkage method under full load. (a) Position estimation. (b) Rotor speed and q-axis current.

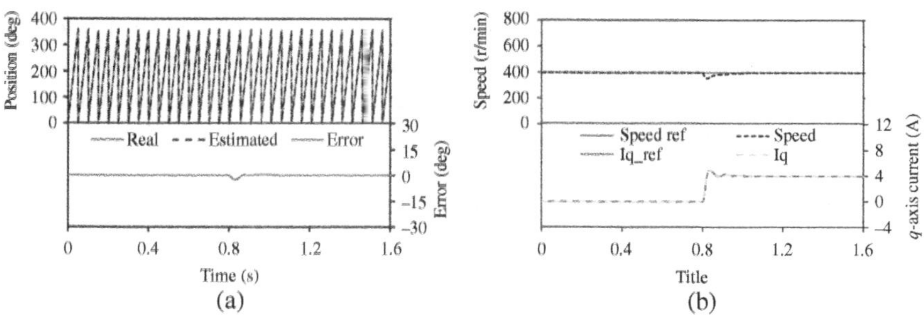

Figure 2.10 Dynamic-state position estimation with active flux-linkage method under step load. (a) Position estimation. (b) Rotor speed and q-axis current.

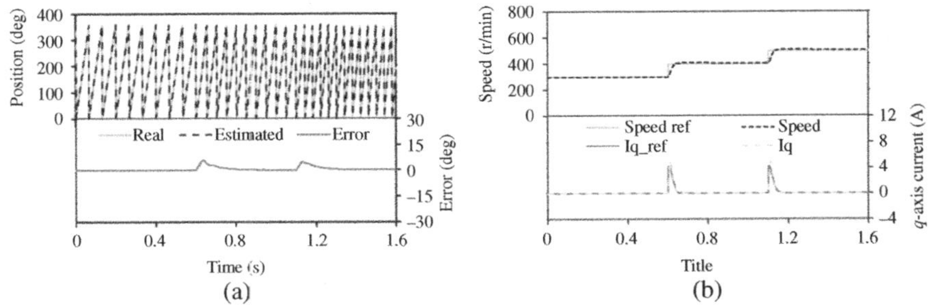

Figure 2.11 Dynamic-state position estimation with active flux-linkage method under step speed. (a) Position estimation. (b) Rotor speed and q-axis current.

full load) and a step speed test (300–400–500 r/min). Overall, good position estimation performance can be obtained.

2.3 Back-EMF-Based Method

When there is a relative motion between the stator and the magnetic field produced by the rotor PMs in the airgap, a voltage is introduced in stator windings, viz. back-EMF, and hence, back-EMF is proportional to the rotor speed. However, back-EMF is difficult to directly measure during machine operation. Instead, based on the fundamental model, back-EMF can be estimated from the stator voltage and current. The back-EMF in the stationary reference frame can be calculated by:

$$\begin{bmatrix} e_\alpha \\ e_\beta \end{bmatrix} = \begin{bmatrix} v_\alpha \\ v_\beta \end{bmatrix} - \begin{bmatrix} R_s + pL_\alpha & 0 \\ 0 & R_s + pL_\beta \end{bmatrix} \begin{bmatrix} i_\alpha \\ i_\beta \end{bmatrix} = \omega_r \psi_m \begin{bmatrix} -\sin\theta_r \\ \cos\theta_r \end{bmatrix} \tag{2.10}$$

Different methods have been developed to estimate the back-EMF. An indirect position estimation method is proposed in [5] by transforming (2.10) into the estimated synchronous reference frame. A concept of extended back-EMF is developed in [6], [7]. Using this concept, salient machines can be considered as non-salient machines. These classical methods are introduced in this section.

2.3.1 Back-EMF Method for Non-salient PMSMs

The voltage equations of a non-salient PMSM are given by:

$$\begin{bmatrix} v_d \\ v_q \end{bmatrix} = \begin{bmatrix} R_s + pL_s & -\omega_r L_s \\ \omega_r L_s & R_s + pL_s \end{bmatrix} \begin{bmatrix} i_d \\ i_q \end{bmatrix} + \begin{bmatrix} 0 \\ \omega_r \psi_m \end{bmatrix} \tag{2.11}$$

Figure 2.12 Relationship between actual and estimated synchronous reference frames.

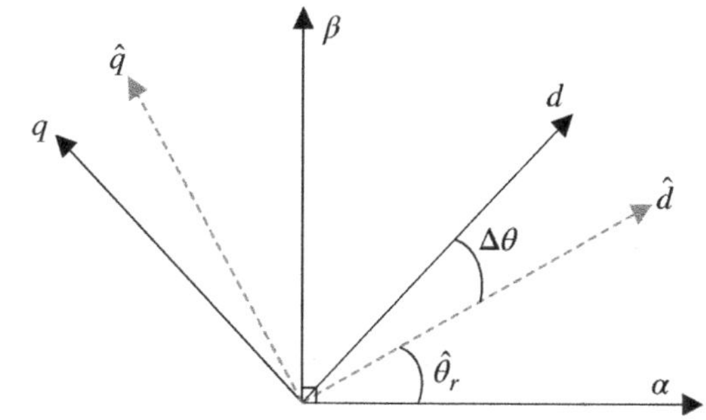

For sensorless rotor position estimation, as the actual rotor position is unknown, the voltages and currents are expressed in the estimated synchronous reference frame [5] as shown in Figure 2.12, where $\hat{\theta}_r$ is the estimated rotor position and $\Delta\theta$ is the error between the actual and estimated rotor positions, i.e. $\Delta\theta = \theta_r - \hat{\theta}_r$. Voltage equations in the estimated frame are obtained from (2.11) by coordinate transformation as:

$$\begin{bmatrix} \hat{v}_d \\ \hat{v}_q \end{bmatrix} = \begin{bmatrix} R_s + L_s p & -\omega_r L_s \\ \omega_r L_s & R_s + L_s p \end{bmatrix} \begin{bmatrix} \hat{i}_d \\ \hat{i}_q \end{bmatrix} + \begin{bmatrix} \hat{E}_d \\ \hat{E}_q \end{bmatrix}, \begin{bmatrix} \hat{E}_d \\ \hat{E}_q \end{bmatrix} = \omega_r \psi_m \begin{bmatrix} -\sin\Delta\theta \\ \cos\Delta\theta \end{bmatrix} \tag{2.12}$$

where ^ means the estimated reference frame. Then, the estimated position error can be derived as:

$$\Delta\hat{\theta} = \tan^{-1}\left(-\frac{\hat{E}_d}{\hat{E}_q}\right) \approx -\frac{\hat{E}_d}{\hat{E}_q} \tag{2.13}$$

As can be seen from (2.12), the rotor position can be estimated by controlling $\Delta\hat{\theta}$ to zero with the phase-locked loop (PLL)-based position observer as shown in Figure 2.13. With this type of observer, the estimated position $\hat{\theta}_r$ is dynamically adjusted until $\Delta\hat{\theta}$ is controlled to zero [5]. The details of the position observer will be discussed in Section 2.4.

Based on the prototype SPMSM-I (its parameters are given in Appendix B), simulation results of position estimation performance with the back-EMF method are shown in Figures 2.14–2.17. In the simulation, the rotor speed is controlled at

Figure 2.13 PLL observer for estimating rotor position.

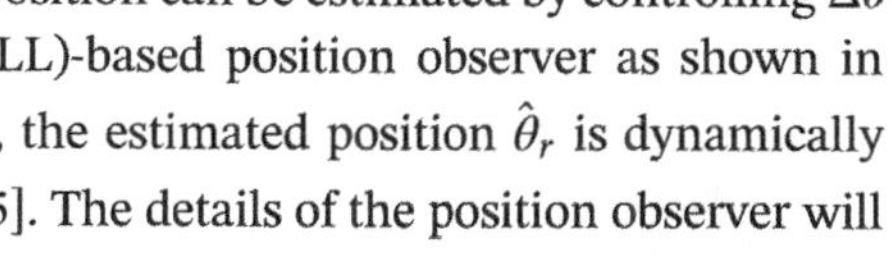

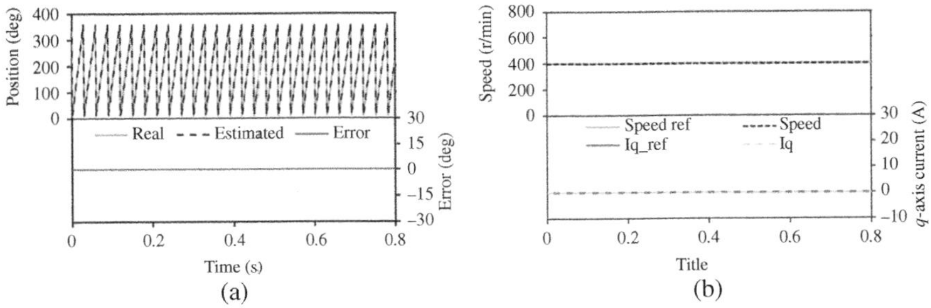

Figure 2.14 Steady-state position estimation with back-EMF method with no load. (a) Position estimation. (b) Rotor speed and q-axis current.

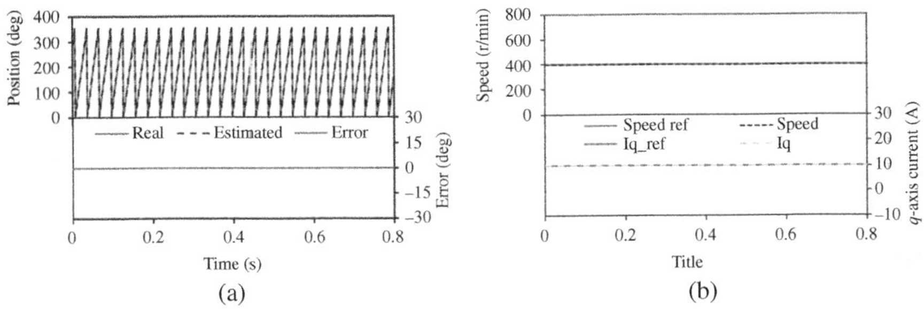

Figure 2.15 Steady-state position estimation performance with back-EMF method with full load. (a) Position estimation. (b) Rotor speed and q-axis current.

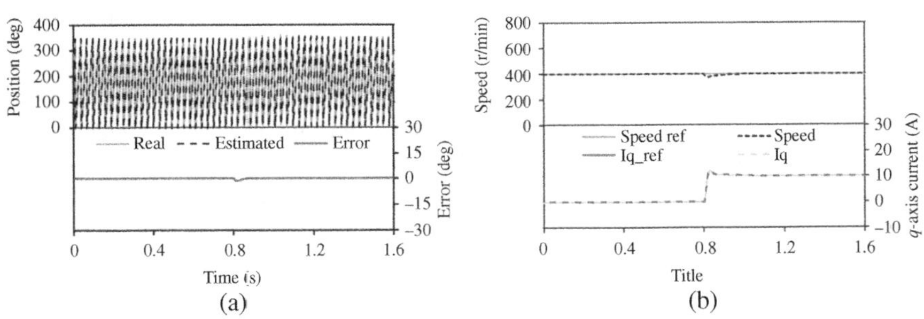

Figure 2.16 Dynamic-state position estimation with back-EMF method with step load. (a) Position estimation. (b) Rotor speed and q-axis current.

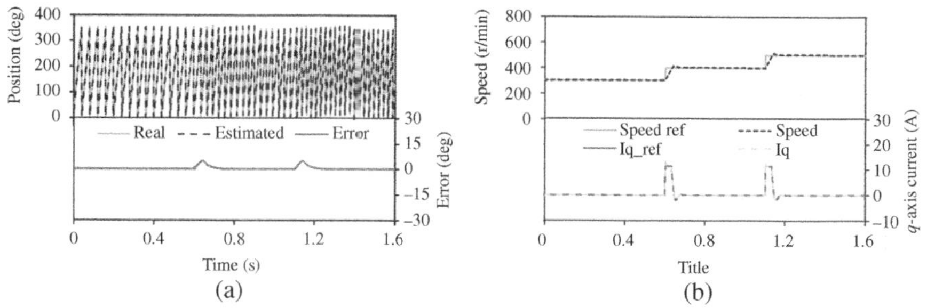

Figure 2.17 Dynamic-state position estimation with back-EMF method with step speed. (a) Position estimation. (b) Rotor speed and q-axis current.

300 r/min. First, in Figures 2.14 and 2.15, the steady-state position estimation performance under no load and full load are shown. In Figures 2.16 and 2.17, the dynamic-state performance is shown including a step load test (from no load to full load) and a step speed test (300–400–500 r/min). The results show that the rotor position can be estimated well in both steady and dynamic states.

2.3.2 Extended Back-EMF Method for Salient PMSMs

An extended back-EMF (EEMF)-based method is introduced in [6], [7]. By using EEMF, salient machines can be treated as non-salient machines, which simplifies the control algorithm in the implementation.

2.3.2.1 In Synchronous Reference Frame

The sensorless position estimation method in Section 2.3.1 is only applicable to non-salient PMSMs. For salient PMSMs, the voltage equations in the dq-axis frame are given by

$$\begin{bmatrix} v_d \\ v_q \end{bmatrix} = \begin{bmatrix} R_s + pL_d & -\omega_r L_q \\ \omega_r L_d & R_s + pL_q \end{bmatrix} \begin{bmatrix} i_d \\ i_q \end{bmatrix} + \begin{bmatrix} 0 \\ \omega_r \psi_m \end{bmatrix} \tag{2.14}$$

where L_d and L_q are the dq-axis apparent inductances. Due to the asymmetry of the dq-axis inductance matrix, the back-EMF cannot be directly extracted in the estimated dq-axis frame. To solve this problem, the extended back-EMF concept is proposed for salient machines [5] as:

$$\begin{bmatrix} v_d \\ v_q \end{bmatrix} = \begin{bmatrix} R_s + pL_d & -\omega_r L_q \\ \omega_r L_q & R_s + pL_d \end{bmatrix} \begin{bmatrix} i_d \\ i_q \end{bmatrix} + \begin{bmatrix} 0 \\ E_{ex} \end{bmatrix} \tag{2.15}$$

$$E_{ex} = \omega_r \psi_m + (L_d - L_q)\left(\omega_r i_d - p i_q\right) \tag{2.16}$$

where E_{ex} is the amplitude of the extended back-EMF. With the rearrangement in (2.15), the inductance matrix becomes symmetrical and the voltage model in the estimated synchronous reference frame can be derived as:

$$\begin{bmatrix} \hat{v}_d \\ \hat{v}_q \end{bmatrix} = \begin{bmatrix} R_s + pL_d & -\omega_r L_q \\ \omega_r L_q & R_s + pL_d \end{bmatrix} \begin{bmatrix} \hat{i}_d \\ \hat{i}_q \end{bmatrix} + \begin{bmatrix} \hat{E}_{ex,d} \\ \hat{E}_{ex,q} \end{bmatrix} \tag{2.17}$$

$$\begin{bmatrix} \hat{E}_{ex,d} \\ \hat{E}_{ex,q} \end{bmatrix} = E_{ex} \begin{bmatrix} -\sin\Delta\theta \\ \cos\Delta\theta \end{bmatrix} + \Delta\omega L_d \begin{bmatrix} \hat{i}_q \\ -\hat{i}_d \end{bmatrix} \tag{2.18}$$

where $\Delta\omega$ is the error between estimated and actual speeds and normally can be assumed as zero. Then, the estimated position error can be expressed as follows:

$$\Delta\hat{\theta} = \tan^{-1}\left(-\frac{\hat{E}_{ex,d}}{\hat{E}_{ex,q}}\right) \approx -\frac{\hat{E}_{ex,d}}{\hat{E}_{ex,q}} \tag{2.19}$$

Subsequently, as shown in Figure 2.18, a position observer is applied to adjust the estimated position to minimize the estimated position error $\Delta\hat{\theta}$ to make the estimated reference frame align with the actual one.

Similarly, based on the prototype IPMSM-I (its parameters are given in Appendix B), the extended back-EMF-based method is also simulated. As shown in Figures 2.19 and 2.20, the rotor position can be estimated accurately in the steady state under no load and full load conditions. In the simulation, the rotor speed is 300 r/min. Then, for the dynamic-state performance, a step load test (from no load to full load) is shown in Figure 2.21 and a step speed test (300–400–500 r/min) is shown in Figure 2.22. From the simulated results, the estimated position can quickly converge to the actual position during the transient.

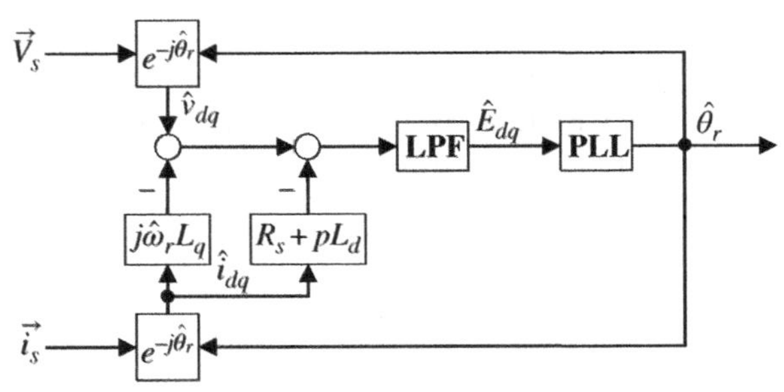

Figure 2.18 Extended back-EMF-based position observer. (*Source:* Adapted from [6].)

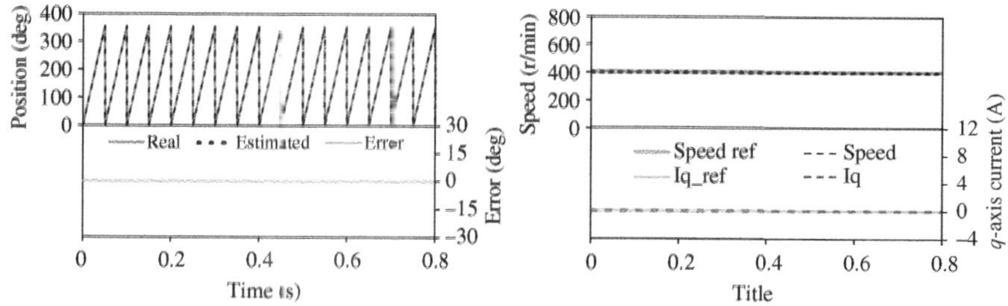

(a) (b)

Figure 2.19 Steady-state position estimation with extended back-EMF method under no load. (a) Position estimation. (b) Rotor speed and q-axis current.

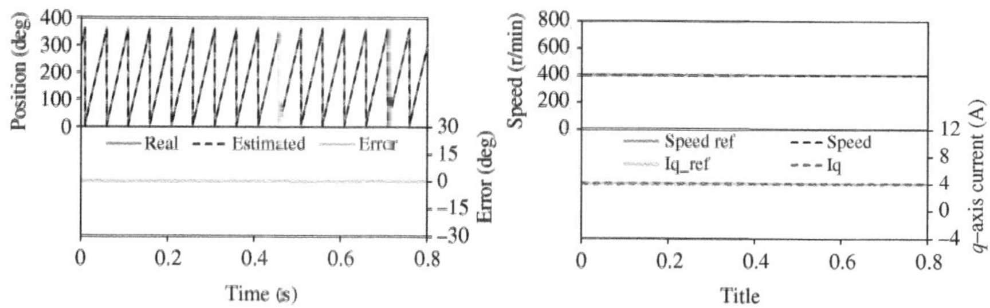

(a) (b)

Figure 2.20 Steady-state position estimation with extended back-EMF method under full load. (a) Position estimation. (b) Rotor speed and q-axis current.

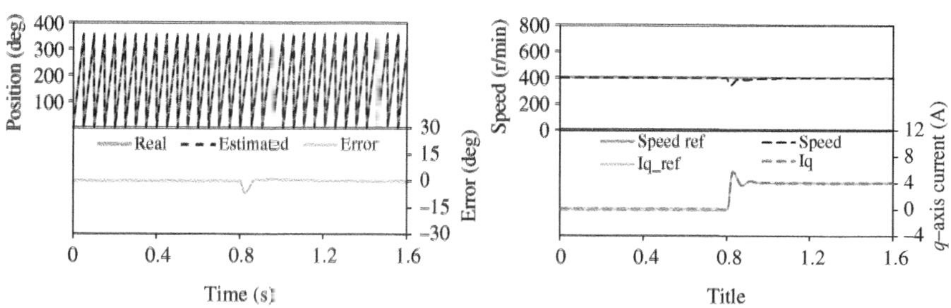

(a) (b)

Figure 2.21 Dynamic-state position estimation with extended back-EMF method under step load. (a) Position estimation. (b) Rotor speed and q-axis current.

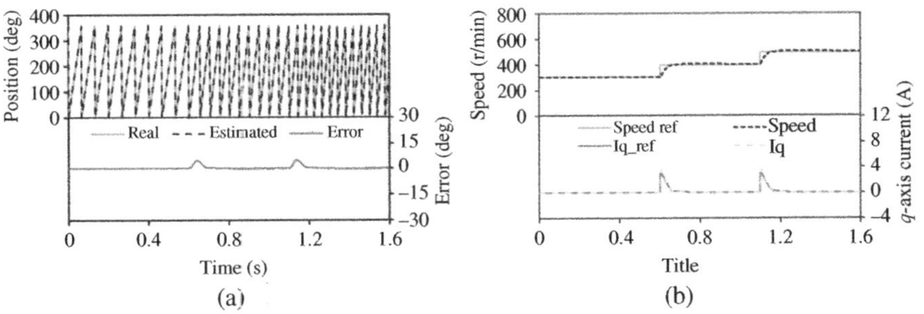

Figure 2.22 Dynamic-state position estimation with extended back-EMF method under step speed. (a) Position estimation. (b) Rotor speed and q-axis current.

2.3.2.2 In Stationary Reference Frame

The extended back-EMF based method can also be implemented in the stationary reference frame [7]. The extended back-EMF model in the stationary reference frame is expressed as:

$$\begin{bmatrix} v_\alpha \\ v_\beta \end{bmatrix} = \begin{bmatrix} R_s + pL_d & -\omega_r(L_q - L_d) \\ \omega_r(L_q - L_d) & R_s + pL_d \end{bmatrix} \begin{bmatrix} i_\alpha \\ i_\beta \end{bmatrix} + \begin{bmatrix} e_{ex,\alpha} \\ e_{ex,\beta} \end{bmatrix} \tag{2.20}$$

$$\begin{bmatrix} e_{ex,\alpha} \\ e_{ex,\beta} \end{bmatrix} = E_{ex} \begin{bmatrix} -\sin\theta_r \\ \cos\theta_r \end{bmatrix} \tag{2.21}$$

Hence, the rotor position can be obtained by:

$$\theta_r = \tan^{-1}\left(-\frac{e_{ex,\alpha}}{e_{ex,\beta}}\right) \tag{2.22}$$

An alternative is to utilize the PLL to estimate the rotor position that can suppress the noises caused by the "arctan" operation. The details of the PLL will be discussed in Section 2.4.

2.3.3 Comparison

Back-EMF and flux-linkage methods for position estimation are all based on the fundamental machine model. They can both be expressed in the stationary reference frame, i.e. $\alpha\beta$ frame, and the synchronous reference frame, i.e. dq frame. Comparisons between the back-EMF model and the flux-linkage model are made in this section.

2.3.3.1 Comparison Between Back-EMF and Flux-Linkage Methods

In the previous sections, the back-EMF and flux-linkage-based sensorless methods were introduced. These two methods are compared and summarized in Table 2.1.

Table 2.1 Comparison between back-EMF and flux-linkage-based sensorless methods.

	Back-EMF	Flux-linkage
General equations	$e_x = v_x - R_s i_x - L_y \dfrac{di_x}{dt}$	$\psi_x = \int (v_x - R_s i_x)dt - L_y i_x$
Speed dependency	Dependent	Independent
Calculation	Differential	Integral
Parameters required	Resistance, inductance	Resistance, inductance
Robustness to noise	Low	High
Dynamic	High	Low

where x, y refer to the corresponding reference frames, i.e., α, β, d, q. The flux-linkage-based method calculation is more robust to noise compared to the back-EMF-based method as the integration behaves as a low-pass filter to the noise, while differential operation may introduce noise for the back-EMF-based method. However, the integration for the flux-linkage-based method also introduces delay and results in decreased dynamic performance. Moreover, the flux-linkage is independent of the rotor speed, which makes it potentially better at low speed range than the back-EMF-based method. However, since flux-linkage must be obtained by the integration of back EMF, the improvement at low speed is limited.

2.3.3.2 Comparison of Active Flux and Extended Back-EMF

The extended back-EMF [7] and active flux [3] are similar in that they transform a salient machine into a non-salient machine. This significantly simplifies the position estimation for salient machines and enables a unified sensorless theory at medium and high speeds for PMSMs. However, they also have some differences. The active flux concept refers to the torque-producing flux in the electromagnetic torque formulas of an ac machine [3], while the extended back-EMF concept does not have a clear physical meaning. A detailed comparison is given in Table 2.2.

The main difference between the extended back-EMF and the active flux is that the extended back-EMF is aligned with the q-axis as shown in Figure 2.23a, while the active flux model-based back-EMF has an additional back-EMF component $(L_d - L_q)pi_d$ on the d-axis as shown in Figure 2.23b.

Table 2.2 Comparison between extended back-EMF and active flux.

	Extended back-EMF (subscript ex)	**Active flux (subscript a)**
Definition	$E_{ex} = \omega_r \psi_m + (L_d - L_q)(\omega_r i_d - p i_q)$	$\psi_a = \psi_m + (L_d - L_q) i_d$
***Dq*-frame voltages**	$\begin{bmatrix} v_d \\ v_q \end{bmatrix} = \begin{bmatrix} R_s + pL_d & -\omega_r L_q \\ \omega_r L_q & R_s + pL_d \end{bmatrix} \begin{bmatrix} i_d \\ i_q \end{bmatrix} + \begin{bmatrix} E_{ex,d} \\ E_{ex,q} \end{bmatrix}$	$\begin{bmatrix} v_d \\ v_q \end{bmatrix} = \begin{bmatrix} R_s + pL_q & -\omega_r L_q \\ \omega_r L_q & R_s + pL_q \end{bmatrix} \begin{bmatrix} i_d \\ i_d \end{bmatrix} + \begin{bmatrix} E_{a,d} \\ E_{a,q} \end{bmatrix}$
***Dq*-frame back EMF**	$\begin{bmatrix} E_{ex,d} \\ E_{ex,q} \end{bmatrix} = \begin{bmatrix} 0 \\ E_{ex} \end{bmatrix}$ (*d*-axis back-EMF is zero)	$\begin{bmatrix} E_{a,d} \\ E_{a,q} \end{bmatrix} = \begin{bmatrix} (L_d - L_q) p i_d \\ \omega_r \psi_m + (L_d - L_q) \omega_r i_d \end{bmatrix}$ (*d*-axis back-EMF is nonzero)

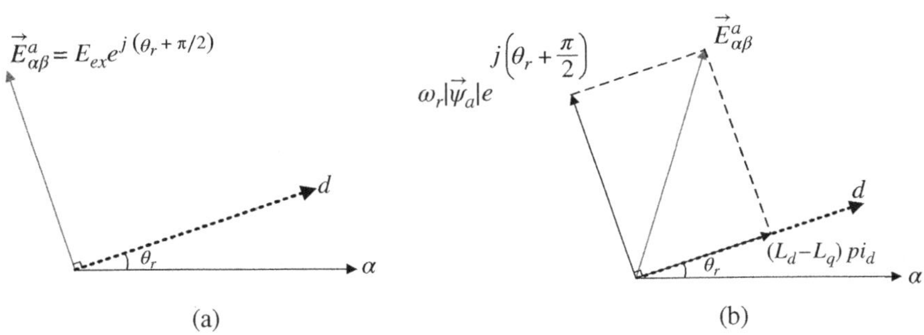

Figure 2.23 Vector diagram for extended back-EMF and active flux. (a) Extended back-EMF. (b) Active flux.

2.4 Position Observer

From the previous sections, after the back-EMF and flux-linkage are obtained, a position observer is required to estimate the rotor position. In this section, several commonly used approaches to position estimation are introduced.

2.4.1 Arctangent Method

The simplest way is the direct calculation using the arctangent function as expressed by:

$$\theta_r = \tan^{-1}\left(-\frac{e_\alpha}{e_\beta}\right), \theta_r = \tan^{-1}\left(\frac{\psi_{m\beta}}{\psi_{m\alpha}}\right) \tag{2.23}$$

This method has a fast dynamic. However, due to the existence of noisy signals, the arctangent function may have a large estimation error when the back-EMF or flux-linkage crosses zero. An LPF should be applied, however, which adds phase lag to the estimation [8].

2.4.2 Phase-Locked Loop

Normally, (2.23) can be replaced by a PLL (phase-locked loop) [9] as shown in Figure 2.24. The back-EMF and flux-linkage in stationary reference frame are the input of the PLL.

Using the back-EMF-based sensorless control as an example, the error input of PLL ε can be expressed as:

$$\varepsilon = -e_\alpha \cos\hat{\theta}_r - e_\beta \sin\hat{\theta}_r \approx k(\theta_r - \hat{\theta}_r) \tag{2.24}$$

where $k = \omega_r \psi_m$. As can be seen from (2.24), the magnitude of back-EMF changes with the rotor speed, i.e. $k = \omega_r \psi_m$. To fix the poles of the PLL, the error signal ε of the PLL should be normalized [9] as:

$$\varepsilon_n = \frac{\varepsilon}{\sqrt{e_\alpha^2 + e_\beta^2}} \tag{2.25}$$

Then, the estimated position can be expressed as:

$$\hat{\theta}_r = \frac{1}{s}\left(K_p + \frac{K_i}{s}\right)\varepsilon_n \tag{2.26}$$

where K_p, K_i are the coefficients of the PI (proportional integral) controller in the PLL.

Furthermore, the closed-loop transfer function of the PLL can be given by:

$$H(s) = \frac{2\xi\omega_n s + \omega_n^2}{s^2 + 2\xi\omega_n s + \omega_n^2} \tag{2.27}$$

$$\xi = \sqrt{K_i}, \omega_n = \frac{K_p}{2}\sqrt{\frac{1}{K_i}} \tag{2.28}$$

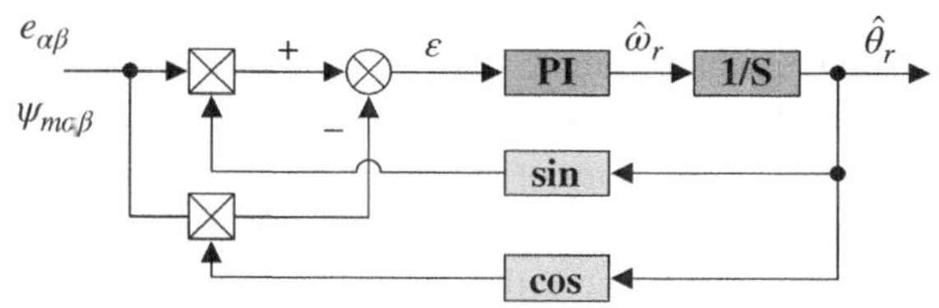

Figure 2.24 PLL block diagram.

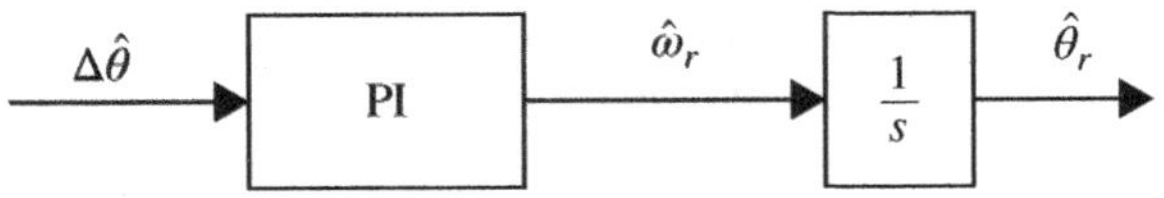

Figure 2.25 Block diagram of PLL-based position observer.

where ω_n is the natural frequency and can be used as the bandwidth of PLL. ξ is the damping ratio [6]. A higher damping ratio can reduce the overshoot but may also sacrifice the dynamic performance. Then, the parameters of PLL can be initially designed based on classical control theory and tuned in the real test.

Moreover, for sensorless control in the estimated synchronous reference frame, the estimated position error is expressed by:

$$\Delta\hat{\theta} \approx -\frac{\hat{E}_d}{\hat{E}_q} \tag{2.29}$$

Then, the PLL-based position observer shown in Figure 2.25 can be used to estimate the rotor speed and position [6].

2.4.3 Simplified Extended Kalman Filter

In [10], a simplified extended Kalman filter (EKF) algorithm is proposed to extract the position information from the noisy output of a resolver. It is further extended to sensorless control to estimate the rotor position [11].

First, a discrete-time nonlinear dynamic system can be expressed in state-space form as:

$$x(k+1) = f(x(k), k) + g(u(k), k) + w(k) \tag{2.30}$$

$$y(k) = h(x(k), k) + v(k) \tag{2.31}$$

where $u(k)$ and $y(k)$ are the input and output signals, respectively, $w(k)$ and $v(k)$ are the process noise and measurement noise, respectively. $x(k)$ is the state vector, which can be estimated by the extended Kalman filter and is given by:

$$\hat{x}(k+1) = f(\hat{x}(k), k) + g(u(k), k) + K_k[y(k) - h(\hat{x}(k), k)] \tag{2.32}$$

The Kalman gain K_k is determined through the Riccati difference equation. However, it requires relatively complex matrix calculations and, therefore, takes a significant amount of computing time. In a sensorless PMSM drive, the output variables of the EKF may be chosen as PM flux-linkage in (2.33) or back-EMF in (2.34). In this section, the PM flux-linkage is taken as an example.

$$\begin{bmatrix} y_1(k) \\ y_2(k) \end{bmatrix} = \begin{bmatrix} \psi_{m\beta} \\ \psi_{m\alpha} \end{bmatrix} \tag{2.33}$$

$$\begin{bmatrix} y_1(k) \\ y_2(k) \end{bmatrix} = \begin{bmatrix} e_\beta \\ e_\alpha \end{bmatrix} \tag{2.34}$$

while the state variables are chosen as the rotor speed, ω_r, and position, θ_r, and the double integration of the noise, w'. Since $\psi_{m\beta}$ and $\psi_{m\alpha}$ are the sine and cosine terms of the stator flux-linkage vector position, to simplify the model, in particular the computation of the Kalman gain, the output variables in (2.23) are expressed in normalized form as:

$$\begin{bmatrix} y_1(k) \\ y_2(k) \end{bmatrix} = \begin{bmatrix} \cos\theta(k) \\ \sin\theta(k) \end{bmatrix} + \begin{bmatrix} v_1(k) \\ v_2(k) \end{bmatrix} \tag{2.35}$$

Therefore, with a state vector $x = [\theta_r, \omega_r, w']^T$ and input $u(k) = 0$, the state-space model, (2.30) and (2.31), can be written as:

$$x(k+1) = Fx(k) + w(k) \tag{2.36}$$

$$y(k) = h(x(k)) + v(k) \tag{2.37}$$

$$F = \begin{bmatrix} 1 & T_s & 0 \\ 0 & 1 & 1 \\ 0 & 0 & 1 \end{bmatrix}, \quad h(x(k)) = \begin{bmatrix} \cos\theta(k) \\ \sin\theta(k) \end{bmatrix} \tag{2.38}$$

where T_s is the sampling time.

The Kalman filter gain can now be significantly simplified and is given by:

$$K = \begin{bmatrix} 0 & k_1 \\ 0 & k_2 \\ 0 & k_3 \end{bmatrix} \cdot \begin{bmatrix} \cos\hat{\theta}_r & \sin\hat{\theta}_r \\ -\sin\hat{\theta}_r & \cos\hat{\theta}_r \end{bmatrix} \tag{2.39}$$

where k_1, k_2, and k_3 are the tuning parameters. These can be pre-computed from simulations.

From the foregoing, it can be shown that the rotor speed and position can be estimated from the following equations:

$$\varepsilon(k) = y_2(k)\cos\hat{\theta}_r(k) - y_1(k)\sin\hat{\theta}_r(k) \tag{2.40}$$

$$\hat{\theta}_r(k+1) = \left[\hat{\theta}_r(k) + T_s\hat{\omega}_r(k) + k_1\varepsilon(k)\right] \tag{2.41}$$

$$\hat{\omega}_r(k+1) = \hat{\omega}_r(k) + w'(k) + k_2\varepsilon(k) \tag{2.42}$$

$$w'(k+1) = w'(k) + k_3\varepsilon(k) \tag{2.43}$$

where $\hat{\theta}_r$ and $\hat{\omega}_r$ are the estimated position and speed, respectively.

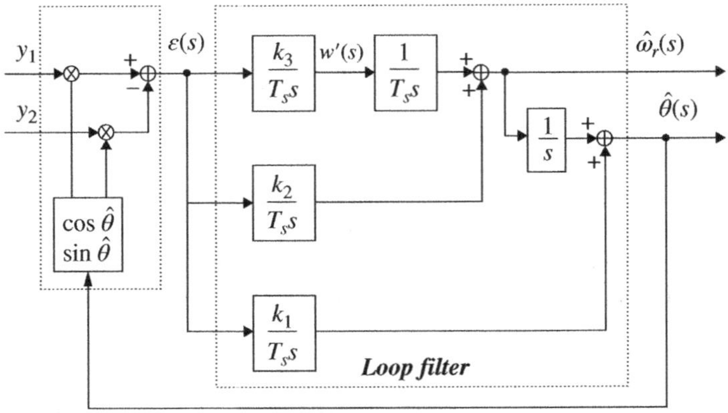

Figure 2.26 Block diagram of simplified EKF. (*Source:* [11]/IET).

Figure 2.26 shows a block diagram of the simplified EKF that is used for the sensorless PMSM system.

Compared to the conventional arctangent-based method, the simplified EKF can have a better steady-state position estimation performance; its estimation accuracy is less affected by the harmonic errors since it is less sensitive to such noises [12].

2.4.4 Simulation Results

Based on the prototype SPMSM-I (its parameters are given in Appendix B), simulation results of position estimation performance with different position calculation methods are shown in Figures 2.27–2.29. In the simulation, the rotor speed is controlled at 300 r/min under full load. Moreover, for both methods, the simulation is carried out based on an ideal PM machine model and a practical PM

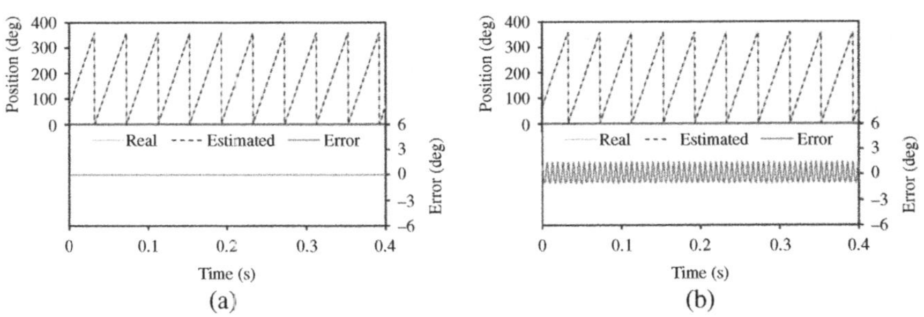

Figure 2.27 Position estimation performance with arctangent method. (a) Ideal model. (b) Practical model.

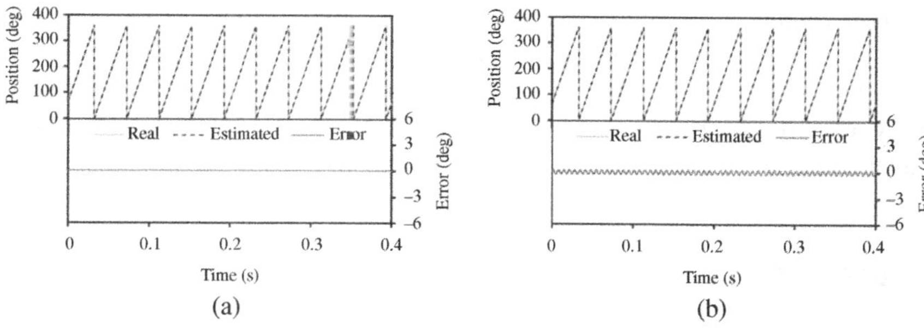

Figure 2.28 Position estimation performance with PLL. (a) Ideal model. (b) Practical model.

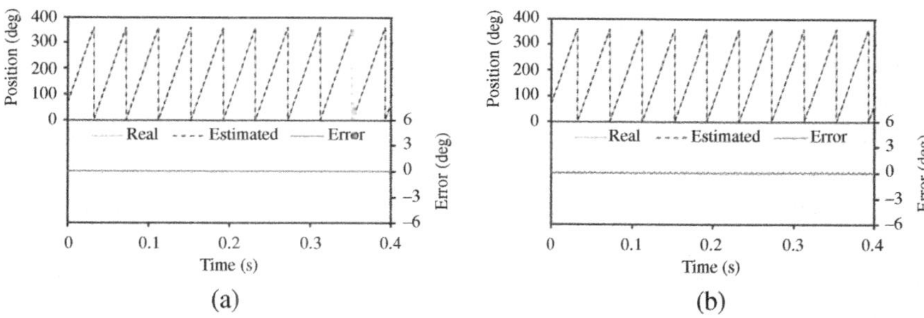

Figure 2.29 Position estimation performance with simplified EKF. (a) Ideal model. (b) Practical model.

machine model with high-frequency harmonics and noises. First, it can be seen that both methods work perfectly based on the ideal model. For the practical model case, as shown in Figure 2.27, the arctangent method has the largest harmonic errors in the position estimation. Then, as shown in Figure 2.28, PLL can achieve a better performance by tuning its control parameters as described in Section 2.4.2. For the simplified EKF shown in Figure 2.29, it has the least harmonic error.

2.5 Summary

In this chapter, basic model-based sensorless control methods, including flux-linkage and back-EMF-based methods, are described for both salient and non-salient PMSMs. In addition, different position observers are included and compared. It should be noted that the methods introduced in this chapter are mostly under ideal conditions. However, in real-world applications, there will be some non-ideal scenarios, e.g. integration drift, filter delays, cross-saturation effect, back-EMF and current harmonics, parameter mismatch, parameter asymmetry, etc. Hence, the next chapter will discuss the common practical issues in implementation and the corresponding solutions.

References

1 R. Wu and G. R. Slemon, "A permanent magnet motor drive without a shaft sensor," *IEEE Trans. Ind. Appl.*, vol. 27, no. 5, pp. 1005–1011, Sep. 1991.

2 J. Hu and B. Wu, "New integration algorithms for estimating motor flux over a wide speed range," *IEEE Trans. Power Electron.*, vol. 13, no. 5, pp. 969–977, Sep. 1998.

3 I. Boldea, M. C. Paicu, and G. Andreescu, "Active flux concept for motion-sensorless unified ac drives," *IEEE Trans. Power Electron.*, vol. 23, no. 5, pp. 2612–2618, Sep. 2008.

4 J. X. Shen, Z. Q. Zhu, and D. Howe, "Improved speed estimation in sensorless PM brushless ac drives," *IEEE Trans. Ind. Appl.*, vol. 38, no. 4, pp. 1072–1080, Jul. 2002.

5 N. Matsui, "Sensorless PM brushless dc motor drives," *IEEE Trans. Ind. Electron.*, vol. 43, no. 2, pp. 300–308, Apr. 1996.

6 S. Morimoto, K. Kawamoto, M. Sanada, and Y. Takeda, "Sensorless control strategy for salient-pole PMSM based on extended EMF in rotating reference frame," *IEEE Trans. Ind. Appl.*, vol. 38, no. 4, pp. 1054–1061, Jul. 2002.

7 Z. Chen, M. Tomita, S. Doki, and S. Okuma, "An extended electromotive force model for sensorless control of interior permanent-magnet synchronous motors," *IEEE Trans. Ind. Electron.*, vol. 50, no. 2, pp. 288–295, Apr. 2003.

8 H. Kim, M. C. Harke, and R. D. Lorenz, "Sensorless control of interior permanent-magnet machine drives with zero-phase lag position estimation," *IEEE Trans. Ind. Appl.*, vol. 39, no. 6, pp. 1726–1733, Nov. 2003.

9 G. Wang, R. Yang, and D. Xu, "DSP-based control of sensorless IPMSM drives for wide-speed-range operation," *IEEE Trans. Ind. Electron.*, vol. 60, no. 2, pp. 720–727, Feb. 2013.

10 L. Harnefors, "Speed estimation from noisy resolver signals," in *1996 6th Int. Conf. Power Electron. Variable Speed Drives (Conf. Publ. No. 429)*, pp. 279–282, Sep. 1996.

11 Y. Liu, Z. Q. Zhu, Y. F. Shi, and D. Howe, "Sensorless direct torque control of a permanent magnet brushless ac drive via an extended Kalman filter," in *2nd Int. Conf. Power Electron, Mach. Drives (PEMD 2004)*, vol. 1, Mar. 2004, pp. 303–307.

12 A. H. Almarhoon and Z. Q. Zhu, "Influence of back-EMF and current harmonics on position estimation accuracy of permanent magnet synchronous machine," in *2014 17th Int. Conf. Elect. Mach. Syst. (ICEMS)*, pp. 2728–2733, Oct. 2014.

3

Fundamental Model-Based Sensorless Control—Issues and Solutions

3.1 Introduction

In Chapter 2, the fundamental model-based sensorless control methods were introduced. However, the sensorless control methods described in Chapter 2 are mainly based on an ideal machine model and drive system. In practice, non-ideal characteristics exist that affect the position estimation of fundamental model-based methods. Hence, in this chapter, some common practical issues are discussed, and the corresponding solutions are also provided. Some examples of common issues that exist in a PM machine drive are as follows:

- Integration initial value and drift issues
- Filter phase delay
- Back-EMF and current harmonics
- Cross-coupling magnetic saturation
- Parameter mismatch
- Parameter asymmetry

3.2 Integration and Filter

The flux-linkage-based sensorless control methods were introduced in Chapter 2. In order to obtain the flux-linkage, the integration of the estimated back-EMF is required. However, in practice, the implementation of a pure integration has several issues that need to be considered, such as those involving initial value and drift. These issues and solutions are introduced in this section.

Sensorless Control of Permanent Magnet Synchronous Machine Drives, First Edition.
Zi Qiang Zhu and Xi Meng Wu.

3.2.1 Initial Value

The initial value issue of the pure integration can be illustrated as follows: When a sine-wave signal is integrated at its positive and negative peaks, the output can be a cosine-wave signal; otherwise, a constant dc offset will appear. A transient change in the input of an integrator also introduces this dc offset [1]. A measured result is shown in Figure 3.1, where there is an offset in the flux-linkage estimation that causes a first order harmonic in the position estimation [3].

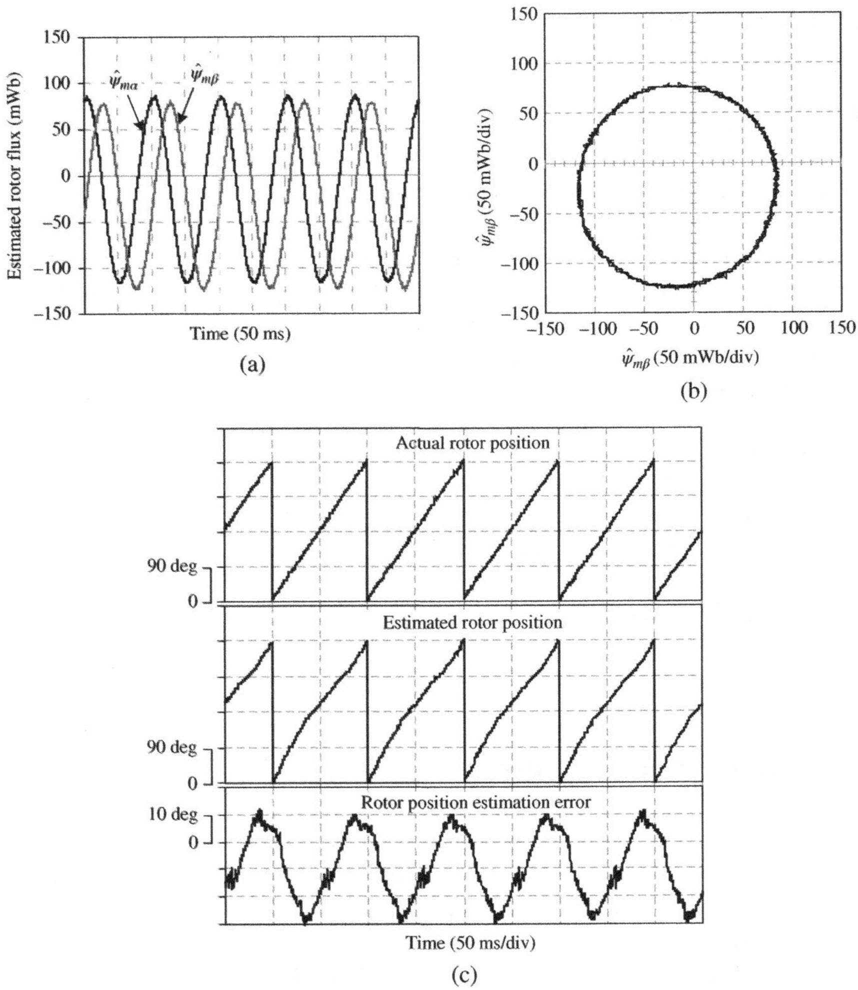

Figure 3.1 Measured waveforms with flux observer without considering initial values. (a) Estimated PM flux-linkages. (b) PM flux-linkage locus. (c) Position estimation. (*Source:* [2]/IEEE.)

Hence, the stator flux-linkages calculation should involve the initial values of the flux-linkages $\psi_\alpha(0)$ and $\psi_\beta(0)$ as:

$$\begin{aligned} \psi_\alpha &= \int (v_\alpha - i_\alpha R_s)dt + \psi_\alpha(0) \\ \psi_\beta &= \int (v_\beta - i_\beta R_s)dt + \psi_\beta(0) \end{aligned} \tag{3.1}$$

When there is no current in the windings, the total flux-linkage is the same as the PM excitation flux-linkage, and $\psi_\alpha(0)$ and $\psi_\beta(0)$ can be expressed as:

$$\begin{cases} \psi_\alpha(0) = \psi_{m\alpha}(0) \\ \psi_\beta(0) = \psi_{m\beta}(0) \end{cases} \tag{3.2}$$

Therefore, if the rotor is aligned to a certain position before the flux observer begins to work, $\psi_\alpha(0)$ and $\psi_\beta(0)$ are pre-known. With the correct initial values, the measured PM flux-linkages are shown in Figure 3.2. Clearly, the offset and position error are eliminated compared to those in Figure 3.1.

3.2.2 Drift

Figure 3.3a shows the measured locus of the observed PM flux-linkage vector $\psi_{m\alpha\beta}$. As can be seen, while it is almost circular, it is continuously being displaced. This is due to the use of pure integration, since any dc offset or error in the measured currents is continuously amplified by the pure integration until saturation is reached [3], [4].

To solve these problems, in [5], a high-pass filter (HPF), for which the transfer function is $s/(s+\omega_c)$, is applied to the variables to be integrated. As the transfer function of a pure integrator is $1/s$, the combined transfer function becomes $1/(s+\omega_c)$, which is equivalent to the replacement of the integrator by a low-pass filter (LPF). ω_c is the cutoff frequency. Hence, (3.1) is modified to (3.3) in the s-domain and (3.4) in the time domain.

$$\begin{aligned} \psi_\alpha(s) &= \frac{1}{s+\omega_c}(v_\alpha - i_\alpha R_s) + \psi_\alpha(0) \\ \psi_\beta(s) &= \frac{1}{s+\omega_c}(v_\beta - i_\beta R_s) + \psi_\beta(0) \end{aligned} \tag{3.3}$$

$$\begin{aligned} \psi_\alpha &= \int [-\omega_c\psi_\alpha + (v_\alpha - i_\alpha R_s)]dt + \psi_\alpha(0) \\ \psi_\beta &= \int [-\omega_c\psi_\beta + (v_\beta - i_\beta R_s)]dt + \psi_\beta(0) \end{aligned} \tag{3.4}$$

With the modified integration of back-EMF voltages, the measured PM flux-linkage vector $\psi_{m\alpha\beta}$ is shown in Figure 3.3b. All loci are circular and overlapped

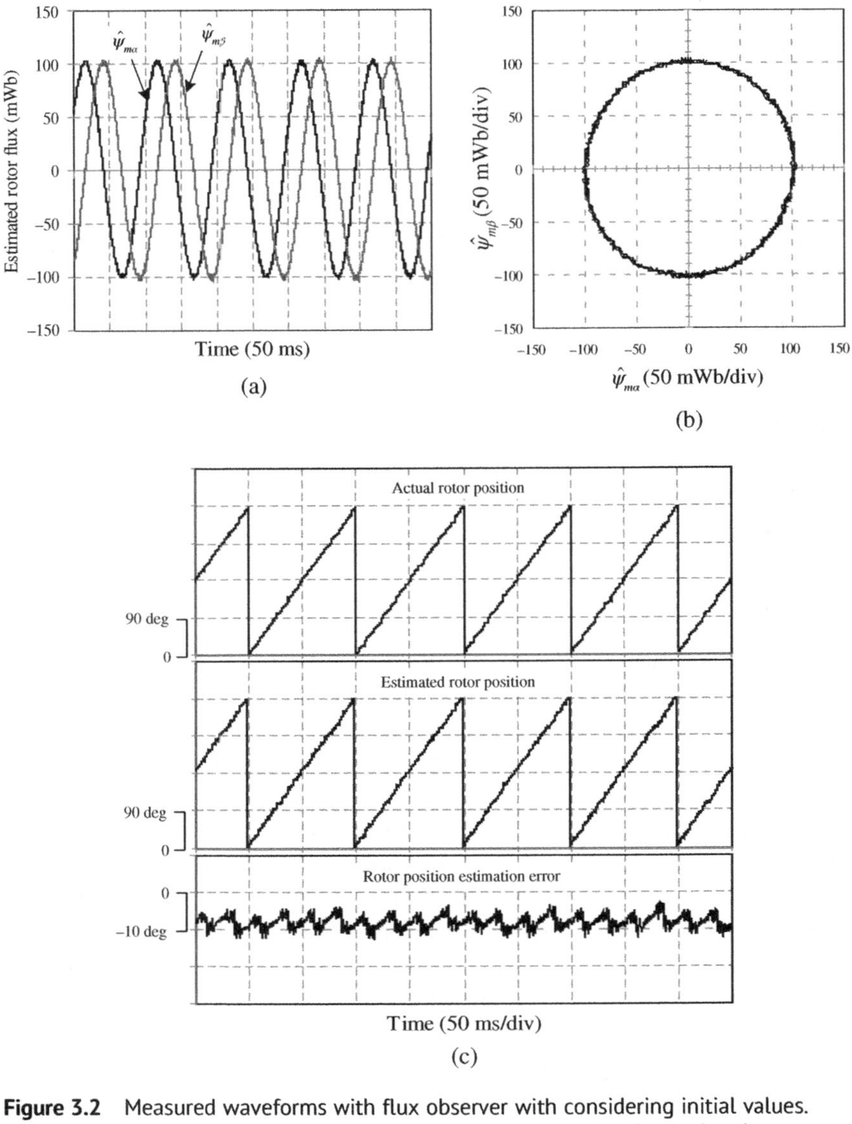

Figure 3.2 Measured waveforms with flux observer with considering initial values. (a) Estimated PM flux-linkages. (b) PM flux-linkage locus. (c) Position estimation. (*Source:* [2]/IEEE.)

with each other. In Figure 3.3b, three captures are presented; the time interval between the first and second captures is about 20 minutes, while the interval between the second and third captures is about 1 hour. Clearly, the experimental results prove that the circular locus always remains stable.

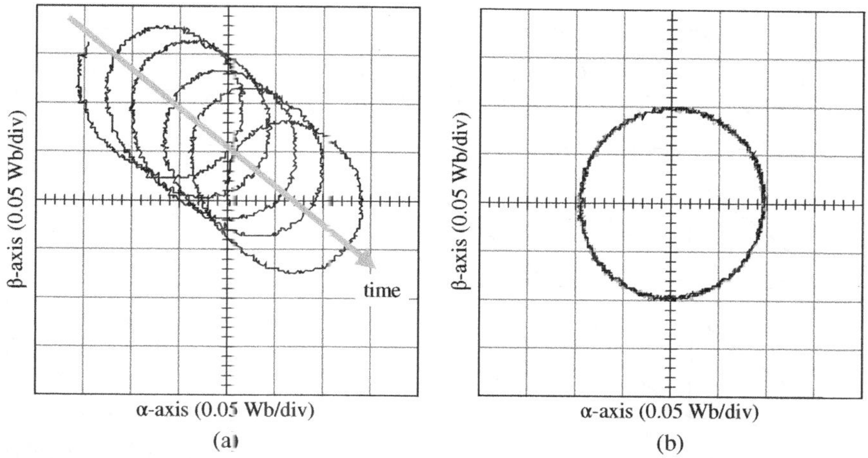

Figure 3.3 Locus of observed PM flux-linkage vector with different strategies. (a) Pure integration. (b) LPF. (*Source:* [4]/IEEE.)

3.2.3 Delay

Although the LPF can effectively address the drift issue, it will cause inevitable phase delay during the position estimation [1]. In [2], a flux-linkage estimator with a delay compensation is applied. The block diagram of the modified flux-linkage estimator is shown in Figure 3.4,

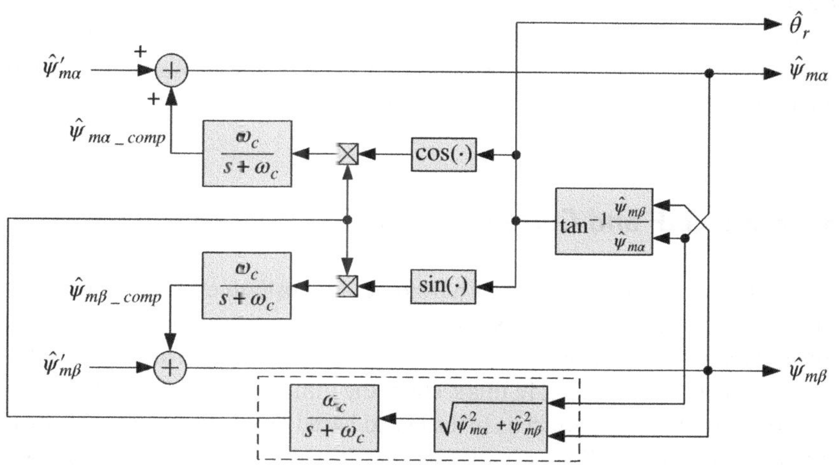

Figure 3.4 Modified flux observer. (*Source:* [2]/IEEE.)

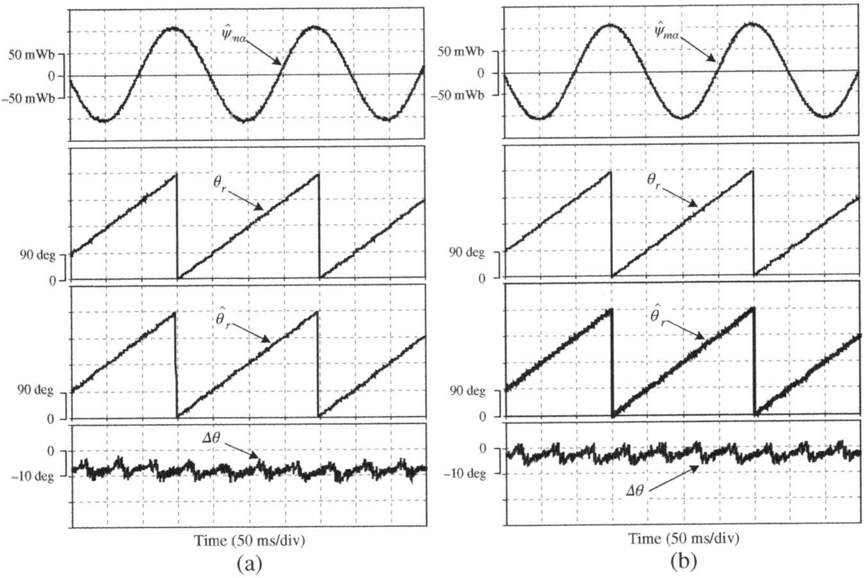

Figure 3.5 Measured results of position estimation performance with different flux-linkage observers. (a) Flux observer without LPF compensation. (b) Flux observer with LPF compensation. (*Source:* [2]/IEEE.)

where $\hat{\psi}'_{m\alpha}$ and $\hat{\psi}'_{m\beta}$ are the outputs of the LPF-based flux-linkage observer by (3.4). With the compensation, the modified integrator can achieve a better performance than LPF while the issues associated with the pure integrator can also be avoided. Based on the prototype SPMSM-II (its parameters are given in Appendix B), experimental results are shown in Figure 3.5 that prove the effectiveness of the modified flux-linkage observer with delay compensation. Clearly, the steady-state position estimation error due to LPF delay is reduced by compensation.

3.3 Back-EMF and Current Harmonics

The influence of voltage and current harmonics on position estimation accuracy should be considered for sensorless control. Such harmonic components lead to a significant reduction in the machine's performance in addition to inaccurate rotor position estimation. Therefore, in this section, the influence of such harmonics on the position estimation accuracy is investigated based on two different sensorless control methods, i.e. flux-linkage with LPF (FO) [4], and flux-linkage with simplified EKF [6]. To compare and evaluate the influence of back-EMF and current harmonics on rotor position estimation accuracy, both sensorless control methods are tested by experiment.

3.3.1 Influence of Back-EMF Harmonics

First, based on the prototype SPMSM-I (its parameters are given in Appendix B), the measured phase back-EMF waveform and spectrum are shown in Figures 3.6a, b. It should be noted that significant third, fifth, and seventh harmonics are observed. However, the most critical harmonics are the fifth and seventh harmonic components.

Then, by applying FO-based sensorless control, the position estimation performance is shown in Figure 3.7. Due to the back-EMF harmonic components, the sixth order fluctuation in the position estimation error occurs. Thereafter, the sensorless control strategy based on simplified EKF is utilized to evaluate and compare its position estimation accuracy with that of FO. Figure 3.8 shows the position estimation performance. It can be seen that the position estimation error is relatively small, and it exhibits fewer fluctuations compared to that of FO. It can be concluded that the rotor position estimation accuracy is less affected by the back-EMF harmonics when the simplified EKF technique is implemented since it is less sensitive to such noise.

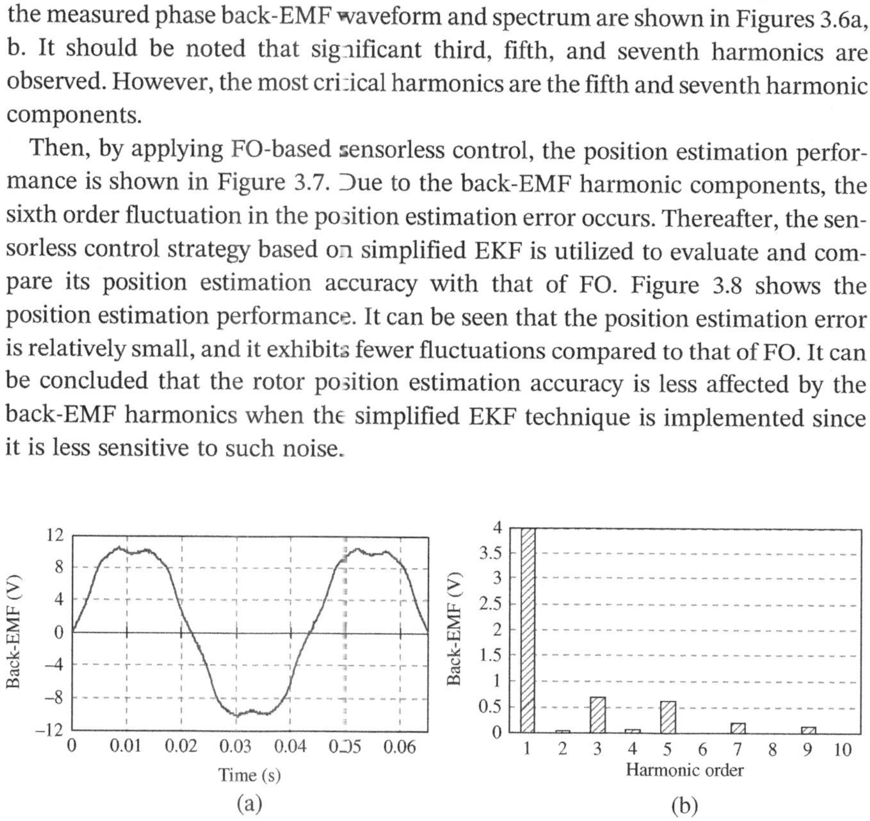

Figure 3.6 Phase back-EMF and spectrum. (a) Phase back-EMF. (b) Spectrum. (*Source:* [7]/IEEE.)

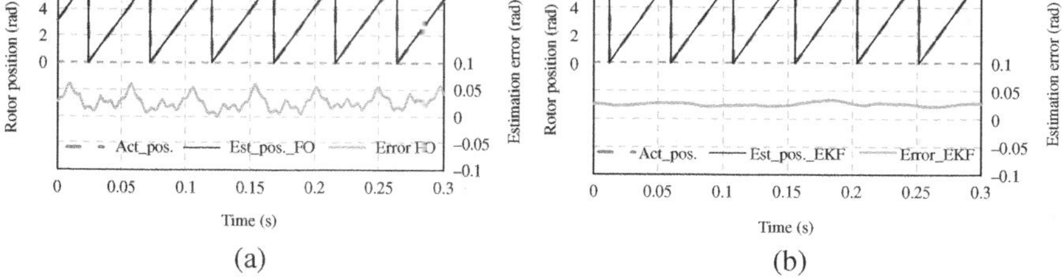

Figure 3.7 Rotor position estimation using different flux-linkage observers. (a) Flux-linkage observer. (b) Simplified EKF. (*Source:* [7]/IEEE.)

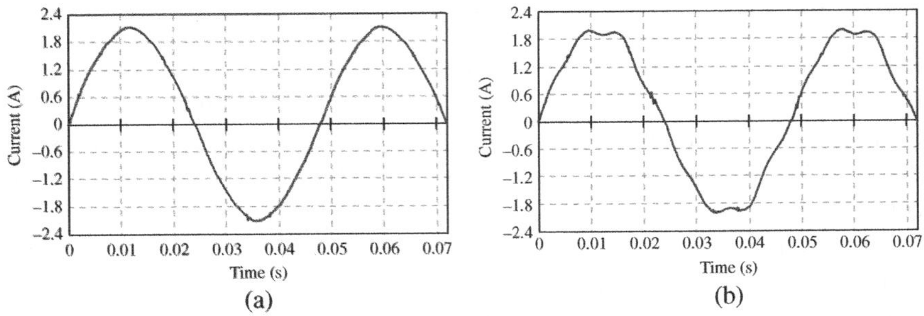

Figure 3.8 Phase A current before/after increasing harmonics. (a) Phase A current before increasing harmonics. (b) Phase A current after increasing harmonics. (*Source:* [7]/IEEE.)

3.3.2 Influence of Current Harmonics

Generally, the main reasons for stator current harmonics ($k = 6n \pm 1$, where $n = 1$, 3, 5, ...) are as follows [7]:

1) The construction of the machine: Permanent magnets can produce non-sinusoidal air-gap flux distribution, and the air-gap flux harmonics might also be caused by the pole shape, slotting, rotor saliency, and possible magnetic saturation.
2) Inverter nonlinearities: The supplied voltage contains harmonics of the order ($k = 6n \pm 1$, where $n = 1, 3, 5, ...$); inverter nonlinearities can cause additional stator current harmonics (mainly the fifth and seventh order harmonics in a stationary reference frame and the sixth order harmonic in a synchronous reference frame).

To verify the influence of current harmonics on the position estimation accuracy, the amplitudes of both fifth and seventh harmonic components in the phase currents have been increased by ~5% of fundamental amplitude by adding a harmonic component to both the d- and q-axis voltage references. Figures 3.8a,b show the currents of phase A before and after increasing the amplitudes of harmonic components, respectively. To evaluate the influence of current harmonics on position estimation accuracy, the comparison of position estimation performance with FO and simplified EKF-based sensorless control techniques is shown in Figure 3.9. It can be seen that the simplified EKF has much smaller harmonic errors in the position estimation than the FO.

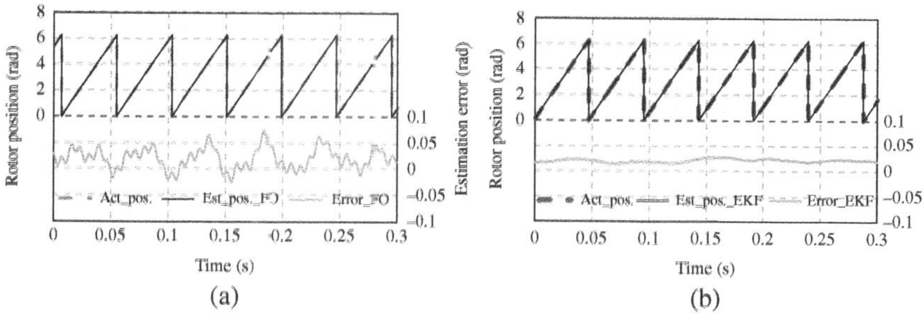

Figure 3.9 Rotor position estimation using different rotor flux-linkage observers. (a) Conventional rotor flux-linkage observer. (b) Simplified EKF. (*Source:* [7]/IEEE.)

3.4 Cross-Coupling Magnetic Saturation

This section analyses the position estimation errors in the extended back-EMF-based sensorless control method due to the cross-coupling magnetic saturation. The cross-coupling effect, i.e. apparent mutual-inductance, can be predicted and modeled from the finite element results of the dq-axis flux-linkages. The predicted value can be used in the machine model to improve the extended back-EMF sensorless method. When the cross-coupling magnetic saturation and the apparent mutual-inductance are accounted for in the model, significant improvement has been achieved in the back-EMF-based sensorless control.

3.4.1 Impact on Position Estimation

In [8]–[10], it is shown that the motor load introduces a cross-coupling magnetic saturation between d- and q-axes. It is proved by [9], [10] that the cross-coupling magnetic saturation will cause rotor position estimation error for extended back-EMF-based sensorless control. For the prototype IPMSM-I (its parameters are given in Appendix B), the d-axis flux-linkage and q-axis flux-linkage are predicted by finite element analyses at various dq-axis currents and are shown in Figure 3.10.

From Figure 3.10, both the dq-axis flux-linkages depend on dq-axis currents, and they can be rewritten as:

$$\begin{aligned} \psi_q(i_d, i_q) &= \psi_{q0} + L_q i_q + L_{qd} i_d, \quad \psi_{q0} = \psi_q(0,0) = 0 \\ \psi_d(i_d, i_q) &= \psi_{d0} + L_d i_d + L_{dq} i_q, \quad \psi_{d0} = \psi_d(0,0) \end{aligned} \tag{3.5}$$

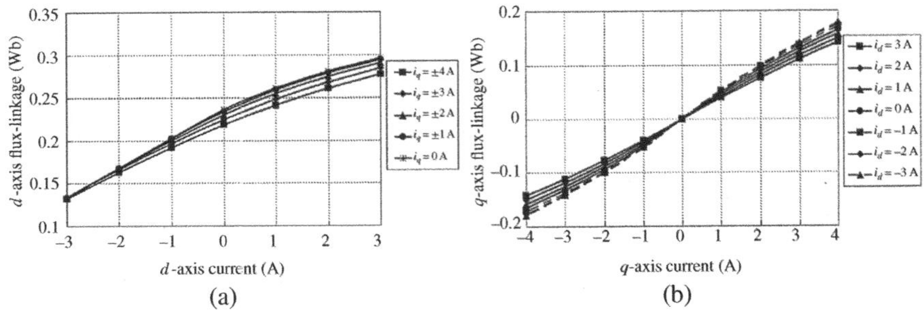

Figure 3.10 *dq*-axis flux-linkages at various currents. (a) *d*-axis flux-linkage. (b) *q*-axis flux-linkage. (*Source:* [9]/IEEE.)

The apparent self- and mutual-inductances then can be calculated as:

$$\begin{aligned} L_d &= \frac{\psi_d(i_d,0)-\psi_d(0,0)}{i_d}, L_q = \frac{\psi_q(0,i_q)-\psi_q(0,0)}{i_q} \\ L_{dq} &= \frac{\psi_d(i_d,i_q)-\psi_d(i_d,0)}{i_d}, L_{qd} = \frac{\psi_q(i_d,i_q)-\psi_q(0,i_q)}{i_d} \end{aligned} \tag{3.6}$$

Next, the *dq*-axis apparent self- and mutual-inductances can be calculated by finite element software as shown in Figure 3.11. Clearly, due to magnetic saturation, L_d, L_q, L_{dq}, and L_{qd} vary with both i_d and i_q.

Then, based on the conventional extended back-EMF model, the position estimation error due to cross-coupling magnetic saturation can be derived as:

$$\Delta\theta = \frac{\Delta L_q \hat{i}_q + L_{qd}\hat{i}_d}{\left(E_{ex,q}/\omega_r\right)} \tag{3.7}$$

It can be seen that mainly L_q and L_{qd} will produce the position estimation error. For *q*-axis inductance mismatch, it can be simply compensated as [11]:

$$L_q\left(i_q\right) = L_q(0) + \frac{L_q\left(i_q\right)-L_q(0)}{i_q}\left|i_q\right| \tag{3.8}$$

where $L_q(i_q)$ and $L_q(0)$ can be obtained from the finite element prediction or experiment.

Considering that the magnetic saturation on the *q*-axis inductance is well compensated with (3.8), the position error due to *dq*-axis cross-coupling is given by [9]:

$$\Delta\theta = \frac{L_{qd}\hat{i}_d}{\left(E_{ex,q}/\omega_r\right)} \tag{3.9}$$

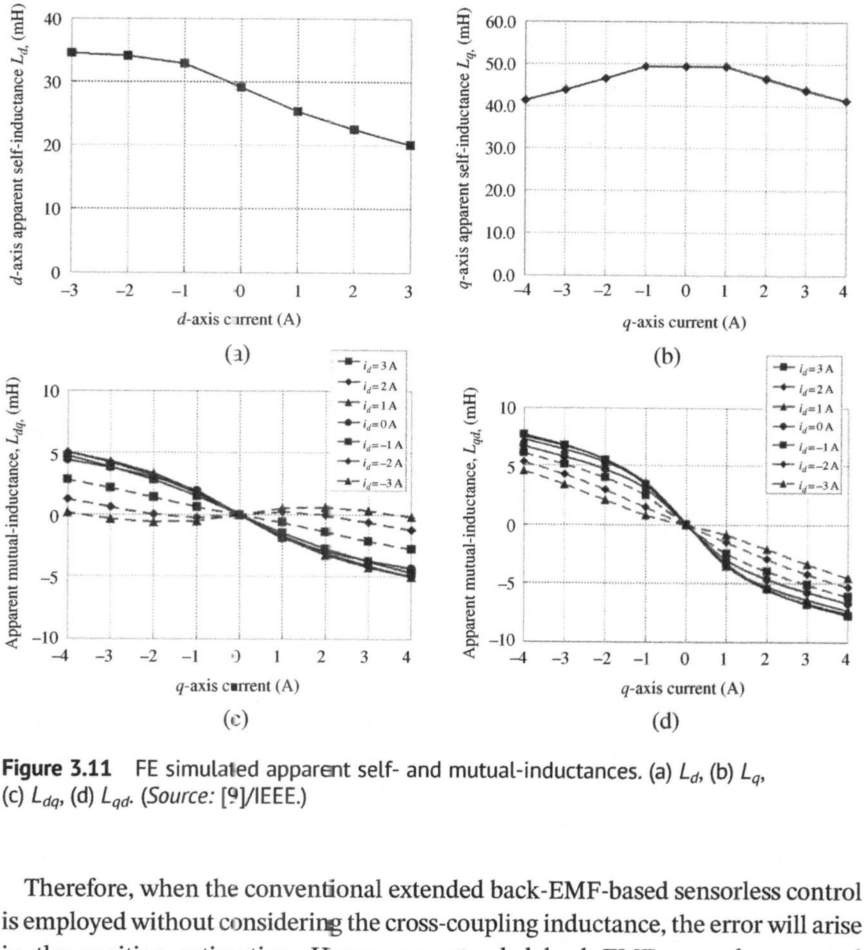

Figure 3.11 FE simulated apparent self- and mutual-inductances. (a) L_d, (b) L_q, (c) L_{dq}, (d) L_{qd}. (*Source:* [9]/IEEE.)

Therefore, when the conventional extended back-EMF-based sensorless control is employed without considering the cross-coupling inductance, the error will arise in the position estimation. Hence, an extended back-EMF sensorless control method accounting for cross-coupling [9] is introduced in the next section.

3.4.2 Sensorless Control Accounting for Cross-Coupling Inductance

To account for the mutual-inductances, L_{dq} and L_{qd}, they are considered in the machine model, from which an improved extended back-EMF sensorless control [9] is developed as:

$$\begin{bmatrix} v_d \\ v_q \end{bmatrix} = \begin{bmatrix} R_s + pL_d - \omega_r L_{qd} & -\omega_r L_q + pL_{dq} \\ \omega_r L_d + pL_{qd} & R_s + pL_q + \omega_r L_{dq} \end{bmatrix} \begin{bmatrix} i_d \\ i_q \end{bmatrix} + \begin{bmatrix} 0 \\ \omega_r \psi_m \end{bmatrix} \tag{3.10}$$

Similarly, as in the conventional extended EMF sensorless control, its impedance matrix is transformed into a symmetrical matrix between estimated and actual rotor position reference frames:

$$\begin{bmatrix} v_d \\ v_q \end{bmatrix} = \begin{bmatrix} R_s + pL_d - \omega_r L_{qd} & -\omega_r L_q + pL_{dq} \\ \omega_r L_q - pL_{dq} & R_s + pL_d + \omega_r L_{qd} \end{bmatrix} \begin{bmatrix} i_d \\ i_q \end{bmatrix} + \begin{bmatrix} 0 \\ E_{ex\_imp} \end{bmatrix}$$
$$\begin{aligned} E_{ex\_imp} = \omega_r \psi_m &+ \left(\omega_r L_d - \omega_r L_q + pL_{qd} + pL_{dq}\right) i_d \\ &+ \left(pL_q - pL_d + \omega_r L_{dq} + \omega_r L_{qd}\right) i_q \end{aligned} \tag{3.11}$$

The voltage in Eq. (3.11) is based on the actual rotor position, which is unavailable in the sensorless control. Therefore, it is transformed into the estimated synchronous reference frame as:

$$\begin{aligned} \begin{bmatrix} \hat{v}_d \\ \hat{v}_q \end{bmatrix} &= \begin{bmatrix} R_s + pL_d - \hat{\omega}_r L_{qd} & -\hat{\omega}_r L_q + pL_{dq} \\ \hat{\omega}_r L_q - pL_{dq} & R_s + pL_d - \hat{\omega}_r L_{qd} \end{bmatrix} \begin{bmatrix} \hat{i}_d \\ \hat{i}_q \end{bmatrix} \\ &+ E_{ex\_imp} \begin{bmatrix} -\sin(\Delta\theta) \\ \cos(\Delta\theta) \end{bmatrix} + (\hat{\omega}_r - \omega_r) \begin{bmatrix} L_q \hat{i}_q + L_{qd} \hat{i}_d \\ -L_q \hat{i}_d + L_{qd} \hat{i}_q \end{bmatrix} \end{aligned} \tag{3.12}$$

In (3.12), the mutual-inductance is much smaller than the self-inductance, and the current derivative is much smaller than the current value in steady or low acceleration operation. Therefore, the term pL_{dq} can be neglected in (3.12). The estimated speed error $(\hat{\omega}_r - \omega_r)$ is corrected quickly compared with the estimated position error, and this term can also be neglected. Then, the voltage equation in the estimated rotor position reference frame is simplified to:

$$\begin{bmatrix} \hat{v}_d \\ \hat{v}_q \end{bmatrix} = \begin{bmatrix} R_s + pL_d - \hat{\omega}_r L_{qd} & -\hat{\omega}_r L_q \\ \hat{\omega}_r L_q & R_s + pL_d - \hat{\omega}_r L_{qd} \end{bmatrix} \begin{bmatrix} \hat{i}_d \\ \hat{i}_q \end{bmatrix} + E_{ex\_imp} \begin{bmatrix} -\sin(\Delta\theta) \\ \cos(\Delta\theta) \end{bmatrix} \tag{3.13}$$

From the above simplified voltage equation, the error in the estimated rotor position is calculated from the improved extended EMF of dq-axes as:

$$\begin{aligned} &\begin{cases} \hat{E}_{ex\_imp\_d} = \hat{v}_d - \left(R_s + pL_d - \hat{\omega}_r L_{qd}\right)\hat{i}_d + \hat{\omega}_r L_q \hat{i}_q \\ \hat{E}_{ex\_imp\_q} = \hat{v}_q - \left(R_s + pL_d - \hat{\omega}_r L_{qd}\right)\hat{i}_q - \hat{\omega}_r L_q \hat{i}_d \end{cases} \\ &\Delta\hat{\theta} \approx \left(-\hat{E}_{ex\_imp\_d}/\hat{E}_{ex\_imp\_q}\right) \end{aligned} \tag{3.14}$$

The value of $\Delta\hat{\theta}$ is then used to correct the estimated rotor position and speed. Compared with the conventional extended back-EMF-based sensorless control in

Chapter 2, the cross-coupling effect is accounted for in (3.14) to improve the estimation accuracy.

The value of L_{qd} can be measured or simulated by FE software. Since the variation of L_{qd} with i_d and i_q can be a complex function that may be difficult to implement on low-cost applications, a simple approximation function, L_{qds}, can be used to represent the variation of L_{qd} with i_d and i_q, viz.

$$L_{qds} = \begin{cases} 0, & \hat{i}_d \leq 0 \\ k_c \hat{i}_q, & \hat{i}_d > 0 \end{cases} \tag{3.15}$$

where k_c is the compensation factor.

Alternatively, the error in the estimated rotor position due to the neglect of the dq-axis cross-coupling effect can also be compensated for indirectly, by first assuming L_{qd} to be zero and then compensating for the resulting rotor position error according to the measured error. For simplicity, it is approximated by:

$$\theta_c = \begin{cases} 0, & \hat{i}_d \leq 0 \\ k_r \hat{i}_d \hat{i}_q, & \hat{i}_d > 0 \end{cases} \tag{3.16}$$

where θ_c is the value of the rotor position compensation and k_r is the compensation factor. In this case, information regarding L_{qd} is not required.

Based on the prototype IPMSM-I (its parameters are given in Appendix B), the conventional and improved extended back-EMF sensorless methods are compared. For the conventional method, the inductances are fixed as $L_q = 43$ mH and $L_{qd} = 0$ mH. Then, the conventional method considering the q-axis inductance variation as (3.8) is also tested, with $L_q = (51 - 2|i_q|)$ mH and $L_{qd} = 0$ mH. For the improved methods, two compensation approaches are examined. For the first improved method with cross-coupling inductance in the model, the cross-coupling inductance is set by (3.15) with $k_c = -0.002$. For the position error compensation-based improved method, k_r is approximated as 0.9.

The results are given in Figure 3.12. When the q-axis apparent inductance, L_q, is regarded as constant and L_{qd} is neglected in the model, Figure 3.12a, the maximum estimated error is large, e.g. 16.6° at $i_q = 4$ A. When the magnetic saturation effect in L_q is considered, Figure 3.12b, there is a small improvement in the rotor position estimation accuracy, e.g. 15.5° at $i_q = 4$ A. By accounting for L_{qd}, the maximum error in the estimated rotor position in Figure 3.12c is significantly reduced to 6.2°. The small errors are due to the inaccuracy of finite element predicted dq-axis inductances, and the approximation within R_s, L_q, and L_{qd}. For the position error compensation shown in Figure 3.12d, the error is the same as in Figure 3.12c.

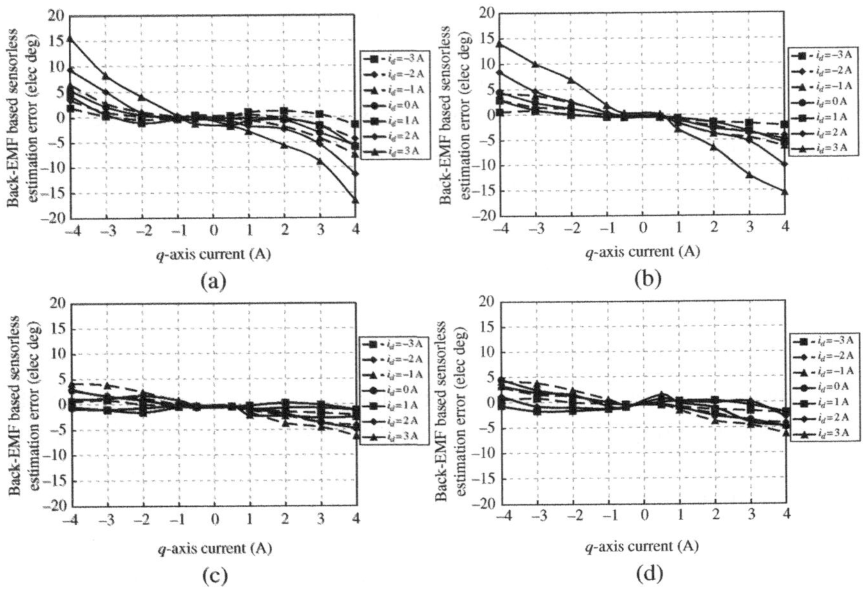

Figure 3.12 Position estimation errors of extended back-EMF-based sensorless. (a) Conventional extended back-EMF method. (b) Conventional extended back-EMF method considering q-axis inductance variation. (c) Improved extended back-EMF method with L_{qd} accounted in the model. (d) Improved extended back-EMF method with position error compensation. (*Source:* [9]/IEEE.)

3.5 Parameter Mismatch

In real-world applications it is challenging to obtain accurate parameter values of an electrical machine, since the actual parameters vary with temperature, saturation effect, load condition, etc. For conventional model-based sensorless methods, only nominal parameter values are used in the back-EMF or flux-linkage estimator. Therefore, when actual parameters vary, an issue of the parameter mismatch is introduced, which may not only cause position estimation error but also deteriorates the performance of the position observer and the overall control system [12].

In this section the influence of the parameter mismatch on rotor position estimation will be investigated based on the extended back-EMF methods. To be more specific, these parameters include phase resistance, d-axis inductance, and q-axis inductance. The cross-coupling inductance L_{qd} was discussed in the previous section, and thus, it is not included here. The relationship between the nominal ($\tilde{R}_s$, $\tilde{L}_d$, $\tilde{L}_q$) and actual (R_s, L_d, L_q) values of the parameters are defined as:

$$\begin{aligned} R_s &= \tilde{R}_s + \Delta R_s \\ L_d &= \tilde{L}_d + \Delta L_d \\ L_q &= \tilde{L}_q + \Delta L_q \end{aligned} \tag{3.17}$$

where the nominal values are those parameters used in the position observer and the parameters with "Δ" represent the mismatched values. It should be mentioned that the PM flux-linkage ψ_m does not need to be considered here since it is not required when an extended back-EMF-based sensorless control method is applied [12].

3.5.1 Impact on Position Estimation

In this subsection, the position estimation error due to parameter mismatches will be analyzed. First, the analysis is based on the extended back-EMF-based sensorless control as introduced in Chapter 2. The estimated position error can be obtained as:

$$\Delta\hat{\theta} = -\frac{\hat{E}_{ex,d}}{\hat{E}_{ex,q}} = \frac{\hat{v}_d - R_s\hat{i}_d - pL_d\hat{i}_d + \omega_r L_q\hat{i}_q}{\hat{v}_q - R_s\hat{i}_q - pL_d\hat{i}_q - \omega_r L_q\hat{i}_d} \tag{3.18}$$

Considering that the current is controlled to track the reference well in steady or low acceleration operation, then pL_d can be neglected, and (3.18) is then simplified as:

$$\Delta\hat{\theta} = \frac{\hat{v}_d - R_s\hat{i}_q - \omega_r L_q\hat{i}_q}{\hat{v}_q - R_s\hat{i}_q - \omega_r L_q\hat{i}_d} \tag{3.19}$$

By substituting (3.17) into (3.19), the position estimation error due to parameter mismatches can be derived as:

$$\begin{aligned} \Delta\theta_{\Delta R_s} &= \tan^{-1}\left(\frac{\Delta R_s\hat{i}_d}{E_{ex} + \Delta R_s\hat{i}_q}\right) \\ \Delta\theta_{\Delta L_d} &= 0 \\ \Delta\theta_{\Delta L_q} &= \tan^{-1}\left(\frac{-\Delta L_q\hat{i}_q}{E_{ex}/\hat{\omega}_r + \Delta L_q\hat{i}_d}\right) \end{aligned} \tag{3.20}$$

where $\Delta\theta_{\Delta R_s}$, $\Delta\theta_{\Delta L_d}$, and $\Delta\theta_{\Delta L_q}$ represent the estimated position errors caused by the mismatch in phase resistance, and d- and q-axis inductances, respectively.

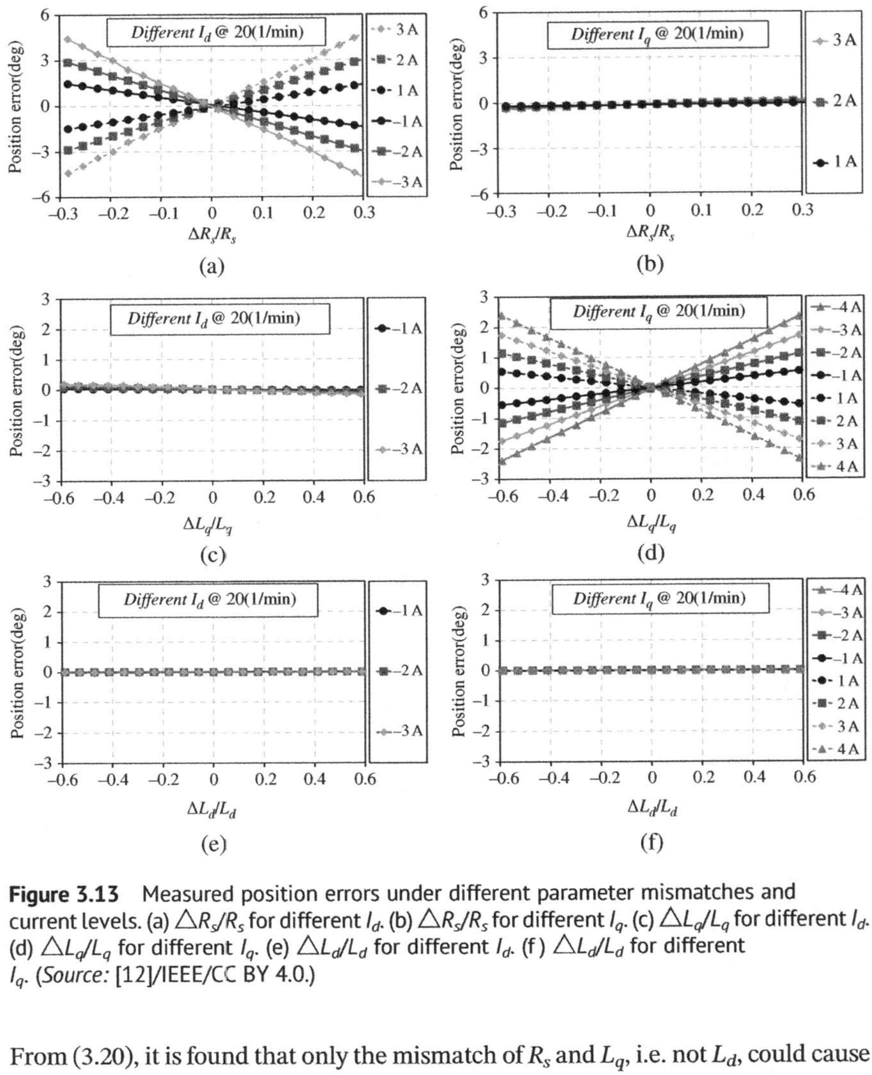

Figure 3.13 Measured position errors under different parameter mismatches and current levels. (a) $\triangle R_s/R_s$ for different I_d. (b) $\triangle R_s/R_s$ for different I_q. (c) $\triangle L_q/L_q$ for different I_d. (d) $\triangle L_q/L_q$ for different I_q. (e) $\triangle L_d/L_d$ for different I_d. (f) $\triangle L_d/L_d$ for different I_q. (*Source:* [12]/IEEE/CC BY 4.0.)

From (3.20), it is found that only the mismatch of R_s and L_q, i.e. not L_d, could cause rotor position error (dc offset). These position errors are related to current and speed levels.

Figure 3.13 shows the experimental results of SPMSM-III (Appendix B) at 20 r/min of the rotor position error due to $\triangle R_s$, $\triangle L_q$, and $\triangle L_d$ when one axis current is varied and the other axis current is controlled to zero. For resistance mismatch, as shown in Figure 3.13a, at a given I_d value, the error is proportional to $\triangle R_s$, while in Figure 3.13b, it is observed that $\triangle R_s$ has barely introduced any position errors for different I_q values. Then, for q-axis inductance mismatch, as shown in

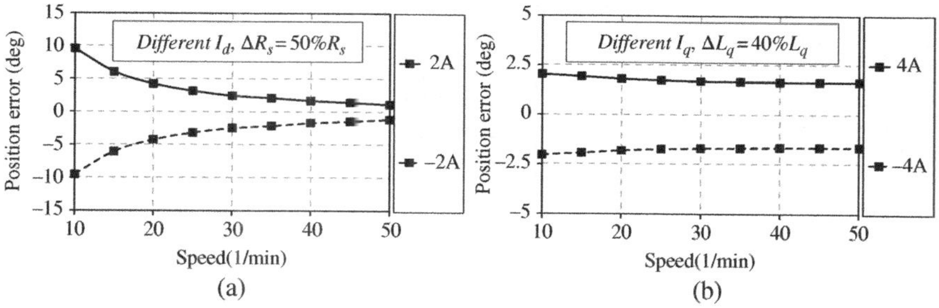

Figure 3.14 Measured position errors under different speeds. (a) R_s mismatch. (b) L_q mismatch. (*Source:* [12]/IEEE/CC BY 4.0.)

Figure 3.13c, at a given I_q value, the error is proportional to ΔL_q, while in Figure 3.13d, when L_q is mismatched, a different I_d level has little effect on position estimation accuracy.

From another perspective, at a given ΔR_s (or ΔL_q), the error is proportional to the *d*- (or *q*-) axis current. However, if there is no mismatch in the parameters, the change in current levels should make little difference in terms of position error. Moreover, from Figures 3.13e,f, it is found that ΔL_d does not affect position estimation. Along with (3.20), these features can be concluded as a position error mechanism. That is, the position error is mainly caused by the parameter mismatches of R_s and L_q, and the level of position error is varied according to I_d for ΔR_s and I_q for ΔL_q. Furthermore, the effect of ΔR_s is hardly coupled with I_q level, nor ΔL_q with I_d level.

On the other hand, from (3.20) and Figure 3.14a, the position error due to ΔR_s decreases as the speed increases, while from (3.20) and Figure 3.14b, the error due to ΔL_q is independent of speed variation. Again, if there is no parameter mismatch, the position error is zero and the speed becomes irrelevant.

Furthermore, the dynamic performance of the position estimation is tested by applying current impulses in the *dq*-axis. Figures 3.15a–c are the results of position error with *q*-axis current impulse (no load to half load) when the mismatches of L_d, L_q, and R_s are considered, respectively. It can be seen that the position estimation is more sensitive to L_q mismatch under the I_q impulse condition. In Figure 3.15c, an accurate L_q can minimize the influence on the dynamic of the position estimation. In Figure 3.15d, the dynamic performance is tested with *d*-axis current impulse under R_s mismatch. Similarly, the results show that any impact on the dynamic of the position estimation can be minimized when accurate R_s is applied. Therefore, parameter accuracy is important for estimation not only at the steady state but also at the transient state.

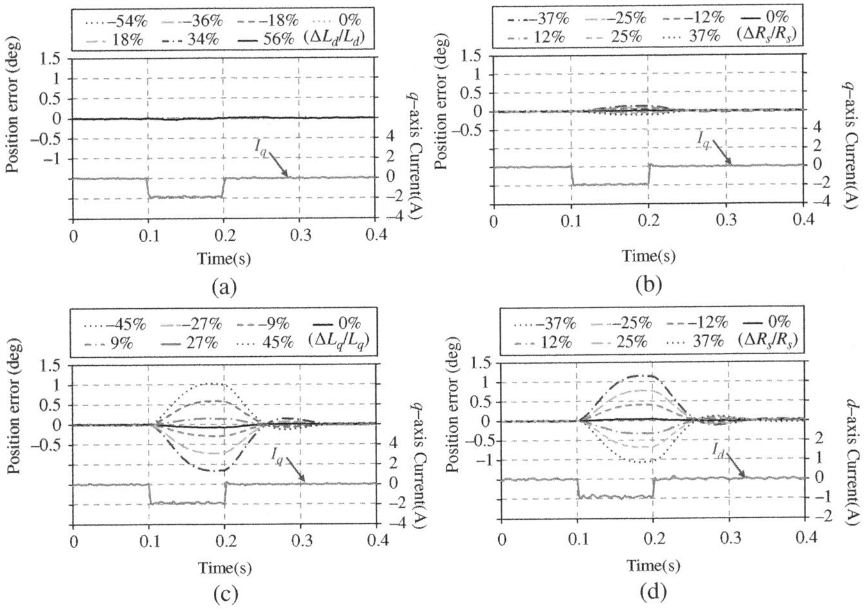

Figure 3.15 Measured position error dynamic response under current impulse. (a) L_d mismatch with q-axis current impulse. (b) R_s mismatch with q-axis current impulse. (c) L_q mismatch with q-axis current impulse. (d) R_s mismatch with d-axis current impulse. (*Source:* [12]/IEEE/CC BY 4.0.)

3.5.2 Position Correction Method Under Parameter Mismatches

Since phase resistance and q-axis inductance mismatch cause position error in the estimation, the error due to parameter mismatches can be generally written as:

$$\Delta\theta_{par} = \Delta\theta_{\Delta L_q} + \Delta\theta_{\Delta R_s} \approx K_{\Delta L_q}\hat{i}_q + K_{\Delta R_s}\hat{i}_d \tag{3.21}$$

where $K_{\Delta R_s}$ and $K_{\Delta L_q}$ are the mismatch factors and defined as:

$$K_{\Delta R_s} = \frac{\Delta R_s}{E_{ex} + \Delta R_s \hat{i}_q},\ K_{\Delta L_q} = -\frac{\Delta L_q}{E_{ex}/\hat{\omega}_r + \Delta L_q \hat{i}_d} \tag{3.22}$$

The error in (3.21) is composed of two parts: one is related to $\triangle R_s$, and the other to $\triangle L_q$. If each part of the mismatched parameter can be corrected independently, the whole position error can be eliminated. Normally, accurate parameters are achieved by parameter identifications. All the parameters need to be identified based on the mathematical model of PMSMs, which may have the issue of ill convergence and rank deficiency [13]. However, if the position error mechanism (3.22) is exploited, only two parameters are related to the position error, and they can be independently corrected. Based on this principle, a position error correction method [12] is introduced in this section.

According to (3.22), if there is no parameter mismatch, the error will not respond to the current variation. Otherwise, in the presence of a parameter mismatch, the error will take the same trend as the current variation. Therefore, by superimposing a sinusoidal current signal (3.23) with the frequency ω_{ac} and the amplitude A_{ac} onto d- or q-axis current reference, the position error should vary at the same frequency of the current signal, unless there is no mismatch.

$$i_{ac}^* = A_{ac} \sin \omega_{ac} t \tag{3.23}$$

3.5.2.1 q-Axis Injection for q-Axis Inductance Mismatch

The sinusoidal current signal can be superimposed onto q-axis current command i_q^* for L_q correction, and the estimated q-axis current can be expressed as:

$$\hat{i}_q = \hat{i}_{q,dc} + \hat{i}_{q,ac} \tag{3.24}$$

where the subscript dc or ac denotes the dc or ac component, respectively, $\hat{i}_{q,ac}$ is defined as the corresponding q-axis ac current component, and due to the current control, its frequency will stay the same as the reference current. Therefore, (3.21) can be rewritten as (3.25) when the extra signal is injected.

$$\begin{aligned} \Delta\theta_{\Delta L_q,ac} &\approx K_{\Delta L_q}\left(\hat{i}_{q,dc} + \hat{i}_{q,ac}\right) + \frac{\Delta R_s}{E_{ex} + \Delta R_s\left(\hat{i}_{q,dc} + \hat{i}_{q,ac}\right)}\hat{i}_{d,dc} \\ &= K_{\Delta L_q}\hat{i}_{q,dc} + K_{\Delta L_q}\hat{i}_{q,ac} - K_{\Delta R_s}\hat{i}_{d,dc} + K_{\Delta R_s,ac}\hat{i}_{d,dc} \\ &= \Delta\theta_{par} + K_{\Delta L_q}\hat{i}_{q,ac} + K_{\Delta R_s,ac}\hat{i}_{d,dc} \end{aligned} \tag{3.25}$$

where

$$\begin{cases} K_{\Delta R_s} = \dfrac{\Delta R_s}{E_{ex} + \Delta R_s \hat{i}_q} \\ K_{\Delta R_s,ac} = \dfrac{\Delta {R_s}^2 \hat{i}_{q,ac}}{\left(E_{ex} + \Delta R_s \hat{i}_{q,dc} + \Delta R_s \hat{i}_{q,ac}\right)\left(E_{ex} + \Delta R_s \hat{i}_{q,dc}\right)} \end{cases} \tag{3.26}$$

In (3.26), two extra ac components $K_{\Delta L_q}\hat{i}_{q,ac}$ and $K_{\Delta R_s,ac}\hat{i}_{d,dc}$ appear in the position error due to sinusoidal current signal injection, and the amplitude $\left|K_{\Delta L_q}\hat{i}_{q,ac}\right| = K_{\Delta L_q}A'_{q,ac}\left(A'_{q,ac} = \left|\hat{i}_{q,ac}\right|\right)$ has a close relationship with L_q mismatch (ΔL_q) while $\left|K_{\Delta R_s,ac}\hat{i}_{d,dc}\right|$ is independent of ΔL_q. To be more specific, if the degree of ΔL_q decreases ($K_{\Delta L_q}$ decreases), the amplitude of the ac component $K_{\Delta L_q}A'_{q,ac}$ should decrease while the amplitude $\left|K_{\Delta R_s,ac}\hat{i}_{d,dc}\right|$ remains unchanged. Furthermore, if there were no L_q mismatch, $K_{\Delta L_q}A'_{q,ac}$ would be zero, and so would this part of the ac component. Therefore, for a given injected current signal, the accurate q-axis inductance can be determined when the amplitude of the ac component is minimized, and the correction of L_q can be independently achieved without considering the effect due to R_s deviations on position estimation. With accurate L_q, the related position error $\Delta\theta_{\Delta L_q}$ can be eliminated.

3.5.2.2 *d*-Axis Injection for Resistance Mismatch

Similarly, to correct R_s, a sinusoidal current signal is superimposed onto d-axis current command i_d^*, and the corresponding d-axis current can be expressed as:

$$\hat{i}_d = \hat{i}_{d,dc} + \hat{i}_{d,ac} \tag{3.27}$$

Therefore, by substituting (3.27) into (3.21), it can be rewritten as:

$$\begin{aligned}\Delta\theta_{\Delta R_s,ac} &\approx K_{\Delta L_q}\hat{i}_{q,dc} + K_{\Delta L_q,ac}\hat{i}_{q,dc} + K_{\Delta R_s}\hat{i}_{d,dc} + K_{\Delta R_s}\hat{i}_{d,ac} \\ &= \Delta\theta_{\text{par}} + K_{\Delta R_s}\hat{i}_{d,ac} + K_{\Delta L_q,ac}\hat{i}_{q,dc}\end{aligned} \tag{3.28}$$

where

$$K_{\Delta L_q,ac} = \frac{\Delta L_q{}^2\hat{i}_{d,ac}}{\left(E_{ex}/\hat{\omega}_r + \Delta L_q\hat{i}_{d,dc} + \Delta L_q\hat{i}_{d,ac}\right)\left(E_{ex}/\hat{\omega}_r + \Delta L_q\hat{i}_{d,dc}\right)} \tag{3.29}$$

In (3.29), two extra terms $K_{\Delta R_s}\hat{i}_{d,ac}$ and $K_{\Delta L_q,ac}\hat{i}_{q,dc}$ appear due to d-axis current signal injection, and only the magnitude of $K_{\Delta R_s}\hat{i}_{d,ac}$ is related to the resistance mismatch. Therefore, by correcting the resistance parameter, the amplitude of the ac component in the position error can be minimized as well.

3.5.2.3 Amplitude Calculation Technique

From the above analysis, the amplitude of the ac component in the position error is the critical information to the parameter mismatch. Therefore, the ac component has to be extracted. Since directly accessing the position error $\Delta\theta_{\text{par}}$ is impossible, it is more convenient to extract the ac component from the estimated speed signal $\hat{\omega}_r$ instead, since it is the derivative of the estimated position. Then, with the help of a peaking filter, the ac component $\hat{\omega}_r^{ac}$ contained in $\hat{\omega}_r$ can be extracted. The transfer function of a peaking filter is expressed as:

$$G(s) = \frac{\mu s}{s^2 + \mu s + \omega_{ac}^2} \tag{3.30}$$

where μ is the bandwidth of the peaking filter. As μ increases, the response is faster, but the peaking performance becomes worse. Then, the amplitude $\left|\hat{\omega}_r^{ac}\right|$ of the ac component $\hat{\omega}_r^{ac}$ can be calculated through the orthogonal generation system, which is based on a second order generalized integrator (SOGI). The structure

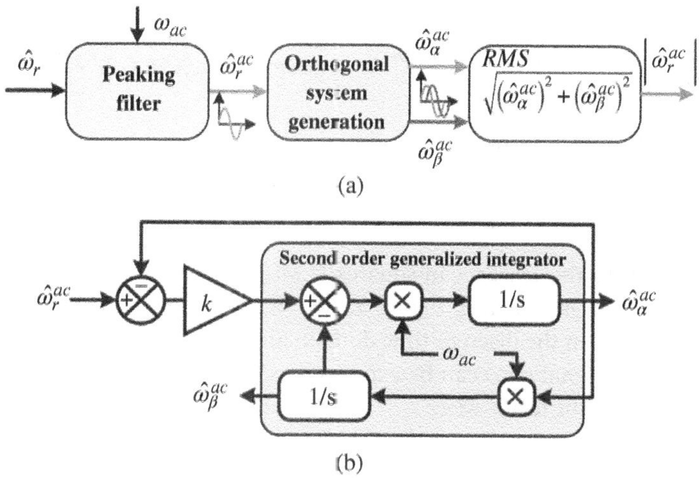

Figure 3.16 General structure. (a) An ac component amplitude calculator. (b) Orthogonal system generation. (*Source:* [12]/IEEE/CC BY 4.0.)

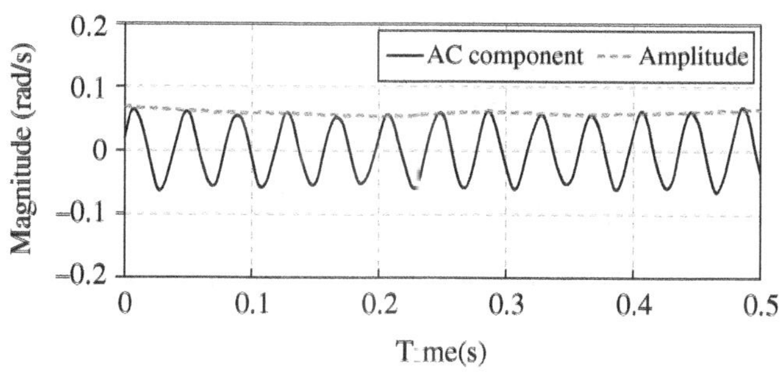

Figure 3.17 Experimental result of amplitude calculator. (*Source:* [12]/IEEE/CC BY 4.0.)

block diagrams of the amplitude calculator and orthogonal generation system are shown in Figures 3.16a,b, respectively, and the experimental result of amplitude calculation under the injection frequency of 25 Hz is shown in Figure 3.17.

3.5.2.4 Position Error Correction with LMS Algorithm

As previously mentioned, $|\hat{\omega}_r^{ac}|$ has a close relationship with the parameter mismatch. To observe the accurate parameter, the minimum value of $|\hat{\omega}_r^{ac}|$ should be acquired by adjusting the parameter that is set in the position observer.

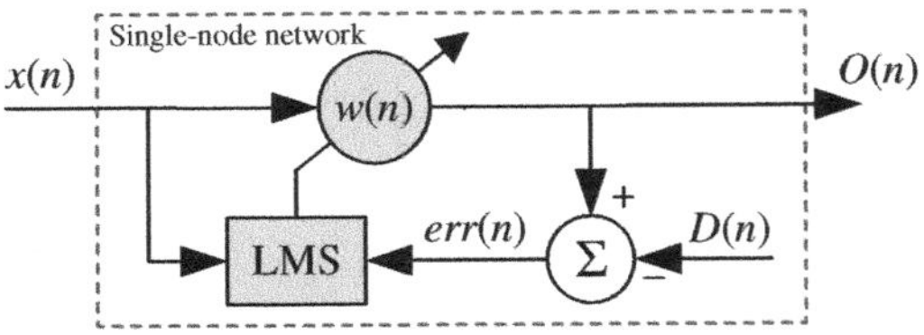

Figure 3.18 Block diagram of LMS algorithm. (*Source:* [12]/IEEE/CC BY 4.0.)

To accomplish this task, the least-mean-squares (LMS) algorithm is used. The structure of the algorithm is shown in Figure 3.18.

The algorithm operates in the discrete-time domain and n denotes the nth sampling point. The output signal $O(n)$ can be expressed as the multiplication of the input signal $x(n)$ and weight factor $w(n)$:

$$O(n) = x(n)w(n) \tag{3.31}$$

and the error $err(n)$, which is the difference between the output signal $O(n)$ and the desired response $D(n)$, can be defined as:

$$err(n) = D(n) - O(n) \tag{3.32}$$

The objective of this algorithm is to find a proper weight factor $w(n)$ to produce the least mean squares of $err(n)$. Thus, the parameters (L_q or R_s) of the sensorless observer are considered as the weight factors and $|\hat{\omega}_r^{ac}|$ is the input signal of the algorithm. This can be rewritten as (take L_q correction, for example):

$$O(n) = x(n)w(n) = |\hat{\omega}_r^{ac}|(n)L_q(n) \tag{3.33}$$

To minimize $|\hat{\omega}_r^{ac}|$, the desired response $D(n)$ can be set to zero, and then from (3.32) can be expressed as:

$$err(n) = D(n) - O(n) = 0 - |\hat{\omega}_r^{ac}|(n)L_q(n) \tag{3.34}$$

In (3.34), with a proper $L_q(n)$, $|\hat{\omega}_r^{ac}|(n)$ will be minimized when $|err(n)|$ is minimized. Thus, when the objective function of the LMS algorithm $J(n)$ is defined as half of the squared error criterion:

$$J(n) = \frac{1}{2}[err(n)]^2 = \frac{1}{2}\left[-|\hat{\omega}_r^{ac}|(n)L_q(n)\right]^2 \tag{3.35}$$

L_q can be trained by applying a gradient descent method to minimize the objective function. The gradient of the objective function $J(n)$ with respect to $L_q(n)$ can be expressed as:

$$\nabla J(n) = \frac{\partial J(n)}{\partial L_q(n)} = -err(n)|\hat{\omega}_r^{ac}|(n) \tag{3.36}$$

and the updated rule of the weight factor $L_q(n)$ is expressed as:

$$L_q(n+1) = L_q(n) + \xi[-\nabla J(n)] \tag{3.37}$$

where ξ is the training constant, which is a positive coefficient that determines the speed of convergence: the bigger the ξ, the faster the convergence. In (3.37), the weight factors are updated in the reverse direction of the gradient $\nabla J(n)$ since the value of the objective function needs to be reduced. Therefore, from (3.34), (3.36), and (3.37), the parameter can be updated as:

$$L_q(n+1) = L_q(n) - \xi\left[\left|\hat{\omega}_r^{ac}\right|(n)\right]^2 L_q(n) \tag{3.38}$$

Therefore, with the help of the LMS algorithm, $L_q(n)$ can be trained to accurate value adaptively and then the position error can be corrected. The procedure for resistance mismatch correction is basically the same; the only difference is that the sinusoidal current signal should be superimposed onto the d-axis current reference instead of the q-axis.

In (3.38), it is assumed that $L_q(n) > L_q(n+1)$, and the parameter is trained in the direction of decrease. However, the initial value of the parameter in the position observer may be smaller or larger than the actual value before the correction, and the right direction of correction (decrease or increase) can be solved with the direction judgment strategy.

The technique used for initial direction judgment is based on the following criterion: When the adapting parameter is perturbed in a given direction and if $\left|\hat{\omega}_r^{ac}\right|$ decreases, it means that the parameter has moved toward the accurate value. Otherwise, if $\left|\hat{\omega}_r^{ac}\right|$ increases, the parameter has moved away from the accurate value, and therefore, the direction of the perturbation must be reversed. Take the very first cycle ($n = 1$) as an example, where the adjustment rule is defined as:

$$L_q(1) = L_q(0) - \Delta L_{\text{step}}(0), \left(\Delta L_{\text{step}}(0) > 0\right) \tag{3.39}$$

where $L_q(0) = \tilde{L}_q$ represents the q-axis inductance value set in the observer initially, while $L_q(1)$ is the value after the first cycle of training and $\Delta L_{step}(0) = \xi\left[\left|\hat{\omega}_r^{ac}\right|(0)\right]^2 L_q(0)$. After the first adjustment cycle, the training direction can be determined with the following logical judgment.

$$\begin{aligned} &\textit{if } \left|\hat{\omega}_r^{ac}\right|_{L_q(0)} > \left|\hat{\omega}_r^{ac}\right|_{L_q(1)} \Rightarrow \textit{Correct Direction} \\ &\textit{else if } \left|\hat{\omega}_r^{ac}\right|_{L_q(0)} < \left|\hat{\omega}_r^{ac}\right|_{L_q(1)} \Rightarrow \textit{Incorrect Direction} \end{aligned} \tag{3.40}$$

where $\left|\hat{\omega}_r^{ac}\right|_{L_q(0)}$ and $\left|\hat{\omega}_r^{ac}\right|_{L_q(1)}$ represent the amplitudes of ac components when L_q is set to $L_q(0)$ and $L_q(1)$, respectively.

The above procedure can be repeated once or twice to confirm that the correction is in the right direction. Moreover, various techniques can be applied to secure direction judgment in practical application. For example, the step of the adjustment can be set bigger than the defined value to reduce the effect of noise and unexpected disturbance, or by using the moving average value of $|\hat{\omega}_r^{ac}|$ to improve the accuracy of the direction judgment. With the help of these techniques, the correct direction of judgment can be guaranteed.

3.5.2.5 Experimental Results

To validate the position error correction technique, experiments are carried out based on the prototype SPMSM-III (its parameters are given in Appendix B). Since the feedback current is required to track the current reference variation properly with the same frequency as the injected signal, the frequency of the injected current signal ω_{ac} must be lower than the current control loop bandwidth. Moreover, to avoid introducing unexpected oscillations, the frequency should be higher than the outer loop bandwidth, whether for speed, torque, or power.

For the extended EMF position observer in Chapter 2, the estimated position error $\Delta\hat{\theta}$ is derived from the estimated back-EMFs, and then the estimated speed and position are obtained by the observer to force $\Delta\hat{\theta}$ to zero.

If the sinusoidal current signal is superimposed onto the q-axis current reference, the amplitude of the signal must be small enough to avoid introducing torque ripples. The amplitude and frequency are chosen as 0.2 A and 25 Hz, respectively, as an example for the test system. Since the prototype machine has a relatively high inertia, with the selection of injection, the real speed hardly varies. On the other hand, for the low-inertia systems that may react to the q-axis current injection, the magnitude and frequency of the injected current signal can be selected so that its impact on the system can be minimized. For example, the frequency could be chosen to avoid the mechanical resonance of the system, and the instant of injection can also be managed to prevent the possible interference between speed/torque harmonics and the correction procedure. Moreover, the magnitude can be set small, as only the relative change in response during parameter correction is used. The capability of fast correction and, thus, the need for only a short time of injection, would also help. Moreover, to avoid unexpected effects, the correction procedure should be applied in a relatively stable state of speed and load.

The overall block diagram of the sensorless control system with the parameter correction method is shown in Figure 3.19. The injection procedure controller is used to control which axis the signal is injected into based on the mismatch situation. The correction for L_q and R_s mismatch could be taken independently or in series.

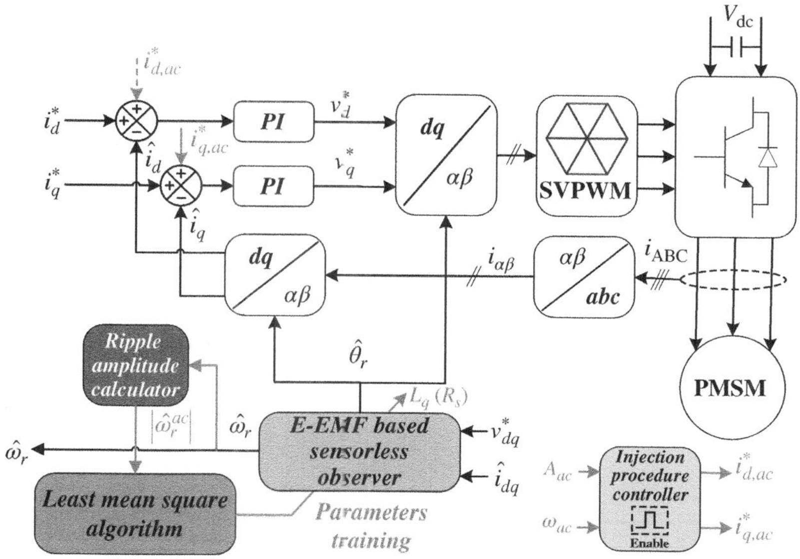

Figure 3.19 Overall block diagram of position correction method for parameter mismatch. (*Source:* [12]/IEEE/CC BY 4.0.)

Initially, with the help of an extended back-EMF sensorless observer, the estimated position and speed can be acquired. The parameters in the observer can be set to their nominal values. When the correction procedure starts, the sinusoidal current signal is injected into the system (q- or d-axis of current), and then, the ac component should appear in the estimated speed. The corresponding amplitude can be acquired from the amplitude calculator. Finally, the LMS algorithm can train the parameters adaptively to more accurate values compared with the original ones. With the updated parameters, the estimated position error caused by parameter mismatch can be corrected.

Figure 3.20 shows the extracted ac component $\hat{\omega}_r^{ac}$ and the amplitude $|\hat{\omega}_r^{ac}|$ variation for the q-axis inductance mismatch ratio ($\Delta L_q/L_q$) under the half load condition. The sinusoidal current signal (0.2 A, 25 Hz) is superimposed onto the q-axis. From Figure 3.20, the minimum point of $\hat{\omega}_r^{ac}$ corresponds to the actual L_q, and $\hat{\omega}_r^{ac}$ decreases as the level of mismatch decreases.

Furthermore, in Figures 3.21a,b, the influence of the load variation on $\hat{\omega}_r^{ac}$ has been evaluated for L_q and R_s mismatches at 40 r/min, respectively. Figure 3.21 shows that the load variation has very little effect on the $\hat{\omega}_r^{ac}$ value.

However, for the change in speed, as illustrated in Figure 3.22, since $K_{\Delta L_q}$ is independent of speed whereas $K_{\Delta R_s}$ is inversely proportional to the speed as (3.22), thus,

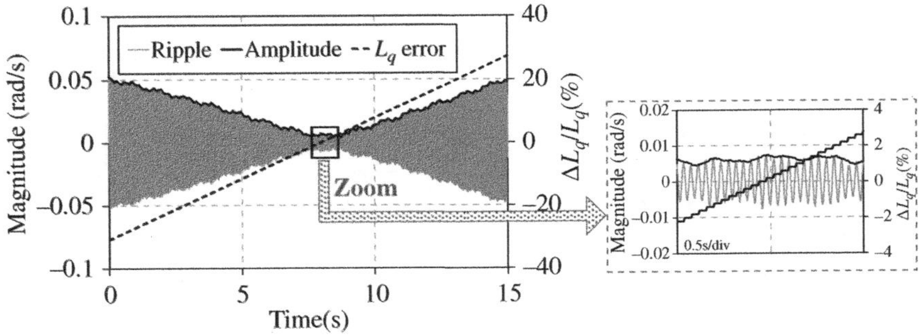

Figure 3.20 Extracted $\hat{\omega}_r^{ac}$, amplitude $|\hat{\omega}_r^{ac}|$ according to L_q variation with signal injected into q-axis (50% load, 40 r/min). (*Source:* [12]/IEEE/CC BY 4.0.)

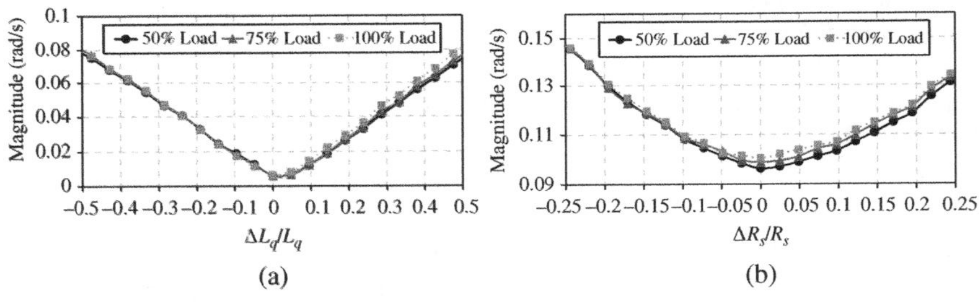

Figure 3.21 Measured $|\hat{\omega}_r^{ac}|$ against mismatch ratio under different load conditions at 40 r/min. (a) L_q mismatch with signal injected into q-axis at I_d = 0 A. (b) R_s mismatch with signal injected into d-axis at I_d = −2 A. (*Source:* [12]/IEEE/CC BY 4.0.)

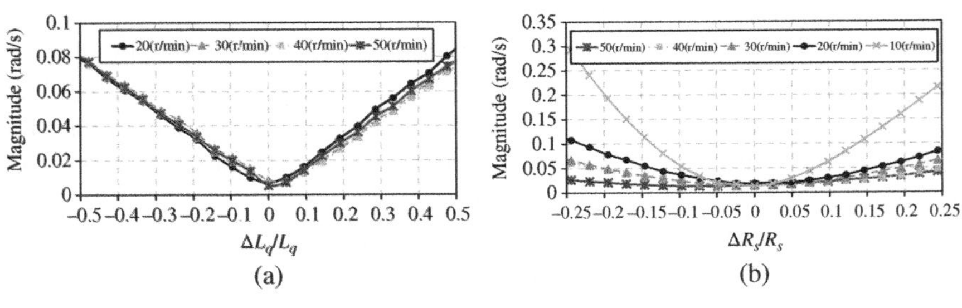

Figure 3.22 Measured $|\hat{\omega}_r^{ac}|$ against mismatch ratio under different speeds with 50% load condition. (a) L_q mismatch with signal injected into q-axis at I_d = 0 A. (b) R_s mismatch with signal injected into d-axis at I_d = −2 A. (*Source:* [12]/IEEE/CC BY 4.0.)

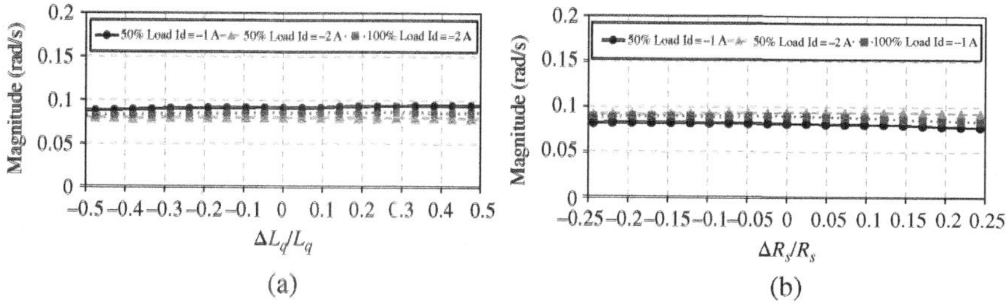

Figure 3.23 Measured $|\hat{\omega}_r^{ac}|$ against mismatch ratio under different load conditions at 40 r/min. (a) L_q mismatch with signal injected into the q-axis. (b) R_s mismatch with signal injected into the d-axis. (*Source:* [12]/IEEE/CC BY 4.0.)

$|\hat{\omega}_r^{ac}|$ due to resistance mismatch reduces as the speed increases. Therefore, it is more effective to correct the resistance in the low-speed rather than the high-speed region.

In Figure 3.23a, for the case of L_q mismatch, instead of injecting the current into the q-axis, it is injected into the d-axis to see the influence of the variation. It can be seen that the L_q variation has little effect on $|\hat{\omega}_r^{ac}|$. Similarly, Figure 3.23b shows that the resistance variation also has little effect on $|\hat{\omega}_r^{ac}|$, when the current signal is injected into the q-axis instead of the d-axis. This confirms that the position errors caused by mismatched L_q and R_s can be decoupled and can be independently corrected.

By applying the LMS algorithm, the correction results can be shown in Figures 3.24 and 3.25 for L_q and R_s mismatches, respectively, where the signals of feedback current, actual position error, and corrected parameters are included. In Figure 3.24a, the machine is running at 40 r/min, while in Figure 3.24b the

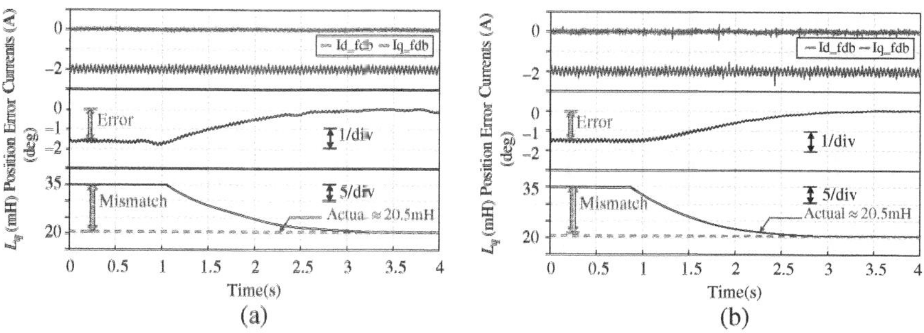

Figure 3.24 Position correction of L_q mismatch at 50% load condition with signal injected into q-axis. (a) 40 r/min. (b) 60 r/min. (*Source:* [12]/IEEE/CC BY 4.0.)

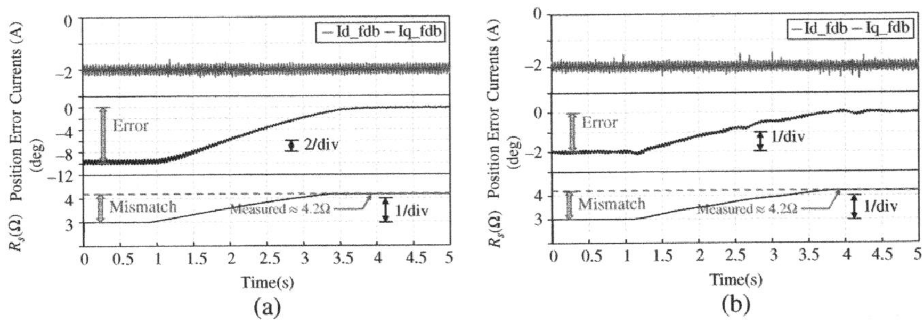

Figure 3.25 Position correction of L_q mismatch at 50% load condition with signal injected into *d*-axis. (a) 10 r/min. (b) 40 r/min. (*Source:* [12]/IEEE/CC BY 4.0.)

speed is at 60 r/min, and the initial nominal L_q is set at 35 mH in the position observer and the current is injected into the *q*-axis. After the LMS algorithm is enabled, L_q is trained to approach the actual value, which is approximately 20.5 mH, and then the position error caused by L_q mismatch can be removed.

Similarly, for the case of R_s mismatch, the test machine is rotating at 10 r/min in Figure 3.25a, while the speed is set to 40 r/min for Figure 3.25b. For both conditions, the nominal value is set to 3 Ω initially, and by applying the proposed technique, the mismatched R_s can be corrected to the actual value of 4.2 Ω, and the position error can be eliminated. From (3.20), when $\hat{i}_d = 0$, there should be no position error due to resistance mismatch. Therefore, the magnitude of the *d*-axis current in the test (Figure 3.25) is only used for validation purpose since the position error and the correction procedure can be seen during the test when $\hat{i}_d \neq 0$. For SPMSM under a $\hat{i}_d = 0$ control strategy, the extra sinusoidal current signal injection is only needed when R_s correction is required. It is noticed that with the same R_s mismatch, the position error at low speed, Figure 3.25a, is much greater than in Figure 3.25b when the speed is high. This is because the position error due to the R_s mismatch decreases as the speed increases.

3.6 Parameter Asymmetry

Among various fundamental model-based sensorless methods, the majority were developed by considering that the machine parameters are symmetrical. However, it is very common that an electrical machine may have unbalanced or asymmetric parameters including resistance [14], [15], inductance [16], [17], and back-EMF [17], [18] in three phases. Specifically, cabling, inverter, or stator windings may cause resistance asymmetry [14]. Inductance asymmetry may come from rough mechanical manufacturing, rotor eccentricity, and winding fault [16]. For back-EMF asymmetry, it can be due to rotor eccentricity, winding fault or specific machine topologies, etc. [17], [18]. If the conventional methods are employed without considering the parameter asymmetry, the unbalanced parameters will result in the second order harmonic (2*f*) error in the estimated position in model-based sensorless control [15], [17]. Therefore, in this section, the parameter asymmetry impact on position estimation with model-based sensorless control is described.

3.6.1 Asymmetric Modeling

3.6.1.1 Resistance Asymmetry

The mathematical model of three-phase asymmetric resistances [15] can be expressed as:

$$R_{\mathrm{ABC}} = \begin{bmatrix} \overline{R}_s + \Delta R_{\mathrm{A}} & 0 & 0 \\ 0 & \overline{R}_s + \Delta R_{\mathrm{B}} & 0 \\ 0 & 0 & \overline{R}_s + \Delta R_{\mathrm{C}} \end{bmatrix} \tag{3.41}$$

where ΔR_{A}, ΔR_{B}, ΔR_{C} are the asymmetric resistances and are assumed to be positive values, and $\overline{R}_s$ denotes the balanced part of the phase resistances.

By transforming into the synchronous reference frame, the resistance matrix is given as:

$$R_{dq} = \begin{bmatrix} R_d & R_{dq} \\ R_{dq} & R_q \end{bmatrix} \tag{3.42}$$

$$\begin{aligned}
R_d &= \overline{R}_s + \frac{\Delta R_{\mathrm{A}} + \Delta R_{\mathrm{B}} + \Delta R_{\mathrm{C}}}{3} + \frac{1}{3}\left[R_{\mathrm{A}}\cos(2\theta_r) + R_{\mathrm{B}}\cos\left(2\theta_r - \frac{4\pi}{3}\right) + R_{\mathrm{C}}\cos\left(2\theta_r + \frac{4\pi}{3}\right)\right] \\
R_q &= \overline{R}_s + \frac{\Delta R_{\mathrm{A}} + \Delta R_{\mathrm{B}} + \Delta R_{\mathrm{C}}}{3} - \frac{1}{3}\left[R_{\mathrm{A}}\cos(2\theta_r) + R_{\mathrm{B}}\cos\left(2\theta_r - \frac{4\pi}{3}\right) + R_{\mathrm{C}}\cos\left(2\theta_r + \frac{4\pi}{3}\right)\right] \\
R_{dq} &= \frac{\sin(2\theta_r)}{6}(-2R_{\mathrm{A}} + R_{\mathrm{B}} + R_{\mathrm{C}}) - \frac{\sqrt{3}\cos(2\theta_r)}{6}(R_{\mathrm{B}} - R_{\mathrm{C}})
\end{aligned} \tag{3.43}$$

Then, the voltage equation can be expressed as:

$$\begin{bmatrix} v_d \\ v_q \end{bmatrix} = \begin{bmatrix} \overline{R}_s + pL_d & -\omega_r L_q \\ \omega_r L_q & \overline{R}_s + pL_d \end{bmatrix} \cdot \begin{bmatrix} i_d \\ i_q \end{bmatrix} + \begin{bmatrix} 0 \\ \omega_r \psi_m \end{bmatrix} + \begin{bmatrix} \varepsilon_{d\Delta R} \\ \varepsilon_{q\Delta R} \end{bmatrix} \tag{3.44}$$

$$\begin{bmatrix} \varepsilon_{\Delta R} \\ \varepsilon_{\Delta R} \end{bmatrix} = \begin{bmatrix} \Delta R_d & R_{dq} \\ R_{qd} & \Delta R_q \end{bmatrix} \begin{bmatrix} i_d \\ i_q \end{bmatrix} \tag{3.45}$$

$$\Delta R_d = \Delta R_{\text{ave}} + \frac{1}{3}\left[R_{\text{A}}\cos(2\theta_r) + R_{\text{B}}\cos\left(2\theta_r - \frac{4\pi}{3}\right) + R_{\text{C}}\cos\left(2\theta_r + \frac{4\pi}{3}\right)\right]$$
$$\Delta R_q = \Delta R_{\text{ave}} - \frac{1}{3}\left[R_{\text{A}}\cos(2\theta_r) + R_{\text{B}}\cos\left(2\theta_r - \frac{4\pi}{3}\right) + R_{\text{C}}\cos\left(2\theta_r + \frac{4\pi}{3}\right)\right]$$
$$\Delta R_{\text{ave}} = \frac{\Delta R_{\text{A}} + \Delta R_{\text{B}} + \Delta R_{\text{C}}}{3} \tag{3.46}$$

Therefore, it can be noted that the asymmetric resistances of three phases will introduce two additional components in the voltage model: a resistance mismatch of ΔR_{ave} and a $2f$ disturbance term.

3.6.1.2 Inductance Asymmetry

First, for simplicity, assuming that one-phase self-inductance is unbalanced (the result will be the same when three-phase self-inductances are unbalanced), the inductance matrix can be expressed as:

$$L_{\text{ABC}} = \begin{bmatrix} L_{\text{AA}} & M_{\text{AB}} & M_{\text{AC}} \\ M_{\text{BA}} & L_{\text{BB}} & M_{\text{BC}} \\ M_{\text{CA}} & M_{\text{CB}} & L_{\text{CC}} \end{bmatrix} + \begin{bmatrix} \Delta L & 0 & 0 \\ 0 & 0 & 0 \\ 0 & 0 & 0 \end{bmatrix} \tag{3.47}$$

where ΔL presents the asymmetry part for the self-inductance. With coordinate transformation, the inductances in the d-/q-axis can be expressed as:

$$L_{dq} = \begin{bmatrix} L_d + \frac{\Delta L}{3}[1 + \cos(2\theta_r)] & -\frac{\Delta L}{3}\sin(2\theta_r) \\ -\frac{\Delta L}{3}\sin(2\theta_r) & L_q + \frac{\Delta L}{3}[1 - \cos(2\theta_r)] \end{bmatrix} \tag{3.48}$$

Based on (3.48), the voltage model [17] is modified as:

$$\begin{bmatrix} v_d \\ v_q \end{bmatrix} = \begin{bmatrix} R_s + pL_d & -\omega_r L_q \\ \omega_r L_d & R_s + pL_q \end{bmatrix} \cdot \begin{bmatrix} i_d \\ i_q \end{bmatrix} + \begin{bmatrix} 0 \\ \omega_r \psi_m \end{bmatrix} + \begin{bmatrix} \varepsilon_{d\Delta L} \\ \varepsilon_{q\Delta L} \end{bmatrix} \tag{3.49}$$

$$\begin{bmatrix} \varepsilon_{d\Delta L} \\ \varepsilon_{q\Delta L} \end{bmatrix} = \underbrace{\frac{\Delta L}{3}\begin{bmatrix} 1 & 0 \\ 0 & 1 \end{bmatrix} \cdot p \begin{bmatrix} i_d \\ i_q \end{bmatrix} + \frac{\omega_r \Delta L}{3}\begin{bmatrix} 0 & -1 \\ 1 & 0 \end{bmatrix} \cdot \begin{bmatrix} i_d \\ i_q \end{bmatrix}}_{\text{Inductance mismatch}}$$
$$\underbrace{+ \frac{\Delta L}{3}\begin{bmatrix} \cos 2\theta_r & -\sin 2\theta_r \\ -\sin 2\theta_r & -\cos 2\theta_r \end{bmatrix} \cdot p \begin{bmatrix} i_d \\ i_q \end{bmatrix} + \frac{\omega_r \Delta L}{3}\begin{bmatrix} -\sin 2\theta_r & -\cos 2\theta_r \\ -\cos 2\theta_r & \sin 2\theta_r \end{bmatrix} \cdot \begin{bmatrix} i_d \\ i_q \end{bmatrix}}_{2f\text{ disturbance}} \tag{3.50}$$

Clearly, from (3.49) and (3.50), the inductance asymmetry also introduces two additional components in the voltage model: an inductance mismatch and a $2f$ disturbance term.

3.6.1.3 Back-EMF Asymmetry

For the unbalanced case, the amplitudes of three-phase back-EMFs can be different. Then, the three-phase back-EMFs in the three-phase reference frame are given by:

$$\begin{bmatrix} e_{\mathrm{A}} \\ e_{\mathrm{B}} \\ e_{\mathrm{C}} \end{bmatrix} = -\omega_r \begin{bmatrix} (\psi_m + \Delta\psi_{\mathrm{A}})\sin\theta_r \\ (\psi_m + \Delta\psi_{\mathrm{B}})\sin\left(\theta_r - \dfrac{2\pi}{3}\right) \\ (\psi_m + \Delta\psi_{\mathrm{C}})\sin\left(\theta_r + \dfrac{2\pi}{3}\right) \end{bmatrix} \tag{3.51}$$

where $\Delta\psi_{\mathrm{A}}$, $\Delta\psi_{\mathrm{B}}$, and $\Delta\psi_{\mathrm{C}}$ represent the undesired unbalance flux-linkage terms. By transforming (3.51) into the synchronous reference frame, the voltage model [17] can be modified as:

$$\begin{bmatrix} v_d \\ v_q \end{bmatrix} = \begin{bmatrix} R_s + pL_d & -\omega_r L_q \\ \omega_r L_d & R_s + pL_q \end{bmatrix} \cdot \begin{bmatrix} i_d \\ i_q \end{bmatrix} + \begin{bmatrix} 0 \\ \omega_r \psi_m \end{bmatrix} + \begin{bmatrix} \varepsilon_{d2,\Delta\psi} \\ \varepsilon_{q2,\Delta\psi} \end{bmatrix} \tag{3.52}$$

$$\begin{bmatrix} \varepsilon_{d2,\Delta\psi} \\ \varepsilon_{q2,\Delta\psi} \end{bmatrix} = \begin{bmatrix} 0 \\ \omega_r \Delta\psi_{\mathrm{ave}} \end{bmatrix} - \frac{\omega_r}{3} \begin{bmatrix} \Delta\psi_{\mathrm{A}}\sin 2\theta_r - \Delta\psi_{\mathrm{B}}\sin\left(2\theta_r - \dfrac{2\pi}{3}\right) - \Delta\psi_{\mathrm{C}}\sin\left(2\pi_r + \dfrac{2\pi}{3}\right) \\ \Delta\psi_{\mathrm{A}}\cos 2\theta_r - \Delta\psi_{\mathrm{B}}\cos\left(2\theta_r - \dfrac{2\pi}{3}\right) - \Delta\psi_{\mathrm{C}}\cos\left(2\pi_r + \dfrac{2\pi}{3}\right) \end{bmatrix} \tag{3.53}$$

$$\Delta\psi_{\mathrm{ave}} = \frac{\Delta\psi_{\mathrm{C}} + \Delta\psi_{\mathrm{B}} + \Delta\psi_{\mathrm{C}}}{3} \tag{3.54}$$

Similarly, the back-EMF asymmetry introduces a PM flux-linkage mismatch $\Delta\psi_{\mathrm{ave}}$ and a $2f$ disturbance term in the voltage model.

3.6.2 Impacts on Position Estimation

Based on the derived mathematical model in Section 3.6.1, it is found that parameter asymmetry will introduce parameter mismatches and $2f$ disturbances in the voltage model. The parameter mismatch can be compensated for by the correction method discussed in Section 3.5. The $2f$ disturbance term will be considered in this section.

For simplicity, the dq-axis estimated back-EMFs with parameter asymmetry can be generally modeled [15] as:

$$\begin{aligned}\hat{E}_d &= \hat{E}_{d0} + \hat{E}_{d,2f} = \hat{E}_{d0} + A_{d,2f}\sin\left(2\theta_r + \varphi_{d,2f}\right)\\ \hat{E}_q &= \hat{E}_{q0} + \hat{E}_{q,2f} = \hat{E}_{q0} + A_{q,2f}\sin\left(2\theta_r + \varphi_{q,2f}\right)\end{aligned} \tag{3.55}$$

where $\hat{E}_{d0}$ and $\hat{E}_{q0}$ are the dc components, $\hat{E}_{d,2f}$ and $\hat{E}_{q,2f}$ denote the $2f$ component, and $A_{d,2f}$ and $A_{q,2f}$ are the amplitudes of the d- and q-axis $2f$ components, respectively. $\varphi_{d,2f}$ and $\varphi_{q,2f}$ are the phase shifts of the d- and q-axis $2f$ components, respectively.

As introduced in Chapter 2, the d-axis estimated back-EMF will be forced to zero by the position observer to track the rotor position. According to (3.55), a $2f$ position estimation error will arise due to parameter asymmetries. A position observer such as the PLL is used to extract the rotor position. PLL is a practical and simple phase and frequency estimation method due to its insensitivity to noise and distortion [19]. However, to fulfill the dynamic performance, the bandwidth of PLL is tuned relatively wide. Hence, it may be difficult to eliminate the low-order harmonics ($2f$) in (3.55) effectively. Therefore, some filtering techniques can be simply applied to the estimated back-EMFs to eliminate this $2f$ disturbance, such as an adaptive notch filter (ANF) [17], [20], an adaptive linear neuron (Adaline) network-based filter [21], [22], a synchronous reference frame filter [16], etc.

3.6.3 Harmonic Suppression

In this section, an ANF is designed to filter out the undesirable $2f$ disturbance due to parameter asymmetry [17]. Based on the $2f$ disturbance, the ANF is designed and shown in Figure 3.26. The transfer function of this ANF is given by:

$$G(s) = \frac{s^2 + \omega_{h,\mathrm{ANF}}^2}{s^2 + \mu s + \omega_{h,\mathrm{ANF}}^2} \tag{3.56}$$

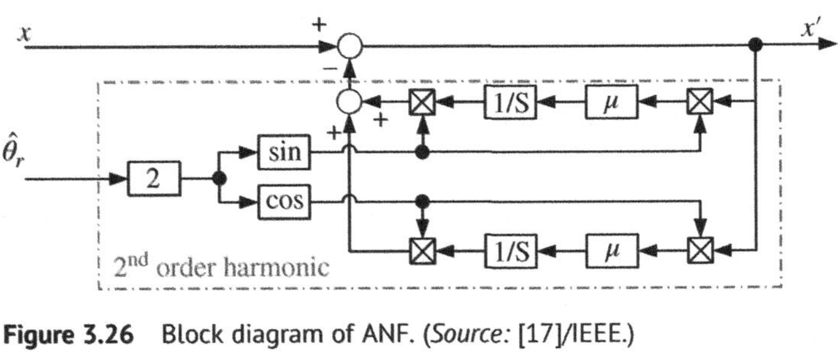

Figure 3.26 Block diagram of ANF. (*Source:* [17]/IEEE.)

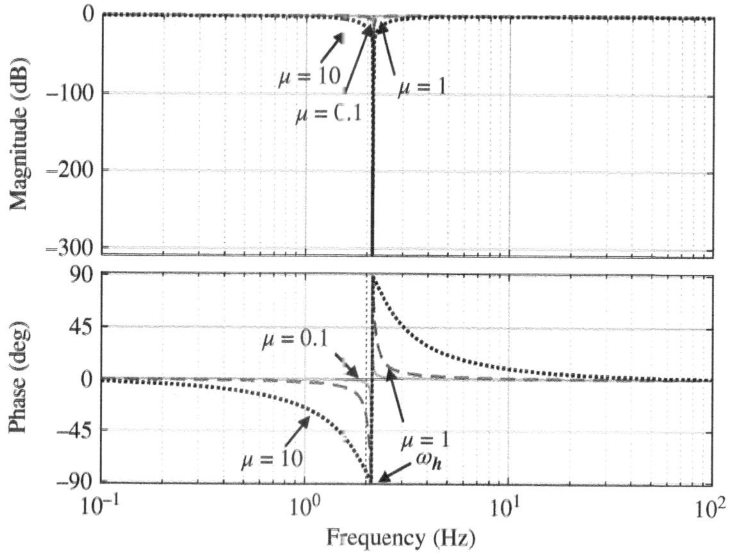

Figure 3.27 Frequency response of ANF with different bandwidths. (*Source:* [17]/IEEE.)

where $\omega_{h,\ \mathrm{ANF}}$ is the center frequency of the system, which specifies the order of harmonic to be suppressed. The bandwidth of ANF is μ, which is the only parameter to be tuned. This parameter should be selected carefully, since as μ becomes larger, the dynamic response is faster but the steady-state error becomes larger too. The frequency response of ANF is shown in Figure 3.27. Then, the conventional back-EMF estimation method introduced in Chapter 2 is modified by adding an ANF in the position estimation. The overall block diagram of the modified back-EMF estimator is shown in Figure 3.28.

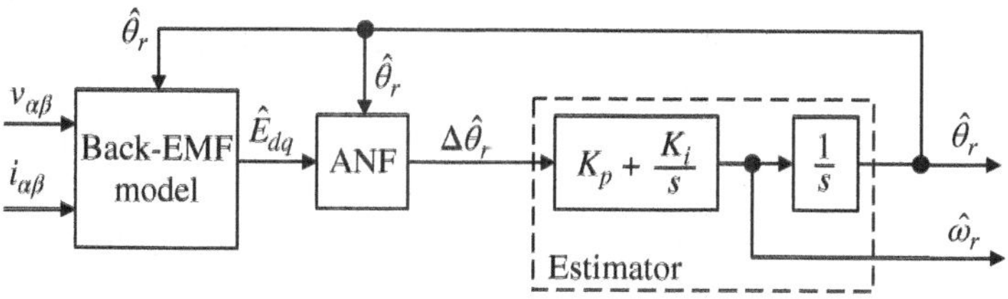

Figure 3.28 Block diagram of back-EMF position observer with ANF. (*Source:* [17]/IEEE.)

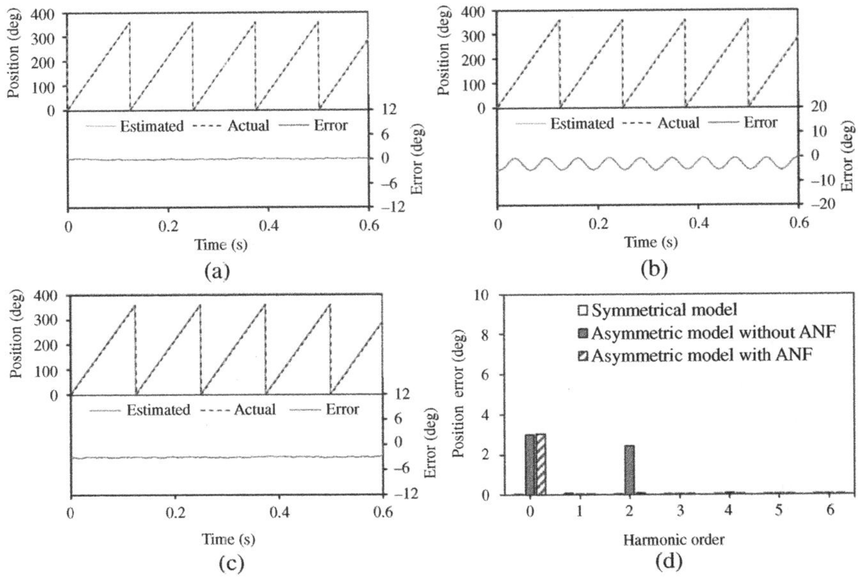

Figure 3.29 Position estimation performance under phase resistance asymmetry. (a) Symmetrical model. (b) Asymmetric model with ANF filter. (c) Asymmetric model without ANF filter. (d) Spectra.

To further illustrate the influence of parameter asymmetry and harmonic suppression performance, parameter asymmetries and ANF are implemented based on the simulation system (SPMSM-III, Appendix B). In the simulation, the d- and q-axis currents are all controlled at $-$ 4 A and the rotor speed is controlled at 30 r/min. First, taking the phase resistance asymmetry as an example, an extra 2 Ω resistance is added to phase A. The harmonic error and compensation performance are shown in Figure 3.29. There are obvious second order harmonic errors due to resistance asymmetry. Meanwhile, a dc offset error appears in the position estimation under resistance asymmetry. These phenomena match well with the analysis in (3.44)–(3.46). Then, with ANF, the second order harmonic error can be suppressed significantly. Moreover, similar results are obtained for inductance and back-EMF asymmetries as shown in Figure 3.30. For the inductance asymmetry, an extra 10 mH inductance is added to phase A self-inductance. The spectra of position error shows that there are $2f$ and dc offset errors in the estimation, which is the same as in (3.49) and (3.50). For back-EMF asymmetry, an extra 0.3 Wb flux-linkage is added to phase A, as indicated by (3.52)–(3.54), only a $2f$ disturbance is

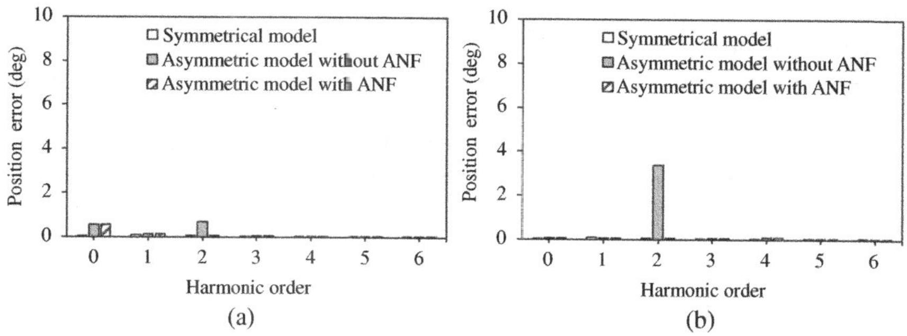

Figure 3.30 Spectra of position estimation errors under parameter asymmetries. (a) Inductance asymmetry. (b) Back-EMF asymmetry.

introduced in the position estimation. After applying ANF, the $2f$ error can be suppressed effectively for all the asymmetry cases.

3.7 Summary

In this chapter, some common parasitic effects of the fundamental model-based sensorless control methods were described. First, for the back-EMF and current harmonics, they mainly introduce sixth order harmonic errors in position estimation. Second, the cross-coupling inductance will cause a dc offset position estimation error, which can be simply compensated for by an improved machine model or offline measured angles. Third, apart from cross-coupling inductance, resistance and q-axis inductance will cause position estimation error. Based on the relationship between the parameter mismatches and the current level, a simple position correction method was presented. Fourth, analysis shows that a second order harmonic disturbance and a mismatch in the parameters will be produced by parameter asymmetries. The second order harmonic error can be suppressed by adding a notch filer to the estimated back-EMFs. Finally, these parasitic effects and corresponding solutions are summarized in Table 3.1.

As described in Chapter 2 and this chapter, the model-based methods rely on back-EMF or flux-linkage, which is unobservable at zero and low speed. Hence, in Chapters 4 and 5, the saliency tracking-based methods will be introduced, which are independent of the rotor speed.

Table 3.1 Summary of parasitic effect on position estimation of model-based sensorless control.

		Impact on position estimation	Solution
Voltage and current harmonics		Sixth order harmonic error	Filter [20]–[22], simplified EKF [7]
Cross-coupling effect	L_{qd}	DC offset position error	Offline measured data [9]
Parameter mismatch	L_q, R_s	DC offset position error	Current injection method [12]
Parameter asymmetry	Resistance, inductance	DC offset position error	Current injection method [12]
	Resistance, inductance, back-EMFs	Second order harmonic error	Filter [16], [20]–[22]

References

1 J. Hu and B. Wu, "New integration algorithms for estimating motor flux over a wide speed range," *IEEE Trans. Power Electron.*, vol. 13, no. 5, pp. 969–977, Sep. 1998.

2 P. Wipasuramonton, Z. Q. Zhu, and D. Howe, "Sensorless vector control of non-salient BLAC machines based on a modified rotor flux-linkage observer," in *2009 Int. Conf. Power Elect. Drive Syst. (PEDS)*, pp. 1056–1061, Nov. 2009.

3 G. R. Chen, J. Y. Chen, and S. C. Yang, "Implementation issues of flux linkage estimation on permanent magnet machine position sensorless drive at low speed," *IEEE Access*, vol. 7, pp. 164641–164649, 2019.

4 J. X. Shen, Z. Q. Zhu, and D. Howe, "Improved speed estimation in sensorless PM brushless ac drives," *IEEE Trans. Ind. Appl.*, vol. 38, no. 4, pp. 1072–1080, Jul. 2002.

5 H. Tajima and Y. Hori, "Speed sensorless field-orientation control of the induction machine," *IEEE Trans. Ind. Appl.*, vol. 29, no. 1, pp. 175–180, Jan. 1993.

6 Y. Liu, Z. Q. Zhu, Y. F. Shi, and D. Howe, "Sensorless direct torque control of a permanent magnet brushless ac drive via an extended Kalman filter," in *2nd Int. Conf. Power Elect., Mach. Drives (PEMD 2004)*, vol. 1, pp. 303–307, Mar. 2004.

7 A. H. Almarhoon and Z. Q. Zhu, "Influence of back-EMF and current harmonics on position estimation accuracy of permanent magnet synchronous machine," in *2014 17th Int. Conf. Elect. Mach. Syst. (ICEMS)*, pp. 2728–2733, Oct. 2014.

8 J. X. Shen and K. J. Tseng, "Analyses and compensation of rotor position detection error in sensorless PM brushless dc motor drives," *IEEE Trans. Energy Convers.*, vol. 18, no. 1, pp. 87–93, Mar. 2003.

9 Y. Li, Z. Q. Zhu, D. Howe, and C. M. Bingham, "Improved rotor position estimation in extended back-emf based sensorless PM brushless ac drives with magnetic saliency," in *2007 IEEE Int. Elect. Mach. Drives Conf.*, 2007, vol. 1, pp. 214–219.

10 Z. Q. Zhu, Y. Li, D. Howe, C. M. Bingham, and D. Stone, "Influence of machine topology and cross-coupling magnetic saturation on rotor position estimation accuracy in extended back-emf based sensorless pm brushless ac drives," in *2007 IEEE Ind. Appl. Annu. Meeting*, pp. 2378–2385, Sep. 2007.

11 S. Morimoto, K. Kawamoto, M. Sanada, and Y. Takeda, "Sensorless control strategy for salient-pole PMSM based on extended EMF in rotating reference frame," *IEEE Trans. Ind. Appl.*, vol. 38, no. 4, pp. 1054–1061, Jul. 2002.

12 T. Y. Liu, Z. Q. Zhu, B. Shuang, Z. Y. Wu, D. A. Stone, and M. P. Foster, "An online position error correction method for sensorless control of permanent magnet synchronous machine with parameter mismatch," *IEEE Access*, vol. 9, pp. 135708–135722, 2021.

13 K. Liu, Q. Zhang, J. Chen, Z. Q. Zhu, and J. Zhang, "Online multi-parameter estimation of nonsalient-pole PM synchronous machines with temperature variation tracking," *IEEE Trans. Ind. Electron.*, vol. 58, no. 5, pp. 1776–1788, May 2011.

14 D. Reigosa, P. Garcia, F. Briz, D. Raca, and R. D. Lorenz, "Modeling and adaptive decoupling of high-frequency resistance and temperature effects in carrier-based sensorless control of PM synchronous machines," *IEEE Trans. Ind. Appl.*, vol. 46, no. 1, pp. 139–149, Jan./Feb. 2010.

15 T. Y. Liu, Z. Q. Zhu, Z. Y. Wu, D. A. Stone, and M. P. Foster, "Improved sensorless control method and asymmetric phase resistances determination for permanent magnet synchronous machines," *IEEE Trans. Ind. Appl.*, vol. 58, no. 3, pp. 3624–3636, May 2022.

16 M. W. Degner and R. D. Lorenz, "Using multiple saliencies for the estimation of flux, position, and velocity in AC machines," *IEEE Trans. Ind. Appl.*, vol. 34, no. 5, pp. 1097–1104, 1998.

17 X. M. Wu, Z. Q. Zhu, Z. Y. Wu, T. Y. Liu, and Y. X. Li, "Analysis and suppression of rotor eccentricity effects on fundamental model based sensorless control of permanent magnet synchronous machine," *IEEE Trans. Ind. Appl.*, vol. 56, no. 5, pp. 4896–4905, Sep. 2020.

18 T. C. Lin, Z. Q. Zhu, and J. M. Liu, "Improved rotor position estimation in sensorless-controlled permanent-magnet synchronous machines having asymmetric-EMF with harmonic compensation," *IEEE Trans. Ind. Elect.*, vol. 62, no. 10, pp. 6131–6139, Oct. 2015.

19 G. Wang, R. Yang, and D. Xu, "DSP-based control of sensorless IPMSM drives for wide-speed-range operation," *IEEE Trans. Ind. Electron.*, vol. 60, no. 2, pp. 720–727, Feb. 2013.

20 G. Wang, H. Zhan, G. Zhang, X. Gui, and D. Xu, "Adaptive compensation method of position estimation harmonic error for emf-based observer in sensorless IPMSM drives," *IEEE Trans. Power Electron.*, vol. 29, no. 6, pp. 3055–3064, Jun. 2014.

21 G. Zhang, G. Wang, D. Xu, and N. Zhao, "ADALINE-network-based PLL for position sensorless interior permanent magnet synchronous motor drives," *IEEE Trans. Power Electron.*, vol. 31, no. 2, pp. 1450–1460, Feb. 2016.

22 L. Wang, Z. Q. Zhu, H. Bin, and L. M. Gong, "Current harmonics suppression strategy for PMSM with non-sinusoidal back-emf based on adaptive linear neuron method," *IEEE Trans. Ind. Electron.*, vol. 67, no. 11, pp. 9164–9173, Nov. 2020.

4

Saliency Tracking-Based Sensorless Control Methods

4.1 Introduction

In Chapter 2 it was shown that fundamental model-based sensorless control methods fail at zero and low speed due to unobservable models. Hence, at zero and low speed, the so-called "saliency tracking-based sensorless control method" can be employed to estimate the rotor position as described in this chapter.

Due to the anisotropy property of PMSMs, there is geometric (e.g. in IPMSM) or magnetic (e.g. in SPMSM) inductance saliency, which is a function of rotor position. Since the machine saliency is independent of the rotor speed, it can be exploited for sensorless position estimation at zero and low speed. Figure 4.1 illustrates the inductance variation according to the electrical rotor position for an IPMSM. In the cross-section view of the machine, the shaded rotor area is characterized by high permeability (high inductance), while the white area between the rotor and stator indicates the relative air gap, where the permeability is assumed to be low (low inductance). It confirms that the machine inductance is modulated by the position-dependent spatial saliency.

To exploit these machine saliencies at zero and in low-speed regions, the rotor position and speed information are often obtained by persistent high-frequency signal injections (HFSI) [1]–[6]. The basic principle is that with the interaction between the high-frequency (HF) excitations and the machine saliency, the rotor-position-dependent HF responses can be produced. With the appropriate signal demodulation process and position observers, the rotor position and speed can then be estimated from these HF responses.

The majority of HFSI-based sensorless control methods can fit into the frame in Figure 4.2. As shown in Figure 4.2, there are five main parts of a saliency tracking-based method. For each part, the relevant methods are also demonstrated.

Sensorless Control of Permanent Magnet Synchronous Machine Drives, First Edition.
Zi Qiang Zhu and Xi Meng Wu.

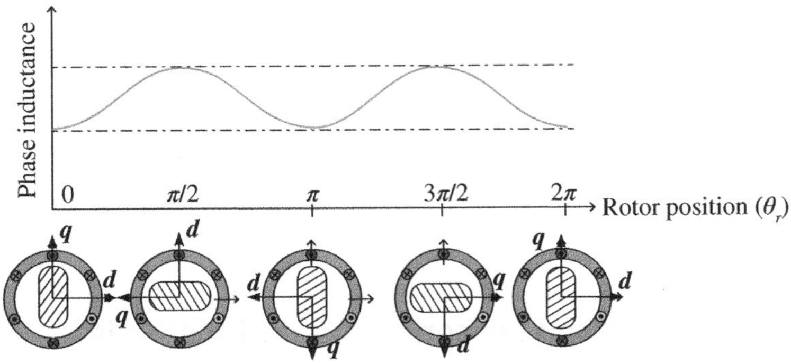

Figure 4.1 Inductance variation according to electrical rotor position.

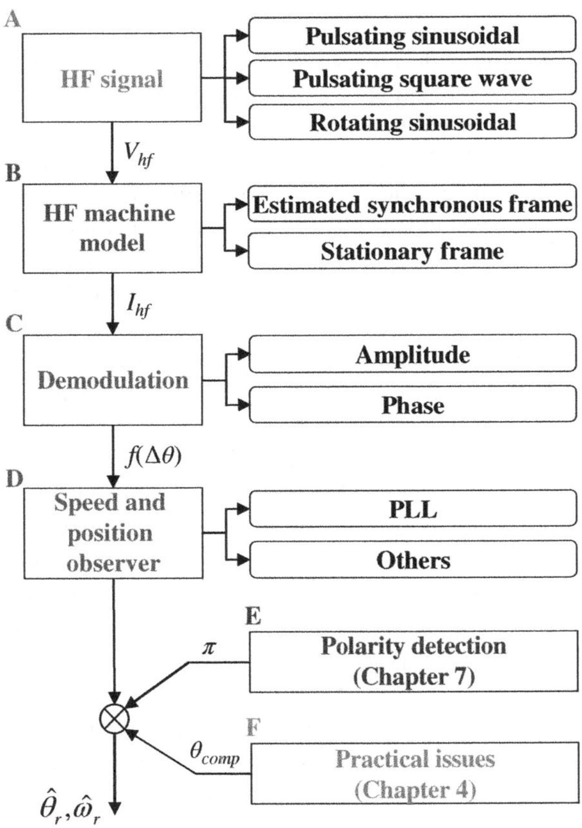

Figure 4.2 Classification of high-frequency signal injection-based sensorless control methods.

In this chapter, fundamentals (Parts A–D) of the HFSI-based sensorless control method will be described and demonstrated. For the injected HF signal V_{hf} in Part A, pulsating sinusoidal signal, pulsating square-wave signal, and rotating sinusoidal signal are included, which are normally used for position estimation. Part B represents the HF machine model when a high-frequency signal is injected. HF signals can be injected into the stationary reference frame and the estimated synchronous reference frame. For different HF machine models, the HF current responses I_{hf} will be different. For Part C, after injecting HF signals, the HF current responses should be demodulated to remove the high frequency component and obtain the position dependent signal $f(\Delta\theta)$. The demodulation process can vary depending on the injection methods used. In this chapter, different demodulation processes are illustrated with respect to different injection methods. In Part D, the demodulated position-dependent signal $f(\Delta\theta)$ is used as an input of a position observer to generate the rotor speed and position information. A commonly used phase-locked loop (PLL)-based position observer will be introduced in this chapter.

For Part E, as indicated in Figure 4.1, the inductance varies twice in one electrical period and hence an extra polarity detection is necessary, which will be illustrated in Chapter 8. Since the preceding parts A–E were based on an ideal model, Part F discusses some common issues faced during practical implementation and which will be detailed further in Chapter 5.

4.2 High-Frequency Model of PM Machines

In HF voltage signal injection-based sensorless control methods, the position-dependent HF current response is subject to the impedance of the machine, and hence, the high-frequency model of a PM machine needs to be analyzed first.

4.2.1 Model in Synchronous Reference Frame

The voltage equations of a PM machine in the synchronous reference frame are given as:

$$\begin{bmatrix} v_d \\ v_q \end{bmatrix} = \begin{bmatrix} R_s & 0 \\ 0 & R_s \end{bmatrix} \cdot \begin{bmatrix} i_d \\ i_q \end{bmatrix} + p \begin{bmatrix} \psi_d \\ \psi_q \end{bmatrix} + \begin{bmatrix} -\omega_r \psi_q \\ \omega_r \psi_d \end{bmatrix} \tag{4.1}$$

where p denotes the differential operator, and ψ_d, ψ_q represent the d- and q-axis flux-linkages, respectively. The flux-linkages determined by load condition (i_d, i_q) are shown in (4.2).

$$\begin{cases} \psi_d = \psi_d(i_d, i_q) \\ \psi_q = \psi_q(i_d, i_q) \end{cases} \tag{4.2}$$

If the injected frequency is sufficiently higher than the fundamental frequency (normally the case at standstill and low-speed operation), back-EMF and resistive voltage drop are negligible. Hence, only the high-frequency components are considered, and the high-frequency voltage equation can be derived as:

$$\begin{bmatrix} v_{dh} \\ v_{qh} \end{bmatrix} = p \begin{bmatrix} \psi_d(i_d, i_q) \\ \psi_q(i_d, i_q) \end{bmatrix} = \begin{bmatrix} \frac{\partial \psi_d}{\partial i_d} \cdot \frac{di_d}{dt} + \frac{\partial \psi_d}{\partial i_q} \cdot \frac{di_q}{dt} \\ \frac{\partial \psi_q}{\partial i_d} \cdot \frac{di_d}{dt} + \frac{\partial \psi_q}{\partial i_q} \cdot \frac{di_q}{dt} \end{bmatrix} \tag{4.3}$$

where $\partial \psi / \partial i$ is the partial derivative of the flux-linkage with respect to the current, and di/dt is the derivative of the current with respect to the time.

According to [7], the incremental self- and mutual-inductances can be defined as:

$$\begin{cases} L_{dh}(i_d, i_q) = \partial \psi_d / \partial i_d \approx [\psi_d(i_d + \Delta i_d, i_q) - \psi_d(i_d, i_q)] / \Delta i_d \\ L_{qh}(i_d, i_q) = \partial \psi_q / \partial i_q \approx [\psi_q(i_d, i_q + \Delta i_q) - \psi_q(i_d, i_q)] / \Delta i_q \\ L_{dqh}(i_d, i_q) = \partial \psi_d / \partial i_q \approx [\psi_d(i_d, i_q + \Delta i_q) - \psi_d(i_d, i_q)] / \Delta i_q \\ L_{qdh}(i_d, i_q) = \partial \psi_q / \partial i_d \approx [\psi_q(i_d + \Delta i_d, i_q) - \psi_q(i_d, i_q)] / \Delta i_d \end{cases} \tag{4.4}$$

With the d-axis as an example, the d-axis self-incremental inductance is illustrated in Figure 4.3 and is defined as the slope at the operating point, while the apparent inductance is defined as the flux-linkage divided by the current. In the HFSI sensorless control, since the frequency of the injection signal is relatively high compared to the fundamental frequency, the slope (incremental inductance) can be approximated as the flux-linkage change divided by the high-frequency current change in (4.4).

With the aid of a definition of incremental inductance, the high-frequency voltage equation can be simplified as:

$$\begin{bmatrix} v_{dh} \\ v_{qh} \end{bmatrix} = \begin{bmatrix} L_{dh} & L_{dqh} \\ L_{qdh} & L_{qh} \end{bmatrix} \cdot p \begin{bmatrix} i_{dh} \\ i_{qh} \end{bmatrix} \tag{4.5}$$

where L_{dqh} and L_{qdh} are the high-frequency incremental mutual-inductances due to the cross-saturation effect between the d- and q-axes. In most practical applications [7], L_{dqh} is considered equal to L_{qdh} for simplifying the analysis, which is

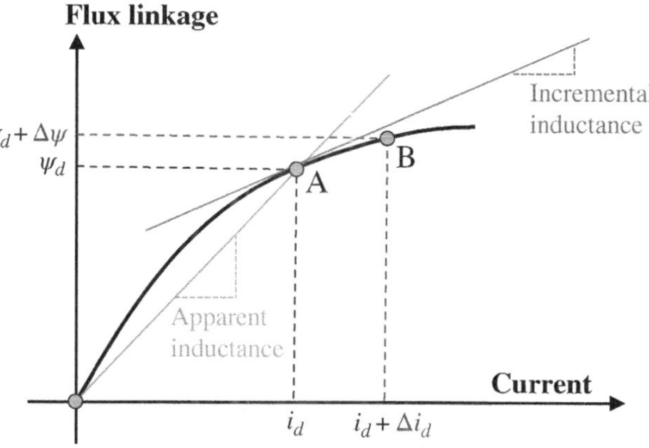

Figure 4.3 Illustration of incremental and apparent inductances.

adopted in this chapter. Consequently, the high-frequency voltage model of a PM machine in the synchronous reference frame can be finally simplified as:

$$\begin{bmatrix} v_{dh} \\ v_{qh} \end{bmatrix} = \begin{bmatrix} L_{dh} & L_{dqh} \\ L_{dqh} & L_{qh} \end{bmatrix} \cdot p \begin{bmatrix} i_{dh} \\ i_{qh} \end{bmatrix} \tag{4.6}$$

From the above equation, it can be seen that only incremental inductances are contained in the high-frequency voltage model. Generally, the saliency tracking-based method can be realized in either the estimated synchronous reference frame or the stationary reference frame, which will be derived in the next sections.

4.2.2 Model in Estimated Synchronous Reference Frame

In a sensorless control-based system, since the real rotor position is unknown, the estimated rotor position is used in the control algorithm. Hence, (4.6) can be transformed into the estimated synchronous reference frame from the real synchronous reference frame by the transformation matrix $T(\Delta\theta)$ as:

$$T(\Delta\theta) = \begin{bmatrix} \cos(\Delta\theta) & -\sin(\Delta\theta) \\ \sin(\Delta\theta) & \cos(\Delta\theta) \end{bmatrix} \tag{4.7}$$

$$T^{-1}(\Delta\theta) = \begin{bmatrix} \cos(\Delta\theta) & \sin(\Delta\theta) \\ -\sin(\Delta\theta) & \cos(\Delta\theta) \end{bmatrix} \tag{4.8}$$

where $\Delta\theta$ is the estimated position error, $\Delta\theta = \theta_r - \hat{\theta}_r$, as shown in Figure 4.4.

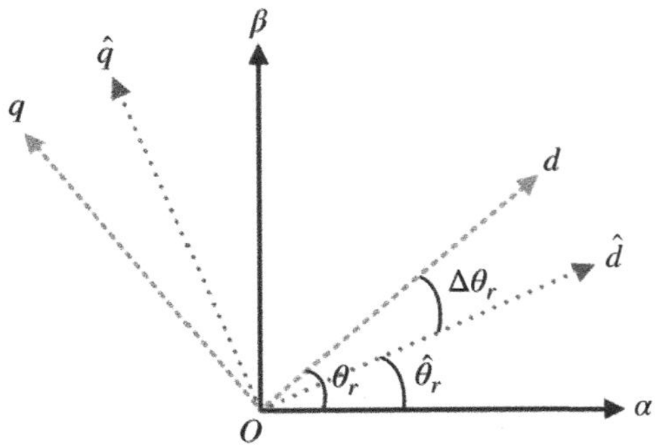

Figure 4.4 Estimated and real synchronous reference frames.

In the estimated synchronous reference frame, the high-frequency voltage model of the PM machine can be derived as:

$$\begin{bmatrix}\hat{v}_{dh}\\ \hat{v}_{qh}\end{bmatrix} = T(\Delta\theta)\begin{bmatrix}L_{dh} & L_{dqh}\\ L_{dqh} & L_{qh}\end{bmatrix}T^{-1}(\Delta\theta)\cdot p\begin{bmatrix}\hat{i}_{dh}\\ \hat{i}_{qh}\end{bmatrix}$$
$$= \begin{bmatrix}L_{sa}-L_{sd}\cos 2\Delta\theta - L_{dqh}\sin 2\Delta\theta & -L_{sd}\sin 2\Delta\theta + L_{dqh}\cos 2\Delta\theta\\ -L_{sd}\sin 2\Delta\theta + L_{dqh}\cos 2\Delta\theta & L_{sa}+L_{sd}\cos 2\Delta\theta + L_{dqh}\sin 2\Delta\theta\end{bmatrix}\cdot p\begin{bmatrix}\hat{i}_{dh}\\ \hat{i}_{qh}\end{bmatrix} \tag{4.9}$$

where L_{sa} and L_{sd} are the average and difference of d- and q-axis incremental inductances, defined by:

$$\begin{cases} L_{sa} = (L_{qh} + L_{dh})/2\\ L_{sd} = (L_{qh} - L_{dh})/2\end{cases} \tag{4.10}$$

Equation (4.9) is solved to calculate the differential of the high-frequency current response in the estimated synchronous reference frame as:

$$p\begin{bmatrix}\hat{i}_{dh}\\ \hat{i}_{qh}\end{bmatrix} = \begin{bmatrix}\dfrac{L_{sa}+L_{sd}\cos 2\Delta\theta + L_{dqh}\sin 2\Delta\theta}{L_{sa}^2-L_{sd}^2-L_{dqh}^2} & \dfrac{L_{sd}\sin 2\Delta\theta - L_{dqh}\cos 2\Delta\theta}{L_{sa}^2-L_{sd}^2-L_{dqh}^2}\\ \dfrac{L_{sd}\sin 2\Delta\theta - L_{dqh}\cos 2\Delta\theta}{L_{sa}^2-L_{sd}^2-L_{dqh}^2} & \dfrac{L_{sa}-L_{sd}\cos 2\Delta\theta - L_{dqh}\sin 2\Delta\theta}{L_{sa}^2-L_{sd}^2-L_{dqh}^2}\end{bmatrix}\cdot\begin{bmatrix}\hat{v}_{dh}\\ \hat{v}_{qh}\end{bmatrix}$$
$$= \begin{bmatrix}\dfrac{L_{sa}+\sqrt{L_{sd}^2+L_{dqh}^2}\cos(2\Delta\theta+\theta_m)}{L_{dh}L_{qh}-L_{dqh}^2} & \dfrac{\sqrt{L_{sd}^2+L_{dqh}^2}\sin(2\Delta\theta+\theta_m)}{L_{dh}L_{qh}-L_{dqh}^2}\\ \dfrac{\sqrt{L_{sd}^2+L_{dqh}^2}\sin(2\Delta\theta+\theta_m)}{L_{dh}L_{qh}-L_{dqh}^2} & \dfrac{L_{sa}-\sqrt{L_{sd}^2+L_{dqh}^2}\cos(2\Delta\theta+\theta_m)}{L_{dh}L_{qh}-L_{dqh}^2}\end{bmatrix}\cdot\begin{bmatrix}\hat{v}_{dh}\\ \hat{v}_{qh}\end{bmatrix}$$
$$= \begin{bmatrix}\dfrac{1}{L_p}+\dfrac{1}{L_n}\cos(2\Delta\theta+\theta_m) & \dfrac{1}{L_n}\sin(2\Delta\theta+\theta_m)\\ \dfrac{1}{L_n}\sin(2\Delta\theta+\theta_m) & \dfrac{1}{L_p}-\dfrac{1}{L_n}\cos(2\Delta\theta+\theta_m)\end{bmatrix}\cdot\begin{bmatrix}\hat{v}_{dh}\\ \hat{v}_{qh}\end{bmatrix} \tag{4.11}$$

where θ_m is referred to as the cross-saturation angle due to L_{dqh}.

$$\theta_m = \tan^{-1}\left(\frac{-L_{dqh}}{L_{sd}}\right), L_p = \frac{L_{dh}L_{qh} - L_{dqh}^2}{L_{sd}}, L_n = \frac{L_{dh}L_{qh} - L_{dqh}^2}{\sqrt{L_{sd}^2 + L_{dqh}^2}} \tag{4.12}$$

Normally, $L_p \ll L_n$.

4.2.3 Model in Stationary Reference Frame

From (4.6), with the aid of the transformation matrix $T(\theta_r)$, the high-frequency voltage model of a PMSM in the stationary reference frame can also be derived as:

$$T(\theta_r) = \begin{bmatrix} \cos(\theta_r) & -\sin(\theta_r) \\ \sin(\theta_r) & \cos(\theta_r) \end{bmatrix} \tag{4.13}$$

$$\begin{aligned} \begin{bmatrix} v_{\alpha h} \\ v_{\beta h} \end{bmatrix} &= T(\theta_r) \begin{bmatrix} L_{dh} & L_{dqh} \\ L_{dqh} & L_{qh} \end{bmatrix} T^{-1}(\theta_r) \cdot p \begin{bmatrix} i_{\alpha h} \\ i_{\beta h} \end{bmatrix} \\ &= \begin{bmatrix} L_{sa} - L_{sd}\cos 2\theta_r - L_{dqh}\sin 2\theta_r & -L_{sd}\sin 2\theta_r + L_{dqh}\cos 2\theta_r \\ -L_{sd}\sin 2\theta_r + L_{dqh}\cos 2\theta_r & L_{sa} + L_{sd}\cos 2\theta_r + L_{dqh}\sin 2\theta_r \end{bmatrix} \cdot p \begin{bmatrix} i_{\alpha h} \\ i_{\beta h} \end{bmatrix} \end{aligned} \tag{4.14}$$

In a similar way, the differential of the high-frequency current response in the stationary reference frame can be obtained as:

$$p\begin{bmatrix} i_{\alpha h} \\ i_{\beta h} \end{bmatrix} = \begin{bmatrix} \frac{1}{L_p} - \frac{1}{L_n}\cos(2\theta_r + \theta_m) & \frac{1}{L_m}\sin(2\theta_r + \theta_m) \\ \frac{1}{L_n}\sin(2\theta_r + \theta_m) & \frac{1}{L_p} - \frac{1}{L_n}\cos(2\theta_r + \theta_m) \end{bmatrix} \cdot \begin{bmatrix} v_{\alpha h} \\ v_{\beta h} \end{bmatrix} \tag{4.15}$$

In summary, based on the HF PM machine model in the estimated synchronous reference and stationary reference frames, different HF voltage signals can be injected to realize the sensorless control, which will be introduced in the next sections.

4.3 High-Frequency Signal Injection in Estimated Synchronous Reference Frame

Based on the HF PM machine model introduced in (4.11), pulsating sinusoidal [2] and pulsating square-wave signals [3] can be injected into either the d-axis or the q-axis of the estimated reference frame to realize the sensorless operation. Injection in the estimated d-axis is normally used since q-axis injection generates more torque ripples and additional losses [8]. Hence, in this section, the commonly used

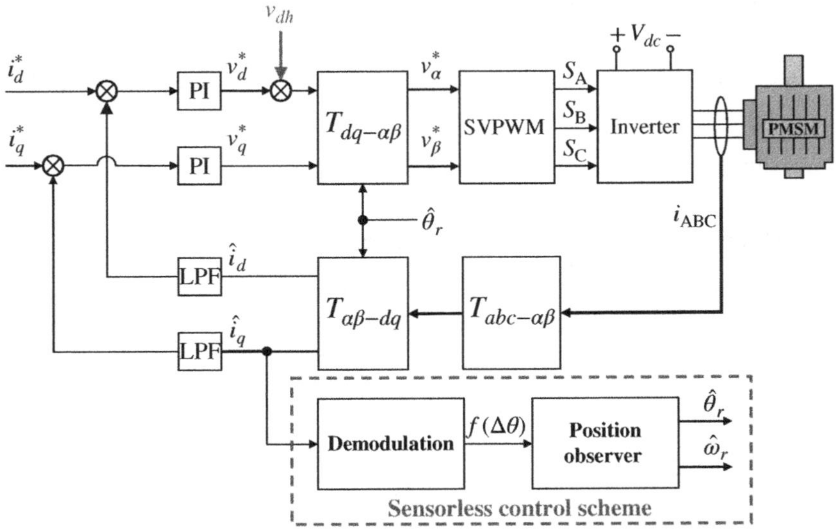

Figure 4.5 Block diagram of *d*-axis pulsating signal injection-based sensorless control.

d-axis injection-based methods are described. With the *d*-axis injection as an example, based on a field-oriented control (FOC) system, the block diagram of the injection in the estimated synchronous reference frame is shown in Figure 4.5. As shown in Figure 4.5, the sensorless control scheme consists of a demodulation process and a position observer. A position error signal $f(\Delta\theta)$ is obtained by a demodulation process depending on the injected signals and used as the input of a position observer to generate the estimated rotor position and speed.

4.3.1 Pulsating Sinusoidal Signal

As depicted in Figure 4.6, a high-frequency pulsating sinusoidal voltage signal [2] is injected into the estimated *d*-axis and the injected voltage signals are expressed by (4.16):

$$\begin{bmatrix} \hat{v}_{dh} \\ \hat{v}_{qh} \end{bmatrix} = V_h \begin{bmatrix} \cos\alpha \\ 0 \end{bmatrix}, \alpha = \omega_h t \tag{4.16}$$

where V_h and ω_h are the amplitude and angular speed of the injected HF voltage signal.

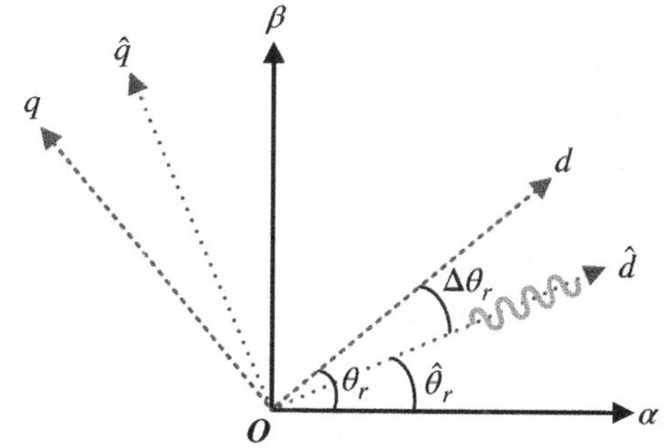

Figure 4.6 Pulsating sinusoidal signal injection in estimated *d*-axis.

After injection, the HF current responses in the estimated synchronous reference frame can be expressed by:

$$p\begin{bmatrix}\hat{i}_{dh}\\ \hat{i}_{qh}\end{bmatrix}=\begin{bmatrix}\frac{1}{L_p}+\frac{1}{L_n}\cos(2\Delta\theta+\theta_m) & \frac{1}{L_n}\sin(2\Delta\theta+\theta_m)\\ \frac{1}{L_n}\sin(2\Delta\theta+\theta_m) & \frac{1}{L_p}-\frac{1}{L_n}\cos(2\Delta\theta+\theta_m)\end{bmatrix}\cdot V_h\begin{bmatrix}\cos\alpha\\ 0\end{bmatrix} \tag{4.17}$$

$$\begin{bmatrix}\hat{i}_{dh}\\ \hat{i}_{qh}\end{bmatrix}=\begin{bmatrix}\frac{V_h}{\omega_h L_p}+\frac{V_h}{\omega_h L_n}\cos(2\Delta\theta+\theta_m)\\ \frac{V_h}{\omega_h L_n}\sin(2\Delta\theta+\theta_m)\end{bmatrix}\cdot\sin\alpha=\begin{bmatrix}I_p+I_n\cos(2\Delta\theta+\theta_m)\\ I_n\sin(2\Delta\theta+\theta_m)\end{bmatrix}\cdot\sin\alpha \tag{4.18}$$

$$I_p=\frac{V_h}{\omega_h L_p},I_n=\frac{V_h}{\omega_h L_n} \tag{4.19}$$

Clearly, as shown in (4.18), the HF current responses have the rotor position information $\Delta\theta$. In order to extract the rotor position information, a demodulation process is required, which is shown in Figure 4.7.

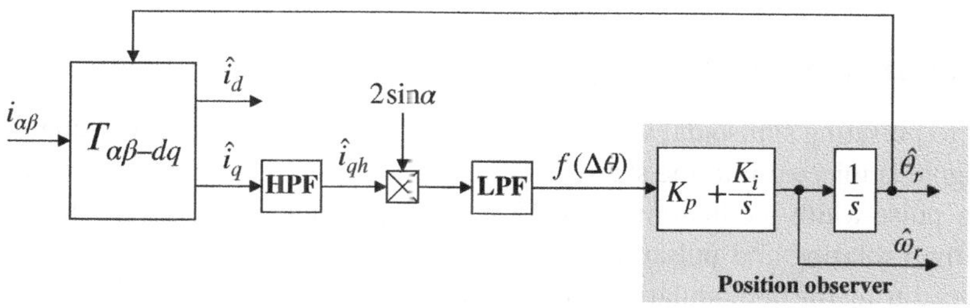

Figure 4.7 Demodulation process of pulsating sinusoidal signal injection method.

In the demodulation process, first, the stator currents in the stationary reference frame are transformed into the estimated synchronous reference frame with the estimated rotor position. Then, the high-frequency component $\hat{i}_{dqh}$ in the estimated dq-axis currents $\hat{i}_{dq}$ is obtained by a high-pass filter (HPF). The cutoff frequency of the HPF should be lower than the injected frequency to obtain the injected frequency component in the current response.

Then, the high-frequency component $\hat{i}_{dqh}$ is demodulated as:

$$\left|\hat{i}_{dqh}\right| = \begin{bmatrix} \left|\hat{i}_{dh}\right| \\ \left|\hat{i}_{qh}\right| \end{bmatrix} = LPF\left(\begin{bmatrix} \hat{i}_{dh} \\ \hat{i}_{qh} \end{bmatrix} \cdot 2\sin\alpha\right) = \begin{bmatrix} I_p + I_n\cos(2\Delta\theta + \theta_m) \\ I_n\sin(2\Delta\theta + \theta_m) \end{bmatrix} \tag{4.20}$$

As indicated in (4.20), a low-pass filter (LPF) is used to filter out the high-frequency component in the current response. The cutoff frequency of the LPF should be lower than the injection frequency. Meanwhile, a lower cutoff frequency will affect the dynamic performance, which will be discussed in Section 4.5.2. Hence, the cutoff frequency of LPF should be selected properly.

From (4.20), the amplitude of the q-axis HF current response is related to the position error $\Delta\theta$, which can be used as the input of a position observer as shown in Figure 4.5.

$$f(\Delta\theta) = \left|\hat{i}_{qh}\right| = I_n\sin(2\Delta\theta + \theta_m) \tag{4.21}$$

With the position observer, $f(\Delta\theta)$ will be forced to zero as:

$$f(\Delta\theta) = I_n\sin(2\Delta\theta + \theta_m) = 0 \tag{4.22}$$

After $f(\Delta\theta)$ is minimized by the position observer, the estimated position aligns with the actual rotor position. In (4.22), an error angle θ_m is introduced by cross-saturation effect; the amplitude of this error mainly depends on the load conditions. Several compensation methods have been proposed and will be introduced in Chapter 5.

4.3.2 Pulsating Square-Wave Signal

Similar to pulsating sinusoidal voltage signal injection, in [3], a pulsating square-wave signal is injected into the stator windings. The injected frequency can be as high as pulse-width-modulation (PWM) switching frequency, which is much higher than rotating and pulsating sinusoidal injection schemes [4], and hence, the filters during the demodulation process are eliminated, the bandwidth of position estimation is improved, and the dynamic performance is greatly enhanced. In this section, similar to pulsating sinusoidal signal injection, pulsating square-wave signal injection at the d-axis in the estimated synchronous reference frame is illustrated.

A pulsating square-wave signal injected into the estimated d-axis is shown in Figure 4.8.

The injected pulsating square-wave signal is expressed by:

$$\begin{bmatrix} \hat{v}_{dh} \\ \hat{v}_{qh} \end{bmatrix} = \begin{bmatrix} V_h \cdot (-1)^n \\ 0 \end{bmatrix}, \quad n = 0, 1, 2... \tag{4.23}$$

where n is an integer and the number of the injected voltages. The injection sequence of a pulsating square-wave signal is demonstrated in Figure 4.9,

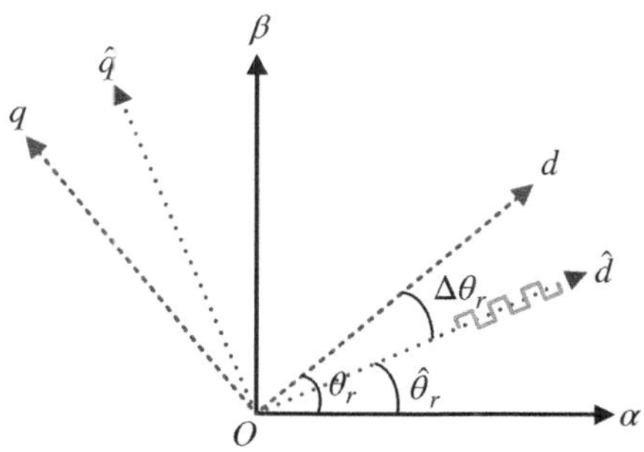

Figure 4.8 Pulsating square signal injection in estimated d-axis.

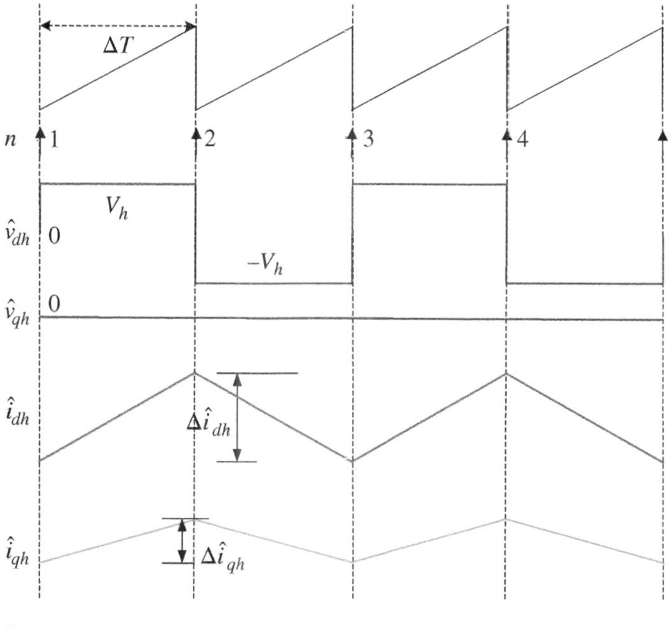

Figure 4.9 Pulsating square-wave voltage injection sequence.

where ΔT is the period of an injected square-wave voltage signal, and V_h is the amplitude of the injection square-wave voltage. By substituting the injected voltages (4.23) into the voltage equations (4.11) in the estimated reference frame, the dq-axis HF current responses can be solved as:

$$\begin{bmatrix} \hat{i}_{dh} \\ \hat{i}_{qh} \end{bmatrix} = \begin{bmatrix} \frac{1}{L_p} + \frac{1}{L_n}\cos(2\Delta\theta + \theta_m) \\ \frac{1}{L_n}\sin(2\Delta\theta + \theta_m) \end{bmatrix} \cdot \int \begin{bmatrix} \hat{v}_{dh} \\ \hat{v}_{qh} \end{bmatrix} dt \tag{4.24}$$

The current variations between adjacent sampling points can be solved as:

$$\begin{bmatrix} \Delta\hat{i}_{dh} \\ \Delta\hat{i}_{qh} \end{bmatrix} = V_h \Delta T \begin{bmatrix} \frac{1}{L_p} + \frac{1}{L_n}\cos(2\Delta\theta + \theta_m) \\ \frac{1}{L_n}\sin(2\Delta\theta + \theta_m) \end{bmatrix} \cdot (-1)^{n-1}, n = 1, 2, 3... \tag{4.25}$$

The signal processing is demonstrated in Figure 4.10.

Similar to the pulsating sinusoidal signal injection, the currents in the stationary reference frame are first transformed into the estimated synchronous reference frame with the estimated position. Then, the high-frequency current responses are obtained by an HPF, and the HF current variations due to pulsating square-wave voltage signal injection can be calculated by:

$$\begin{bmatrix} \Delta\hat{i}_{dh}(k) \\ \Delta\hat{i}_{qh}(k) \end{bmatrix} \approx \begin{bmatrix} \hat{i}_{dh}(k) \\ \hat{i}_{qh}(k) \end{bmatrix} - \begin{bmatrix} \hat{i}_{dh}(k-1) \\ \hat{i}_{qh}(k-1) \end{bmatrix} \tag{4.26}$$

Moreover, if the injection frequency is the same as the PWM frequency, the fundamental frequency components in the dq-axis currents can be assumed to be

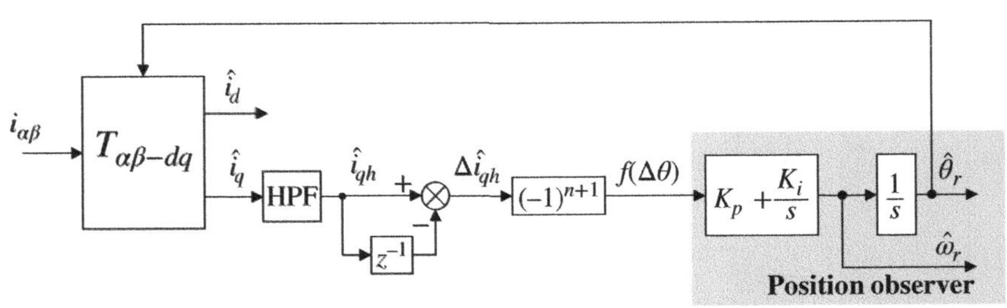

Figure 4.10 Demodulation process of pulsating square-wave signal injection-based method.

constant in one sampling period. The calculation of HF current variations can then be further simplified without using filters as:

$$\begin{bmatrix} \hat{i}_{dh}(k) \\ \hat{i}_{qh}(k) \end{bmatrix} \approx \begin{bmatrix} \hat{i}_{d}(k) \\ \hat{i}_{q}(k) \end{bmatrix} - \begin{bmatrix} \hat{i}_{d}(k-1) \\ \hat{i}_{q}(k-1) \end{bmatrix} \tag{4.27}$$

$$\begin{bmatrix} \Delta\hat{i}_{dh}(k) \\ \Delta\hat{i}_{qh}(k) \end{bmatrix} \approx \begin{bmatrix} \hat{i}_{dh}(k) \\ \hat{i}_{qh}(k) \end{bmatrix} - \begin{bmatrix} \hat{i}_{dh}(k-1) \\ \hat{i}_{qh}(k-1) \end{bmatrix} \tag{4.28}$$

After obtaining the HF current variations, similar to pulsating sinusoidal signal injection methods, the HF current variation response $\Delta\hat{i}_{qh} \cdot (-1)^{n+1}$ is regarded as the input of the position observer as:

$$f(\Delta\theta) = \Delta\hat{i}_{qh} \cdot (-1)^{n+1} = \frac{V_h \Delta T}{L_n} \sin(2\Delta\theta + \theta_m) \tag{4.29}$$

By controlling $f(\Delta\theta)$ with the position observer, the rotor position can be estimated. The minimum sampling period can be half of the switching frequency if sampling at both the bottom and peak of the PWM carrier waveform can be performed. In that case, the injection frequency is the same as the switching frequency and the conventional HPF and LPF can be eliminated. It is worth noting that in (4.29), the same as the pulsating sinusoidal signal injection method, due to cross-saturation effect (L_{dqh}), an error angle θ_m exists in the estimated rotor position.

4.4 High-Frequency Signal Injection in Stationary Reference Frame

Based on the high-frequency PM machine model in (4.15), a rotating sinusoidal signal [1] or pulsating signals [5], [6] can be injected into the stationary reference frame to realize the sensorless operation. For the HF signal injection into the stationary reference frame, the overall block diagram is shown in Figure 4.11.

4.4.1 Rotating Sinusoidal Signal

For the rotating sinusoidal signal injection, a balanced high-frequency sinusoidal voltage rotating vector is injected into the stationary reference, as expressed by:

$$\begin{bmatrix} v_{\alpha h} \\ v_{\beta h} \end{bmatrix} = V_h \begin{bmatrix} \cos\alpha \\ \sin\alpha \end{bmatrix}, \alpha = \omega_h t \tag{4.30}$$

where V_h and ω_h are the amplitude and angular speed of the injected HF voltage signal, respectively. The injection is depicted in Figure 4.12.

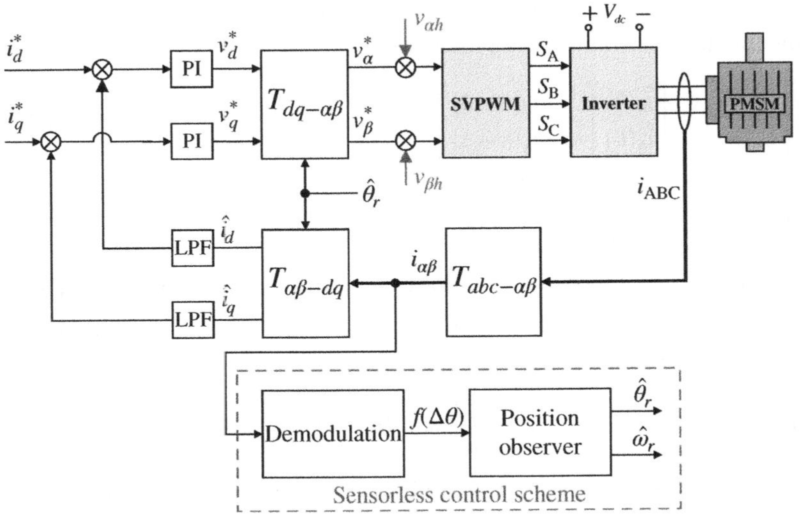

Figure 4.11 Block diagram of rotating sinusoidal signal injection-based sensorless control.

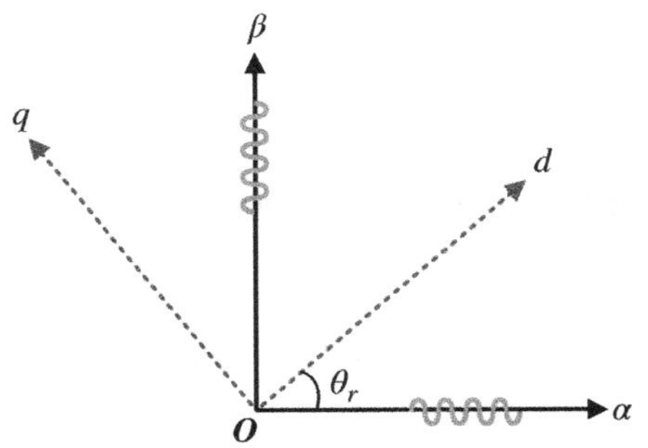

Figure 4.12 Conventional rotating signal injection injection in stationary reference frame.

After injection, the differential of HF current responses in the stationary reference frame can be expressed by:

$$p\begin{bmatrix} i_{\alpha h} \\ i_{\beta h} \end{bmatrix} = \begin{bmatrix} \frac{1}{L_p} + \frac{1}{L_n}\cos(2\theta_r + \theta_m) & \frac{1}{L_n}\sin(2\theta_r + \theta_m) \\ \frac{1}{L_n}\sin(2\theta_r + \theta_m) & \frac{1}{L_p} - \frac{1}{L_n}\cos(2\theta_r + \theta_m) \end{bmatrix} \cdot V_h \begin{bmatrix} \cos\alpha \\ \sin\alpha \end{bmatrix} \tag{4.31}$$

and then, the resultant HF current response in the stationary reference frame can be derived as:

$$\begin{bmatrix} i_{\alpha h} \\ i_{\beta h} \end{bmatrix} = \frac{V_h}{\omega_h} \begin{bmatrix} \frac{1}{L_p} + \frac{1}{L_n}\cos(2\theta_r + \theta_m) & \frac{1}{L_n}\sin(2\theta_r + \theta_m) \\ \frac{1}{L_n}\sin(2\theta_r + \theta_m) & \frac{1}{L_p} - \frac{1}{L_n}\cos(2\theta_r + \theta_m) \end{bmatrix} \cdot \begin{bmatrix} \sin\alpha \\ -\cos\alpha \end{bmatrix} \tag{4.32}$$

$$\begin{bmatrix} i_{\alpha h} \\ i_{\beta h} \end{bmatrix} = \begin{bmatrix} I_p \cos\left(\alpha - \frac{\pi}{2}\right) \\ I_p \sin\left(\alpha - \frac{\pi}{2}\right) \end{bmatrix} + \begin{bmatrix} I_n \cos\left(-\alpha + 2\theta_r + \theta_m + \frac{\pi}{2}\right) \\ I_n \sin\left(-\alpha + 2\theta_r + \theta_m + \frac{\pi}{2}\right) \end{bmatrix} \tag{4.33}$$

$$I_p = \frac{V_h}{\omega_h L_p}, I_n = \frac{V_h}{\omega_h L_n} \tag{4.34}$$

The injected HF voltage signal and HF current responses in the stationary reference frame can be expressed by using a complex vector as:

$$v_{\alpha\beta h} = V_h \cdot e^{j\alpha}, \alpha = \omega_h t \tag{4.35}$$

$$i_{\alpha\beta h} = i_p + i_n = I_p \cdot e^{j(\alpha - \pi/2)} + I_n \cdot e^{j(-\alpha + 2\theta_r + \theta_m + \pi/2)} \tag{4.36}$$

Clearly, as shown in (4.36), there are two main components in the HF current response. The first component is a positive sequence i_p, which has the same frequency as the injected voltage signal. The second component is a negative sequence i_n, which has the rotor position information. Therefore, rotor position can be estimated from the negative sequence current.

In order to estimate the rotor position from the negative sequence HF current response, a demodulation process is required and shown in Figure 4.13.

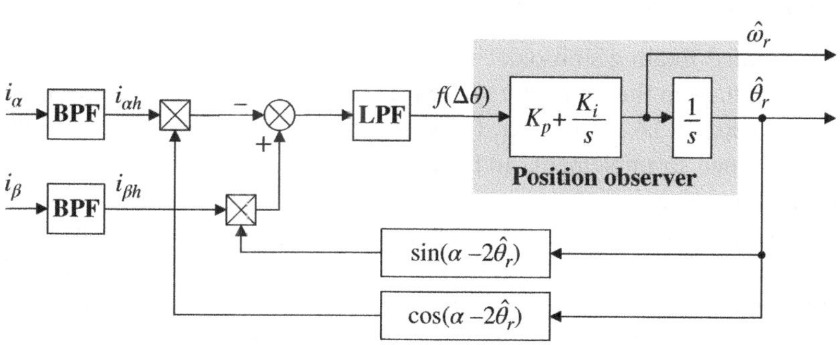

Figure 4.13 Demodulation process of rotating sinusoidal signal injection-based method.

The high-frequency current responses are first obtained by band-pass filters (BPF). Then, the positive sequence component in the HF current response is shifted to the high-frequency domain and can be easily filtered by an LPF, and the negative sequence is obtained for position estimation. The demodulation process of high-frequency current responses is expressed as:

$$i_{\alpha\beta h} \cdot e^{j\left(\alpha - 2\hat{\theta}_r - \pi/2\right)} = I_p \cdot e^{j\left(2\alpha - 2\hat{\theta}_r - \pi\right)} + I_n \cdot e^{j(2\Delta\theta + \theta_m)} \tag{4.37}$$

$$\hat{i}_n = \mathrm{LPF}\left(i_{\alpha\beta h} \cdot e^{j\left(\alpha - 2\hat{\theta}_r - \pi/2\right)}\right) = I_n \cdot e^{j(2\Delta\theta + \theta_m)} \tag{4.38}$$

After the demodulation process, the negative sequence current $\hat{i}_n$ can be expressed by:

$$\hat{i}_n = I_n e^{j(2\Delta\theta + \theta_m)} \tag{4.39}$$

$$\begin{bmatrix} \hat{i}_{nd} \\ \hat{i}_{nq} \end{bmatrix} = I_n \begin{bmatrix} \cos(2\Delta\theta + \theta_m) \\ \sin(2\Delta\theta + \theta_m) \end{bmatrix} \tag{4.40}$$

Then, the magnitude of $\hat{i}_{nq}$ is the input of a position observer as given by:

$$f(\Delta\theta) = \hat{i}_{nq} = I_n \sin(2\Delta\theta + \theta_m) \tag{4.41}$$

After the error signal $f(\Delta\theta)$ is minimized by the position observer, the estimated position aligns with the actual position.

$$f(\Delta\theta) = \hat{i}_{nq} = I_n \sin(2\Delta\theta + \theta_m) = 0 \tag{4.42}$$

In (4.42), the same as the pulsating signal injection method, an error angle θ_m due to cross-saturation exists in the estimated rotor position.

4.4.2 Pulsating Sinusoidal Signal

The conventional rotating sinusoidal signal injection methods inject a balanced three-phase voltage in the stationary reference frame to form a rotating excitation as shown in Section 3.4.1. However, the signal demodulation procedure is relatively complicated to implement, and the utilization of filters will degrade the dynamic performance.

This section introduces a method that injects a pulsating high-frequency voltage signal into the stationary reference frame [5]. Since the high-frequency HF current response is amplitude-modulated the same as the pulsating HF signal injection method, the signal demodulation process can be simplified to obtain a good dynamic performance compared to rotating sinusoidal signal injection.

4.4.2.1 Mathematical Model

First, a pulsating high-frequency sinusoidal voltage signal is injected into the α-axis of the stationary reference frame, as shown in Figure 4.14.

The injection HF voltage signal is expressed as:

$$\begin{bmatrix} v_{\alpha h} \\ v_{\beta h} \end{bmatrix} = V_h \begin{bmatrix} \cos\alpha \\ 0 \end{bmatrix}, \quad \alpha = \omega_h t \tag{4.43}$$

The differential of the HF current response in the stationary reference frame can be derived as:

$$p\begin{bmatrix} i_{\alpha h} \\ i_{\beta h} \end{bmatrix} = \begin{bmatrix} \frac{1}{L_p} + \frac{1}{L_n}\cos(2\theta_r + \theta_m) & \frac{1}{L_n}\sin(2\theta_r + \theta_m) \\ \frac{1}{L_n}\sin(2\theta_r + \theta_m) & \frac{1}{L_p} - \frac{1}{L_n}\cos(2\theta_r + \theta_m) \end{bmatrix} \cdot V_h \begin{bmatrix} \cos\alpha \\ 0 \end{bmatrix} \tag{4.44}$$

The resultant HF current response can be derived as:

$$\begin{bmatrix} i_{\alpha h} \\ i_{\beta h} \end{bmatrix} = \begin{bmatrix} \frac{V_h}{\omega_h L_p} + \frac{V_h}{\omega_h L_n}\cos(2\theta_r + \theta_m) \\ \frac{V_h}{\omega_h L_n}\sin(2\theta_r + \theta_m) \end{bmatrix} \cdot \sin\alpha = \begin{bmatrix} I_p + I_n\cos(2\theta_r + \theta_m) \\ I_n\sin(2\theta_r + \theta_m) \end{bmatrix} \cdot \sin\alpha \tag{4.45}$$

$$I_p = \frac{V_h}{\omega_h L_p}, \quad I_n = \frac{V_h}{\omega_h L_n} \tag{4.46}$$

It is clearly shown that the HF current response is amplitude-modulated by the rotor position.

In order to demodulate the position-dependent amplitude from the HF current response, a demodulation process similar to the pulsating sinusoidal signal injection method is applied, as shown in Figure 4.15.

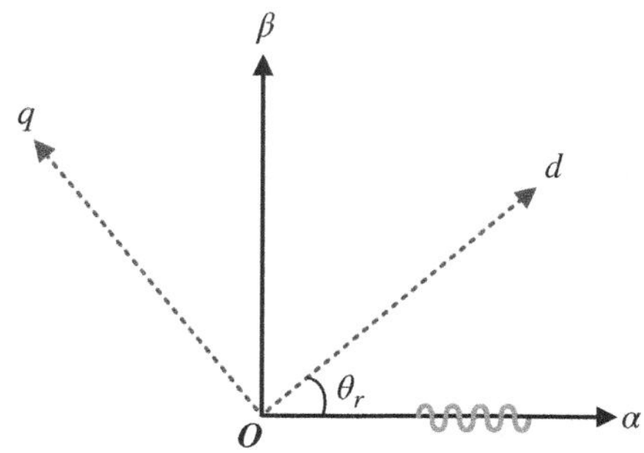

Figure 4.14 Pulsating sinusoidal high-frequency signal injection in stationary reference frame.

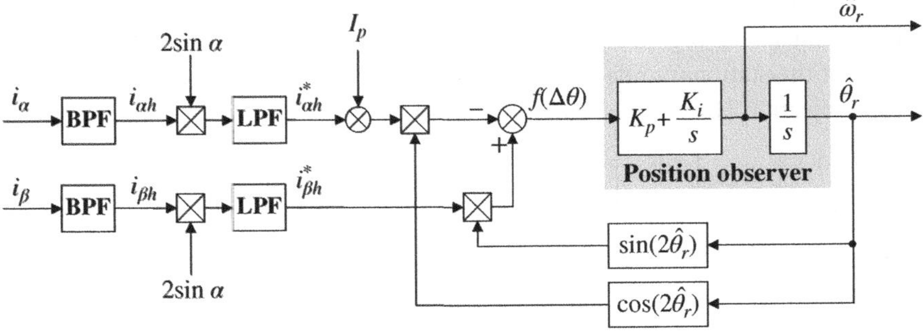

Figure 4.15 Block diagram of stationary pulsating sinusoidal signal injection-based method.

Both sides of (4.45) are multiplied by 2sinα, and then applying the signal to an LPF, the amplitude of the HF current can be obtained as expressed in (4.47).

$$\begin{bmatrix}|i_{\alpha h}|\\|i_{\beta h}|\end{bmatrix} = LPF(\begin{bmatrix}i_{\alpha h}\\i_{\beta h}\end{bmatrix}\cdot 2\sin\alpha) = \begin{bmatrix}I_p + I_n\cos(2\theta_r+\theta_m)\\ I_n\sin(2\theta_r+\theta_m)\end{bmatrix} = \begin{bmatrix}I_p\\0\end{bmatrix} + \begin{bmatrix}i^{**}_{\alpha h}\\ i^{**}_{\beta h}\end{bmatrix} \tag{4.47}$$

If I_p and θ_m could be pre-detected and compensated for, then

$$\begin{bmatrix}i^*_{\alpha h}\\ i^*_{\beta h}\end{bmatrix} = \begin{bmatrix}I_n\cos(2\theta_r)\\ I_n\sin(2\theta_r)\end{bmatrix} \tag{4.48}$$

and the error signal can be calculated from (4.48) as:

$$f(\Delta\theta) = -i^*_{\alpha h}\sin\left(2\hat{\theta}_r\right) + i^*_{\beta h}\cos\left(2\hat{\theta}_r\right) = I_n\sin(2\Delta\theta) \tag{4.49}$$

From (4.49), the rotor position could be derived easily by forcing $f(\Delta\theta)$ to zero by a position observer as shown in Figure 4.15.

If a β-axis HF pulsating sinusoidal voltage signal (4.50) is injected into the stationary reference frame,

$$\begin{bmatrix}v_{\alpha h}\\ v_{\beta h}\end{bmatrix} = V_h\begin{bmatrix}0\\ \cos\alpha\end{bmatrix},\quad \alpha = \omega_h t \tag{4.50}$$

Then, the differential of the HF current response in the stationary reference frame can be expressed as:

$$p\begin{bmatrix}i_{\alpha h}\\ i_{\beta h}\end{bmatrix} = \begin{bmatrix}\frac{1}{L_p} + \frac{1}{L_n}\cos(2\theta_r+\theta_m) & \frac{1}{L_n}\sin(2\theta_r+\theta_m)\\ \frac{1}{L_n}\sin(2\theta_r+\theta_m) & \frac{1}{L_p} - \frac{1}{L_n}\cos(2\theta_r+\theta_m)\end{bmatrix}\cdot V_h\begin{bmatrix}0\\ \cos\alpha\end{bmatrix} \tag{4.51}$$

Then, the resultant HF current response can be derived as:

$$\begin{bmatrix} i_{\alpha h} \\ i_{\beta h} \end{bmatrix} = \begin{bmatrix} I_n \sin(2\theta_r + \theta_m) \\ I_p - I_n \cos(2\theta_r + \theta_m) \end{bmatrix} \cdot \sin\alpha = \left(\begin{bmatrix} 0 \\ I_p \end{bmatrix} + \begin{bmatrix} i_{\beta h}^{**} \\ -i_{\alpha h}^{**} \end{bmatrix} \right) \cdot \sin\alpha \tag{4.52}$$

Similar to the α-axis injection, the HF current response is also amplitude-modulated by the rotor position, and the demodulation process will be the same.

4.4.2.2 I_p Pre-detection and Compensation

As described above, e.g. with α-axis pulsating injection, I_p should be subtracted from $|i_{\alpha h}|$ to obtain $i_{\alpha h}^{**}$ in (4.47) for tracking the rotor position. Based on (4.12) and (4.46), I_p could be expressed as:

$$I_p = \frac{V_h}{\omega_h \cdot L_p} = \frac{V_h \cdot L_{sa}}{\omega_h \cdot \left(L_{dh} L_{qh} - L_{dqh}^2\right)} \tag{4.53}$$

As shown in (4.53), I_p is machine parameter dependent and proportional to the ratio of V_h/ω_h. Since the parameter value of L_p is almost unaffected by i_q, on the condition of fixed V_h and ω_h, the value of I_p can be eliminated by a look-up table that is designed theoretically based on the pre-measured inductances. Based on the prototype IPMSM-I (its parameters are given in Appendix B), the measured I_p against a different fundamental current is shown in Figure 4.16, from which I_p could be treated as constant at different q-axis current for simplicity.

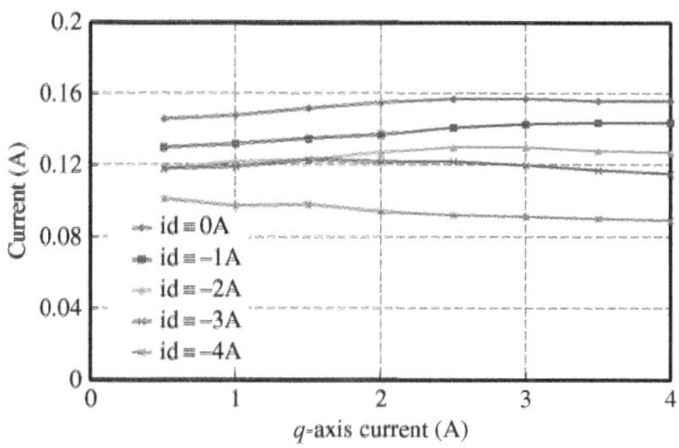

Figure 4.16 Measured I_p versus different fundamental currents. (*Source:* [5]/IEEE.)

For general applications at zero and very low speed, i_d is usually set as negative and kept the same. Hence, it is possible to derive I_p from the look-up table under that condition and apply it for starting.

However, when the rotor speed is higher than a certain value, e.g. 10 r/min, various i_d may apply. Due to the variations of the self- and mutual-inductances at different i_d, it is impractical to calculate every I_p at all different current conditions. However, $i_{\alpha h}^{**}$ in (4.47) will appear as a sinusoidal signal where I_p is the constant dc offset. Therefore, if the $|i_{\alpha h}|$ signal goes through an LPF, the output should be I_p. In this case, the demodulation process is modified as shown in Figure 4.17. To effectively filter the contained ac components, which are 1 Hz minimally without phase lag, the cutoff frequency should be online adjusted with the feedback rotor speed, where the maximum cutoff frequency is 20% of the ac component frequency. Since the derived I_p by LPF will maintain the same value with fixed i_d, the value derived at the starting condition could be used, stored into controller memory and applied as I_p during starting from zero to very low speed. The obtaining procedure is machine parameter free, which makes it superior to the calculation from pre-tested data.

After deriving I_p, the HF current responses are given by:

$$\begin{bmatrix} i_{\alpha h}^{**} \\ i_{\beta h}^{**} \end{bmatrix} = \begin{bmatrix} I_n \cos(2\theta_r + \theta_m) \\ I_n \sin(2\theta_r + \theta_m) \end{bmatrix} \tag{4.54}$$

As shown in (4.54), the phase angle of $i_{\alpha\beta h}^{**}$ is modulated by rotor position θ_r and cross-saturation angle θ_m. Hence, the rotor position can be derived with the load-dependent position estimation error $(-\theta_m/2)$ due to the cross-saturation effect.

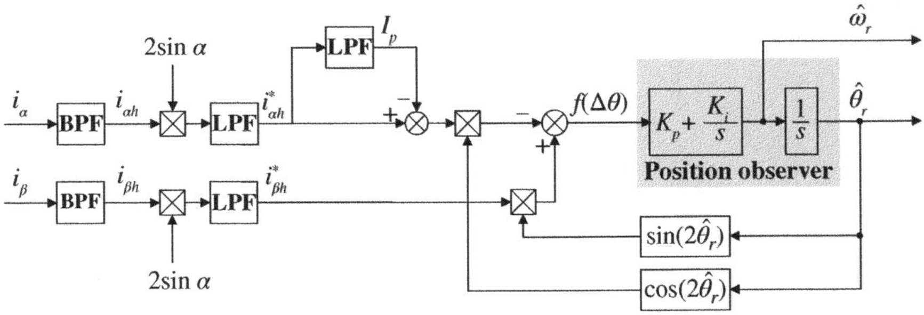

Figure 4.17 Demodulation process of stationary pulsating sinusoidal signal injection method with I_p estimation.

4.4.2.3 Experiment Results

Experiments have been performed to validate the high-frequency sinusoidal pulsating signal injection method based on the stationary reference frame. The specifications of the prototype machine IPMSM-I are given in Appendix B. The magnitude of injected HF voltage is set as 12 V, and the injected frequency is selected as 330 Hz to obtain a satisfied estimation performance. The detailed magnitude and frequency selection method will be introduced in Chapter 5.

As indicated by (4.54), there will be position error due to the cross-saturation effect. Hence, based on the cross-saturation compensation strategy [7], which will be introduced in Chapter 5, the estimated rotor positions before and after compensation under steady-state at 50 r/min with 1 A q-axis current are shown in Figure 4.18a. Figure 4.18b shows the dynamic performance when rotor speed changes from −50 to 50 r/min and then back to −50 r/min. It can be found that there are significant errors before compensation, while after compensation, the estimated rotor positions can match the actual rotor position well and the errors are close to zero.

The dynamic performance test is carried out when the rotor initial speed is zero and step changed to 25 r/min and then 50 r/min, and the estimated and actual rotor positions and the estimation error are shown in Figure 4.19a. Compared with the dynamic performances of pulsating sinusoidal signal injection, Figure 4.20a, and rotating sinusoidal signal injection, Figure 4.20b, it can be concluded that the stationary pulsating sinusoidal signal injection strategy has a similar dynamic performance as the pulsating sinusoidal signal injection strategy, and better performance than the rotating sinusoidal signal injection strategy. The dynamic performance under the step load condition where the q-axis current step change from

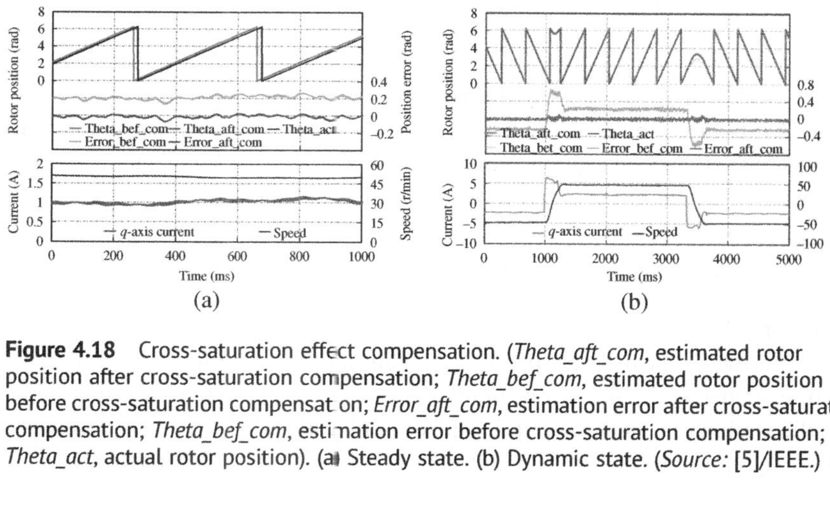

Figure 4.18 Cross-saturation effect compensation. (*Theta_aft_com*, estimated rotor position after cross-saturation compensation; *Theta_bef_com*, estimated rotor position before cross-saturation compensation; *Error_aft_com*, estimation error after cross-saturation compensation; *Theta_bef_com*, estimation error before cross-saturation compensation; *Theta_act*, actual rotor position). (a) Steady state. (b) Dynamic state. (*Source:* [5]/IEEE.)

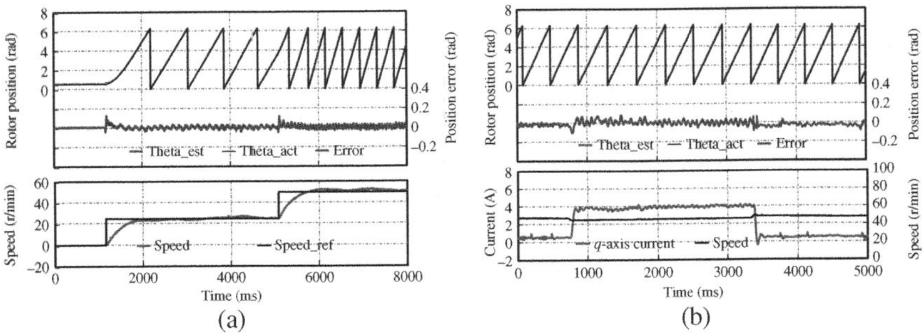

Figure 4.19 Dynamic performance of stationary pulsating sinusoidal signal injection method. (a) Step speed. (b) Step load. (*Source:* [5]/IEEE.)

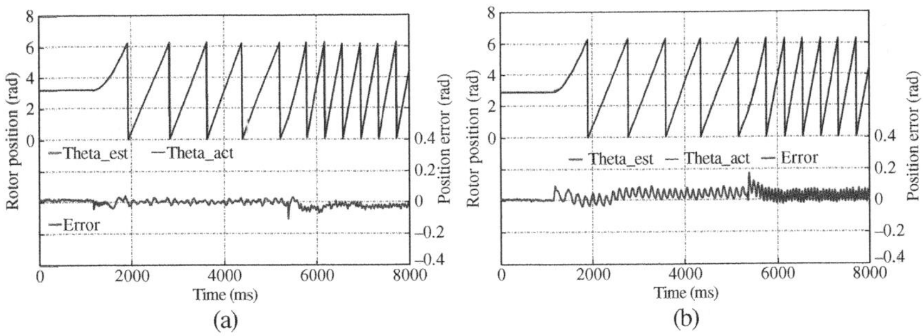

Figure 4.20 Dynamic performance of pulsating and rotating sinusoidal signal injection methods. (a) Pulsating sinusoidal signal in estimated synchronous reference frame. (b) Rotating sinusoidal signal in stationary reference frame. (*Source:* [5]/IEEE.)

0.5 A to full load, which is 4 A, and then back to 0.5 A at 50 r/min is shown in Figure 4.19b. From the experimental results, it can be concluded that the stationary pulsating sinusoidal signal injection strategy has outstanding dynamic performance against different speed and load conditions.

4.4.3 Pulsating Square-Wave Signal

4.4.3.1 Mathematical Model

To extend the application bandwidth, the concept of square waveform injection can be extended to the strategy of an injection of a high-frequency pulsating signal into the stationary reference frame [6]. A high-frequency pulsating square-wave voltage signal in (4.55) is injected into the α-axis as shown in Figure 4.21.

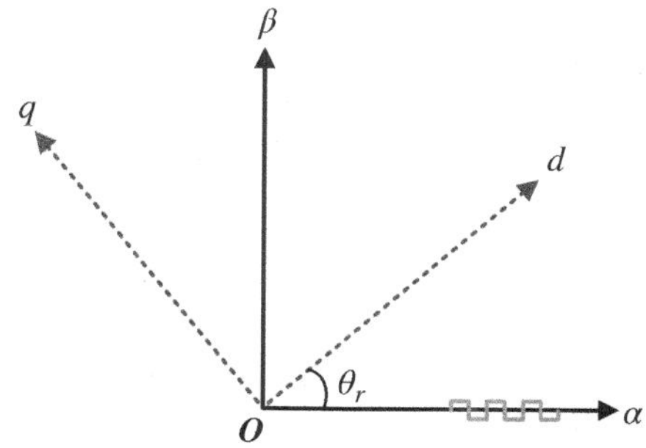

Figure 4.21 Pulsating square-wave high-frequency signal injection in stationary reference frame.

$$\begin{bmatrix} v_{\alpha h} \\ v_{\beta h} \end{bmatrix} = \begin{bmatrix} V_h \cdot (-1)^n \\ 0 \end{bmatrix}, \quad n = 0, 1, 2... \tag{4.55}$$

where n is an integer and the number of injected voltages. After injection, the variation of resultant HF current response in the stationary reference frame is expressed as:

$$\begin{bmatrix} \Delta i_{\alpha h} \\ \Delta i_{\beta h} \end{bmatrix} = V_h \Delta T \begin{bmatrix} \frac{1}{L_p} + \frac{1}{L_n}\cos(2\theta_r + \theta_m) \\ \frac{1}{L_n}\sin(2\theta_r + \theta_m) \end{bmatrix} \cdot (-1)^{n-1}, n = 1, 2, 3... \tag{4.56}$$

From (4.56), it can be found that the sign of $\Delta i_{\alpha\beta h}$ is decided by the polarity of injected HF voltage. Then, $\Delta i_{\alpha\beta h}$ should be normalized by considering the sign of the injected voltage as:

$$\Delta i^*_{\alpha\beta h} = \Delta i_{\alpha\beta h} \cdot (-1)^{n+1} \tag{4.57}$$

$$\begin{bmatrix} \Delta i^*_{\alpha h} \\ \Delta i^*_{\beta h} \end{bmatrix} = \begin{bmatrix} \frac{V_h \Delta T}{L_p} + \frac{V_h \Delta T}{L_n}\cos(2\theta_r + \theta_m) \\ \frac{V_h \Delta T}{L_n}\sin(2\theta_r + \theta_m) \end{bmatrix} = \begin{bmatrix} I_p^{SQ} \\ 0 \end{bmatrix} + \begin{bmatrix} \Delta i^{**}_{\alpha h} \\ \Delta i^{**}_{\beta h} \end{bmatrix} \tag{4.58}$$

$$\begin{bmatrix} \Delta i^{**}_{\alpha h} \\ \Delta i^{**}_{\beta h} \end{bmatrix} = \begin{bmatrix} I_n^{SQ}\cos(2\theta_r + \theta_m) \\ I_n^{SQ}\sin(2\theta_r + \theta_m) \end{bmatrix} \tag{4.59}$$

$$I_p^{SQ} = \frac{V_h \Delta T}{L_p}, \quad I_n^{SQ} = \frac{V_h \Delta T}{L_n} \tag{4.60}$$

From (4.58), it is clearly shown that the normalized current responses $\Delta i^{**}_{\alpha h}$ and $\Delta i^{**}_{\beta h}$ are modulated by the rotor position θ_r. Similar to the pulsating sinusoidal signal injection method in Section 4.4.2, if I_p^{SQ} can be compensated and the cross-saturation effect is considered, the rotor position can be retrieved from the response current precisely. The error signal of the position observer is expressed as:

$$f_1(\Delta\theta) = -\Delta i_{\alpha h}^{**}\sin\left(2\hat{\theta}_r\right) + \Delta i_{\beta h}^{**}\cos\left(2\hat{\theta}_r\right) = I_n\sin(2\Delta\theta) \tag{4.61}$$

From (4.61), the rotor position can be derived easily by forcing $f(\Delta\theta)$ to zero by a position observer. The overall block diagram of the stationary pulsating square-wave signal injection method is shown in Figure 4.22.

If a β-axis square waveform pulsating HF voltage signal (4.62) is injected into the stationary reference frame,

$$\begin{bmatrix} v_{\alpha h} \\ v_{\beta h} \end{bmatrix} = \begin{bmatrix} 0 \\ V_h \cdot (-1)^n \end{bmatrix}, \quad n = 0, 1, 2\ldots \tag{4.62}$$

Then, the amplitude difference in the stationary reference frame between the present and the previous values of the HF current responses can be derived as:

$$\begin{bmatrix} \Delta i_{\alpha h} \\ \Delta i_{\beta h} \end{bmatrix} = V_h \Delta T \begin{bmatrix} \dfrac{1}{L_n}\sin(2\theta_r + \theta_m) \\ \dfrac{1}{L_p} - \dfrac{1}{L_n}\cos(2\theta_r + \theta_m) \end{bmatrix} \cdot (-1)^{n-1}, n = 1, 2, 3\ldots \tag{4.63}$$

Then, the amplitude variation of HF current responses can be derived considering the polarity of the injected voltage, i.e.

$$\begin{bmatrix} \Delta i_{\alpha h}^{*} \\ \Delta i_{\beta h}^{*} \end{bmatrix} = \begin{bmatrix} \dfrac{V_h \Delta T}{L_n}\sin(2\theta_r + \theta_m) \\ \dfrac{V_h \Delta T}{L_p} - \dfrac{V_h \Delta T}{L_n}\cos(2\theta_r + \theta_m) \end{bmatrix} = \begin{bmatrix} 0 \\ I_p^{SQ} \end{bmatrix} + \begin{bmatrix} \Delta i_{\beta h}^{**} \\ -\Delta i_{\alpha h}^{**} \end{bmatrix} \tag{4.64}$$

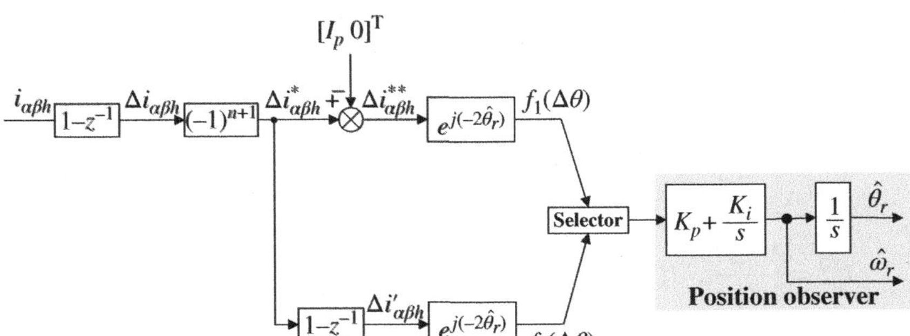

Figure 4.22 Block diagram of stationary pulsating square-wave signal injection-based method.

$$\begin{bmatrix} \Delta i_{\beta h}^{**} \\ -\Delta i_{\alpha h}^{**} \end{bmatrix} = \begin{bmatrix} I_n^{SQ} \sin(2\theta_r + \theta_m) \\ -I_n^{SQ} \cos(2\theta_r + \theta_m) \end{bmatrix} \tag{4.65}$$

Similar to α-axis injection, the HF current response is also amplitude modulated by the rotor position information, and the demodulation process will be the same.

4.4.3.2 I_p^{SQ} Pre-detection and Compensation

As with the sinusoidal waveform injection method, I_p^{SQ} also should be subtracted from $\Delta i_{\alpha h}^{*}$ to obtain $\Delta i_{\alpha h}^{*}$ for tracking the rotor position. Based on (4.12) and (4.60), I_p^{SQ} could be expressed as:

$$I_p^{SQ} = \frac{V_h \Delta T}{L_p} = \frac{V_h \cdot L_{sa} \cdot \Delta T}{L_{dh} L_{qh} - L_{dqh}^2} \tag{4.66}$$

where I_p^{SQ} is parameter dependent and proportional to the ratio of $V_h \Delta T$, which is nearly the same as I_p of sinusoidal signal injection. As mentioned above, L_p is almost unaffected by i_q, and for general applications at zero and very low speed, i_d is usually set as negative and maintains the same value. Furthermore, it is possible to obtain I_p by a look-up table designed from the pre-measured machine inductances. Then, the derived I_p can be stored and applied for starting from zero to very low speed. Finally, subtracting the calculated or measured I_p^{SQ} from $\Delta i_{\alpha h}^{*}$, $\Delta i_{\alpha h}^{**}$, which is modulated by rotor position θ_r and cross-saturation angle θ_m, could be derived as:

$$\begin{bmatrix} \Delta i_{\alpha h}^{**} \\ \Delta i_{\beta h}^{**} \end{bmatrix} = \begin{bmatrix} I_n^{SQ} \cos(2\theta_r + \theta_m) \\ I_n^{SQ} \sin(2\theta_r + \theta_m) \end{bmatrix} \tag{4.67}$$

Once the rotor is rotating at higher speed, different i_d may be applied and L_p will vary. Then, deriving I_p^{SQ} from the pre-tested data at all different operating conditions is no longer practical. An alternative solution is provided here, where derivation calculation is considered at each sampling instant. Then, the difference value $\Delta i'_{\alpha\beta h}$ of the present and previous sampling instants of $\Delta i'_{\alpha\beta h}$ can be calculated as:

$$\begin{aligned} \begin{bmatrix} \Delta i'_{\alpha h} \\ \Delta i'_{\beta h} \end{bmatrix} &= \left.\begin{bmatrix} \dfrac{I_p^{SQ} + I_n^{SQ} \cos(2\theta_r + \theta_m)}{T_s} \\ \dfrac{I_n^{SQ} \sin(2\theta_r + \theta_m)}{T_s} \end{bmatrix}\right|_{(k)} - \left.\begin{bmatrix} \dfrac{I_p^{SQ} + I_n^{SQ} \cos(2\theta_r + \theta_m)}{T_s} \\ \dfrac{I_n^{SQ} \sin(2\theta_r + \theta_m)}{T_s} \end{bmatrix}\right|_{(k-1)} \\ &= \begin{bmatrix} \dfrac{I_p^{SQ}\big|_{(k)} - I_p^{SQ}\big|_{(k-1)}}{T_s} + \dfrac{I_n^{SQ} \cos(2\theta_r + \theta_m)\big|_{(k)} - I_n^{SQ} \cos(2\theta_r + \theta_m)\big|_{(k-1)}}{T_s} \\ \dfrac{I_n^{SQ} \sin(2\theta_r + \theta_m)\big|_{(k)} - I_n^{SQ} \sin(2\theta_r + \theta_m)\big|_{(k-1)}}{T_s} \end{bmatrix} \end{aligned} \tag{4.68}$$

Since I_p^{SQ} can be seen as constant between two sampling instants, the difference of I_p^{SQ} between the two sampling instants can be treated as zero. Hence, (4.68) can be re-written as:

$$\begin{bmatrix} \Delta i'_{\alpha h} \\ \Delta i'_{\beta h} \end{bmatrix} = \begin{bmatrix} \dfrac{I_n^{SQ}\cos(2\theta_r + \theta_m)|_{(k)} - I_n^{SQ}\cos(2\theta_r + \theta_m)|_{(k-1)}}{T_s} \\ \dfrac{I_n^{SQ}\sin(2\theta_r + \theta_m)|_{(k)} - I_n^{SQ}\sin(2\theta_r + \theta_m)|_{(k-1)}}{T_s} \end{bmatrix}$$
$$= \frac{I_n^{SQ}}{T_s}\begin{bmatrix} -\sin(2\theta_r + \theta_m)\cdot\sin(2\Delta\theta_s) \\ \cos(2\theta_r + \theta_m)\cdot\sin(2\Delta\theta_s) \end{bmatrix} \tag{4.69}$$
$$= \frac{I_n^{SQ}}{T_s}\cdot\sin(2\Delta\theta_s)\begin{bmatrix} -\sin(2\theta_r + \theta_m) \\ \cos(2\theta_r + \theta_m) \end{bmatrix}$$

where $\Delta\theta_s$ is the rotor position difference between two sampling instants and is constant at constant speed.

From (4.69), it can be found that $\Delta i'_{\alpha\beta h}$ is also modulated by rotor position θ_r and cross-saturation angle θ_m with one sampling time delay, which is nearly the same as (4.67). Then, by neglecting the cross-saturation angle θ_m, the position error signal can be derived as:

$$f_2(\Delta\theta) = -\Delta i'_{\alpha h}\sin(2\hat{\theta}_r) + \Delta i'_{\beta h}\cos(2\hat{\theta}_r) = I_n\sin(2\Delta\theta) \tag{4.70}$$

From (4.70), the rotor position can be derived easily by forcing $f(\Delta\theta)$ to zero by a position observer as shown in Figure 4.22.

4.4.3.3 Experiment Results

Several experiments have been performed to validate the introduced high-frequency pulsating square-wave signal injection method based on the stationary reference frame. The experiments are carried out based on the prototype IPMSM-I, whose parameters are given in Appendix B. A 12 V, 1 kHz pulsating square-wave signal is injected into the α-axis of the stationary reference frame to obtain a reliable estimation.

The estimated rotor positions before and after cross-saturation compensation [7] under steady-state at 50 r/min can be estimated as shown in Figure 4.23. It can be seen that before compensation of the cross-saturation angle, there exists an approximate 0.1 rad position estimation error, and after compensation, the error can be kept within 0.01 rad. The details of the compensation method will be described in Chapter 5.

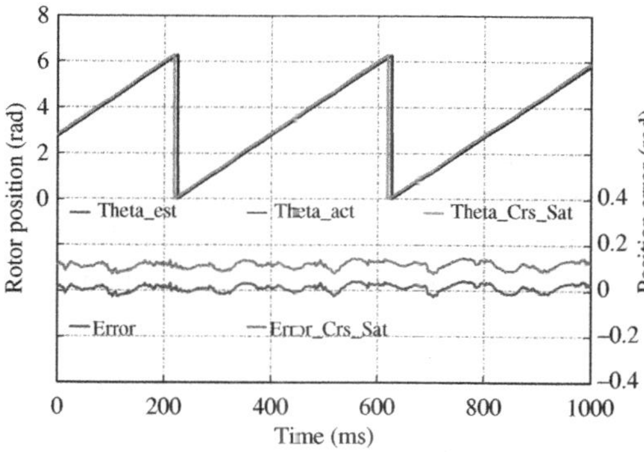

Figure 4.23 Steady-state performance of square waveform injection. (*Theta_est*, estimated rotor position after cross-saturation compensation; *Theta_act*, actual rotor position; *Theta_Crs_Sat*, estimated rotor position before cross-saturation compensation; *Error*, position estimation error after cross-saturation compensation; *Error_Crs_Sat*, position estimation error before cross-saturation compensation). (*Source:* [6]/IEEE.)

Figure 4.24 Dynamic performance of square waveform injection. (*Source:* [6]/IEEE.)

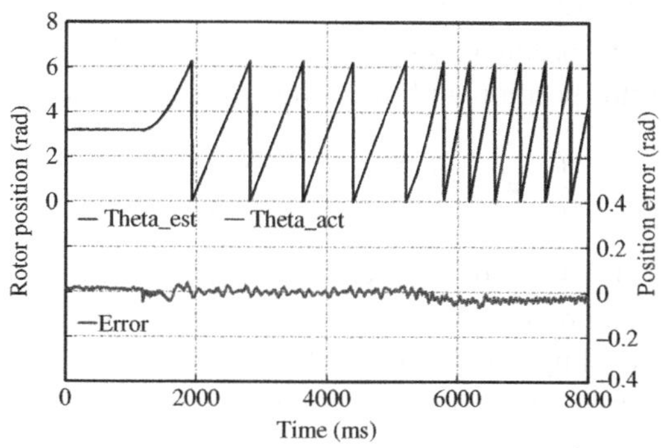

The dynamic performance test is carried out when the rotor initial speed is zero and with step change to 25 r/min and further to 50 r/min. Then, the estimated and actual rotor positions and the estimation error are shown in Figure 4.24. Compared with experimental results by applying the pulsating sinusoidal signal injection in Figure 4.19a, the pulsating square-wave injection method can achieve better dynamic performance due to higher bandwidth caused by the higher injected frequency and elimination of all filters.

4.5 Position Observer

As shown in the block diagrams of five different HFSI methods, a position observer is required to minimize the position error signal $f(\Delta\theta)$ to track the actual position. There are several different position observers that can be used to track the rotor position, including PLL [6], simplified Kalman filter [9], mechanical model based observer [1], etc. In this section, the commonly used PLL-based position observer is introduced, which is simple to implement.

4.5.1 Basic Structure

For all five introduced HF signal injection-based methods, by neglecting the cross-saturation angle θ_m, the error signal $f(\Delta\theta)$ can be expressed as:

$$f(\Delta\theta) = I_n \sin(2\Delta\theta) \tag{4.71}$$

Considering a small error $\Delta\theta$, (4.71) is simplified as:

$$f(\Delta\theta) = I_n \sin(2\Delta\theta) \approx I_n \cdot 2\Delta\theta \tag{4.72}$$

Then, with the error signal $f(\Delta\theta)$ as the input, the block diagram of the position observer is shown in Figure 4.25.

According to the diagram, the closed-loop transfer function between the actual rotor position and the estimated one can be expressed as,

$$\frac{\hat{\theta}_r}{\theta_r} = \frac{2I_nK_p \cdot s + 2I_nK_i}{s^2 + 2I_nK_p \cdot s + 2I_nK_i} \tag{4.73}$$

The value of I_n is load dependent, hence the parameters of K_p and K_i for the position observer should be online varied with I_n to maintain a constant bandwidth. For simplicity in practical implementation, however, the parameters of K_p and K_i can be selected based on the typical value of I_n at a no-load condition.

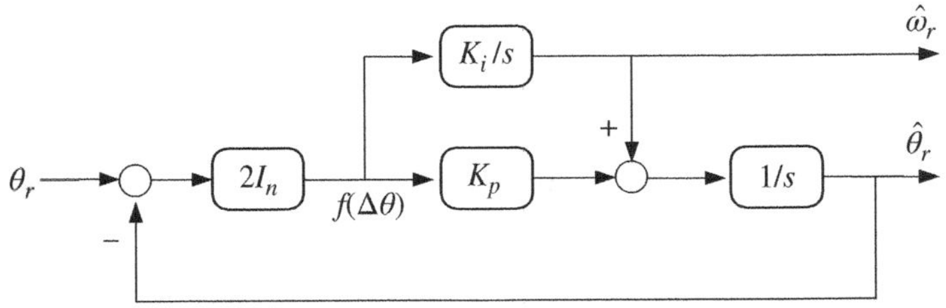

Figure 4.25 Equivalent block diagram of position observer.

4.5.2 Influence of LPF

For pulsating sinusoidal signal injection and rotating sinusoidal signal injection-based methods, an LPF is required to filter out the injected HF component in the current to obtain the position error signal $f(\Delta\theta)$. The bandwidth of the position observer will be limited by this LPF. Accounting for the time lag effect of the LPF, the equivalent model of a position observer for practical applications can be depicted as in Figure 4.26.

For the prototype IFMSM-I (whose parameters are given in Appendix B), the typical value of I_n is considered as $I_n = 70$ mA, and the sampling frequency is configured to be 5 kHz. The parameters of the position observer are selected based on the optimal second order property of ideal closed loop ($K_p = 137$, $K_i = 1250$). Without considering the LPF, the unit step response and Bode plot of the position observer can be drawn analytically, as shown in Figure 4.27. It can be seen that the designed position observer has the rise time of around 17.8 ms in unit step response and loop bandwidth of around 19 Hz.

To evaluate the effect of LPFs on the dynamic behavior of the position observer, Figure 4.28 shows the experimentally measured step response of a position tracking observer, in which LPF (in Figures 4.6, 4.11, and 4.12) with different cutoff frequency is used to remove the HF frequency ripple on $f(\Delta\theta)$. Without fundamental excitation, the prototype machine stays at standstill. It can be seen that when the cutoff frequency of LPF is much higher than the designed bandwidth of the position observer (80 Hz >> 19 Hz), the designed position observer has similar dynamic behavior as the theoretical simulation. The estimated rotor position can steadily track the real rotor position with the rise time of 19 ms, similar to the predicted value of 17.8 ms (Figure 4.28). A decrease in the cutoff frequency of LPF would lead to the dynamic performance of the position observer being degraded, as shown in Figures 4.28b,c.

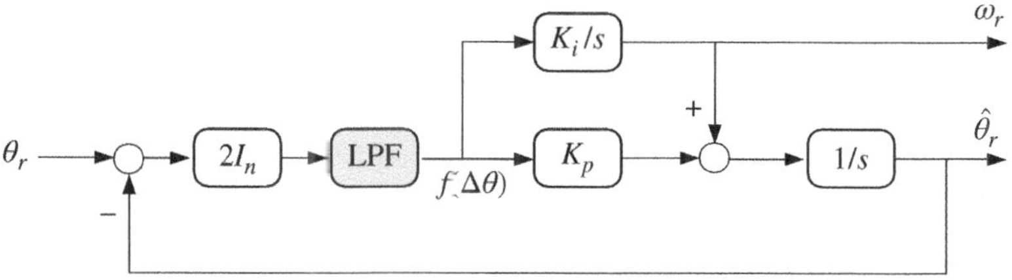

Figure 4.26 Block diagram of position observer including low-pass filter.

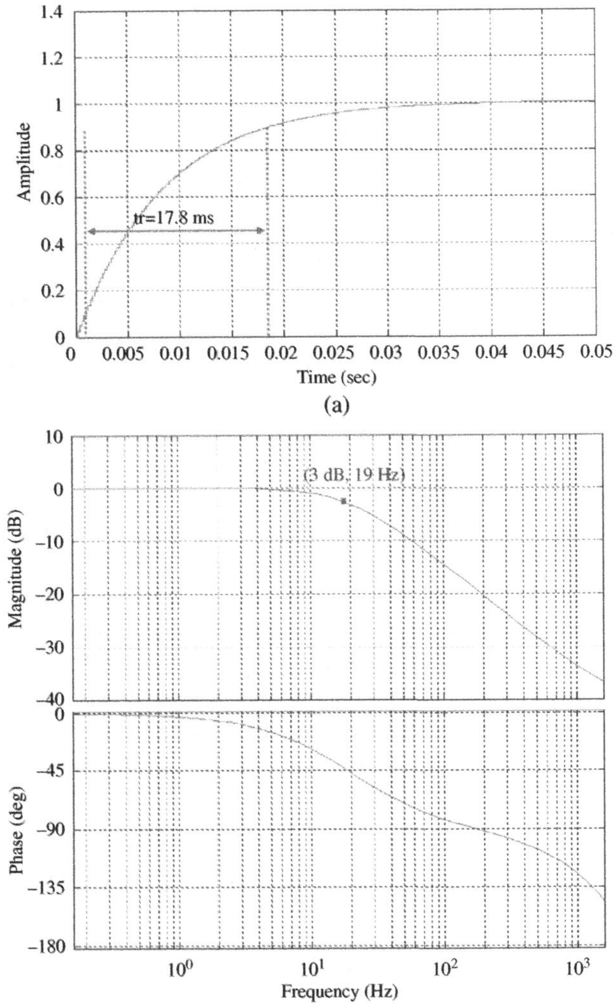

Figure 4.27 Simulated step response of designed position observer. (a) Unit step response. (b) Bode plot.

4.5.3 Convergence Analysis

The criterion for reliable convergence of the PLL is that the error signal ε should have positive slope near the point $\Delta\theta = 0$ [10]. The transfer function of the position observer is given in (4.73). The error signal $f(\Delta\theta)$ with position error $\Delta\theta$ is plotted in Figure 4.29 with the stable regions highlighted.

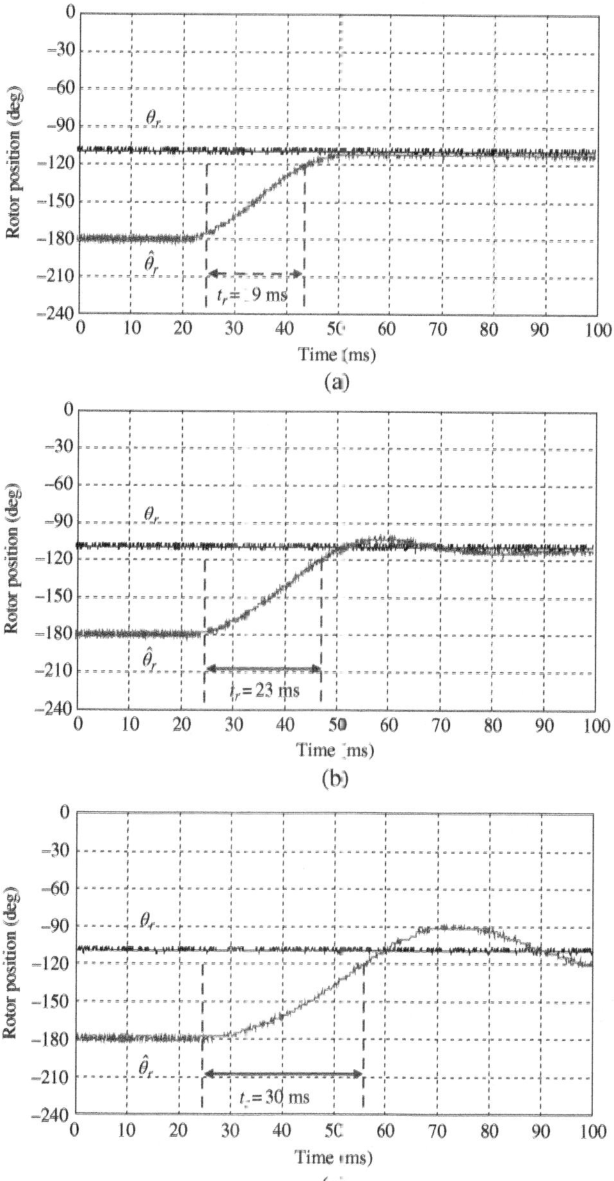

Figure 4.28 Measured position step response with different cutoff frequency of LPF. (a) f_c = 80 Hz. (b) f_c = 40 Hz. (c) f_c = 20 Hz.

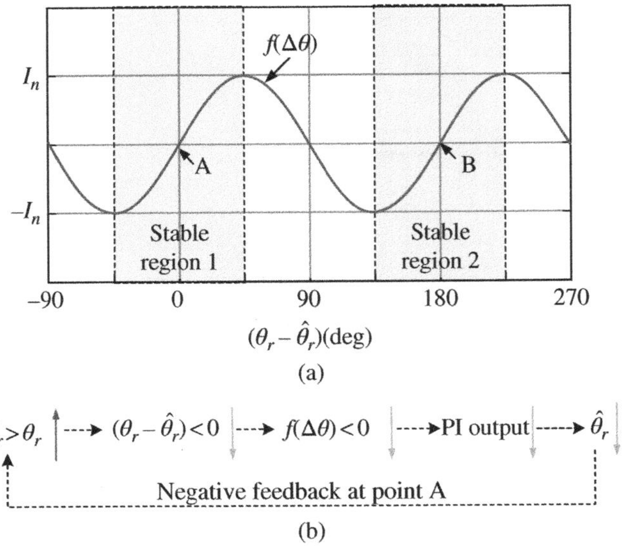

Figure 4.29 Closed-loop control block diagram of position observer. (a) Stable region for convergence. (b) Dynamic process for convergence.

It is observed that A and B are stable convergence points due to positive slopes of the error signal. For instance, the dynamic process of convergence at point A is shown in Figure 4.29b. In one electrical period, there are two convergence points. The initial polarity detection will be introduced later in Chapter 9. The dynamic performance of the position estimation performance is dependent on I_n, which varies with the working currents. Thus, the sufficient saliency level (or I_n) should be guaranteed for reliable convergence.

4.6 Other Saliency Tracking-Based Methods

Apart from the previously introduced persistent HFSI-based sensorless control methods, the rotor saliency can also be employed by other types of excitations including transient voltage vector injection and inherent PWM excitations.

4.6.1 Transient Voltage Vector-Based Method

In the zero- and low-speed region, apart from the persistent HF signal injection methods introduced in previous sections, the transient voltage vector injection in the PWM periods can also be utilized for the position estimation [11]–[13].

To use the voltage pulse injection method, the PWM switching patterns should be modified so that the resultant current variations can be detected and demodulated for rotor position estimation. By way of example, for the INFORM (indirect flux detection by on-line reactance measurement) method [11], the transient vectors are injected during the zero vector dwelling of the PWM as shown in Figure 4.30. As shown in Figure 4.31, three pairs of voltage vectors are injected within three continuous PWM periods, and then the rotor position information can be obtained by demodulating the current variations.

However, due to the INFORM method being performed in a very short period of time, it may increase the implementation complexity for the practical voltage injection and current sampling, etc. [13]. Besides, the current response may also be sensitive to the parameter variations, measurement delay and noises, etc. [3]. Additionally, as with the persistent HFSI methods, the current distortions will also contribute to torque ripples, noise, and vibration.

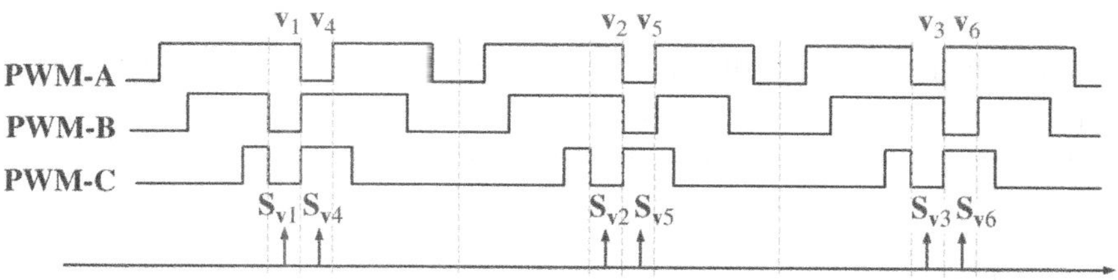

Figure 4.30 Transient voltage vector injection based on INFORM method.
(*Source:* [11]/IEEE.)

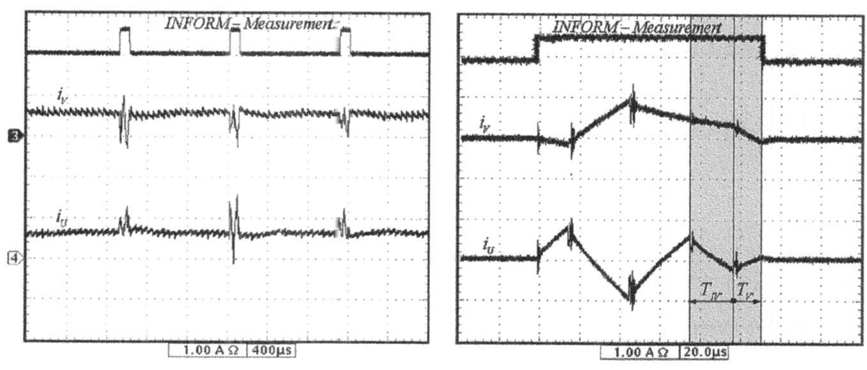

Figure 4.31 Current waveforms of INFORM pulse injection-based sensorless.
(*Source:* [12]/IEEE.)

4.6.2 PWM Excitation-Based Method

Under normal machine operation, the voltage vectors in the inherent PWM switching patterns can also be directly utilized to estimate the rotor position. Inherent PWM methods do not need to inject any additional signals, resulting in reduced switching losses, noise, and torque ripples, compared with persistent HF signal injection and transit voltage vector injection methods [14], [15]. As shown in Figure 4.32, the triangles in the phase currents are due to the PWM switching.

In [14], by neglecting the resistive voltage drop and back-EMF, the high-frequency voltage equation is derived as:

$$v_{hf} = \mathrm{L}\frac{di_{hf}}{dt} \tag{4.74}$$

$$\mathrm{L} = \begin{bmatrix} L_{11} & L_{12} \\ L_{21} & L_{22} \end{bmatrix} = \begin{bmatrix} L_{sa} - L_{sd}\cos 2\theta_r & -L_{sd}\sin 2\theta_r \\ -L_{sd}\sin 2\theta_r & L_{sd} + L_{sd}\cos 2\theta_r \end{bmatrix} \tag{4.75}$$

Based on (4.74), the winding inductances can be derived with the knowledge of applied voltage vectors and current slopes. The rotor position can then be estimated from these calculated winding inductances as:

$$2\theta_r = \tan^{-1}\left(\frac{L_{12} + L_{21}}{L_{11} - L_{22}}\right) \tag{4.76}$$

$$\begin{bmatrix} L_{11} & L_{12} \\ L_{21} & L_{22} \end{bmatrix} = \begin{bmatrix} L_{sa} - L_{sd}\cos 2\theta_r & -L_{sd}\sin 2\theta_r \\ -L_{sd}\sin 2\theta_r & L_{sd} + L_{sd}\cos 2\theta_r \end{bmatrix} \tag{4.77}$$

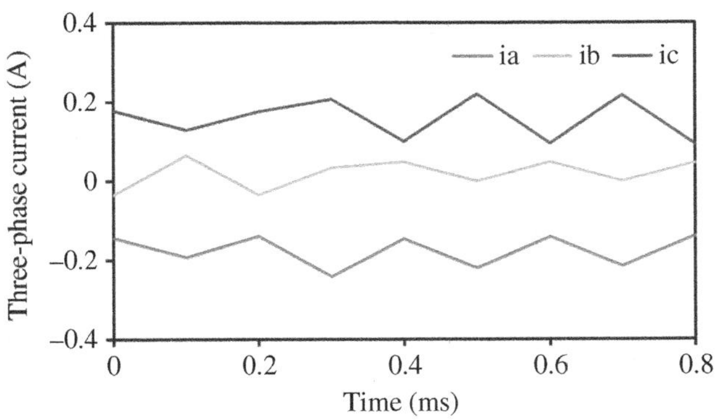

Figure 4.32 Three-phase currents of PWM injection-based sensorless. (*Source:* [14]/IEEE.)

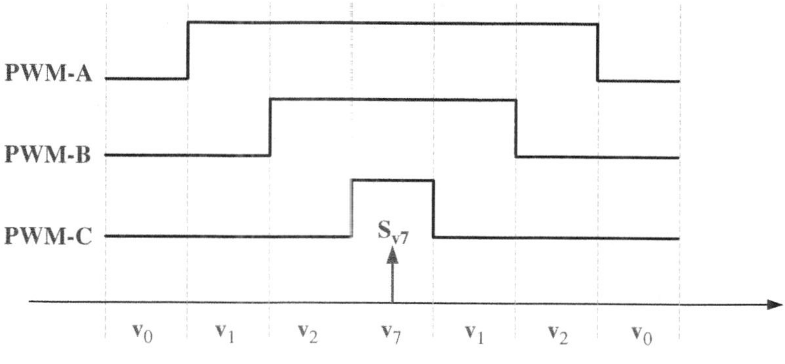

Figure 4.33 Zero vector current derivative method in one PWM period. (*Source:* [15]/IEEE.)

Different from [14], where current derivatives resulting from all the voltage vectors are used for position estimation, a zero vector current derivative method (ZVCD) was proposed in [15], only the current derivative after applying the zero vector is used for position estimation. In this case, the voltage information is not required in the position estimation and calculation becomes simpler. Specifically, the current derivative of the estimated d-axis current during the zero vector dwelling period in Figure 4.33 can be described as:

$$\left.\frac{d\hat{i}_d}{dt}\right|_{v_0/v_7} = -\frac{R_s}{2}\left(\frac{1}{L_q} - \frac{1}{L_d}\right)\hat{i}_q \sin(2\Delta\theta) - \frac{\omega_r \psi_m}{L_q} \sin(\Delta\theta) \tag{4.78}$$

Then, from (4.78), the estimated rotor position and speed can be obtained from the inductive saliency information by neglecting the back-EMF signals in the zero- and low-speed regions.

The three types of methods are compared in Table 4.1. Of the three methods, it can be seen that the persistent HF signal injection-based method has the advantages of high SNR, low cost, and easy implementation, which makes it the most popular choice for zero- and low-speed sensorless control.

4.7 Summary

In this chapter, five traditional HFSI based sensorless control methods were described. The comparison of these methods is given in Table 4.2. Clearly, the square-wave based methods have the best dynamic performance and simplest

Table 4.1 Comparison of saliency tracking-based sensorless methods.

Specifications	Persistent HF signal	Transient voltage vector	PWM excitation
Implementation complexity	Simple	Medium	Medium
Current measurement cost	Low	High	High
Additional signal	Yes	Yes	No
SNR of current response	High	Medium	Low
Bandwidth	Medium	Medium	High
Noise	Medium	Medium	Low
Speed operation range	Narrow	Medium	Wide
Estimated position	Continuous	Discrete	Continuous

Table 4.2 Comparison of persistent signal injection-based methods.

Injection signal	Rotating sinusoidal	Pulsating sinusoidal		Square wave	
Reference frame	Stationary	Estimated synchronous	Stationary	Estimated synchronous	Stationary
Response modulation	Phase-modulated	Amplitude-modulated	Amplitude-modulated	Amplitude-modulated	Amplitude-modulated
Cross-saturation error	$-\theta_m/2$	$-\theta_m/2$	$-\theta_m/2$	$-\theta_m/2$	$-\theta_m/2$
Demodulation	Complex	Medium	Medium	Simple	Simple
Dynamic performance	Low	Medium	Medium	Good	Good
Require parameter	No	No	Yes	No	Yes

demodulation process. Moreover, it can be seen that all of the methods have the same position error due to cross-saturation. It is worth noting that the HFSI methods introduced in this chapter were under ideal conditions. However, in real-world applications, there will be some non-ideal scenarios, e.g. cross-saturation effect, inverter nonlinearity, etc. Hence, the next chapter will discuss some common practical issues in their implementation and the corresponding solutions.

References

1 P. L. Jansen and R. D. Lorenz, "Transducerless position and velocity estimation in induction and salient AC machines," *IEEE Trans. Ind. Appl.*, vol. 31, no. 2, pp. 240–247, Mar. 1995.

2 M. J. Corley and R. D. Lorenz, "Rotor position and velocity estimation for a salient-pole permanent magnet synchronous machine at standstill and high speeds," *IEEE Trans. Ind. Appl.*, vol. 34, no. 4, pp. 784–789, Jul. 1998.

3 Y. Yoon, S. K. Sul, S. Morimoto, and K. Ide, "High-bandwidth sensorless algorithm for AC machines based on square-wave-type voltage injection," *IEEE Trans. Ind. Appl.*, vol. 47, no. 3, pp. 1361–1370, May 2011.

4 S. Kim, J. I. Ha, and S. K. Sul, "PWM switching frequency signal injection sensorless method in IPMSM," *IEEE Trans. Ind. Appl.*, vol. 48, no. 5, pp. 1576–1587, Sep. 2012.

5 J. M. Liu and Z. Q. Zhu, "Novel sensorless control strategy with injection of high-frequency pulsating carrier signal into stationary reference frame," *IEEE Trans. Ind. Appl.*, vol. 50, no. 4, pp. 2574–2583, Jul. 2014.

6 J. M. Liu and Z. Q. Zhu, "Sensorless control strategy by square-waveform high-frequency pulsating signal injection into stationary reference frame," *IEEE J. Emerging Sel. Topics Power Electron.*, vol. 2, no. 2, pp. 171–180, Jun. 2014.

7 Y. Li, Z. Q. Zhu, D. Howe, C. M. Bingham, and D. Stone, "Improved rotor position estimation by signal injection in brushless AC motors, accounting for cross-coupling magnetic saturation," *IEEE Trans. Ind. Appl.*, vol. 45, no. 5, pp. 1843–1849, 2009.

8 J. H. Jang, J. I. Ha, M. Ohto, K. Ide, and S. K. Sul, "Analysis of permanent-magnet machine for sensorless control based on high-frequency signal injection," *IEEE Trans. Ind. Appl.*, vol. 40, no. 6, pp. 1595–1604, Nov. 2004.

9 Y. Liu, Z. Q. Zhu, Y. F. Shi, and D. Howe, "Sensorless direct torque control of a permanent magnet brushless AC drive via an extended Kalman filter," in *2nd Intl. Conf. Power Electron., Mach. Drives (PEMD 2004)*, vol. 1, pp. 303–307, Mar. 2004.

10 S. C. Yang and Y. L. Hsu, "Full speed region sensorless drive of permanent-magnet machine combining saliency-based and back-EMF-based drive," *IEEE Trans. Ind. Electron.*, vol. 64, no. 2, pp. 1092–1101, Feb. 2017.

11 M. Schroedl, "Sensorless control of AC machines at low speed and standstill based on the "INFORM" method," in *IAS '96. Conf. Rec. 1996 IEEE Ind. Appl. Conf. 31st IAS Annu. Meeting*, San Diego, CA, USA, pp. 270–277 vol. 1, 1996.

12 E. Robeischl and M. Schroedl, "Optimized INFORM measurement sequence for sensorless PM synchronous motor drives with respect to minimum current distortion," *IEEE Trans. Ind. Appl.*, vol. 40, no. 2, pp. 591–598, Mar. 2004.

13 G. Xie, K. Lu, S. Dwivedi, R. Riber, and F. Blaabjerg, "Minimum voltage vector injection method for sensorless control of PMSM for low-speed operations," *IEEE Trans. Power Electron.*, vol. 31, no. 2, pp. 1785–1794, Feb. 2016.

14 S. Ogasawara and H. Akagi, "Implementation and position control performance of a position-sensorless IPM motor drive system based on magnetic saliency," *IEEE Trans. Ind. Appl.*, vol. 34, no. 4, pp. 806–812, Jul./Aug. 1998.

15 R. Raute, C. Caruana, J. Cilia, C. S. Staines, and M. Sumner, "A zero speed operation sensorless PMSM drive without additional test signal injection," in *Eur. Conf. Power Electron. Appl.*, pp. 1–10, 2007.

5

Saliency Tracking-Based Sensorless Control Methods—Issues and Solutions

5.1 Introduction

In Chapter 4, saliency tracking-based zero- and low-speed sensorless control methods were introduced. By utilizing the machine saliency, the position and speed can be estimated by high-frequency signal injection (HFSI). However, the sensorless control methods described in Chapter 4 are mainly based on an ideal machine model and drive system. In practice, there are some non-ideal characteristics that will introduce position estimation errors of HFSI-based methods, as summarized in Figure 5.1. Moreover, there are some aspects that need to be considered during the design of sensorless algorithms, which are also depicted in Figure 5.1.

In this chapter, these practical issues of HFSI-based sensorless control methods are introduced and the corresponding solutions are also presented. As shown in Figure 5.1, first, on the machine side, the machine saliency is investigated including the cross-coupling magnetic saturation, the load-dependent saliency, and the multiple saliency effect. Moreover, the asymmetric parameters, which are mainly inductances, are also considered. Second, on the drive side, the inverter nonlinearity effects and signal processing delay are discussed. Third, in terms of the sensorless control design aspect, the selection of an injected HF signal is described and the transition strategy between low speed and high speed is introduced.

5.2 Cross-Coupling Magnetic Saturation

In [1], it was found experimentally that the position estimation error increases with load current, which is due to the cross-coupling saturation between the d- and q-axes that causes the position error in HFSI-based methods [2], [3].

Sensorless Control of Permanent Magnet Synchronous Machine Drives, First Edition.
Zi Qiang Zhu and Xi Meng Wu.

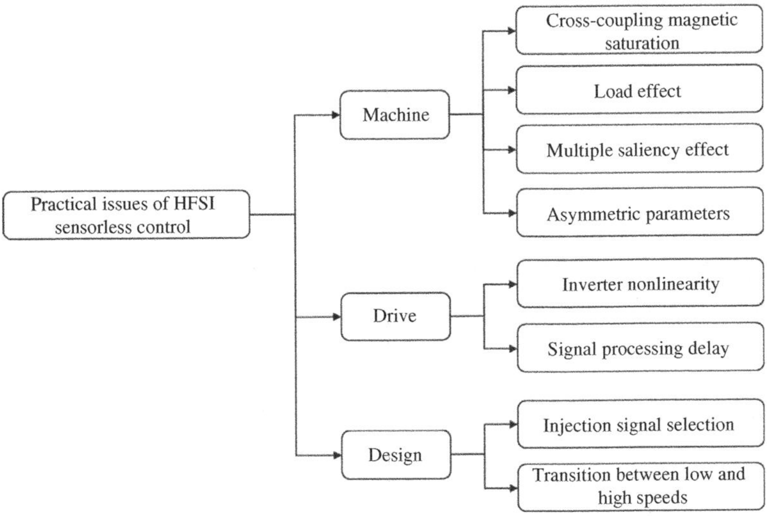

Figure 5.1 Practical issues of HFSI-based sensorless control methods.

By considering the cross-saturation effect, the high-frequency voltage model is expressed by:

$$\begin{bmatrix} v_{dh} \\ v_{qh} \end{bmatrix} = \begin{bmatrix} L_{dh} & L_{dqh} \\ L_{qdh} & L_{qh} \end{bmatrix} \cdot p \begin{bmatrix} i_{dh} \\ i_{qh} \end{bmatrix} \tag{5.1}$$

where L_{dqh} and L_{qdh} are the high-frequency incremental mutual-inductances due to the cross-saturation effect, which are normally regarded as the same [4], [5], i.e. $L_{dqh} \approx L_{qdh}$. The d- and q-axis incremental self- and mutual-inductances are defined by:

$$\begin{cases} L_{dh}(i_d, i_q) = \partial \psi_d / \partial i_d \approx [\psi_d(i_d + \Delta i_d, i_q) - \psi_d(i_d, i_q)] / \Delta i_d \\ L_{qh}(i_d, i_q) = \partial \psi_q / \partial i_q \approx [\psi_q(i_d, i_q + \Delta i_q) - \psi_q(i_d, i_q)] / \Delta i_q \\ L_{dqh}(i_d, i_q) = \partial \psi_d / \partial i_q \approx [\psi_d(i_d, i_q + \Delta i_q) - \psi_d(i_d, i_q)] / \Delta i_q \\ L_{qdh}(i_d, i_q) = \partial \psi_q / \partial i_d \approx [\psi_q(i_d + \Delta i_d, i_q) - \psi_q(i_d, i_q)] / \Delta i_d \end{cases} \tag{5.2}$$

Due to magnetic saturation, L_{dh}, L_{qh}, L_{dqh}, and L_{qdh} vary with both i_d and i_q. By way of example, Figure 5.2 shows the measured incremental inductances for the prototype IPMSM-I, whose parameters are given in Appendix B. Clearly, incremental inductances change with fundamental current excitations. For the cross-coupling inductances, their amplitudes are load dependent. As seen from Figure 5.2c, when there is no load ($i_q = 0$), there is no cross-coupling.

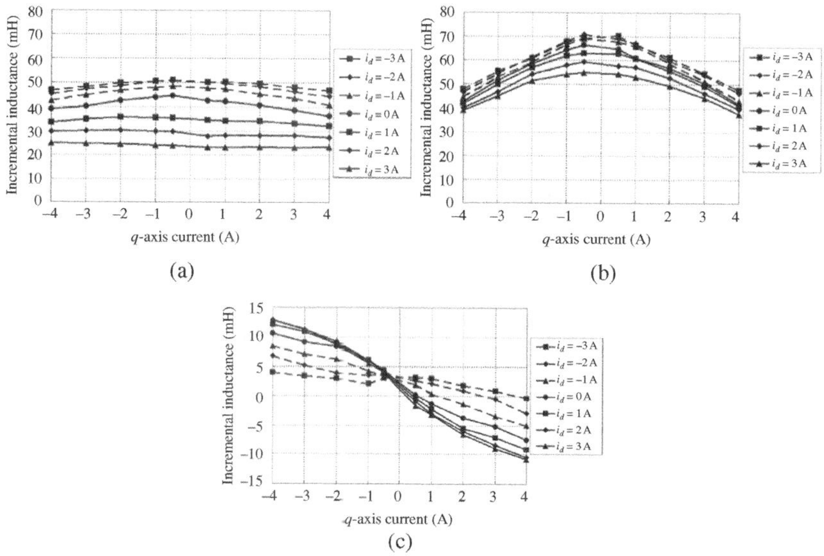

Figure 5.2 Measured incremental self- and mutual-inductances [4]. (a) d-axis incremental self-inductance L_{dh}. (b) q-axis incremental self-inductance L_{qh}. (c) dq-axis incremental mutual-inductances, L_{dqh} and L_{qdh}.

5.2.1 Impact on Position Estimation

As derived in Chapter 4, for conventional HFSI-based methods including pulsating and rotating signal injection methods, the cross-coupling inductances as a result of cross-coupling magnetic saturation between d- and q-axes will cause position estimation errors. The impacts of cross-coupling inductances for different HFSI-based sensorless control methods are summarized in Table 5.1, including the HF current response, the position-dependent signal obtained by demodulating the HF current, and the corresponding position estimation error due to cross-coupling magnetic saturation between d- and q-axes.

Clearly, the position estimation error due to the cross-coupling effect for different methods is the same and can be expressed by:

$$(\hat{\theta}_r - \theta_r) = \Delta\theta = -\theta_m/2 = -\arctan\left(\frac{2L_{dqh}}{L_{qh} - L_{dh}}\right)/2 \tag{5.3}$$

$$\theta_m = \arctan\left(\frac{2L_{dqh}}{L_{qh} - L_{dh}}\right) \tag{5.4}$$

where θ_m is the position error caused by the cross-coupling effect. This position error can be a critical issue as it reduces the torque capability and the reliability of sensorless control. Hence, compensation of the cross-coupling inductances is necessary for HFSI-based sensorless control.

Table 5.1 Impact of cross-coupling inductances on HFSI sensorless control methods.

Sensorless control method	HF current response	Position-dependent signal	Position error
Pulsating sinusoidal	$\begin{bmatrix} \hat{i}_{dh} \\ \hat{i}_{qh} \end{bmatrix} = \begin{bmatrix} I_p + I_n \cos(2\Delta\theta + \theta_m) \\ I_n \sin(2\Delta\theta + \theta_m) \end{bmatrix} \cdot \sin\alpha$	$\lvert\hat{i}_{qh}\rvert = I_n \sin(2\Delta\theta + \theta_m)$	$-\frac{\theta_m}{2}$
Pulsating square wave	$\begin{bmatrix} \Delta\hat{i}_{dh} \\ \Delta\hat{i}_{qh} \end{bmatrix} = V_h \Delta T \begin{bmatrix} \frac{1}{L_p} + \frac{1}{L_n} \cos(2\Delta\theta + \theta_m) \\ \frac{1}{L_n} \sin(2\Delta\theta + \theta_m) \end{bmatrix} \cdot (-1)^{n-1}$	$\Delta\hat{i}_{qh} = \frac{V_h \Delta T}{L_n} \sin(2\Delta\theta + \theta_m)$	$-\frac{\theta_m}{2}$
Rotating sinusoidal	$i_{\alpha\beta h} = i_p + i_n = I_p \cdot e^{j(\alpha - \pi/2)} + I_n \cdot e^{j(-\alpha + 2\theta_r + \theta_m + \pi/2)}$	$\hat{i}_{nq} = I_n \sin(2\Delta\theta + \theta_m)$	$-\frac{\theta_m}{2}$

5.2.2 Compensation Scheme

The cross-coupling inductance can be compensated for both online and offline. Online compensation is quite challenging due to nonlinear characteristics of the cross-coupling effect. For reducing the computation complexity, some assumptions are necessary for online compensation, e.g. fixed d-axis HF inductance [6], which is not feasible for all of the machines. Hence, some methods compensate for the cross-coupling effect by using offline measured data of the machine. The offline data can be position errors [5], [7] or a coupling factor [4].

5.2.2.1 Direct Compensation

The position errors caused by the cross-coupling effect at different working points can be measured offline and directly used for compensation. For HF rotating and pulsating voltage injection-based sensorless control, the direct compensation method is expressed as:

$$\hat{\theta}_r = \hat{\theta}_{r0} + \theta_m\left(\hat{i}_d, \hat{i}_q\right) \tag{5.5}$$

where θ_m is the offline measured error caused by the cross-coupling effect, $\hat{\theta}_{r0}$ is the original estimated position from a position observer, and $\hat{\theta}_r$ is the position after compensation. The error angle θ_m can be obtained from the finite element analysis or measurement. Generally, for the measurement, the error angle can be measured with a sensorless position estimation technique as introduced in Chapter 4. For example, the estimated rotor position can be estimated by injecting pulsating voltage. The error caused by the cross-coupling effect can be obtained from the difference between the actual rotor position and the estimated rotor position. The measured errors can be stored in the memory of a microprocessor. In sensorless operation, the position error caused by the cross-coupling effect is corrected online by adding the error as in (5.5).

For reducing the memory requirement for storing the offline measured errors, the error can be curve-fitted with i_d, i_q. In [5], [7], for simplicity, the error is an approximately linear function of the q-axis current as:

$$\theta_m = k_c i_q \tag{5.6}$$

where k_c can be fitted by the least squares method. Based on the prototype IPMSM-I, the measured results of the compensation performance are shown in Figure 5.3. The compensation is applied to a conventional pulsating sinusoidal signal injection method with $k_c \approx 8.0$ deg /Amp. Moreover, the compensation performance under different load with and without compensation is shown in Figure 5.4. Clearly, the estimation error is reduced under load with the compensation method.

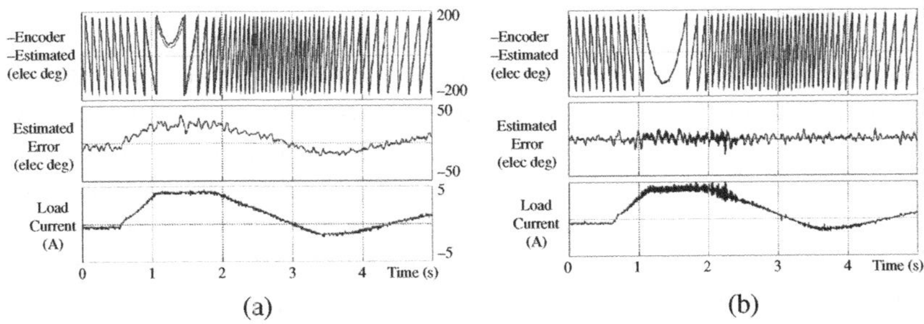

Figure 5.3 Position estimation with/without compensation. (a) Without compensation. (b) With compensation. (*Source:* [5]/IEEE.)

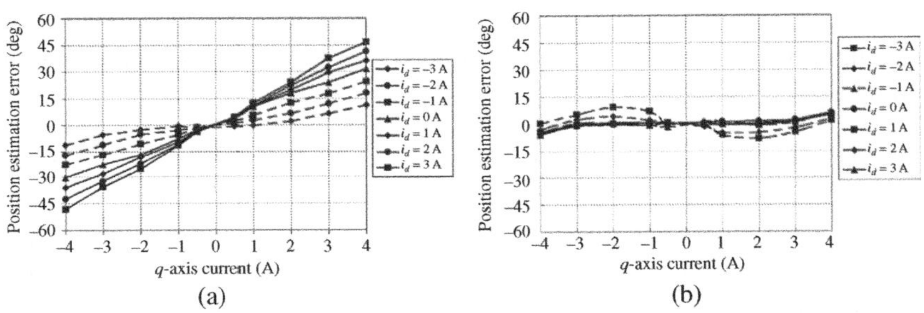

Figure 5.4 Position estimation errors with/without compensation. (a) Without compensation. (b) With compensation. (*Source:* [5]/IEEE.)

However, since the d-axis current is neglected and the error is approximated as a linear function of the q-axis current for simplicity, for the prototype machine at some load conditions, the estimation error is relatively large, which may deteriorate the estimation accuracy and stability of the sensorless control. If it is necessary, other higher order approximation functions may be used.

5.2.2.2 Indirect Compensation

An improved compensation method was proposed in [4]. Instead of directly using the offline measured position errors for compensation, an offline measured coupling factor between the HF inductances is used. Based on the pulsating voltage injection-based sensorless control, the observer input error signal ε is obtained by the linear combination of $\hat{I}_{dh}$ and $\hat{I}_{qh}$ as shown in Figure 5.5.

ε, λ are defined as:

$$\varepsilon = \hat{I}_{qh} + \lambda \hat{I}_{dh}, \quad \lambda = \frac{L_{dqh}}{L_{qh}} \tag{5.7}$$

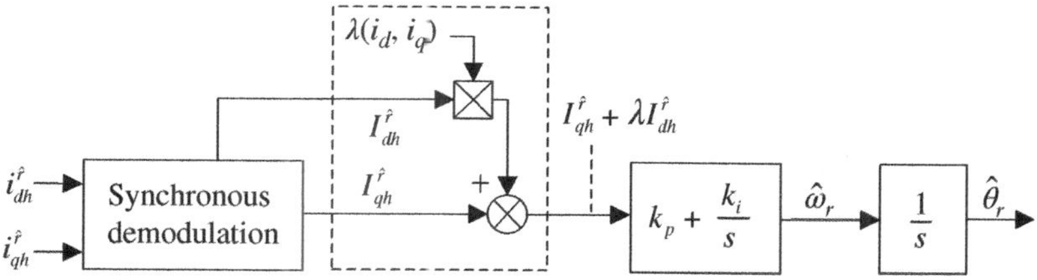

Figure 5.5 Cross-coupling effect compensation with offline measured λ [4].

The modified error signal ε is derived as:

$$\varepsilon = \hat{I}_{qh} + \lambda\hat{I}_{dh} \approx 2I_n \cos\theta_m \Delta\theta \tag{5.8}$$

Thus, by controlling $\varepsilon = 0$, the cross-coupling effect can be indirectly compensated. The coefficient λ can be predicted by FEA or offline measurement. In [4], the coupling factor λ is directly measured by driving the PMSM using the actual rotor position with an encoder. In this way, by injecting HF voltage $v_{dh} = V_h \sin(\omega_h t)$ on the actual d-axis, the high-frequency voltage equation can be expressed as:

$$\begin{bmatrix} v_{dh} \\ 0 \end{bmatrix} = \begin{bmatrix} L_{dh} & L_{dqh} \\ L_{qdh} & L_{qh} \end{bmatrix} p \begin{bmatrix} i_{dh} \\ i_{qh} \end{bmatrix} \tag{5.9}$$

The corresponding high-frequency d- and q-axis current components are obtained by solving (5.9) as:

$$i_{dh} = \frac{V_h L_{qh} \sin(\omega_h t)}{\omega_h \left(L_{dh} L_{qh} - L_{dqh}^2\right)}, \quad i_{qh} = \frac{-V_h L_{dqh} \sin(\omega_h t)}{\omega_h \left(L_{dh} L_{qh} - L_{dqh}^2\right)} \tag{5.10}$$

Then, the value of λ can be obtained from the ratio between measured values of i_{dh}, i_{qh} as:

$$\lambda = -\frac{i_{qh}}{i_{dh}} = \frac{L_{dqh}}{L_{qh}} \tag{5.11}$$

The measured results for the prototype IPMSM-I0 are shown in Figure 5.6 with a 20 V/500 Hz pulsating voltage injection into the d-axis. Considering a low-cost implementation, the value of λ can be simplified by representing the data in Figure 5.6 by the following simple function as:

$$\lambda = \begin{cases} -k_1 \cdot \hat{i}_q, & \hat{i}_d \geq 0 \\ -\left(k_1 + k_2 \cdot \hat{i}_d\right) \cdot \hat{i}_q, & \hat{i}_d < 0 \end{cases} \tag{5.12}$$

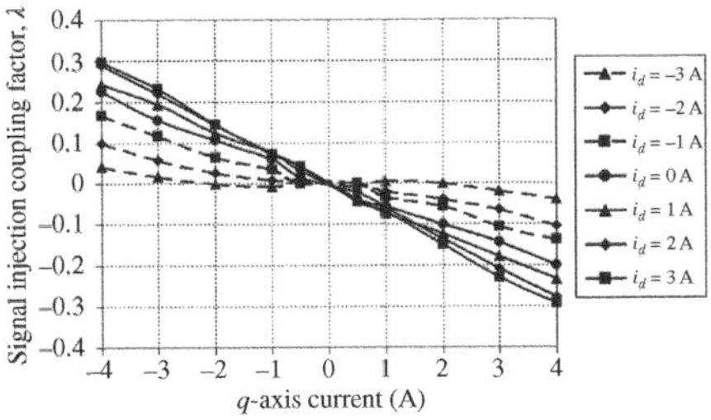

Figure 5.6 Measured λ. (*Source:* [4]/IEEE.)

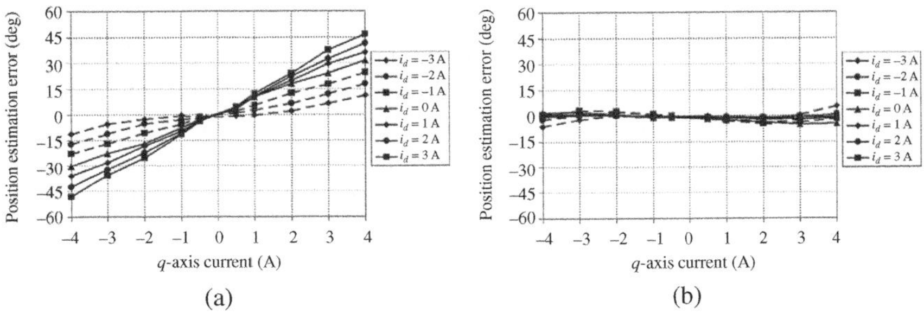

Figure 5.7 Position estimation with/without compensation. (a) Position estimation error without compensation. (b) Position estimation error with compensation. (*Source:* [4]/IEEE.)

where k_1 and k_2 are coefficients. For the prototype IPMSM-I, $k_1 = 0.06/\text{A}$, and $k_2 = 0.011/\text{A}^2$. The measured results in Figure 5.7 show the improved compensation performance compared to the results in Figure 5.4. In the experiment, a 35 V/330 Hz sinusoidal voltage is injected.

5.3 Machine Saliency and Load Effect

Machine saliency is essential for saliency-based sensorless control of permanent magnet synchronous machines. It is well known that the machine's saliency properties result from physical geometric saliency and magnetic saturation as well. The saturation saliency is closely related to machine load condition [4]. Therefore, it is desirable to evaluate the machine saliency information before saliency-based sensorless control is employed. Hence, in this section, the machine saliency is investigated including a saliency measurement approach [8] and a saliency circle-based method [9] to clearly show the saliency information.

5.3.1 Machine Saliency Investigation

In [8], the machine saliency can be measured with the aid of a position sensor. First, the rotor is locked at zero position ($\theta_r = 0$). Then, a pulsating sinusoidal voltage signal with the form of (5.13) is injected into the virtual dq-axes ($d^v q^v$), which rotate at a fixed speed (2 Hz) in clockwise direction, as shown in Figure 5.8. For the prototype IPMSM-I, the injected HF voltage is 35 V/330 Hz. The HF current response in the virtual dq-axis reference frame can be derived as (5.14).

$$\begin{bmatrix} v_{dh}^v \\ v_{qh}^v \end{bmatrix} = V_h \begin{bmatrix} \cos\alpha \\ 0 \end{bmatrix}, \quad \alpha = \omega_h t + \phi \tag{5.13}$$

$$\begin{bmatrix} i_{dh}^v \\ i_{qh}^v \end{bmatrix} = \begin{bmatrix} I_p + I_n \cos(2\Delta\theta + \theta_m) \\ I_n \sin(2\Delta\theta + \theta_m) \end{bmatrix} \cdot \sin\alpha \tag{5.14}$$

where $\Delta\theta = \theta_r - \theta_r^v = -\theta_r^v$.

$$\begin{bmatrix} |i_{dh}^v| \\ |i_{qh}^v| \end{bmatrix} = LPF\left(\begin{bmatrix} i_{dh}^v \\ i_{qh}^v \end{bmatrix} \cdot 2\sin\alpha \right) = \begin{bmatrix} I_p + I_n \cos(2\Delta\theta + \theta_m) \\ I_n \sin(2\Delta\theta + \theta_m) \end{bmatrix} \tag{5.15}$$

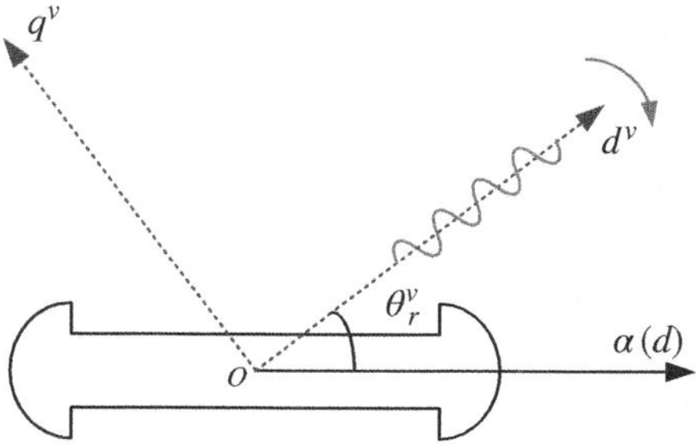

Figure 5.8 Machine saliency evaluation. (*Source:* [8]/IEEE.)

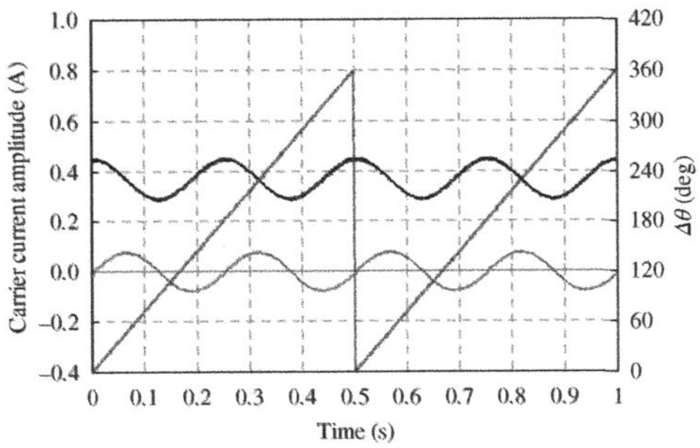

Figure 5.9 Measured variation of HF current amplitude with position (i_d = 0 A, i_q = 0 A) [9].

After demodulation, the position-dependent amplitude of the HF current can be obtained as (5.52) and is shown in Figure 5.9. The experimental result clearly shows that the amplitude of the HF current is modulated by the position difference $\Delta\theta$ between the virtual and real *dq*-axis reference frames. When the position difference is 0° or 180°, i.e. the virtual *d*-axis is aligned with the real *d*-axis, the *d*-axis current amplitude reaches the maximum value, while the *q*-axis current amplitude is close to zero. Furthermore, the amplitude modulation of the HF current undergoes two cycles per single electrical cycle of position, which reveals the angle ambiguity of π in sensorless position estimation.

5.3.2 Machine Saliency Circle

When the HF current amplitude variations are combined in Figure 5.10, the formed circle is designated as the machine's saliency circle, which clearly shows the machine saliency information [9]. Corresponding to (5.15), the center location of the machine saliency circle is determined by the value of I_p, while the radius of the saliency circle is dependent on the value of I_n. According to the definition of I_p and I_n (in Chapter 4), for a given injected HF voltage signal (V_h/ω_h), the location and radius of the saliency circle are only decided by the machine parameters L_p and L_n, respectively. Therefore, the scale of the saliency circle indicates the machine saliency level. The longer the radius of the saliency circles, the higher the saliency level, and vice versa.

In order to investigate the machine saliency information at different load conditions, a specific fundamental excitation can be applied to the machine. In the same way, the machine saliency information under different load conditions can be measured, as shown in Figures 5.11 and 5.12. The measured saliency circles

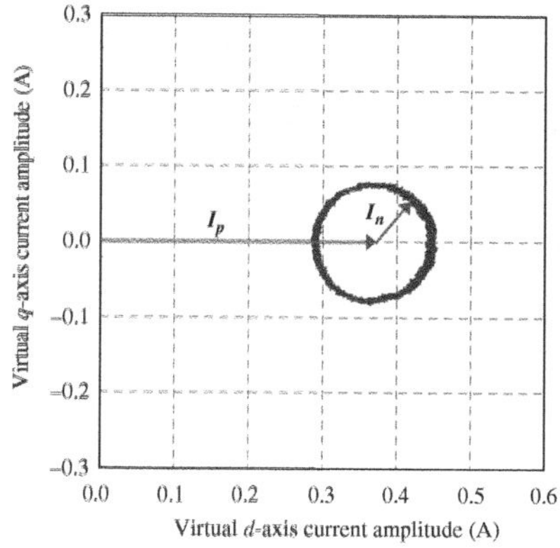

Figure 5.10 Measured machine saliency circle (i_d = 0 A, i_q = 0 A). (*Source:* [9]/IEEE.)

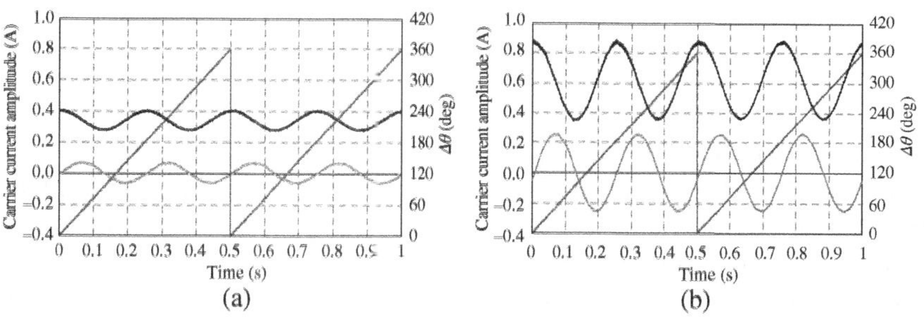

Figure 5.11 Measured variation of HF current amplitude against $\Delta\theta$ at different i_d (i_q = 0 A) [9]. (a) i_d = −4 A, i_q = 0 A. (b) i_d = 4 A, i_q = 0 A.

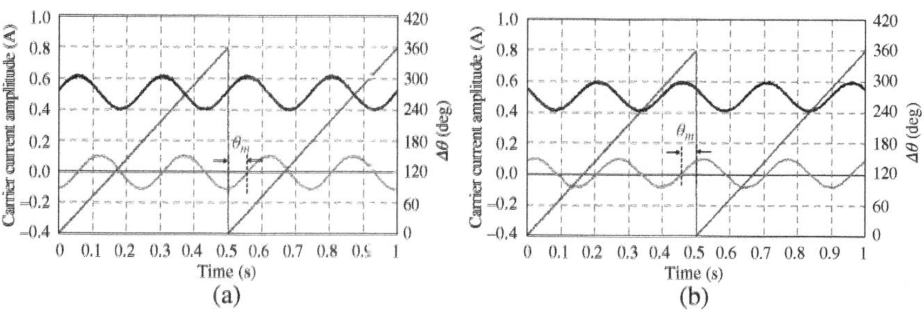

Figure 5.12 Measured variation of HF current amplitude against $\Delta\theta$ at different i_q (i_d = 0) [9]. (a) i_d = 0 A, i_q = −4 A. (b) i_d = 0 A, i_q = 4 A.

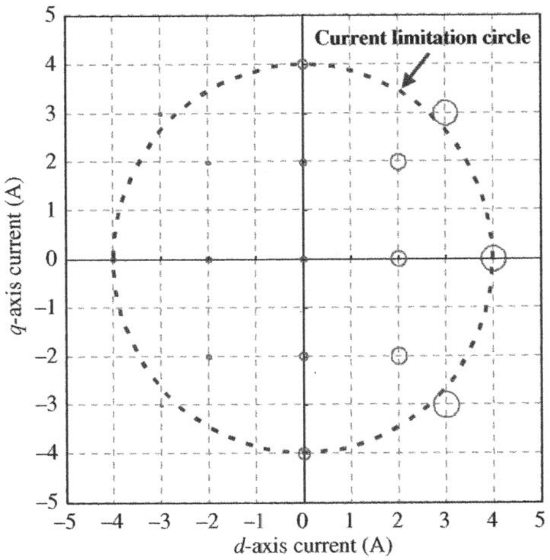

Figure 5.13 Summary of measured saliency circles with different fundamental excitations. (*Source:* [9]/IEEE.)

for the prototype IPMSM-I under different load conditions are summarized in Figure 5.13. From the experimental results in Figures 5.11–5.13, it is known that the radius of the saliency circle changes significantly with the d-axis current, which reveals that d-axis current makes more contributions to the machine saliency level. Positive d-axis current gives rise to an increase in the saliency, which is beneficial for saliency-based sensorless control. Moreover, the measured cross-saturation angle θ_m shows that the cross-saturation effect is mostly associated with q-axis current rather than d-axis current.

5.4 Multiple Saliency Effect

The machine saliency investigation presented above only reveals the existence of a single spatial saliency, i.e. L_{dh} and L_{qh} are constant for a given load condition and are independent of the rotor position. However, non-sinusoidal spatial distribution of the reluctance normally exists in a real machine due to the practical design. Hence, L_{dh} and L_{qh} are associated with the rotor position in practice, which can be modeled as a spatial Fourier series. The spatial saliency for sensorless position tracking is denoted as primary saliency, while other harmonic components are referred to as secondary saliency [10], [11]. The summation of primary and secondary saliencies is known as the multiple saliency effect. The HF inductance model considering the multiple saliency effect can be expressed as:

$$L_{\alpha\beta} = \begin{bmatrix} L_{sa} + \sum_{h=1}^{\infty} L_{\Delta h}\cos(2h\theta_r) & -\sum_{h=1}^{\infty} L_{\Delta h}\sin(2h\theta_r) \\ -\sum_{h=1}^{\infty} L_{\Delta h}\sin(2h\theta_r) & L_{sa} - \sum_{h=1}^{\infty} L_{\Delta h}\cos(2h\theta_r) \end{bmatrix} \tag{5.16}$$

where h is the harmonic order ($h = 1$ represents the primary saliency), and $L_{\Delta h}$ is the amplitude of the spatial inductance.

For rotating voltage injection [12], the HF currents will contain current harmonics caused by the secondary saliencies as

$$i_{\alpha\beta h}^{s} = I_p e^{j(\theta_h - \pi/2)} + \sum_{h=1}^{\infty} I_{nh} e^{-j(\theta_h - 2h\theta_r - \pi/2)}, I_{nh} = \frac{V_h L_{\Delta h}}{2\omega_h \left(L_{sa}^2 - L_{\Delta h}^2\right)} \tag{5.17}$$

For pulsating voltage injection, the estimated q-axis HF current is

$$\hat{i}_{qh} = \left[I_n \sin\left(2\theta_r - 2\hat{\theta}_r\right) + \sum_{h=1}^{\infty} I_{nh}\sin\left(2\theta_r - 2h\hat{\theta}_r\right)\right]\sin(\omega_h t) \tag{5.18}$$

The HF current components with $h > 1$ are called the secondary saliency currents. It is difficult to separate the primary currents with the secondary saliency currents as their frequencies are too close. Hence, there will be ripples in the estimated position. In [12], a neural network is introduced to decouple secondary saliencies. In [13], the multi-saliencies are measured offline with pulse voltage injection and used for online compensation. In [14], with the multi-signal injection method, multi-saliencies can be removed with only the primary saliency left. On the one hand, in general, it is hard to eliminate the influence of multi-saliencies in sensorless control and it is desirable to minimize the multi-saliencies in machine design. On the other hand, in many applications, due to relatively low magnitudes, multi-saliencies can be neglected in machines with high primary saliency.

Based on the pulsating sinusoidal signal injection method, the measured results of the HF current response are shown in Figure 5.14. The measurement is based on the prototype IPMSM-I (Appendix B). The variation of HF current amplitude at different load conditions can be depicted against the rotor position, as shown in Figure 5.14. The ripple term has the frequency of six times the fundamental frequency, which reveals that fourth and/or eighth harmonics are dominant. From Figure 5.14, the dc component of q-axis HF current amplitude is determined by ($I_n \sin\theta_m$) due to cross-coupling. Hence, the positive value indicates positive θ_m, and vice versa. Figure 5.14 shows θ_m has a near zero value at the condition of ($i_d = 0$ A, $i_q = 0$ A), while it has a positive value at $i_q = 2$ A and a negative value at $i_q = -2$ A.

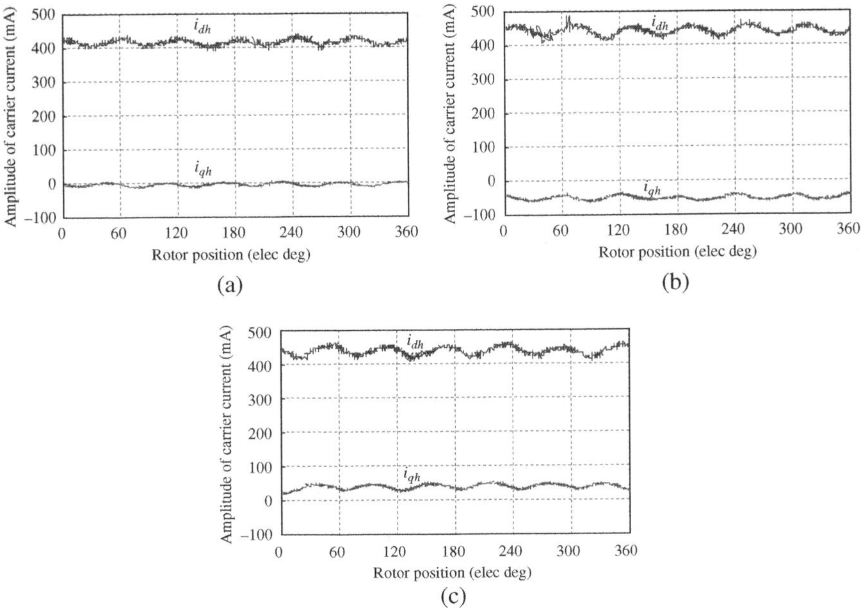

Figure 5.14 Multiple saliency evaluation [15]. (a) (i_d = 0 A, i_q = 0 A). (b) (i_d = 0 A, i_q = −2 A). (c) (i_d = 0 A, i_q = 2 A).

From Figure 5.14, different ripple amplitude can be observed at different load conditions, which reveals that the multiple saliency effect is also connected with load condition.

5.5 Asymmetric Parameters

In the previous chapters, HFSI methods assumed that the machine parameters are symmetrical. However, asymmetries often exist in three-phase PM machines, resulting in machine parameter asymmetries including phase resistances, inductances, and back-EMFs. Since the phase back-EMF and resistance are neglected in the HFSI model, only the inductance asymmetries need to be considered for HFSI methods. Therefore, in this section, the detailed modeling, comparison, and compensation strategies for machine parameter asymmetries are presented for the two typical HF injection methods, i.e. the rotating sinusoidal signal injection method and the pulsating sinusoidal signal injection method. It will be shown that with inductance asymmetry, the second harmonic position errors arise regardless of injection methods. In order to suppress the second harmonic position error due to inductance asymmetry, an online compensation strategy with dual frequency injection [16] is also presented.

5.5.1 High-Frequency Models with Machine Inductance Asymmetry

Practically, the inductance asymmetry may originate from mechanical manufacturing, machine eccentricity, winding fault [10], etc. Assuming that phase A self-inductance is unbalanced, the inductance matrix in the dq-axis can be expressed as:

$$L(d,q) = \begin{bmatrix} L_{dh} + \frac{2}{3}\Delta L \cos^2\theta_r & L_{dqh} - \frac{1}{3}\Delta L \sin 2\theta_r \\ L_{dqh} - \frac{1}{3}\Delta L \sin 2\theta_r & L_{qh} + \frac{2}{3}\Delta L \sin^2\theta_r \end{bmatrix} \tag{5.19}$$

where ΔL presents the asymmetry part for the self-inductance. Thus, with the pulsating signal injection, the HF current response neglecting phase resistance can be obtained as:

$$\begin{aligned} \hat{i}_{qh} \approx \frac{V_h \sin(\omega_h t)}{\omega_h L_{dh} L_{qh}} &\left[L_{sd} \sin 2\Delta\theta - L_{dqh} \cos 2\Delta\theta \right. \\ &\left. - \frac{1}{3}\Delta L \cos 2\theta_r \sin 2\Delta\theta + \frac{1}{3}\Delta L \sin 2\theta_r \cos 2\Delta\theta \right] \end{aligned} \tag{5.20}$$

Accordingly, the position error with one-phase inductance asymmetry for the pulsating injection can be obtained as:

$$\Delta\theta_{p\_L} \approx \frac{1}{2} \arctan \frac{L_{dqh}}{L_{sd}} + \frac{1}{2} \arctan \frac{\Delta L}{3\hat{L}_{sd}} \sin(2\theta_r + \delta_L) \tag{5.21}$$

where δ_L presents the phase angle under various inductance asymmetry conditions and $\hat{L}_{sd} = \sqrt{L_{dqh}^2 + L_{sd}^2}$.

Similarly, the position error with three-phase inductance asymmetry for the pulsating injection can be also obtained as (5.22).

$$\Delta\theta_{p\_L} = \frac{1}{2}\arctan \frac{L_{dqh} - \sum\limits_{i=1,2,3} \left\{ \frac{1}{3}\Delta L_i \sin 2\left(\theta_r - \frac{2}{3}\pi \cdot (i-1)\right) + \frac{2}{3}\Delta M_i \sin 2\left(\theta_r + \frac{2}{3}\pi \cdot i\right) \right\}}{L_{sd} - \sum\limits_{i=1,2,3} \left\{ \frac{1}{3}\Delta L_i \cos 2\left(\theta_r - \frac{2}{3}\pi \cdot (i-1)\right) + \frac{2}{3}\Delta M_i \cos 2\left(\theta_r + \frac{2}{3}\pi \cdot i\right) \right\}} \tag{5.22}$$

where ΔM presents the asymmetry part for the mutual-inductances, $i = 1$–3 for phases A–C, respectively.

Then, for rotating signal injection, the HF current with one-phase inductance asymmetry can be obtained as:

$$i_\alpha + ji_\beta = \frac{V_h}{\omega_h L_{dh} L_{qh}} \left[\left(L_{avg} + \frac{\Delta L}{3}\right) e^{j(\omega_h t - \pi/2)} - \frac{\Delta L}{3} e^{-j(\omega_h t - \pi/2)} + \hat{L}_{sd} e^{j(2\theta_r + \theta_m)} e^{-j(\omega_h t - \pi/2)}\right] \tag{5.23}$$

Similarly, the position error for rotating signal injection can be derived as:

$$\Delta\theta_{r\_L} \approx \frac{1}{2} \arctan \frac{L_{dqh}}{L_{sa}} + \frac{1}{2} \arctan \frac{\Delta L}{3\hat{L}_{sd}} \sin(2\theta_r + \delta_L) \tag{5.24}$$

for one-phase inductance asymmetry and (5.25) for three-phase inductance asymmetry.

$$\Delta\theta_{r\_L} = \frac{1}{2}\arctan \frac{L_{dqh} - \sum\limits_{i=1,2,3} \left\{\frac{1}{3}\Delta L_i \sin 2\left(\theta_r - \frac{2}{3}\pi \cdot (i-1)\right) + \frac{2}{3}\Delta M_i \sin 2\left(\theta_r + \frac{2}{3}\pi \cdot i\right)\right\}}{L_{sd} - \sum\limits_{i=1,2,3} \left\{\frac{1}{3}\Delta L_i \cos 2\left(\theta_r - \frac{2}{3}\pi \cdot (i-1)\right) + \frac{2}{3}\Delta M_i \cos 2\left(\theta_r + \frac{2}{3}\pi \cdot i\right)\right\}} \tag{5.25}$$

Thus, from the above derivations with inductance asymmetry in (5.21) and (5.24), it is noted that the second harmonic position errors are exactly the same for the two injection methods. Moreover, the second harmonic error for inductance asymmetry is not related to HF injection frequency, as can be seen from (5.21) and (5.24) [16].

5.5.2 Suppression of Position Errors Due to Inductance Asymmetry

In order to suppress the influence of machine inductance asymmetry, an online compensation strategy [16] with dual frequency injection is employed. By way of example, for rotating signal injection with one-phase inductance asymmetry, the negative sequence HF current can be expressed as:

$$i_{neg} = \frac{V_h e^{-j(\omega_h t - \pi/2)}}{\omega_h L_{dh} L_{qh}} \left[-\frac{\Delta L}{3} + \hat{L}_{sd} e^{j(2\theta_r + \theta_m)}\right] \tag{5.26}$$

It is noted that the additional part caused by inductance asymmetry presents as the dc component, and the amplitude is only related to the ratio of V_h/ω_h. Therefore, if the dual frequency HF signal is injected, i.e.

$$\begin{bmatrix} v_\alpha \\ v_\beta \end{bmatrix} = V_h \begin{bmatrix} \cos\omega_h t \\ \sin\omega_h t \end{bmatrix} + V_{h1} \begin{bmatrix} \cos\omega_{h1} t \\ \sin\omega_{h1} t \end{bmatrix} \tag{5.27}$$

where $V_h/\omega_h = V_{h1}/\omega_{h1}$, the resultant negative sequence HF currents can be obtained as:

$$
\begin{aligned}
i_{neg} \approx\ & \frac{V_h e^{-j(\omega_h t - \pi/2)}}{\omega_h L_{dh} L_{qh}} \left[-\frac{\Delta L}{3} + \hat{L}_{sd} e^{j(2\theta_r + \theta_m)} \right] \\
& + \frac{V_{h1} e^{-j(\omega_{h1} t - \pi/2)}}{\omega_{h1} L_{dh} L_{qh}} \left[-\frac{\Delta L}{3} + \hat{L}_{sd} e^{j(2\theta_r + \theta_m)} \right]
\end{aligned}
\tag{5.28}
$$

With the signal demodulation processing, the HF signal can be further expressed as:

$$
I_{neg} = \text{LPF1}\left(i_{neg} e^{j(\omega_h t - \pi/2)} \right) = \frac{V_h}{\omega_h L_{dh} L_{qh}} \left(-\frac{\Delta L}{3} + \hat{L}_{sd} e^{j(2\theta_r + \theta_m)} \right) \tag{5.29}
$$

$$
I_{neg1} = \text{LPF2}\left(i_{neg} e^{j(\omega_{h1} t - \pi/2)} \right) = \frac{V_{h1}}{\omega_{h1} L_{dh} L_{qh}} \left(-\frac{\Delta L}{3} + \hat{L}_{sd} e^{j(2\theta_r + \theta_m)} \right) \tag{5.30}
$$

Since the injected second frequency (ω_{h1}) signal is only used for error compensation for the inductance asymmetry, the cutoff frequency of LPF2 can be selected much lower than that of LPF1, i.e. (5.30) can be expressed as:

$$
I_{neg1} = \text{LPF2}\left(i_{neg} e^{j(\omega_{h1} t - \pi/2)} \right) \approx -\frac{V_{h1}}{\omega_{h1} L_{dh} L_{qh}} \frac{\Delta L}{3} \tag{5.31}
$$

Thus, it can be derived that:

$$
I_{neg2} = I_{neg} - I_{neg1} \approx \frac{V_h}{\omega_h L_{dh} L_{qh}} \hat{L}_{sd} e^{j(2\theta_r + \theta_m)} \tag{5.32}
$$

Compared to (5.26), the asymmetry component in (5.32) has been largely suppressed with the compensation strategy. It should be noted that since the added LPF2 is only used for the second frequency (ω_{h1}) HF signals, the bandwidth of the main position estimation loop is unaffected. The block diagram for rotating signal injection with the inductance asymmetry compensation is shown in Figure 5.15.

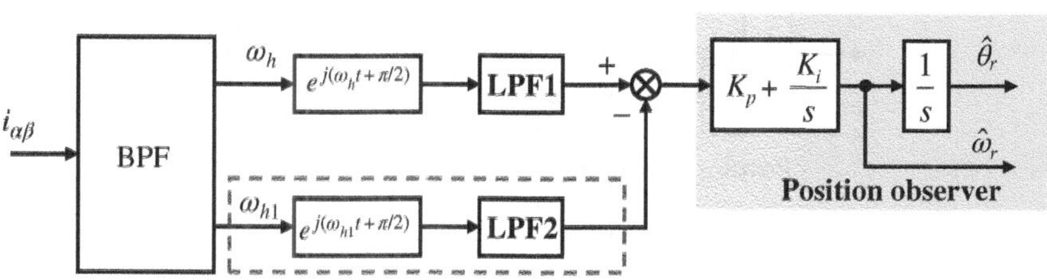

Figure 5.15 Dual frequency injection compensation strategy for rotating signal injection [16].

Similarly, for the pulsating signal injection, the dual frequency signals are injected as:

$$\begin{bmatrix} \hat{v}_{dh} \\ \hat{v}_{qh} \end{bmatrix} = \begin{bmatrix} V_h \cos \omega_h t \\ 0 \end{bmatrix} + \begin{bmatrix} V_{h1} \cos \omega_{h1} t \\ 0 \end{bmatrix} \tag{5.33}$$

Accordingly, the HF current can be expressed as:

$$\begin{aligned} \hat{i}_{dqh} = \hat{i}_{dh} + j\hat{i}_{qh} = \left(\frac{V_h \sin \omega_h t}{\omega_h L_{dh} L_{qh}} + \frac{V_{h1} \sin \omega_{h1} t}{\omega_{h1} L_{dh} L_{qh}} \right) \\ \left[L_{sa} + \frac{1}{3}\Delta L + \hat{L}_{sd} e^{j(2\Delta\theta + \theta_m)} - \frac{1}{3}\Delta L e^{j(-2\theta_r + 2\Delta\theta)} \right] \end{aligned} \tag{5.34}$$

Then, the signal demodulation as shown in Figure 5.16 can be expressed as:

$$\begin{aligned} \hat{I}_{qh} &= \mathrm{Im}\left\{ \left[\mathrm{LPF1}\left(\hat{i}_{dqh} \sin \omega_h t \cdot e^{j2\hat{\theta}_r} \right) - \mathrm{LPF2}\left(\hat{i}_{dqh} \sin \omega_{h1} t \cdot e^{j2\hat{\theta}_r} \right) \right] \cdot e^{j-2\hat{\theta}_r} \right\} \\ &= \frac{V_h \hat{L}_{sd} \sin(2\Delta\theta + \theta_m)}{2\omega_h L_{dh} L_{qh}} \end{aligned} \tag{5.35}$$

where I_m presents the imaginary part of the estimated dq-axis HF current component. From the above derivations, it can be seen that the asymmetry component

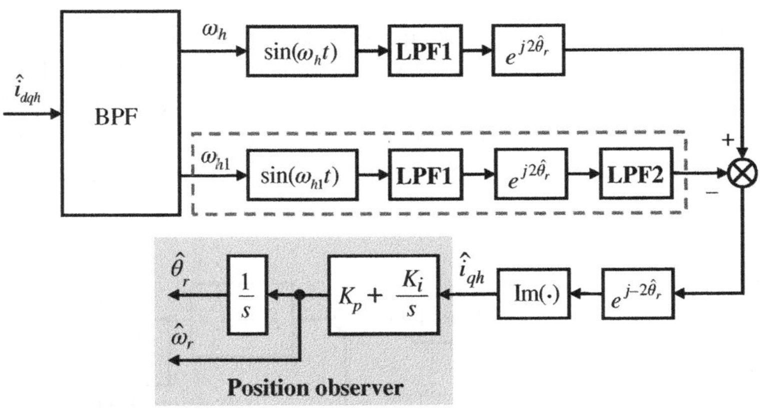

Figure 5.16 Dual frequency injection compensation strategy for pulsating signal injection. (*Source:* [16]/IEEE.)

for pulsating injection is also significantly suppressed with the dual frequency injection strategy compared to (5.21).

5.5.3 Experimental Results

The experiments on the prototype IPMSM-I (Appendix B) are implemented on a dSPACE platform to validate the above theoretical analyses. To simulate the asymmetry conditions discussed in this section, extra phase inductances are connected in series to the phases of the IPMSM.

5.5.3.1 Position Estimation Under Inductance Asymmetry

For inductance asymmetry with added 2 mH inductance (~4% phase inductance asymmetry), the position errors at 500 Hz injection for the two methods are shown in Figures 5.17 and 5.18. It can be clearly seen that the two position errors are approximately the same and also with second harmonic oscillations. The second harmonic position errors at different injection frequencies under different load conditions are also shown in Figure 5.19, and slightly increased errors are observed under loaded condition, which are all consistent with the above analyses.

5.5.3.2 The Second Harmonic Oscillating Error Suppression

For inductance asymmetry, with the compensation methods based on dual frequency injection (12 V/400 Hz and 18 V/600 Hz), the oscillating harmonic errors are significantly reduced under various load conditions, respectively, Figures 5.20 and 5.21. The spectra of these position estimation errors before and after the compensation are also illustrated in Figure 5.22, where the second harmonic errors have been significantly suppressed.

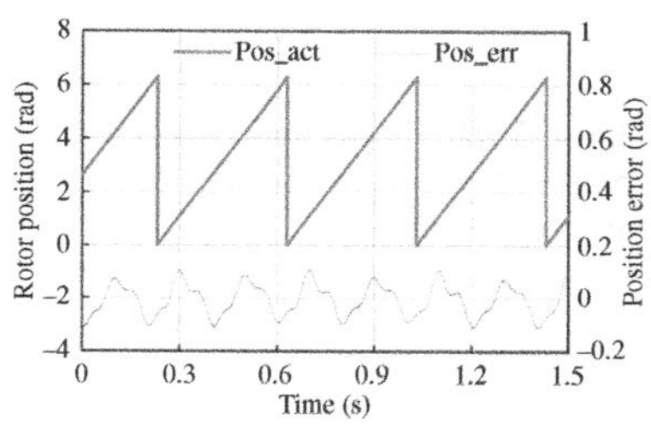

Figure 5.17 Position errors with ~4% self-inductance asymmetry of one phase for rotating signal injection. (*Source:* [16]/IEEE.)

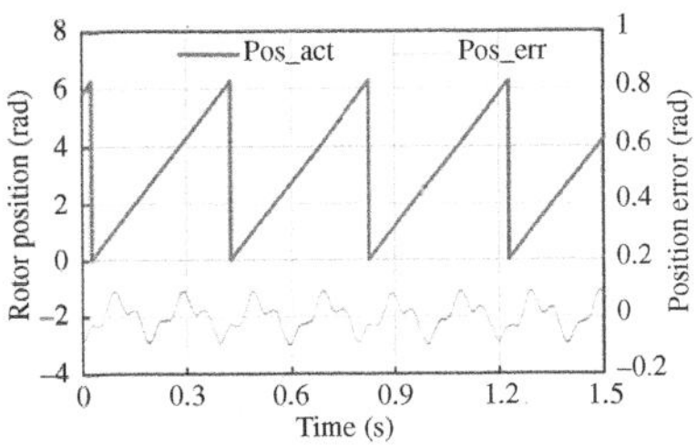

Figure 5.18 Position errors with ~4% self-inductance asymmetry of one phase for pulsating signal injection. (*Source:* [16]/IEEE.)

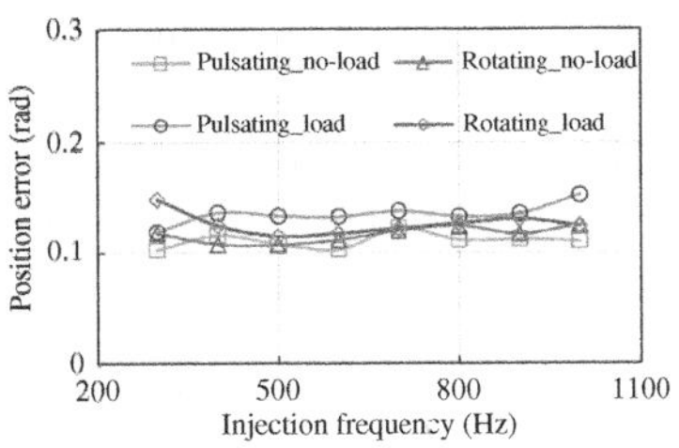

Figure 5.19 The second harmonic position errors with ~4% self-inductance asymmetry at different injection frequencies with/without load. (*Source:* [16]/IEEE.)

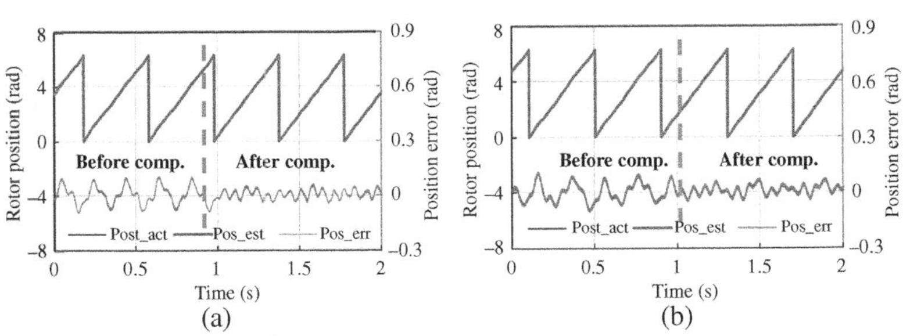

Figure 5.20 The second harmonic error suppression for inductance asymmetry with rotating signal injection. (a) No-load condition. (b) Load condition (i_q = 4 A). (*Source:* [16]/IEEE.)

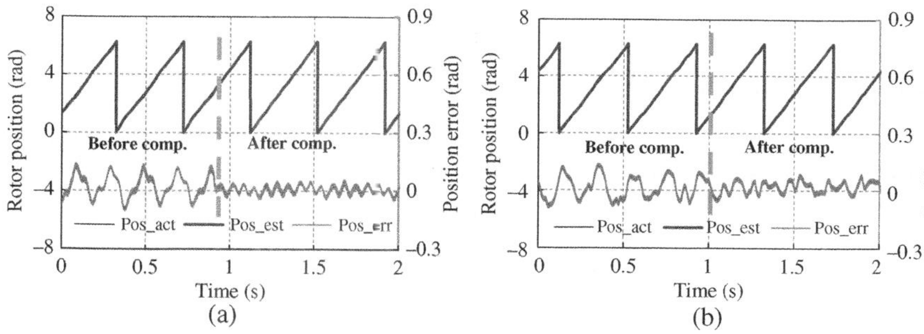

Figure 5.21 The second harmonic error suppression for inductance asymmetry with pulsating signal injection. (a) No-load condition. (b) Load condition (i_q = 4 A). (*Source:* [16]/IEEE.)

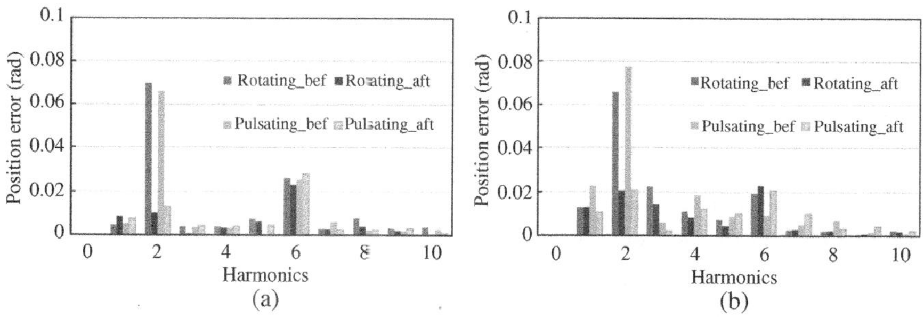

Figure 5.22 Spectra of position error before/after compensation for inductance asymmetry. (a) No-load condition. (b) Load condition (i_q = 4 A). (*Source:* [16]/IEEE.)

5.6 Inverter Nonlinearity Effects

For HFSI-based sensorless control introduced in Chapter 4, an ideal inverter is assumed. However, in practical implementation, nonlinear effects in a real inverter will exert negative impacts on the injected HF voltage introducing additional components in the HF current response, eventually deteriorating the sensorless performance [17], [18]. Therefore, inverter nonlinearity effects on HFSI sensorless control are introduced in this section.

Inverter nonlinear effects play a significant role in the distortion of voltage. They can be viewed as a disturbance voltage generator as shown in Figure 5.23. The distortion of voltage can be caused by several inverter nonlinearities, including [17]

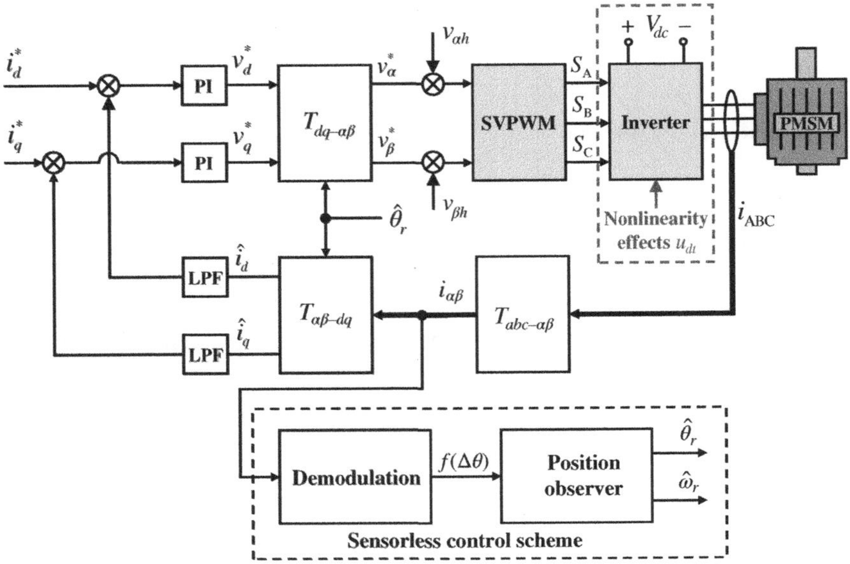

Figure 5.23 Voltage distortion generated by inverter nonlinearity effects.

- Deadtime
- Parasitic capacitance
- Turn-on and turn-off delays
- Gate drive circuit delays
- Voltage drops in the power devices
- Zero-current clamping
- Short-pulse suppression
- Others

The resultant voltage distortion introduced by inverter nonlinear effects consists of fundamental and high-frequency components. The fundamental disturbance voltage, which leads to fundamental current distortion, has been well documented in previous research work [19], [20]. For HFSI based sensorless techniques, the introduced high-frequency disturbance voltage is the more important factor, which results in additional HF current components, eventually giving rise to significant position estimation error [17]. Moreover, it is proved that the deadtime and parasitic capacitance effects are the major source of HF voltage distortion [17], which will be demonstrated later in this section.

Therefore, this section will focus on the inverter nonlinearity effects in HFSI sensorless control methods, including the mechanism of inverter nonlinearity, HF voltage distortion, and HF current distortion models with experimental results. Moreover, corresponding compensation schemes are also included.

5.6.1 Mechanism

For HFSI-based sensorless techniques, the deadtime and parasitic capacitance of semiconductors have been proven to be the major source of HF voltage distortion [17], which are introduced in detail in this section.

5.6.1.1 Deadtime

Considering the inherent turn-off delay of power devices, the deadtime is essential to avoid a shoot-through fault. When the two power devices in one phase leg are switched, the deadtime t_{dt} must be inserted to guarantee that the device commanded to turn off has been switched off effectively, before the counterpart is commanded to turn on. Together with turn-on and turn-off delay, an error voltage pulse would be generated due to the unwanted time delays, as shown in Figures 5.24 and 5.25. It should be noted that the sign of the error voltage pulse

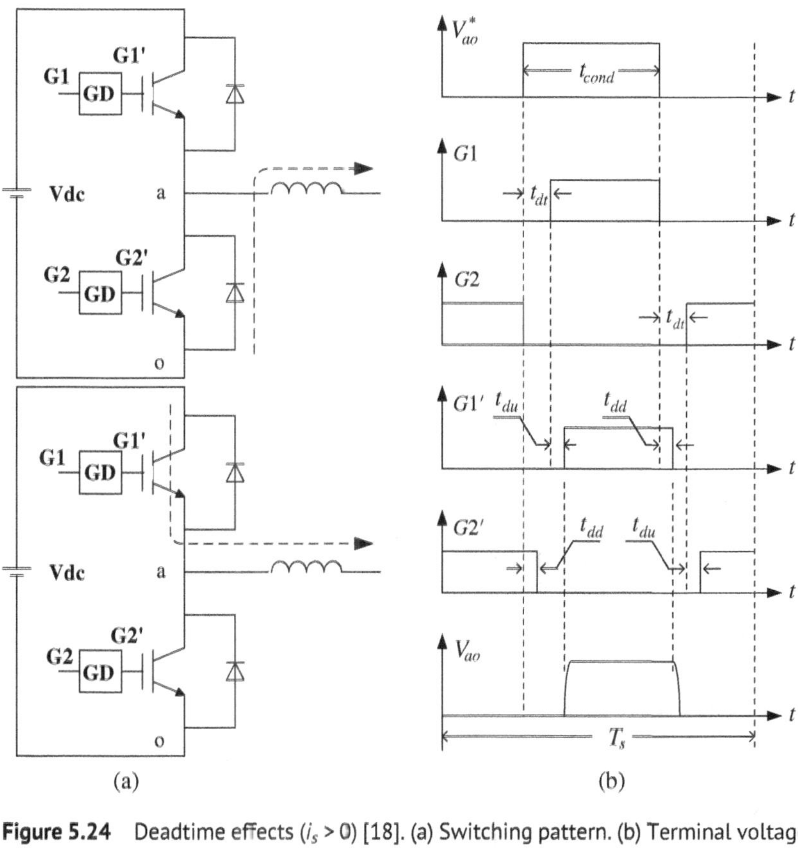

Figure 5.24 Deadtime effects ($i_s > 0$) [18]. (a) Switching pattern. (b) Terminal voltage error.

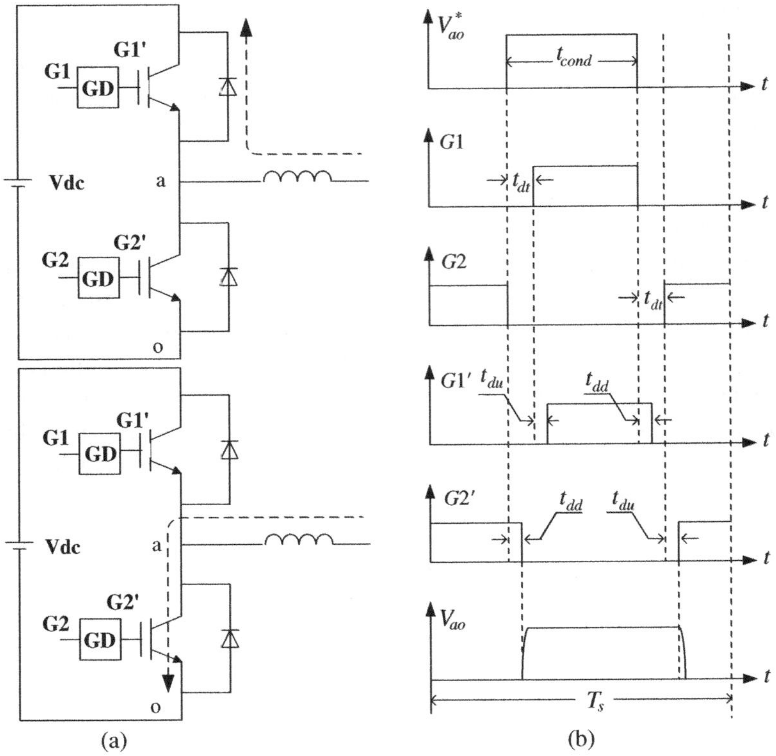

Figure 5.25 Deadtime effects ($i_s < 0$) [18]. (a) Switching pattern. (b) Terminal voltage error.

is opposite to the current polarity. The introduced terminal voltage error can be averaged in one PWM period, as given by

$$V_{dt} = V_{ao} - V_{ao}^* = -\text{sign}(i_s) \cdot V_{dc} \frac{t_{dt} + t_{du} - t_{dd}}{T_s} = -\text{sign}(i_s) \cdot \Delta V \quad (5.36)$$

where i_s is the phase current, V_{dc} is the dc bus voltage, T_s is the PWM period, t_{du} and t_{dd} are the turn-on and turn-off delays, respectively. Obviously, since ΔV is constant for given dc bus voltage and PWM frequency, the average terminal voltage error is only determined by the related phase current polarity, instead of current amplitude.

5.6.1.2 Parasitic Capacitance Effects

The parasitic capacitance of power devices is shown in Figure 5.26, in which the current flow path for positive current is indicated by the dashed line. When the current is transferred from the free-wheel diode to the IGBT, it can be

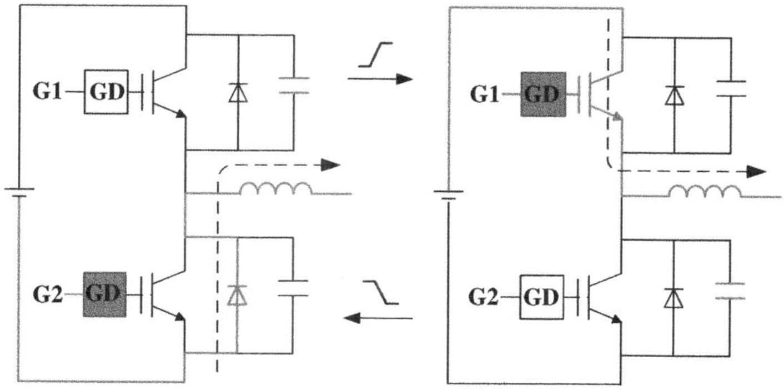

Figure 5.26 Current flowing paths with consideration of parasitic capacitance ($i_s > 0$). (*Source:* [18]/IEEE.)

accomplished immediately. The reason is that IGBTs are active switchers, whose turn-on can provide a low resistance path for immediate charging or discharging of parasitic capacitance. Consequently, as indicated by control signals given in Figures 5.27a and 5.28a, the terminal voltage slope is fairly steep and almost independent of the current level, as shown by the rising edge in Figure 5.27b and the falling edge in Figure 5.28b. However, a different phenomenon can be observed in the current transferring from the IGBT to the free-wheel diode, which is initialized by switching off the IGBT. The falling edge in Figure 5.27b and the rising edge in Figure 5.28b show the terminal voltage slope in this case. Due to the parasitic capacitance effect, the slope of the terminal voltage is significantly dependent on the current level [17].

5.6.2 HF Voltage Distortion

In order to obtain the relationship between terminal voltage error and phase current, the measured data under the condition of 2.0 μs deadtime, 10 kHz PWM switching frequency, and 150 V dc bus voltage can be used to construct the curve, as indicated by the solid line in Figure 5.29.

It can be found that the terminal voltage error tends to saturation at a higher current level. The value of ΔV is determined by the deadtime for a given dc bus voltage and PWM frequency. At a lower phase current, however, the terminal voltage error is significantly dependent on the instantaneous current level and the parasitic capacitance. The lower the parasitic capacitance, the steeper the slope of the terminal voltage error. Without parasitic capacitors, the slope-change behavior of the terminal voltage error at a lower current level becomes a step-change at the zero current point.

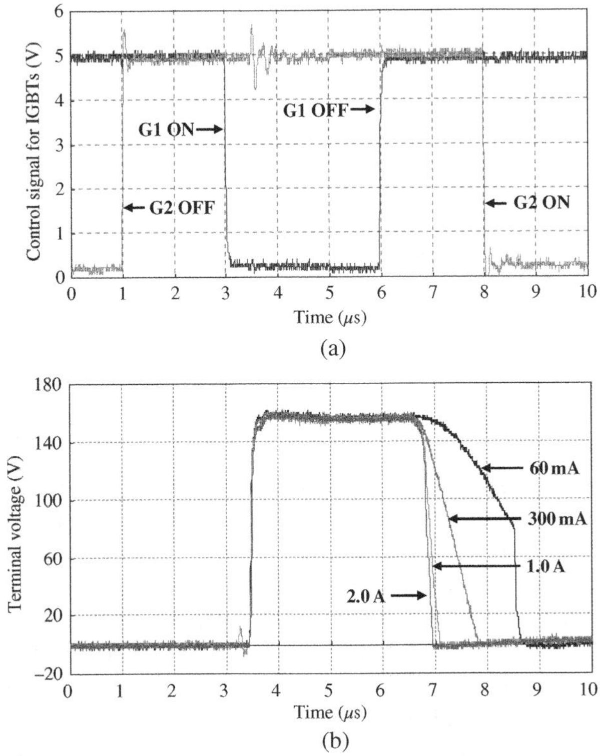

Figure 5.27 Measured terminal voltage (t_{dt} = 2 μs, i_s > 0). (a) Control signals for IGBTs. (b) Terminal voltage. (*Source:* [18]/IEEE.)

From Figure 5.29, the total current response consists of fundamental and HF current components. Hence, the terminal voltage error in the arbitrary phase "X" can be expressed as:

$$\Delta v_{XO} = f(i_X) = f\left(i_{Xf} + i_{Xh}\right) \tag{5.37}$$

where i_X is the total phase current, i_{Xf} is the fundamental current component, and i_{Xh} is the HF current components. When the HF current is much smaller than the fundamental current, the above equation can be approximated as:

$$\Delta v_{XO} \approx f\left(i_{Xf}\right) + f'\left(i_{Xf}\right) \cdot i_{Xh} \approx \mathrm{sign}\left(i_{Xf}\right) \cdot \Delta V + R_{Xh} \cdot i_{Xh} \tag{5.38}$$

where R_{Xh} is designated as the equivalent HF resistance of the inverter.

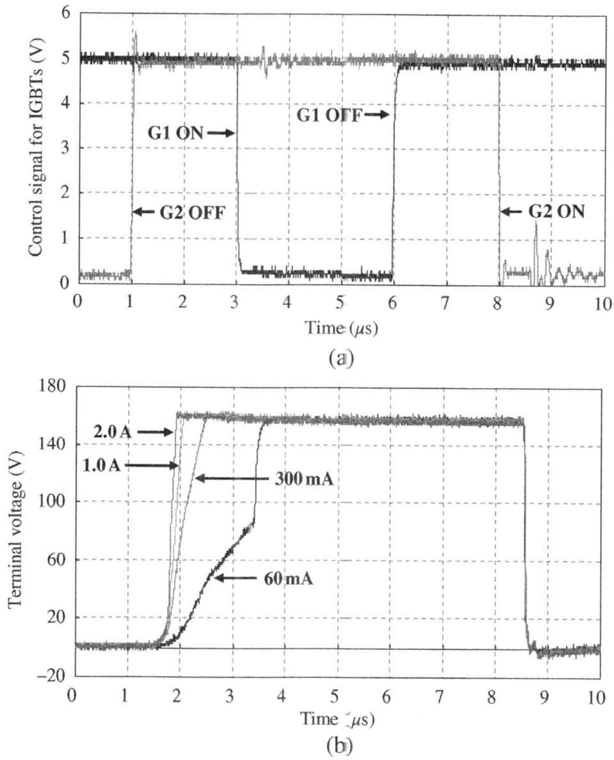

Figure 5.28 Measured terminal voltage (t_{dt} = 2 μs, $i_s < 0$). (a) Control signals for IGBTs. (b) Terminal voltage. (*Source:* [18]/IEEE.)

Based on (5.38), R_{Xh} is calculated and is shown by the dashed line in Figure 5.29. It can be concluded that the equivalent HF resistance of the inverter, determined by the characteristic of terminal voltage error, is an extremely nonlinear resistance. For current higher than 0.5 A, R_{Xh} becomes nearly zero, which implies that the equivalent HF resistance of the inverter only exists near the zero-current region and reaches its maximum value at zero fundamental current. Moreover, the disturbance voltage can be expressed as a phase voltage error vector as

$$\begin{aligned}\Delta v &= \frac{2}{3}\left(\Delta v_{AO} + \Delta v_{EO} \cdot e^{j2\pi/3} + \Delta v_{CO} \cdot e^{j4\pi/3}\right) = \frac{2}{3}\left(\Delta v_A + \Delta v_B \cdot e^{j2\pi/3} + \Delta v_C \cdot e^{j4\pi/3}\right) \\ &= \Delta v_f + \Delta v_h \end{aligned} \tag{5.39}$$

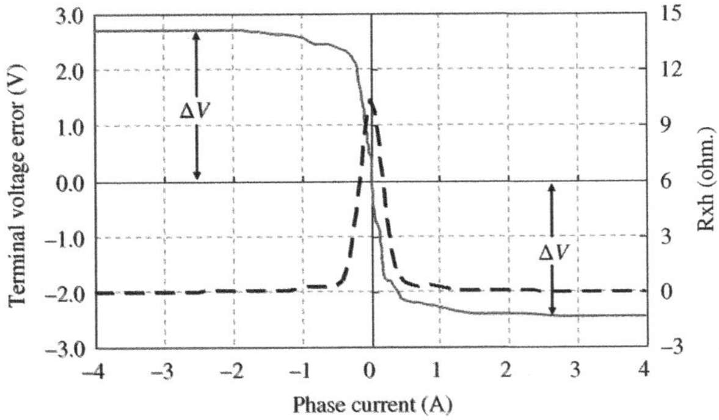

Figure 5.29 Measured terminal voltage error (t_{dt} = 2 μs, V_{dc} = 150 V). (*Source:* [18]/IEEE.)

where v_A, v_B, and v_C are the phase voltages, Δv_f is the fundamental disturbance voltage component, and Δv_h is the high-frequency disturbance voltage component, as defined by

$$\begin{cases} \Delta v_f = -\dfrac{2}{3}\Delta V\left[\text{sign}\left(i_{Af}\right) + \text{sign}\left(i_{Bf}\right)\cdot e^{j2\pi/3} + \text{sign}\left(i_{Cf}\right)\cdot e^{j4\pi/3}\right] \\ \Delta v_h = -\dfrac{2}{3}\left[R_{Ah}\cdot i_{Ah} + R_{Bi}\cdot i_{Bh}\cdot e^{j2\pi/3} + R_{Ch}\cdot i_{Ch}\cdot e^{j4\pi/3}\right] \end{cases} \tag{5.40}$$

In (5.40), for HFSI sensorless control, the introduced HF disturbance voltage Δv_h is the major contributor to HF current distortion [17]. As shown in Figure 5.29, when the fundamental phase current is in the vicinity of zero, the related HF resistance has a nonzero value. Therefore, the injected HF voltage vector Δv_h will be affected six times in one fundamental cycle [17] according to (5.40). As a result, it gives rise to a visible perturbation on the HF current response in the frequency domain, as shown in Figure 5.30. Moreover, in Figure 5.30, a larger deadtime causes more distortion.

5.6.3 HF Current Distortion

When the HF voltage is distorted, the resultant HF current will also have distortion and thus affect the HFSI sensorless control estimation performance. In this section, the HF current distortion due to inverter nonlinearity is described for rotating signal injection [17] and pulsating signal injection [21].

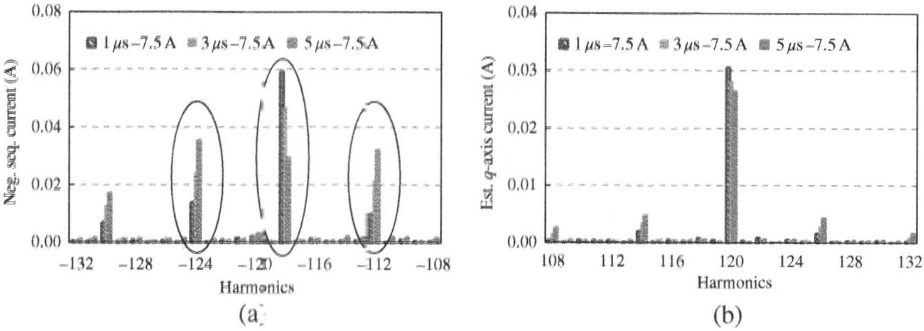

Figure 5.30 Measured spectra of HF current responses with different deadtimes [21]. (a) Rotating signal injection. (b) Pulsating signal injection.

5.6.3.1 Rotating Signal Injection-Based Method

Assuming the phase A current is crossing zero, then R_{Ah} has a nonzero value while $R_{Bh} = R_{Ch} = 0$. In this case, the HF disturbance voltage vector defined by (5.40) can be simplified to:

$$\Delta v_h = -\frac{2}{3} R_{Ah} \cdot i_{Ah} \tag{5.41}$$

The real HF current response becomes different from the ideal expression due to inverter nonlinearity effects. Considering its dominant contribution to the total HF current response, the ideal HF current can be substituted into (5.41),

$$\begin{aligned}\Delta v_h = &-\frac{1}{3} R_{Ah} I_p e^{j(\alpha - \pi/2)} - \frac{1}{3} R_{Ah} I_n e^{j(\alpha - 2\theta_r - \pi/2)} - \frac{1}{3} R_{Ah} I_p e^{j(-\alpha + \pi/2)} \\ &-\frac{1}{3} R_{Ah} I_n e^{j(-\alpha + 2\theta_r + \pi/2)}\end{aligned} \tag{5.42}$$

Equation (5.42) shows that the HF disturbance voltage introduced by the inverter nonlinearity effects consists of four terms. The first two terms are positive sequence vectors, while the last two terms are negative sequence vectors. Each voltage term generates corresponding positive and negative sequence HF currents. Hence, the resultant total HF current response can be calculated and summarized in Table 5.2. For the relatively small amplitudes of I_n, the disturbance HF current components with gray background in Table 5.2 can be neglected.

Table 5.2 Total HF current components ($R_{ah} \neq 0$).

Voltage term	HF current I_{pos} (positive sequence)	HF current I_{neg} (negative sequence)
$V_h e^{j\alpha}$	$I_p e^{j(\alpha - \pi/2)}$	$I_n e^{j(-\alpha + 2\theta_r + \pi/2)}$
$-\frac{R_{Ah} I_p}{3} e^{j(\alpha - \pi/2)}$	$-\frac{R_{Ah} I_p^2}{3V_h} e^{j(\alpha - \pi)}$	$-\frac{R_{Ah} I_p I_n}{3V_h} e^{j(-\alpha + 2\theta_r + \pi)}$
$-\frac{R_{Ah} I_n}{3} e^{j(\alpha - 2\theta_r - \pi/2)}$	$-\frac{R_{Ah} I_p I_n}{3V_h} e^{j(\alpha - 2\theta_r - \pi)}$	$-\frac{R_{Ah} I_n^2}{3V_h} e^{j(-\alpha + 4\theta_r + \pi)}$
$-\frac{R_{Ah} I_p}{3} e^{j(-\alpha + \pi/2)}$	$-\frac{R_{Ah} I_p I_n}{3V_h} e^{j(\alpha + 2\theta_r - \pi)}$	$-\frac{R_{Ah} I_p^2}{3V_h} e^{j(-\alpha + \pi)}$
$-\frac{R_{Ah} I_n}{3} e^{j(-\alpha + 2\theta_r + \pi/2)}$	$-\frac{R_{Ah} I_n^2}{3V_h} e^{j(\alpha - \pi)}$	$-\frac{R_{Ah} I_p I_n}{3V_h} e^{j(-\alpha + 2\theta_r + \pi)}$

(*Source:* [18]/IEEE.)

As introduced in Chapter 4, a parameter of ε constructed from negative sequence HF current is normally used to estimate the position information, as given by

$$\varepsilon = -\sin\left(-\alpha + 2\hat{\theta}_r + \frac{\pi}{2}\right) I_{neg\_\alpha} + \cos\left(-\alpha + 2\hat{\theta}_r + \frac{\pi}{2}\right) I_{neg\_\beta} \tag{5.43}$$

where I_{neg_α} is the α-axis component of the total negative sequence HF current and I_{neg_β} is the β-axis component of the total negative sequence HF current in the stationary reference frame. $\hat{\theta}_r$ is the estimated rotor position.

When the additional HF current components resulting from inverter nonlinearity effects, shown in Table 5.2, are considered, (5.43) can be re-written as

$$\varepsilon \approx I_n \sin(2\Delta\theta) - \frac{R_{Ah} I_p^2}{3V_h} \cos\left(2\hat{\theta}_r\right) - \frac{2R_{Ah} I_p I_n}{3V_h} \cos(2\Delta\theta) \tag{5.44}$$

Furthermore, when the rotor position estimation error $\Delta\theta$ is sufficiently small, considering that the real rotor position is close to zero in this case, (5.44) can be approximated to

$$\varepsilon \approx 2I_n \Delta\theta - \frac{R_{Ah}}{3V_h} \left(I_p^2 + 2I_p I_n\right) \tag{5.45}$$

The same conclusion can be drawn for the cases of $R_{Bh} \neq 0$ and $R_{Ch} \neq 0$. From (5.45), it can be seen that ε is associated with both estimated position error and inverter nonlinearity effects. Inevitably, the inverter nonlinearity effects are adverse to position estimation, which would give rise to a sixth harmonic on the estimated position information due to the nonlinear behavior of the inverter equivalent HF resistance.

5.6.3.2 Pulsating Signal Injection-Based Method

Similarly, assume that the phase A current is crossing zero and then R_{Ah} has a non-zero value while $R_{Bh} = R_{Ch} = 0$. In this case, the dq-axis HF disturbance voltages can be derived as [21]:

$$\begin{aligned} \Delta v_{dh} &= \frac{2}{3} R_{Ah} i_{Ah} \cos\,\theta_r \\ \Delta v_{qh} &= \frac{2}{3} R_{Ah} i_{Ah} \sin\,\theta_r \end{aligned} \tag{5.46}$$

Accordingly, the q-axis HF current can be further described as:

$$\begin{aligned} \hat{i}_{qh} &= \frac{\left(-L_d + L_q\right)}{2\omega_h L_d L_q} \sin(2\Delta\theta)(V_h \sin\alpha + \Delta v_{dh}) \\ &+ \frac{\left(L_d + L_q\right) + \left(-L_d + L_q\right)\cos(2\Delta\theta)}{2\omega_h L_d L_q} \Delta v_{qh} \end{aligned} \tag{5.47}$$

For the pulsating signal injection-based method, the q-axis HF current is normally controlled to zero to track the position. From (5.47), it can be seen that the voltage distortion in the q-axis voltage due to inverter nonlinearity will mainly cause position estimation error.

5.6.3.3 Experiment Results

The inverter nonlinearity gives rise to a visible perturbation on the HF current responses in the frequency domain, as shown by the measured results in Figure 5.30. The measurement is carried out based on the prototype IPMSM-II (Appendix B). The injection frequency is 600 Hz with the fundamental frequency at 5 Hz. The tests are under full load condition with different deadtimes. As shown in Figure 5.30, the inverter nonlinearity effects cause an increase in extra HF current harmonics and reduce the main component that is used for position estimation, thus deteriorating the position estimation. Furthermore, it is noticed that the pulsating-based method is much less sensitive compared to the rotating-based method.

Moreover, the inverter nonlinearity effects on position estimation are also shown in Figure 5.31. Clearly, the sixth order harmonic is introduced in the position estimation error and it becomes larger as deadtime T_{dt} increases. Similar to the

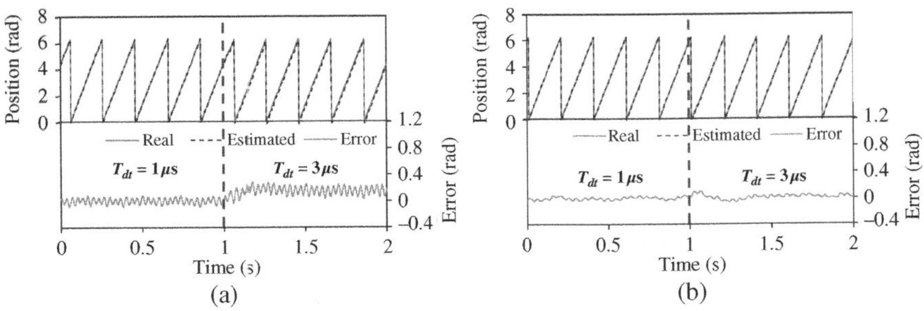

Figure 5.31 Measured position estimation performance for different signal injection methods [21]. (a) Rotating signal injection. (b) Pulsating signal injection.

current response spectra, the pulsating-based method is much less sensitive to inverter nonlinearity effects than the rotating-based method.

5.6.4 Compensation Scheme

5.6.4.1 Pre-compensation

Using the measured phase current, this method estimates the disturbance voltage due to inverter nonlinearity effects based on the inverter's physical characteristics (Figure 5.29), and then superimposes the estimated compensation voltage to the terminal voltage command [22]–[24]. This compensation method can be simply applied to both rotating and pulsating signal injection methods. The overall block diagram is shown in Figure 5.32a and the compensation voltage and phase current are shown in Figure 5.32b. Although the principle of this method is straightforward, to obtain the inverter's physical characteristics from experiments is a time-consuming task. Additionally, the measurement error of instantaneous current especially due to sampling delay makes it impossible to compensate for the disturbance voltage completely with a pre-compensation method.

5.6.4.2 Post-compensation

Specifically, for the rotating HFSI-based sensorless control, only the negative sequence HF current component, which contains position information of the phase angle, is utilized in the sensorless algorithm [18]. In this section, the impacts of HF disturbance voltage on positive sequence HF current are analyzed and the positive sequence HF current response is proven to be useful for the compensation of inverter nonlinearity effects. To analyze the distortion of positive sequence HF current, another important factor λ is introduced as given by,

$$\lambda_{comp} = -\sin\left(\alpha - \frac{\pi}{2}\right)I_{pos\alpha} + \cos\left(\alpha - \frac{\pi}{2}\right)I_{pos\beta} \tag{5.48}$$

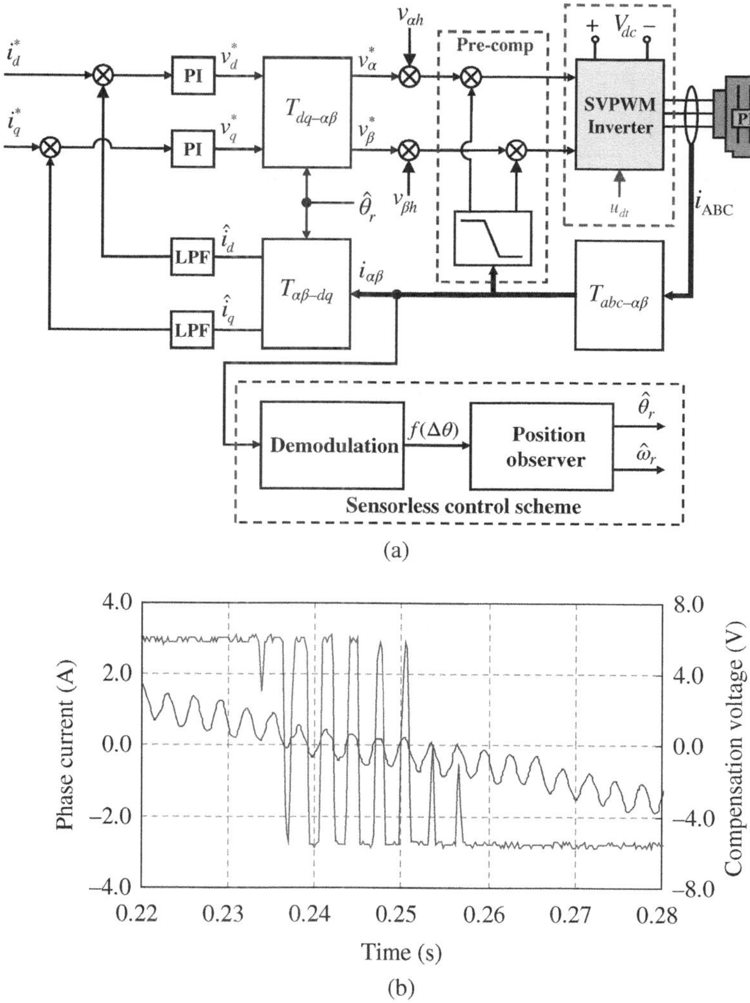

Figure 5.32 Pre-compensation method (t_{dt} = 4 μs, V_{dc} = 150 V) [18]. (a) Block diagram of pre-compensation method. (b) Compensation voltage and measured phase current.

Substituting HF current components shown in Table 5.2 into (5.48), it yields

$$\lambda_{comp} \approx \frac{R_{Ah}}{3V_h}\left(I_p^2 + 2I_pI_n\right) \tag{5.49}$$

Equation (5.49) shows that λ_{comp} is only associated with the inverter nonlinearities. Therefore, it is practical to compensate for the inverter nonlinearity effects based on the distortion of positive sequence HF current. Hence, a simple inverter

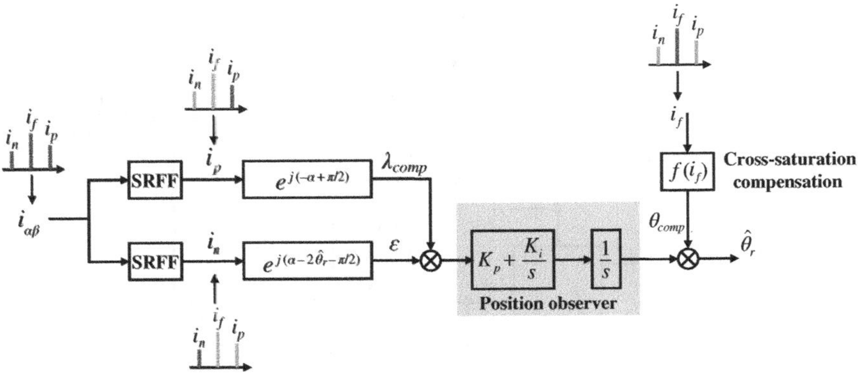

Figure 5.33 Compensation method for inverter nonlinearity effects [18].

nonlinearity compensation method for the rotating signal injection-based sensorless control method is introduced below.

The comparison between (5.45) and (5.49) reveals that without position estimation error ($\Delta\theta = 0$), the signal ε has the opposite phase to that of λ_{comp}. Hence, the combination of them gives:

$$\varepsilon + \lambda_{comp} \approx 2I_n\Delta\theta \tag{5.50}$$

The summation of ε and λ_{comp}, i.e. $(\varepsilon + \lambda_{comp})$, is only dependent on the estimated position error. Hence, it can be employed to estimate the position information accurately [18]. The complete system schematic diagram of the post-compensation method is shown in Figure 5.33, in which $(\varepsilon + \lambda_{comp})$ is utilized as the input signal to the position estimator.

In the compensation scheme, the values of ε and $(\varepsilon + \lambda_{comp})$ can be obtained online from (5.43) and (5.48) without any offline commissioning process. By way of example, the online obtained ε, λ_{comp}, and $(\varepsilon + \lambda_{comp})$ in sensored operation mode ($\Delta\theta = 0$) are shown in Figure 5.34. From them, the sixth harmonic ripple can be observed on both ε and λ_{comp}, while their phases are opposite to each other, which is in good agreement with the foregoing analysis. Although the ripple amplitude of ε and λ_{comp} increases as deadtime increases, the ripple on $(\varepsilon + \lambda_{comp})$ has been significantly suppressed. Consequently, sensorless operation performance is expected to be improved with the post-compensation method. In Figure 5.34, although the dc offsets in ε and λ_{comp} would cause additional constant estimation error for a given load condition, the introduced estimation error can easily be compensated for together with cross-saturation compensation.

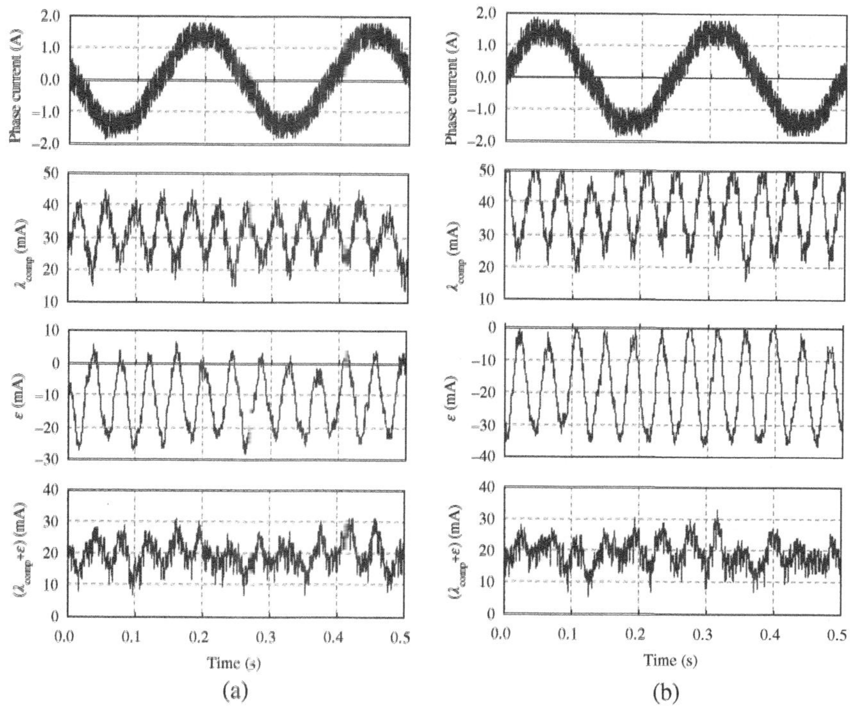

Figure 5.34 Calculated ε, λ_{comp}, and $(\varepsilon + \lambda_{comp})$ in sensored operation mode. (a) t_{dt} = 4 μs. (b) t_{dt} = 6 μs. (*Source:* [18]/IEEE.)

To evaluate the sensorless control performance of the above method, the pre-compensation method as shown in Figure 5.32 is also implemented for comparison. The sensorless operation performances without compensation, with pre-compensation method, and with post-compensation methods are compared at different load conditions, as shown in Figures 5.35 and 5.36. The deadtime is set to 4 μs. For the prototype IPMSM-I (Appendix B), a rotating HF voltage with magnitude of 35 V and a frequency of 330 Hz is determined for the purpose of sensorless effectiveness in the whole operating range. From the experimental results shown in Figures 5.35 and 5.36, the estimated position error without compensation scheme oscillates at six times the fundamental frequency, which results in much distortion on the current response. Although some improvement can be observed with the pre-compensation method, more improvement in the accuracy of the estimated rotor position can be achieved when the post-compensation scheme is employed. Furthermore, the fundamental current waveform is also improved with fewer harmonics due to more accurately estimated position information.

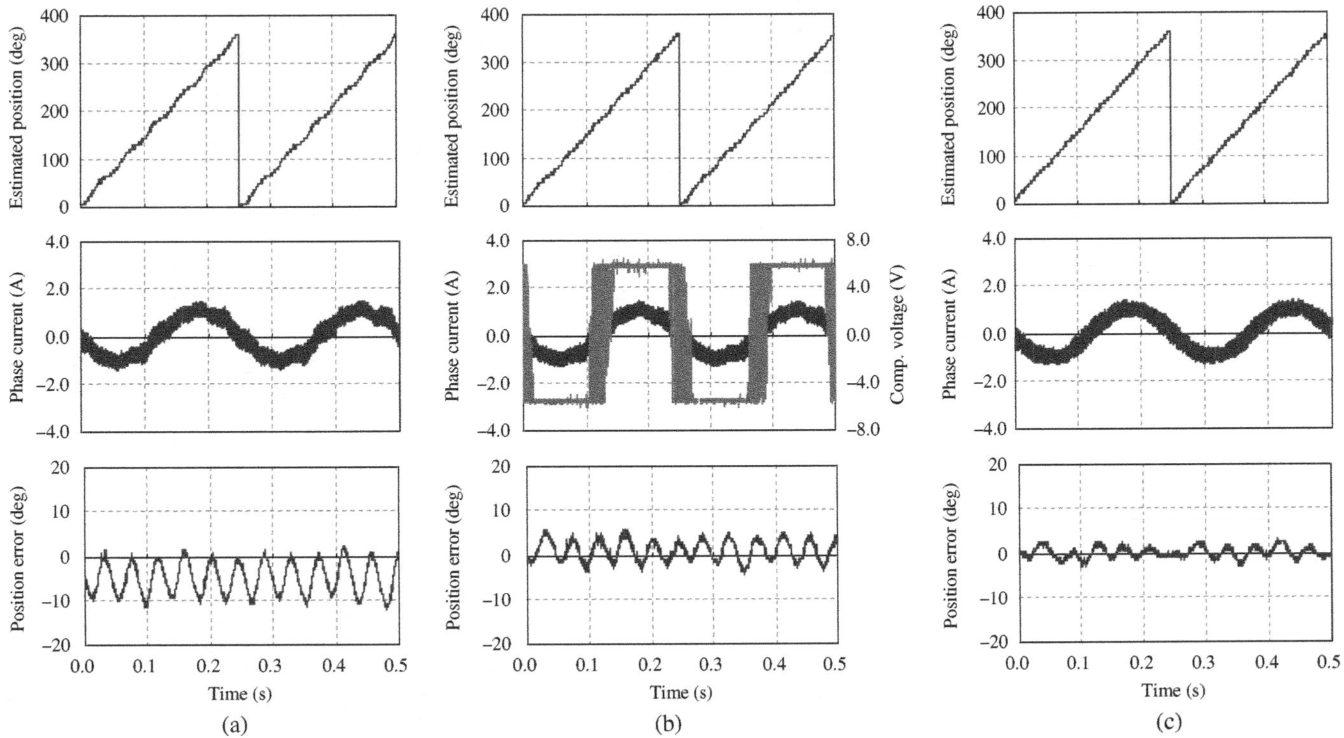

Figure 5.35 Comparison of sensorless operation performance (i_{df} = 0 A, i_{qf} = 1 A, f_e = 4 Hz, t_{dt} = 4 μs, V_{dc} = 150 V). (a) Without compensation. (b) Pre-compensation method. (c) Post-compensation method. (*Source:* [18]/IEEE.)

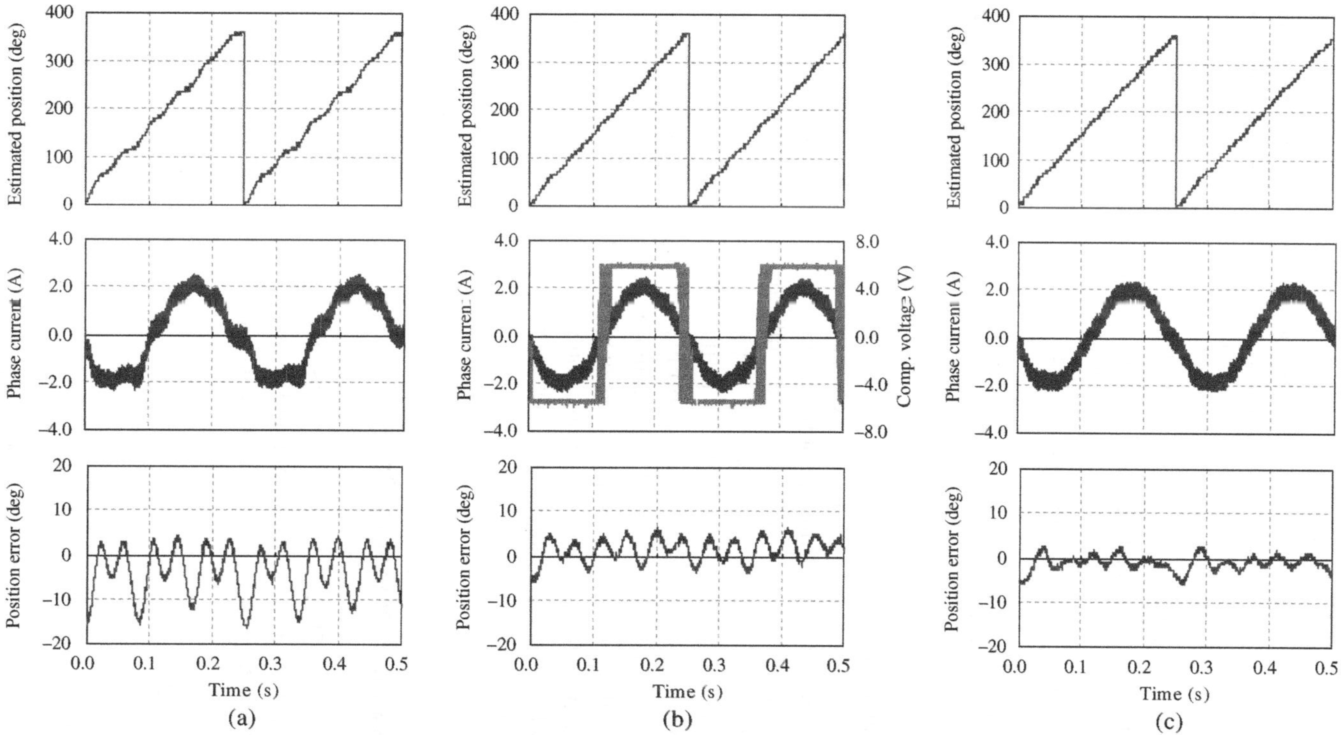

Figure 5.36 Comparison of sensorless operation performance (i_{df} = 0 A, i_{qf} = 2 A, f_e = 4 Hz, t_{dt} = 4 μs, V_{dc} = 150 V). (a) Without compensation. (b) Pre-compensation method. (c) Post-compensation method. (*Source:* [18]/IEEE.)

Table 5.3 Comparison of two inverter nonlinearity effects compensation schemes.

	Pre-compensation	Post-compensation
Compensation manner	Offline measured LUT	Online
Compensation performance	Middle	Good
Compatible sensorless control method	Pulsating and rotating	Rotating
Sensitivity to HF current measurement error	High	Low
Universality of drive system	Poor	Good

5.6.4.3 Comparison

The two compensation schemes are compared in Table 5.3. Since the pre-compensation method requires an offline commissioning process, it is very time consuming and has poor universality for different machines. For the post-compensation method, it can be simply implemented by online compensation without any requirement of knowledge of the drive system, and it has better universality and better estimation performance. Moreover, since the post-compensation method does not directly compensate for the disturbance voltage using instantaneous HF current, it is less sensitive to HF current measurement error. However, the post-compensation is only applicable to the rotating-based method while the pre-compensation method can be applied to both pulsating and rotating methods.

5.7 Signal Processing Delay

As reported in [25]–[28], the digital delays, analog circuit delays, AD sampling delays, etc., can cause the phase shifts of HF current responses. Considering these signal processing delays φ_d, the HF currents can be expressed as:

$$i_{\alpha\beta h} = \frac{V_h}{2\omega_h L_{dh} L_{qh}} \left[\left(L_{dh} + L_{qh}\right) e^{j\left(\omega_h t + \frac{\pi}{2} - \varphi_d\right)} + \left(-L_{dh} + L_{qh}\right) e^{j\left(-\omega_h t + 2\theta_r + \frac{\pi}{2} - \varphi_d\right)} \right] \tag{5.51}$$

for rotating signal injection, and

$$\hat{i}_{qh} = \frac{V_h L_{sd}}{\omega_h L_{dh} L_{qh}} \sin(2\Delta\theta) \sin(\omega_h t - \varphi_d) \tag{5.52}$$

for the synchronous pulsating injection, respectively, where angle φ_d represents the phase shifts due to the signal processing delays. Then, with signal demodulation, the position information signals can be expressed as:

$$\begin{cases} i_{\alpha 1} = \mathrm{Re}\left(\mathrm{LPF}\left(i_{\alpha\beta}^{h} e^{j\left(\omega_h t - \frac{\pi}{2}\right)}\right)\right) = \dfrac{V_h L_{sd}}{\omega_h L_{dh} L_{qh}} \cos(2\theta_r - \varphi_d) \\ i_{\beta 1} = \mathrm{Im}\left(\mathrm{LPF}\left(i_{\alpha\beta}^{h} e^{j\left(\omega_h t - \frac{\pi}{2}\right)}\right)\right) = \dfrac{V_h L_{sd}}{\omega_h L_{dh} L_{qh}} \sin(2\theta_r - \varphi_d) \end{cases} \tag{5.53}$$

$$\hat{I}_{qh} = \mathrm{LPF}\left(\hat{i}_{qh} \cdot 2 \sin \omega_h t\right) = \frac{V_h L_{sd}}{\omega_h L_{dh} L_{qh}} \cos \varphi_d \sin(2\Delta\theta) \tag{5.54}$$

Therefore, it can be seen from (5.53)–(5.54) that the signal processing delays can cause position errors for the rotating signal injection, while not introducing position errors for synchronous pulsating injection since only the amplitudes of (5.54) are affected [27]. In order to illustrate the influence of signal processing delays, average position estimation errors at different injection frequencies (i.e. different signal processing delays) are measured and shown in Figure 5.37. It can be seen that pulsating injection shows great robustness to the variation of injection frequency, while rotating injection is sensitive to the injection frequency.

Since the rotating signal injection method is much affected by signal processing delays, a compensation method is presented by utilizing the positive sequence current [28]. To compensate for the average position errors for rotating signal injection, a PI regulator-based strategy is utilized. According to (5.51), the positive

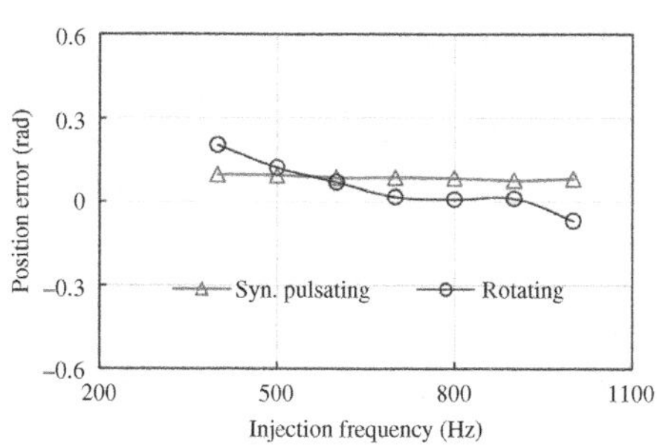

Figure 5.37 Average position errors with sensitivity to injection frequency considering delay effects [27].

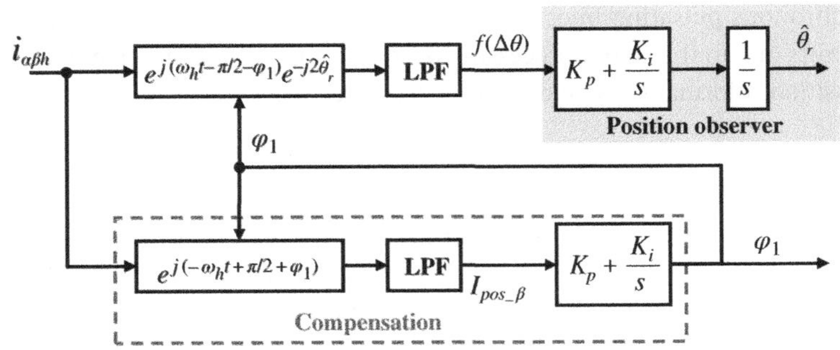

Figure 5.38 Block diagram of the compensation strategy for rotating signal injection [28].

sequence HF current has approximately the same phase shift with the negative sequence HF current due to the delays. Thus, the positive sequence HF current then can be processed as:

$$I_{pos} = \mathrm{LPF}\left(i_{\alpha\beta h}e^{j\left(-\omega_h t+\frac{\pi}{2}+\varphi_1\right)}\right) = I_{pos\_\alpha} + jI_{pos\_\beta} \approx \frac{V_h L_{sd}}{\omega_h L_{dh} L_{qh}}(\cos\Delta\varphi + j\sin\Delta\varphi) \tag{5.55}$$

where I_{pos_β} and φ_1 are the input and output of the added PI regulator, respectively, Figure 5.38, and $\Delta\varphi$ is the difference between the angle φ_d and φ_1. With the PI regulator, I_{pos_β} can be controlled to zero, i.e. $\Delta\varphi = 0$. Then, according to Figure 5.38, the HF current in the negative sequence frame can be described as:

$$I_{\alpha\beta h} = \mathrm{LPF}\left(i_{\alpha\beta h}e^{j\left(\omega_h t-\frac{\pi}{2}-\varphi_1\right)}\right) \approx \frac{V_h L_{sd}}{\omega_h L_{dh} L_{qh}} e^{j(2\theta_r + \Delta\varphi)} \tag{5.56}$$

Therefore, compared to (5.53), it can be seen from (5.56) that the position errors due to the signal processing delay will be significantly suppressed for the rotating signal injection. To illustrate the effectiveness for the compensation method, the position estimation performance before and after the compensation are shown in Figure 5.39, under different load conditions. It can be clearly observed that the average error is significantly suppressed after the delay compensation. Then, with the addition of the cross-coupling effect compensation [5], the position estimation error is further reduced, as shown in Figure 5.39b.

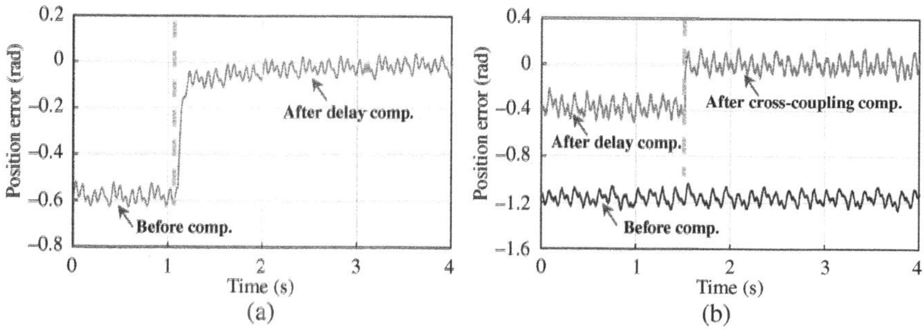

Figure 5.39 Position error compensation for rotating signal injection [28]. (a) No-load condition. (b) 75% load condition.

5.8 Selection of Amplitude and Frequency for Injection Voltage Signal

HFSI sensorless control methods estimate rotor position by the measured HF current response by injection of a HF voltage signal. The amplitude and frequency of the injection voltage signal should be selected properly to obtain a reliable estimation. Therefore, in this section, a sensorless safe operation area (SSOA) is defined, which provides a practical solution to determine the injection voltage signal in terms of amplitude and frequency for a reliable estimation performance [29]. Moreover, by considering the reduction of acoustic noise caused by HFSI, the pseudo-random selection manner of HF signals is introduced.

5.8.1 Quantization Error in AD Conversion

Before defining the SSOA, it is useful to discuss the effect of quantization error in AD conversion. In general, the current sensors designed for fundamental current measurement are also used to measure the HF current in order to reduce the system cost. Therefore, the resolution of the measured HF current component is under the worst condition. Based on the rotating sinusoidal signal injection-based method (introduced in Chapter 4), the effect of quantization error due to AD conversion is introduced below [29].

It is assumed that the current value of $2I_p$ is scaled to be the full range of the AD converter with the resolution of N bits. Then, a parameter of quantum current can be defined as:

$$I_{qu} = 2I_p/2^N \tag{5.57}$$

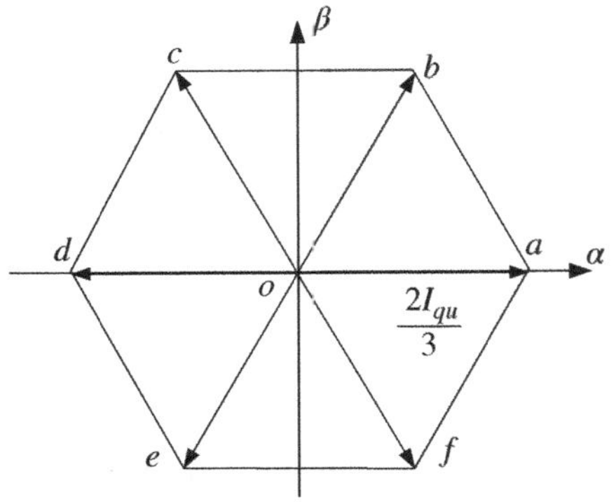

Figure 5.40 Disturbance current vector due to quantization error. (*Source:* [29]/IEEE.)

Therefore, the phase current measurement error due to the quantization error in AD conversion can be expressed as:

$$\varepsilon_s = \pm \frac{I_{qu}}{2} \tag{5.58}$$

Based on the definition of the space vector theory, the disturbance current vector due to current measurement error can be expressed as:

$$I_{AD\_error} = \frac{2}{3}\left(\varepsilon_s + \varepsilon_s e^{j2\pi/3} + \varepsilon_s e^{j4\pi/3}\right) \tag{5.59}$$

It indicates eight vectors (including two zero vectors) in the stationary reference frame, as shown in Figure 5.40. The nonzero disturbance current vectors have an identical magnitude of $2I_{qu}/3$. Due to its random frequency, it is difficult to compensate for the estimation error resulting from AD quantization. Under the worst case condition, the disturbance current vector lies on the q-axis in the estimated synchronous reference frame, and in turn the resultant position estimation error is the maximum, given by:

$$I_n \sin(2\Delta\theta_{\mathrm{AD}}) = 2I_{qu}/3 \tag{5.60}$$

Then, the maximum position estimation error is derived as:

$$\Delta\theta_{\mathrm{AD}} = \pm \frac{1}{2}\sin^{-1}\left[\left(\frac{2I_{qu}}{3}\right)/I_n\right] \tag{5.61}$$

From (5.61), the relationship between $\Delta\theta$ and I_n can be depicted in Figure 5.41, in which I_n is expressed as a per-unit value with reference to I_{qu}. It can be observed that the position estimation error due to AD quantization is decreased considerably with the increase of I_n when I_n is small. However, the estimation error tends to saturation for higher I_n.

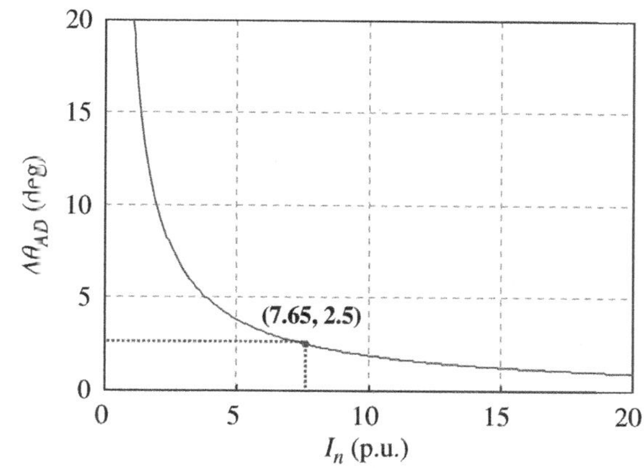

Figure 5.41 Estimated position error due to ADC quantization error. (*Source:* [29]/IEEE.)

5.8.2 Sensorless Safe Operation Area

Providing the position estimation error due to AD quantization is constrained within ±2.5 electric degrees, then from (5.61) the requirement of I_n can be derived as (5.62), which is highlighted in Figure 5.41.

$$I_n = \left(\frac{2I_{qu}}{3}\right) / \sin(2\Delta\theta) \geq 7.65 I_{qu} \tag{5.62}$$

With the definition of I_n (in Chapter 4), (5.62) has an alternative form, as given by:

$$L_n \leq V_h / \left(7.65 \omega_h I_{qu}\right) = L_{lm} \tag{5.63}$$

where L_{lm} is designated as the boundary inductance.

The SSOA in the dq-axis plane for effective sensorless operation can be defined as the area in the current limitation circle, in which $L_n \leq L_{lm}$. Meanwhile, the SSOA can be interpreted as the area in which $I_n \geq 7.65 I_{qu}$ instead of $I_n > 0$ for the feasible region. As a result, the SSOA is determined by the machine saliency level (L_n), injected HF voltage signal (V_h/ω_h), and the resolution of current measurement (I_{qu}).

Since the value of L_n is load-dependent, it is useful to plot its contour map in the dq-axis plane, as shown in Figure 5.42. In the figure, the current locus of the maximum torque per ampere (MTPA) control is indicated by the solid red line, while the current limitation circle is shown by the dashed blue line.

In this work, based on prototype machine IPMSM-I, the current value of 23 A (−11.5 to +11.5 A) is scaled to the full range of the 12 bit AD converter. From the foregoing analysis, it can be obtained that

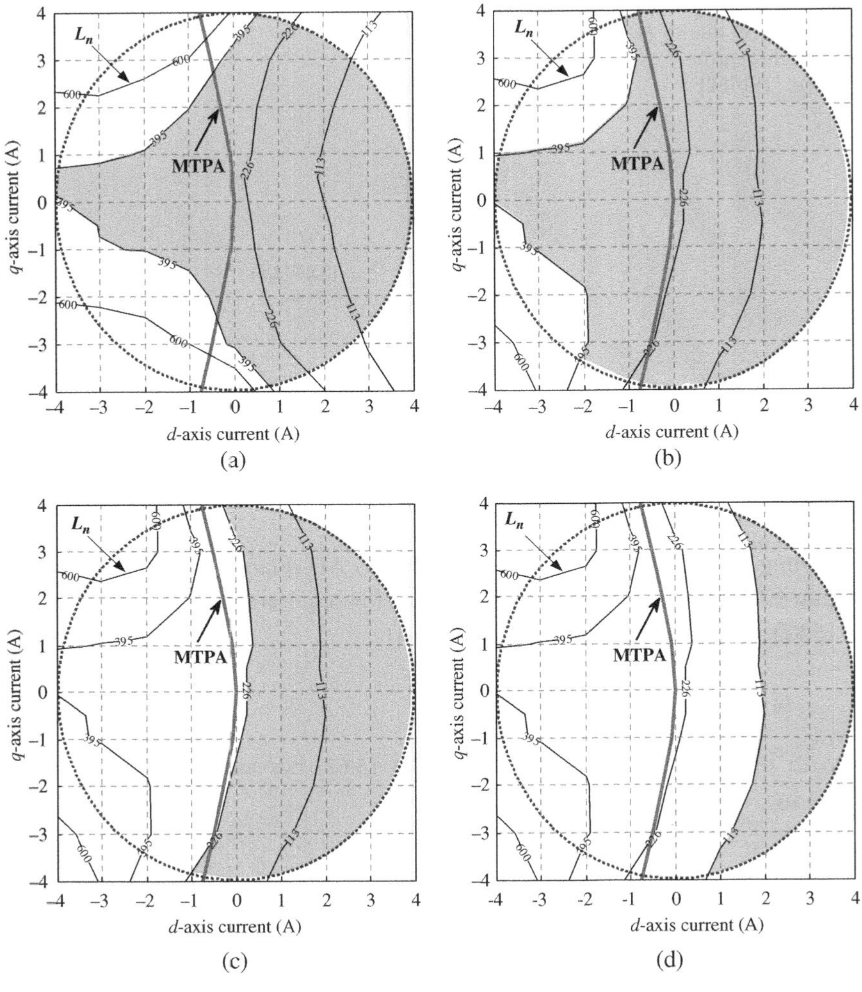

Figure 5.42 Predicted SSOA. (a) Without consideration of cross-saturation effect (V_h/f_h = 35 V/330 Hz). (b) With consideration of cross-saturation effect (V_h/f_h = 35 V/330 Hz). (c) With consideration of cross-saturation effect (V_h/f_h = 20 V/330 Hz). (d) With consideration of cross-saturation effect (V_h/f_h = 10 V/330 Hz). (*Source:* [29]/IEEE.)

$$\begin{cases} I_{qu} = 2 \times 11.5 \times 10^3/2^{12} = 5.62(\text{mA}) \\ I_n \geq 7.65 I_{qu} = 43.0(\text{mA}) \\ L_n \leq V_h/\left(7.65\omega_h I_{qu}\right) = 395(\text{mH}) \end{cases} \tag{5.64}$$

In Figures 5.42a,b, the gray area inside the current limitation circle, in which $L_n \leq 395\,\text{mH}$, is the predicted SSOA for the prototype machine IPMSM-I with

Table 5.4 Predicted SSOA for prototype machine.

Injected signals	I_n (mA)	L_n (mH)	Boundary point at i_q = 2 (A)
$V_h = 35\,\text{V}/f_h = 330\,\text{Hz}$	≥43	≤395	$i_d = -1.0$
$V_h = 20\,\text{V}/f_h = 330\,\text{Hz}$	≥43	≤226	$i_d = 0.3$
$V_h = 10\,\text{V}/f_h = 330\,\text{Hz}$	≥43	≤113	$i_d = 1.8$

(*Source:* [29]/IEEE.)

the injected signal of $V_n = 35$ V and $f_h = 330$ Hz. Figures 5.42a,b show the contour map of L_n without/with consideration of the cross-saturation effect. According to the definition of L_n, the cross-saturation effect decreases the value of L_n, which results in the extension of SSOA to some extent. With the aid of the cross-saturation effect, it can be seen from Figure 5.42b that the whole MTPA trajectory is inside the SSOA. In the same way, the predicted SSOA for different injection signals can be summarized in Table 5.4. The table clearly indicates that the SSOA shrinks with the decrease of HF voltage amplitude, as shown in Figures 5.42c,d. It is worth noting that the SSOA is not completely symmetrical with respect to the line of zero q-axis current, which implies that the estimated rotor position in the case of negative q-axis current is slightly more accurate.

5.8.3 Experimental Results of Determining Amplitude and Frequency

To validate the SSOA for rotating and pulsating HFSI-based techniques, the prototype machine IPMSM-I operates in sensored mode and the q-axis current is fixed to 2 A, while the d-axis current is changed from −3.5 to 3.5 A (inside the current limitation circle). The cross-saturation effect has been compensated for using the methods presented in this chapter. The remaining estimated position error can be attributed to the quantization error in the AD conversion.

In Figures 5.43–5.45, the amplitude of the injected HF voltage is different, while the frequency is identical, being fixed to 330 Hz. The solid red lines indicate the boundary of SSOA defined by ($I_n \geq 43$ mA) in Figure 5.43, and the boundary point of SSOA for ($V_h = 35$ V, $f_h = 330$ Hz) lies at ($i_d = -1$ A, $i_q = 2$ A), the same as the calculated point shown in Table 5.4. The same conclusion can be drawn from Figures 5.44 and 5.45. Regardless of the amplitude of the injected HF voltage, it can be observed that I_n increases nonlinearly with an increase of i_d for a constant q-axis current ($i_q = 2$ A), which is in accordance with the decrease of L_n along the line of $i_q = 2$ A, Figure 5.42b. Meanwhile, it can be observed that the estimated position errors reduce with the increase of I_n for both rotating and pulsating injection-based techniques, which results in the fact that the resultant position error is

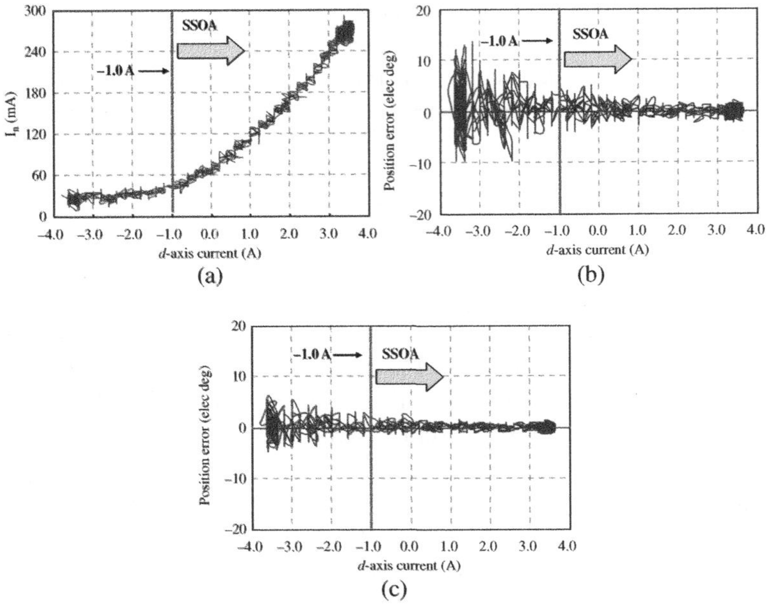

Figure 5.43 Measured SSOA (V_h = 35 V, f_h = 330 Hz). (a) Measured I_n. (b) Estimated position error for rotating injection. (c) Estimated position error for pulsating injection. (*Source:* [29]/IEEE.)

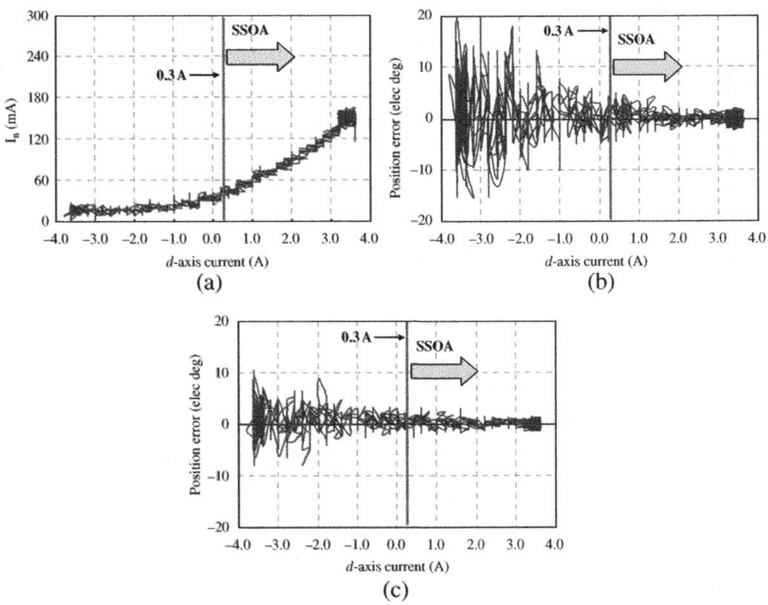

Figure 5.44 Measured SSOA (V_h = 20 V, f_h = 330 Hz). (a) Measured I_n. (b) Estimated position error for rotating injection. (c) Estimated position error for pulsating injection. (*Source:* [29]/IEEE.)

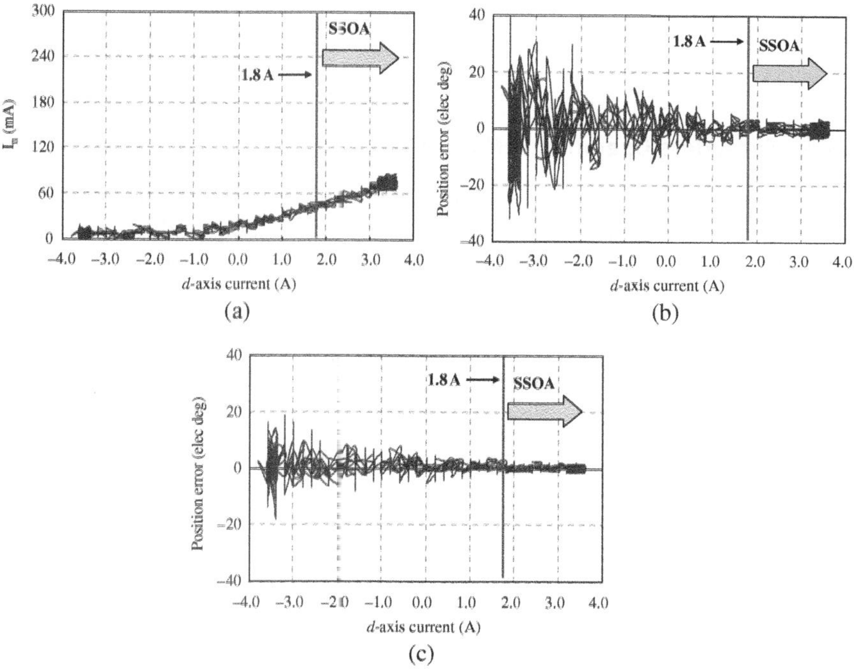

Figure 5.45 Measured SSOA (V_h = 10 V, f_h = 330 Hz). (a) Measured I_n. (b) Estimated position error for rotating injection. (c) Estimated position error for pulsating injection. (*Source:* [29]/IEEE.)

connected with I_n due to quantization error in the AD conversion. On the boundary of SSOA, the estimated position errors for pulsating injection reach nearly the same level of around ±2.5 electric degrees for different injection signals, which is in accordance with the previous analysis. However, the estimated position errors on the boundary of SSOA for rotating injection seem to be a little higher, which may be caused by different signal demodulation processes. Finally, the experimental results show that the SSOA shrinks significantly with the decrease of V_h, as indicated in Figure 5.42.

It is worth pointing out that it does not mean that the machine could not perform with sensorless operation outside the SSOA. Actually, the SSOA for sensorless operation defines a working area in the dq-axis plane, in which the machine can work in sensorless mode with a guaranteed performance in steady state. Besides the machine saliency level (L_n), the definition of SSOA also considers the characteristics of the injected signal (V_h/ω_h) and the resolution of current measurement (I_{qu}). Consequently, for the specific prototype machine and given current measurement resolution, the proper injected HF signal should be selected carefully to make sure that the whole working trajectory is inside the SSOA.

5.8.4 Sensorless Operation Performance

Figures 5.46 and 5.47 show the dynamic performance of the rotating injection-based sensorless scheme with cross-saturation compensation for different magnitudes of injected HF voltage. The reference speed is changed from 0 to 100 and

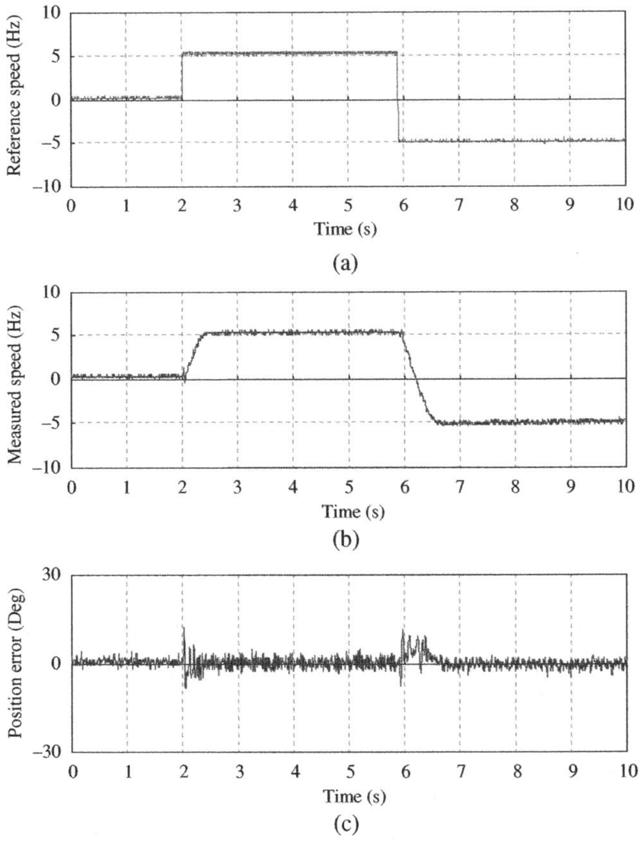

Figure 5.46 Step speed response (35 V, 330 Hz). (a) Speed reference. (b) Speed response. (c) Position estimation error. (*Source:* [29]/IEEE.)

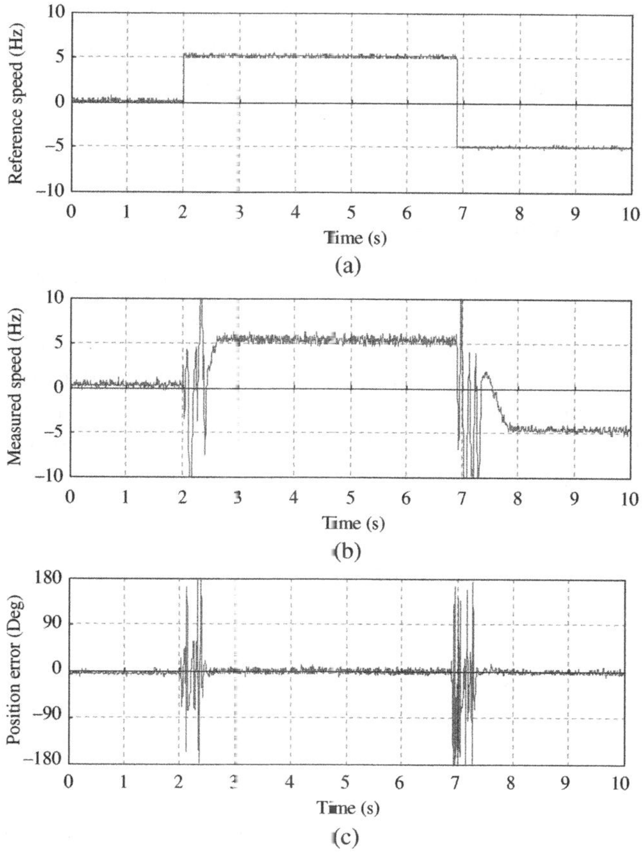

Figure 5.47 Step speed response (10 V, 330 Hz). (a) Speed reference. (b) Speed response. (c) Position estimation error. (*Source:* [29]/IEEE.)

then to −100 r/min. As shown in Fig. 5.46, a rotating HF voltage with a magnitude of 35 V and frequency of 330 Hz is applied. In this case, the whole MTPA trajectory is within the SSOA, as shown in Fig. 5.42(b). The sensorless performance is guaranteed in both transient and steady state. As shown in Fig. 5.47, when the amplitude of injected HF voltage decreases to 10 V, the boundary curve for SSOA is $L_n \leq$ 113 mH. Fig. 5.42(d) shows that the MTPA trajectory lies outside of the SSOA entirely for this case. However, as mentioned before, it does not mean that the machine could not perform with sensorless operation outside the SSOA. The sensorless estimator may still track the rotor position in steady state with relatively large position error and deteriorated performance, although it is more likely to fail to do so in the transient state.

5.8.5 Pseudo-random Selection of Injection Signal

In Sections 5.8.1–5.8.4, the frequency and amplitude of the injected signals were selected to guarantee a reliable estimation performance using fixed values. For conventional HFSI methods, the injected high-frequency voltage signal has fixed frequency and amplitude. However, these methods with fixed signals suffer from acoustic noise [30]–[32], which is not preferred in some applications, e.g. domestic appliances. In order to reduce the acoustic noise due to HFSI, the injected HF signal can be selected in a pseudo-random approach [30], [31], which is introduced in this section.

In [31], based on the conventional pulsating square-wave signal injection method, two injection signals v_{dh1} and v_{dh2} with fixed frequency and fixed amplitude are employed, which are given by:

$$v_{dh1} = \begin{cases} V_{h1} & 0 < t_r \leq T_{i1}/2 \\ -V_{h1} & T_{i1}/2 < t_r \leq T_{i1} \end{cases} \tag{5.65}$$

$$v_{dh2} = \begin{cases} V_{h2} & 0 < t_r \leq T_{i2}/2 \\ -V_{h1} & T_{i2}/2 < t_r \leq T_{i2} \end{cases} \tag{5.66}$$

where V_{h1} and V_{h2} denote the amplitude of the first and second square-wave signals, respectively. T_{i1} and T_{i2} represent the periods of the first and second square-wave signals, respectively. t_r is the remainder of the division of the time t by T_i. It is worth noting that the ratio of voltage amplitude and frequency should be constant, i.e. $V_{h1}T_{i1} = V_{h2}T_{i2}$, so that the current response magnitudes of the two different voltage signals are identical.

As shown in Figure 5.48, two different voltage signals in (5.65) and (5.66) are selected randomly and hence a pseudo-random voltage signal is finally injected in the estimated d-axis. By injecting the pseudo-random signal, the distribution of injected high frequency and the induced current response in the frequency domain are extended. As a consequence, the annoyance of audible noise can be reduced.

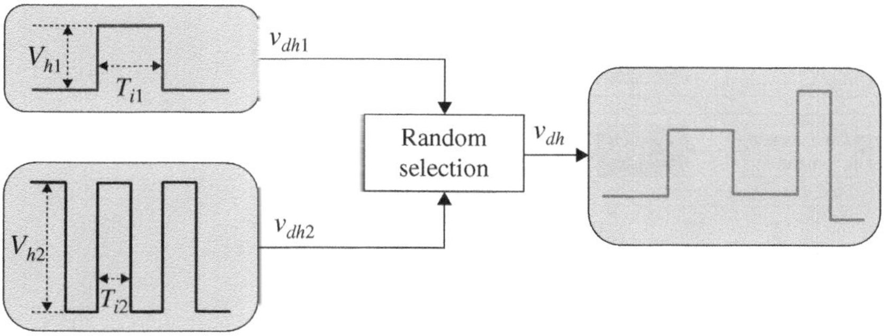

Figure 5.48 Random signal generator [31].

5.9 Transition Between Low Speed and High Speed

To extend the sensorless control performance over a wide speed range, a combination of two sensorless control strategies are introduced in this section. At low-speed range, the HFSI methods are utilized for rotor position estimation, while the fundamental model-based methods are employed to estimate the rotor position at high-speed range. A smooth transition between the two combined methods is required. Between low-speed range and high-speed range, there is an area in which to change the algorithm of the position estimator. It should be changed smoothly to ensure system stability and also high dynamic performance.

Most of the transition strategies can be classified as position fusion or error signal fusion. For position fusion [33]–[35], the saliency model-based position observer and fundamental model-based position observer are running in parallel as shown in Figure 5.49a. With this structure, the dynamic performance of position estimation at low- and high-speed ranges can be adjusted separately. However, this will increase the computation burden due to two closed-loop observers. The estimated positions of two models are combined. For error signal fusion [36], the position error signals from the saliency model and the fundamental model are normalized and combined by the error fusion module as shown in Figure 5.49b. With a normalized error signal, the saliency model and fundamental model can share the same PI parameters with unified dynamic performance. The two fusion methods are summarized in Table 5.5.

By way of example, the position fusion based method is introduced below. To achieve smooth transition between the combined methods, the fusion module shown in Figure 5.50 is applied for a certain speed range with optimum weighted gains for both estimated positions as expressed in (5.67):

$$\hat{\theta}_r = k \times \hat{\theta}_{r\_hf} + (1 - k) \times \hat{\theta}_{r\_fo} \tag{5.67}$$

where $\hat{\theta}_r$ is referred to as final estimated rotor position, $\hat{\theta}_{r\_hf}$ denotes the estimated rotor position of the HFSI method. The estimated rotor position of the

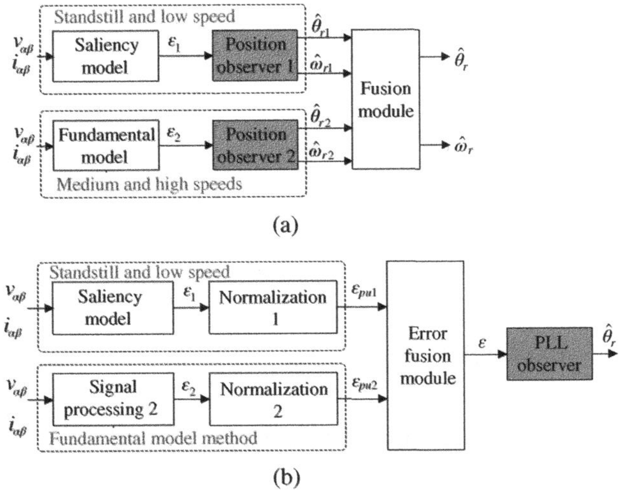

Figure 5.49 Fusion strategies for whole speed range sensorless operation. (a) Position fusion. (b) Error signal fusion.

Table 5.5 Summary of two sensorless fusion strategies.

	Position fusion	**Error signal fusion**
Combined signal	Position and speed	Position error signal
Observer number	2	1
Computation cost	High	Low
Modification of existing sensorless methods	Not required	Required

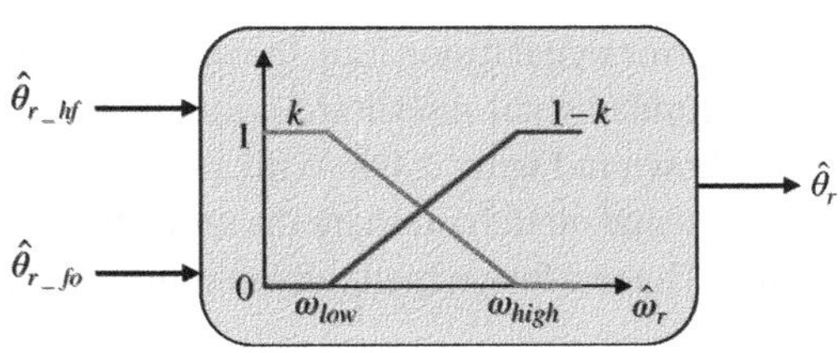

Figure 5.50 Fusion module [35].

fundamental model-based method is represented by $\hat{\theta}_{r\_fo}$, and k is the weighting gain, which can be calculated as:

$$k = \begin{cases} 1 & \text{if } \hat{\omega}_r < \omega_{low} \\ (\hat{\omega}_r - \omega_{high})/\omega_{low} - \omega_{high} & \text{if } \omega_{low} < \hat{\omega}_r < \omega_{high} \\ 0 & \text{if } \hat{\omega}_r > \omega_{high} \end{cases} \tag{5.68}$$

where $\hat{\omega}_r$ is the estimated speed, ω_{low} is referred to as the low-speed border of the area, and the high-speed border of the area is denoted by ω_{high}. When the estimated speed is lower than the low-speed border ω_{low}, k is 1 and the estimated position only comes from the HFSI sensorless method. If the estimated speed is higher than the high-speed border ω_{high}, k is 0 and the estimated position is only contributed by the fundamental model-based sensorless control method. When the estimated speed is between ω_{low} and ω_{high}, a linear function in (5.68) is used for a smooth transition. For error signal fusion, the position signal in (5.67) can be simply changed to the normalized error signals. Moreover, it is worth mentioning that the injection of a high-frequency signal should be gradually reduced until it reaches zero to reduce the adverse effects of the injected signal to the system when the speed is higher than the high-speed border.

In order to illustrate the effectiveness of the combined sensorless control method, experimental tests of position fusion are carried out on prototype IPMSM-II (Appendix B). In the tests, ω_{low} is chosen as 60 r/min, and ω_{high} is chosen as 150 r/min. In the low-speed range, a high-frequency pulsating voltage signal (V_h = 8 V, f_h = 500 Hz) is injected into the system. It will disappear when the machine runs in high-speed range and the estimator of the flux-linkage observer starts to work. In the test, a conventional flux-linkage observer introduced in Chapter 2 is combined with the conventional pulsating sinusoidal signal injection method introduced in Chapter 4. When the rotor's initial speed is zero and step-changed to 80 r/min and then to 250 r/min, the rotor speed response is shown in Figure 5.51a, and the comparison of the actual and estimated rotor positions together with the estimation error are illustrated in Figure 5.51b.

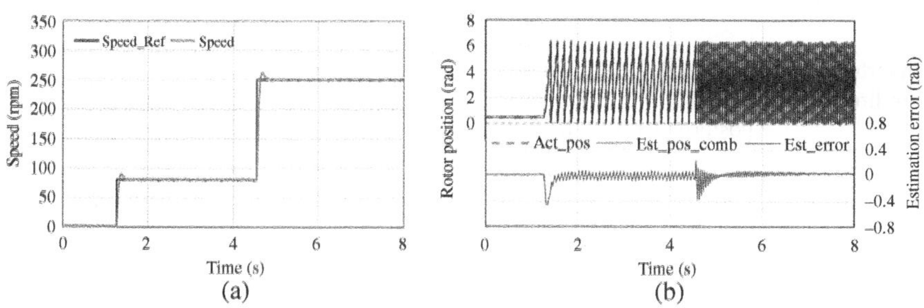

Figure 5.51 Sensorless control performance from low- to high-speed range, 0–80–250 r/min. (a) Speed response. (b) Rotor position estimation.

5.10 Summary

In this chapter, several common practical issues for HFSI sensorless control were introduced including their impacts on position estimation and the corresponding solutions. The practical issues and their impacts on position estimation are summarized in Table 5.6. It can be seen that inverter nonlinearity effect and multiple saliency effect are the main causes of the sixth order (or multiple of sixth) harmonics in the position estimation error, while inductance asymmetry will mainly introduce a second order harmonic position error. The cross-coupling effect will produce a dc offset in the position estimation, which can be simply compensated

Table 5.6 Summary of practical issues impacts on position estimation of HFSI sensorless control.

	Pulsating sinusoidal/square-wave signal injection methods		Rotating sinusoidal signal injection method	
	Issues	**Solution**	**Issues**	**Solution**
Cross-coupling effect	dc offset position error	Offline measured data [4], [5]	dc offset position error	Offline measured data [4], [5]
Multiple saliency effect	Sixth (or multiple) order harmonic position error	Neglectable	Sixth (or multiple) order harmonic position error	Neglectable
Parameter asymmetry	Second order harmonic position error	Filter [16]	Second order harmonic position error	Filter [16]
Inverter nonlinearity effect	Sixth order harmonic position error	Offline measured LUT [22]–[24]	Sixth order harmonic position error	1) Offline measured LUT [22]–[24] 2) Compensation with positive sequence current [18]
Signal processing delay	None	None	dc offset position error	Compensation with positive sequence current [28]
ADC quantization	dc offset position error	SSOA [29]	dc offset position error	SSOA [29]

for by offline measured data. For the signal processing delays, no position error is introduced for the pulsating based methods, while a dc offset error appears in the rotating signal injection method. This dc offset can be compensated by utilizing the positive sequence HF current response. Moreover, A/D conversion (ADC) quantization will introduce dc offset position error for HFSI-based methods; by considering this error, a SSOA is presented that provides a practical solution to select the amplitude and frequency of the injected voltage signal. In addition, to reduce the audible noise caused by HFSI, the injected HF signal can be selected in a pseudo-random manner. Finally, the transition strategy between low-speed (HFSI-based method) and high-speed (fundamental model-based method) sensorless control strategies was discussed.

References

1 M. J. Corley and R. D. Lorenz, "Rotor position and velocity estimation for a salient-pole permanent magnet synchronous machine at standstill and high speeds," *IEEE Trans. Ind. Appl.*, vol. 34, no. 4, pp. 784–789, Jul. 1998.

2 P. Guglielmi, M. Pastorelli, and A. Vagati, "Cross-saturation effects in IPM motors and related impact on sensorless control," *IEEE Trans. Ind. Appl.*, vol. 42, no. 6, pp. 1516–1522, Nov. 2006.

3 N. Bianchi and S. Bolognani, "Influence of rotor geometry of an IPM motor on sensorless control feasibility," *IEEE Trans. Ind. Appl.*, vol. 43, no. 1, pp. 87–96, Jan. 2007.

4 Y. Li, Z. Q. Zhu, D. Howe, C. M. Bingham, and D. Stone, "Improved rotor position estimation by signal injection in brushless AC motors, accounting for cross-coupling magnetic saturation," *IEEE Trans. Ind. Appl.*, vol. 45, no. 5, pp. 1843–1849, 2009.

5 Z. Q. Zhu, Y. Li, D. Howe, and C. M. Bingham, "Compensation for rotor position estimation error due to cross-coupling magnetic saturation in signal injection based sensorless control of PM brushless ac motors," *IEEE Int. Elect. Mach. Drives Conf.*, vol. 1, pp. 208–213, May 2007.

6 G. Wang, D. Xiao, G. Zhang, C. Li, X. Zhang, and D. Xu, "Sensorless control scheme of IPMSMS using HF orthogonal square-wave voltage injection into a stationary reference frame," *IEEE Trans. Power Electron.*, vol. 34, no. 3, pp. 2573–2584, Mar. 2019.

7 G. El-Murr, D. Giaouris, and J. W. Finch, "Online cross-coupling and self-incremental inductances determination of salient permanent magnet synchronous machines," *5th IET Int. Conf. Power Electron. Mach. Drives* (PEMD 2010), Brighton, UK, 2010, pp. 1–4.

8 T. C. Lin, L. M. Gong, J. M. Liu, and Z. Q. Zhu, "Investigation of saliency in a switched-flux permanent-magnet machine using high-frequency signal injection," *IEEE Trans. Ind. Electron.*, vol. 61, no. 9, pp. 5094–5104, Sep. 2014.

9 T. C. Lin and Z. Q. Zhu, "Sensorless operation capability of surface-mounted permanent-magnet machine based on high-frequency signal injection methods," *IEEE Trans. Ind. Appl.*, vol. 51, no. 3, pp. 2161–2171, May 2015.

10 M. W. Degner and R. D. Lorenz, "Using multiple saliencies for the estimation of flux, position, and velocity in AC machines," *IEEE Trans. Ind. Appl.*, vol. 34, no. 5, pp. 1097–1104, 1998.

11 D. Raca, M. C. Harke, and R. D. Lorenz, "Robust magnet polarity estimation for initialization of PM synchronous machines with near-zero saliency," *IEEE Trans. Ind. Appl.*, vol. 44, no. 4, pp. 1199–1209, 2008.

12 D. Reigosa, P. Garcia, D. Raca, F. Briz, and R. D. Lorenz, "Measurement and adaptive decoupling of cross-saturation effects and secondary saliencies in sensorless-controlled IPM synchronous machines," *IEEE Trans. Ind. Appl.*, vol. 44, no. 6, pp. 1758–1767, 2008.

13 L. Chen, G. Götting, S. Dietrich, and I. Hahn, "Self-sensing control of permanent-magnet synchronous machines with multiple saliencies using pulse-voltage-injection," *IEEE Trans. Ind. Appl.*, vol. 52, no. 4, pp. 3480–3491, Jul./Aug. 2016.

14 Z. Chen, J. Gao, F. Wang, Z. Ma, Z. Zhang, and R. Kennel, "Sensorless control for SPMSM with concentrated windings using multi-signal injection method," *IEEE Trans. Ind. Electron.*, vol. 61, no. 12, pp. 6624–6634, Dec. 2014.

15 L. M. Gong, "Carrier signal injection based sensorless control of permanent magnet brushless ac machines," Ph.D. dissertation, Department of Electronic and Electrical Engineering, University of Sheffield, UK, 2012.

16 P. L. Xu and Z. Q. Zhu, "Carrier signal injection-based sensorless control for permanent-magnet synchronous machine drives considering machine parameter asymmetry," *IEEE Trans. Ind. Electron.*, vol. 63, no. 5, pp. 2813–2824, May 2016.

17 J. M. Guerrero, M. Leetmaa, F. Briz, A. Zamarron, and R. D. Lorenz, "Inverter nonlinearity effects in high-frequency signal-injection-based sensorless control methods," *IEEE Trans. Ind. Appl.*, vol. 41, no. 2, pp. 618–626, 2005.

18 L. M. Gong and Z. Q. Zhu, "A novel method for compensating inverter nonlinearity effects in carrier signal injection-based sensorless control from positive-sequence carrier current distortion," *IEEE Trans. Ind. Appl.*, vol. 47, no. 3, pp. 1283–1292, May 2011.

19 J. W. Choi and S. K. Sul, "Inverter output voltage synthesis using novel dead time compensation," *IEEE Trans. Power Electron.*, vol. 11, no. 2, pp. 221–227, 1996.

20 S. H. Hwang and J. M. Kim, "Dead time compensation method for voltage-fed PWM inverter," *IEEE Trans. Energy Convers.*, vol. 25, no. 1, pp. 1–10, 2010.

21 P. L. Xu and Z. Q. Zhu, "Analysis of parasitic effects in carrier signal injection methods for sensorless control of PM synchronous machines," *IET Elect. Power Appl.*, vol. 12, no. 2, pp. 203–212, 2018.

22 C. H. Choi and J. K. Seok, "Compensation of zero-current clamping effects in high-frequency-signal-injection-based sensorless PM motor drives," *IEEE Trans. Ind. Appl.*, vol. 43, no. 5, pp. 1258–1265, Sep. 2007.

23 Y. Inoue, K. Yamada, S. Morimoto, and M. Sanada, "Effectiveness of voltage error compensation and parameter identification for model-based sensorless control of IPMSM," *IEEE Trans. Ind. Appl.*, vol. 45, no. 1, pp. 213–221, 2009.

24 D. E. Salt, D. Drury, D. Holliday, A. Griffo, P. Sangha, and A. Dinu, "Compensation of inverter nonlinear distortion effects for signal-injection-based sensorless control," *IEEE Trans. Ind. Appl.*, vol. 47, no. 5, pp. 2084–2092, Sep. 2011.

25 F. Cupertino, A. Guagnano, A. Altomare, and G. Pellegrino, "Position estimation delays in signal injection-based sensorless PMSM drives," *Proc. IEEE SLED 2012 Symp.*, 2012, pp. 1–6.

26 M. A. G. Moghadam and F. Tahami, "Sensorless control of PMSMs with tolerance for delays and stator resistance uncertainties," *IEEE Trans. Power Electron.*, vol. 28, no. 3, pp. 1391–1399, Mar. 2013.

27 P. L. Xu and Z. Q. Zhu, "Comparison of carrier signal injection methods for sensorless control of PMSM drives," in *2015 IEEE Energy Convers. Congr. Expo. (ECCE)*, Sep. 2015, pp. 5616–5623.

28 P. L. Xu and Z. Q. Zhu, "Carrier signal injection-based sensorless control for permanent magnet synchronous machine drives with tolerance of signal processing delays," *IET Elect. Power Appl.*, vol. 11, no. 6, pp. 1140–1149, 2017.

29 Z. Q. Zhu and L. M. Gong, "Investigation of effectiveness of sensorless operation in carrier-signal-injection-based sensorless-control methods," *IEEE Trans. Ind. Electron.*, vol. 58, no. 8, pp. 3431–3439, Aug. 2011.

30 G. Wang, G. Zhang, and D. Xu, *Position Sensorless Control Techniques for Permanent Magnet Synchronous Machine Drives*. Singapore: Springer, 2020.

31 G. Wang, L. Yang, B. Yuan, B. Wang, G. Zhang, and D. Xu, "Pseudo-random high-frequency square-wave voltage injection based sensorless control of IPMSM drives for audible noise reduction," *IEEE Trans. Ind. Electron.*, vol. 63, no. 12, pp. 7423–7433, Dec. 2016.

32 G. Wang, L. Yang, G. Zhang, X. Zhang, and D. Xu, "Comparative investigation of pseudo-random high-frequency signal injection schemes for sensorless IPMSM drives," *IEEE Trans. Power Electron.*, vol. 32, no. 3, pp. 2123–2132, Mar. 2017.

33 Z. Ma, J. Gao, and R. Kennel, "FPGA implementation of a hybrid sensorless control of SMPMSM in the whole speed range," *IEEE Trans. Ind. Inform.*, vol. 9, no. 3, pp. 1253–1261, Aug. 2013.

34 A. Yousefi-Talouki, P. Pescetto, G. Pellegrino, and I. Boldea, "Combined active flux and high-frequency injection methods for sensorless direct-flux vector control of synchronous reluctance machines," *IEEE Trans. Ind. Inform.*, vol. 33, no. 3, pp. 2447–2457, Mar. 2018.

35 C. Silva, G. M. Asher, and M. Sumner, "Hybrid rotor position observer for wide speed-range sensorless PM motor drives including zero speed," *IEEE Trans. Ind. Electron.*, vol. 53, no. 2, pp. 373–378, Apr. 2006.

36 S. C. Yang and Y. L. Hsu, "Full speed region sensorless drive of permanent-magnet machine combining saliency-based and back-EMF-based drive," *IEEE Trans. Ind. Electron.*, vol. 64, no. 2, pp. 1092–1101, Feb. 2017.

6

Saliency Tracking-Based Sensorless Control Method Using Zero Sequence Voltage

6.1 Introduction

Chapter 4 described the high-frequency signal injection (HFSI)-based rotor position sensorless control by utilizing the machine saliency and high-frequency (HF) current response. Apart from the HF current response, the HF zero sequence voltage (ZSV) response can also be used to estimate the rotor position [1] and will be introduced in this chapter. Generally, to measure the zero sequence HF voltage, a balanced resistor network and access to the machine neutral point are required [1], as shown in Figure 6.1.

Although additional hardware is required, utilizing the ZSV response has several benefits over the current response. According to [1]–[3], in terms of total harmonic distortion (THD), the zero sequence HF voltage response is less sensitive to the distortions caused by injection signal errors and HF response measurement errors compared to the HF current response [1], [2]. Moreover, it is observed that the system bandwidth can be significantly enhanced by using ZSV, and consequently position estimation accuracy and stability are improved [1].

In Chapter 4, to estimate the rotor position from current responses, several HF signals were introduced including the rotating sinusoidal signal, pulsating sinusoidal signal, and pulsating square-wave signal. These HF signals can also be employed with zero sequence HF voltage responses for position estimation as classified in Figure 6.2. The ZSV-based sensorless control is first used with a rotating sinusoidal signal injection that, however, is sensitive to delays [4]. In order to solve this problem, the pulsating signal injection can be used [4]. However, the conventional pulsating signal injection method with ZSV sensing deteriorates the position estimation performance, as will be shown later in this chapter. Thus, this chapter also introduces two improved pulsating HF signal injection strategies (sinusoidal [5] and square-wave signals [6]) that utilize zero

Sensorless Control of Permanent Magnet Synchronous Machine Drives, First Edition.
Zi Qiang Zhu and Xi Meng Wu.

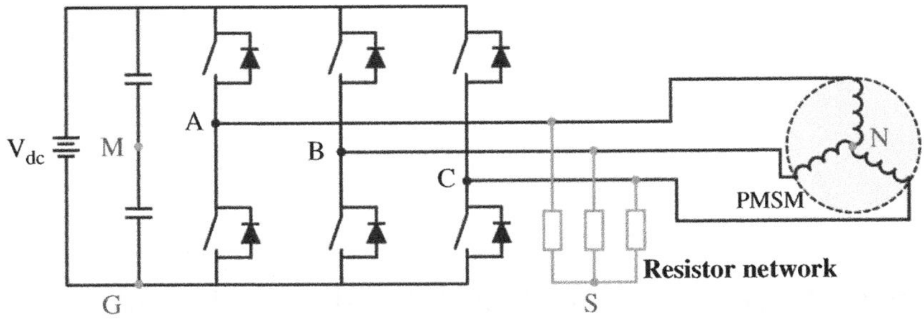

Figure 6.1 Zero sequence HF voltage measurement.

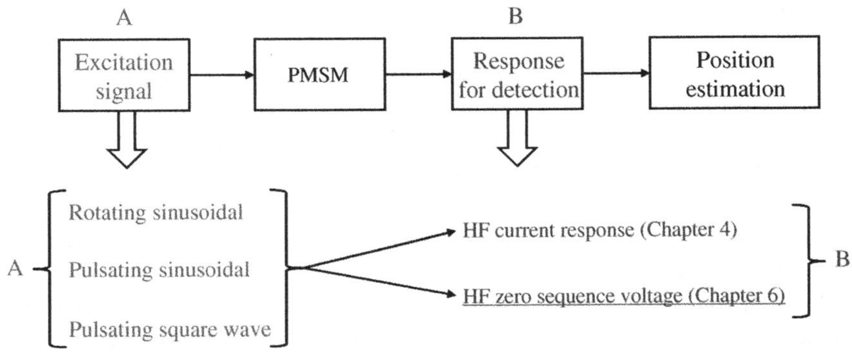

Figure 6.2 Classification of HF current and zero sequence voltage responses with different injection signals.

sequence HF voltage response for sensorless control. Different from the commonly used pulsating signal injection in the estimated synchronous reference frame, the improved methods will be injected on the estimated reference frame, anti-rotating at twice the rotor's estimated electrical speed. Then, the position estimation accuracy can be greatly improved. Additionally, the cross-saturation effect for the zero sequence HF voltage sensing method is also investigated, together with a proposed compensation method.

6.2 Rotating Sinusoidal Signal Injection

In this section, with a rotating sinusoidal signal injection at the stationary reference frame, the ZSV-based sensorless control method [1] is described including the ZSV model and position estimation.

6.2.1 Zero Sequence Voltage Model

First, considering high frequency, three-phase voltage equations neglecting resistances and back-EMFs can be expressed as:

$$\begin{bmatrix} v_{AN} \\ v_{BN} \\ v_{CN} \end{bmatrix} = \begin{bmatrix} L_{AA} & M_{AB} & M_{AC} \\ M_{BA} & L_{BB} & M_{BC} \\ M_{CA} & M_{CB} & L_{CC} \end{bmatrix} \begin{bmatrix} \frac{di_A}{dt} \\ \frac{di_B}{dt} \\ \frac{di_C}{dt} \end{bmatrix} \tag{6.1}$$

where v_{AN}, v_{BN}, v_{CN}, i_A, i_B, i_C are the voltages and currents, respectively, of phases A, B, C; L_{AA}, L_{BB}, L_{CC} are the three-phase self-inductances; and M_{AB}, M_{BA}, M_{AC}, M_{CA}, M_{BC}, M_{CB} are the three-phase mutual-inductances. The self- and mutual-inductances can generally be described as:

$$\begin{cases} L_{AA} = L_{s0} - L_{s2}\cos(2\theta_r) \\ L_{BB} = L_{s0} - L_{s2}\cos\left(2\theta_r + \frac{2}{3}\pi\right) \\ L_{CC} = L_{s0} - L_{s2}\cos\left(2\theta_r - \frac{2}{3}\pi\right) \end{cases} \tag{6.2}$$

$$\begin{cases} M_{AB} = M_{BA} = M_{s0} - M_{s2}\cos\left(2\theta_r - \frac{2}{3}\pi\right) \\ M_{BC} = M_{CB} = M_{s0} - M_{s2}\cos(2\theta_r) \\ M_{CA} = M_{AC} = M_{s0} - M_{s2}\cos\left(2\theta_r + \frac{2}{3}\pi\right) \end{cases} \tag{6.3}$$

where L_{s0}, L_{s2} are the dc and second harmonic amplitudes of self-inductances, and M_{s0}, M_{s2} are the dc and second harmonics amplitudes of mutual-inductances, respectively.

For the rotating sinusoidal signal injection with ZSV sensing [1], [3], the three-phase injection voltages can be expressed as:

$$\begin{aligned} v_{AM} &= V_h \cos \omega_h t \\ v_{BM} &= V_h \cos\left(\omega_h t - \frac{2}{3}\pi\right) \\ v_{CM} &= V_h \cos\left(\omega_h t + \frac{2}{3}\pi\right) \end{aligned} \tag{6.4}$$

where V_h and ω_h are the amplitude and frequency of the injected HF signal. The injection is illustrated in Figure 6.3.

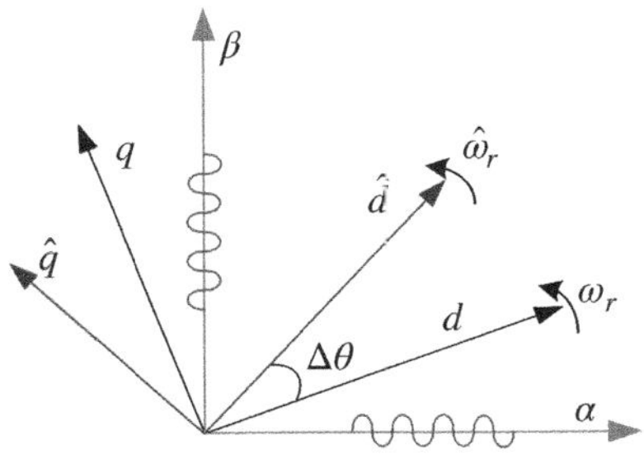

Figure 6.3 Rotating signal injection method in stationary reference frame. (*Source:* [5]/IEEE.)

As shown in Figure 6.1, the relation of the voltages for the machine neutral point N, the resistor network central point S, and the capacitor mid-point M can be expressed as:

$$\begin{aligned} v_{AM} &= v_{AN} - v_{SN} + v_{SM} \\ v_{BM} &= v_{BN} - v_{SN} + v_{SM} \\ v_{CM} &= v_{CN} - v_{SN} + v_{SM} \end{aligned} \tag{6.5}$$

where v_{SN} is the ZSV between the network central point S and the machine neutral point N, and v_{SM} is the voltage between the network central point S and the capacitor mid-point M, respectively.

It can be easily proved that v_{SM} equals to zero around the injection frequency. To be more specific, according to Figure 6.1, the relation of the voltages for the resistor network central point S, the capacitor mid-point M, and the point G of the dc bus can be expressed as:

$$\begin{aligned} v_{AG} &= v_{AS} + v_{SM} + v_{MG} \\ v_{BG} &= v_{BS} + v_{SM} + v_{MG} \\ v_{CG} &= v_{CS} + v_{SM} + v_{MG} \end{aligned} \tag{6.6}$$

Then, from (6.6) and Figure 6.1, the voltage of v_{SM} can be obtained as:

$$v_{SM} = \frac{v_{AG} + v_{BG} + v_{CG}}{3} - \frac{V_{dc}}{2} \tag{6.7}$$

where V_{dc} is the dc-link voltage. By only considering the components around HF injection frequency, v_{AG}, v_{BG}, and v_{CG} can be expressed as:

$$\begin{aligned} v_{AG} &= \frac{V_{dc}}{2} + V_h \cos \omega_h t \\ v_{BG} &= \frac{V_{dc}}{2} + V_h \cos\left(\omega_h t - \frac{2\pi}{3}\right) \\ v_{CG} &= \frac{V_{dc}}{2} + V_h \cos\left(\omega_h t + \frac{2\pi}{3}\right) \end{aligned} \tag{6.8}$$

Therefore, with (6.8), it can be easily derived from (6.7) that v_{SM} equals zero at HF injection frequency. Then, for three-phase Y-connected windings, it is derived that:

$$\frac{di_A}{dt} + \frac{di_B}{dt} + \frac{di_C}{dt} = 0 \tag{6.9}$$

With (6.1)–(6.9), the zero sequence HF voltage v_{SN} in Figure 6.1 can be obtained as:

$$\begin{aligned} v_{\mathrm{SN}} &= v_{o1}\cos(\omega_h t + 2\theta_r) - v_{o2}\cos(\omega_h t - 4\theta_r) \\ v_{o1} &= V_h \frac{2(-M_{s0} + L_{s0})}{4L_{s0}{}^2 - 8L_{s0}M_{s0} - L_{s2}{}^2 - 4M_{s2}{}^2 - 4L_{s2}M_{s2} + 4M_{s0}{}^2}(-M_{s2} + L_{s2}) \\ v_{o2} &= V_h \frac{(-2M_{s2} + L_{s2})}{4L_{s0}{}^2 - 8L_{s0}M_{s0} - L_{s2}{}^2 - 4M_{s2}{}^2 - 4L_{s2}M_{s2} + 4M_{s0}{}^2}(-M_{s2} + L_{s2}) \end{aligned} \tag{6.10}$$

From (6.10), the zero sequence HF voltage for the rotating sinusoidal signal injection consists of two frequency components at $\omega_h t + 2\theta_r$ and $\omega_h t - 4\theta_r$ [1], [3].

The finite element simulation results of the incremental self- and mutual-inductances for the prototype IPMSM-II (Appendix B) are shown in Figure 6.4. From Figure 6.4, since typically $2(-M_0 + L_0)$ is much larger than $(-2M_2 + L_2)$, the first term of (6.10) is normally utilized for rotor position estimation [1], i.e.

$$v_{\mathrm{SN}} = V_h \frac{2(-M_{s2} + L_{s2})(-M_{s0} + L_{s0})}{4L_{s0}{}^2 - 8L_{s0}M_{s0} - L_{s2}{}^2 - 4M_{s2}{}^2 - 4L_{s2}M_{s2} + 4M_{s0}{}^2}\cos(\omega_h t + 2\theta_r) \tag{6.11}$$

It is worth noting that the signal amplitude in (6.11) for position tracking is not related to the injection frequency. Thus, compared with the negative sequence HF current response introduced in Chapter 4, where the signal amplitude is inversely proportional to the injection frequency, a larger injection frequency can be employed for the ZSV method, leading to boosted system bandwidths, which consequently increases the stability of the system [2].

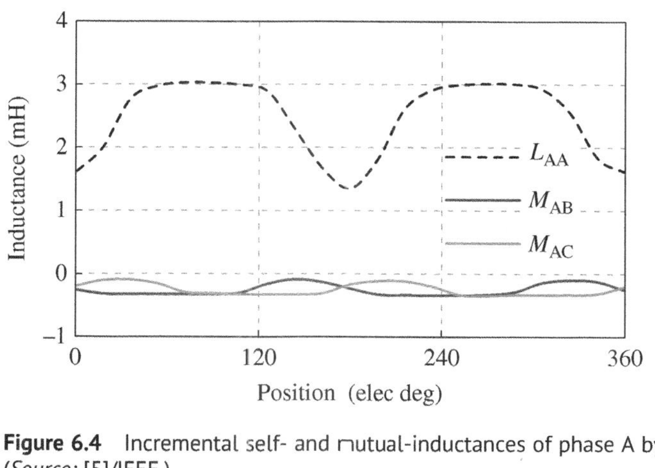

Figure 6.4 Incremental self- and mutual-inductances of phase A by FE simulation. (*Source:* [5]/IEEE.)

6.2.2 Signal Demodulation

In order to obtain the position information from (6.11), the zero sequence HF voltage is further processed as:

$$v_{\text{SN\_cos}} = \text{LPF}(v_{\text{SN}} \cdot (-\cos(\omega_h t))) = -\frac{1}{2}v_{o1}\cos(2\theta_r) \tag{6.12}$$

$$v_{\text{SN\_sin}} = \text{LPF}(v_{\text{SN}} \cdot (-\sin(\omega_h t))) = \frac{1}{2}v_{o1}\sin(2\theta_r) \tag{6.13}$$

Then, based on (6.12) and (6.13), the estimated rotor position can be obtained with the position observer as shown in Figure 6.5.

From the above analyses, it can be seen that the ZSV response is based on rotating signal injection and phase-modulated by the saliency position in (6.11) [1]. However, as reported in [4], the rotating signal injection is sensitive to signal processing delays since the phase of saliency position can be shifted by the delays and consequently these delays can cause larger position estimation errors. In contrast, it is found that the pulsating sinusoidal signal injection in the estimated synchronous reference frame introduced in Chapter 4 is more accurate and robust to these delays since the resultant HF current response is amplitude-modulated by saliency position with the intrinsic cancelation of the phase shifts caused by signal processing delays [4]. Besides, the signal demodulation is less computationally intensive compared to that of rotating sinusoidal signal injection [4]. Thus, it is expected that the ZSV sensing method with pulsating sinusoidal signal injection should be more advantageous since both the merits can be integrated. However, conventional pulsating sinusoidal signal injection in the estimated synchronous reference frame using ZSV degrades the position estimation performance; this is analyzed in the next section.

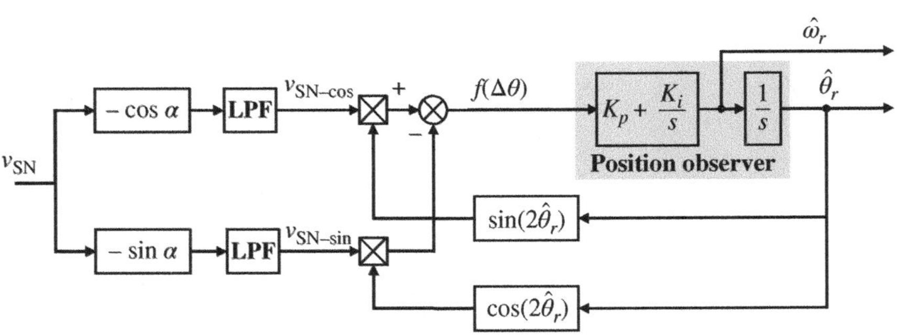

Figure 6.5 Zero sequence HF signal demodulation for rotating sinusoidal signal injection sensorless control [5].

6.3 Conventional Pulsating Sinusoidal Signal Injection

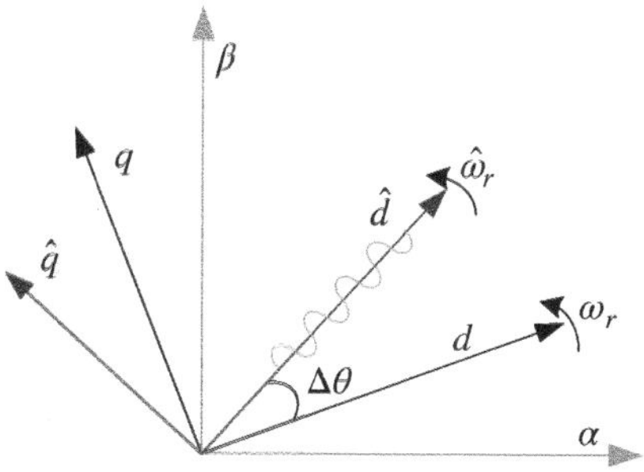

Figure 6.6 Pulsating signal injection method in estimated synchronous reference frame. (*Source:* [5]/IEEE.)

Specifically, if the conventional pulsating signal injection on the estimated synchronous reference frame is employed, as shown in Figure 6.6, the injected voltages of three phases can be expressed as:

$$\begin{aligned} v_{\rm AM} &= V_h \cos\omega_h t \cos(\hat{\theta}_r) \\ v_{\rm BM} &= V_h \cos\omega_h t \cos\left(\hat{\theta}_r - \frac{2}{3}\pi\right) \\ v_{\rm CM} &= V_h \cos\omega_h t \cos\left(\hat{\theta}_r + \frac{2}{3}\pi\right) \end{aligned} \tag{6.14}$$

Similarly, with (6.1)–(6.3), (6.5)–(6.9), and (6.14), the resultant zero sequence HF voltage can be obtained as:

$$v_{\rm SN} = V_h \frac{2(-M_{s2}+L_{s2})(-M_{s0}+L_{s0})}{4L_{s0}{}^2 - 8L_{s0}M_{s0} - L_{s2}{}^2 - 4M_{s2}{}^2 - 4L_{s2}M_{s2} + 4M_{s0}{}^2} \cos\omega_h t \cos(2\theta_r + \hat{\theta}_r) \tag{6.15}$$

Then, the signal demodulation can be expressed as:

$$\begin{aligned} &v_{{\rm SN}-sum} = {\rm LPF}(v_{\rm SN}\cdot 4\sin(\omega_h t - 3\hat{\theta}_r)) = \\ &\frac{2V_h(-M_{s2}+L_{s2})(-M_{s0}+L_{s0})}{4L_{s0}{}^2 - 8L_{s0}M_{s0} - L_{s2}{}^2 - 4M_{s2}{}^2 - 4L_{s2}M_{s2} + 4M_{s0}{}^2}(\sin(2\Delta\theta) + \sin(2\theta_r + 4\hat{\theta}_r)) \end{aligned} \tag{6.16}$$

where $v_{{\rm SN}-sum}$ is the final position information signal. From (6.16), it can be seen that additional sixth harmonics having larger amplitudes arise in the steady condition, which is expected to affect the stability of sensorless control and significantly deteriorate the position estimation performance.

Based on the prototype IPMSM-II, an experiment example is shown in Figure 6.7. In the experiment, an 800 Hz, 4 V pulsating sinusoidal signal is injected for position estimation. With the zero sequence HF voltage response for the pulsating sinusoidal signal injection, the estimated rotor position and position estimation error are obtained as shown in Figure 6.7. It can be clearly

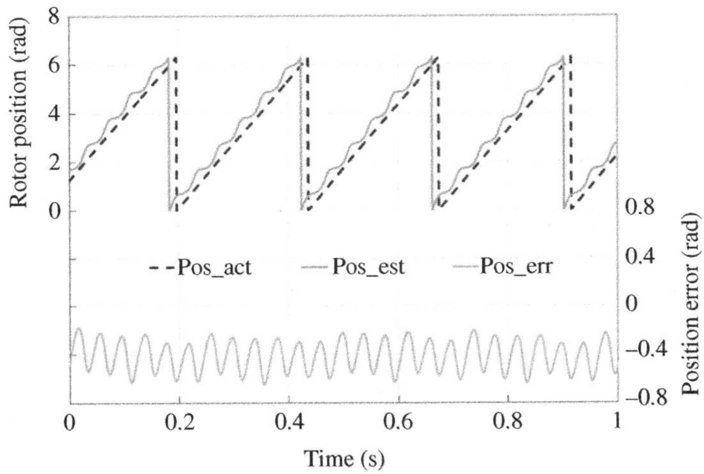

Figure 6.7 Position estimation with conventional pulsating signal injection in estimated synchronous reference frame. (*Source:* [5]/IEEE.)

observed that large dc and oscillation errors arise, which indicates that the ZSV response for the conventional pulsating signal injection is not suitable for rotor position tracking.

6.4 Anti-rotating Pulsating Sinusoidal Signal Injection

From the above analyses, it can be seen that the conventional pulsating injection using ZSV has undesirable dc and large sixth harmonic errors. In order to utilize the synergies of both ZSV sensing and pulsating signal injection, an improved anti-rotating pulsating signal injection method utilizing zero sequence HF voltage is presented for sensorless control of PMSMs. The HF signal is injected on the estimated reference frame, anti-rotating at twice the rotor's estimated electrical angular speed. The anti-rotating pulsating sinusoidal signal injection method is of high bandwidth and stability due to the use of zero sequence HF voltage and of high accuracy and fast dynamic response. Furthermore, the cross-saturation effect due to magnetic saturation for ZSV-based sensorless control is investigated and compensated for. All the theoretical analyses have been compared with the experimental results in both steady and dynamic conditions.

6.4.1 Anti-rotating Signal Injection

From (6.15), it is noted that the ZSV response is modulated by the component of $(2\theta_r + \hat{\theta}_r)$, where $2\theta_r$ appears because the phase inductances vary twice in one electrical period, while $\hat{\theta}_r$ is there because the injection axis is in the estimated synchronous reference frame. Therefore, it is expected that if the injection axis is anti-rotating in the estimated reference frame, it is possible for the resultant zero sequence response to be amplitude-modulated by $\sin(2\Delta\theta)$, which will be exactly the same for the HF current response of the conventional pulsating sinusoidal signal injection, resulting in high accuracy.

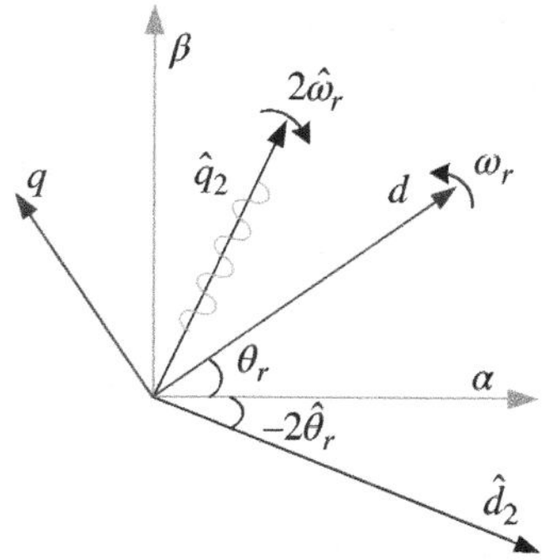

Figure 6.8 Improved anti-rotating pulsating signal injection strategy. (*Source:* [5]/IEEE.)

With the above analyses, an improved HF signal injection method is presented as shown in Figure 6.8. It can be seen that the injection is performed on the q_2-axis in the estimated reference frame, anti-rotating at twice the rotor's estimated electrical speed. The injected signal can be expressed as:

$$\begin{aligned} \hat{v}_{2d} &= 0 \\ \hat{v}_{2q} &= V_h \cos \omega_h t \end{aligned} \tag{6.17}$$

Then, with a coordinate transformation matrix, the injection voltage in the stationary reference frame is described as:

$$\begin{bmatrix} v_\alpha \\ v_\beta \end{bmatrix} = \begin{bmatrix} \cos(-2\hat{\theta}_r) & -\sin(-2\hat{\theta}_r) \\ \sin(-2\hat{\theta}_r) & \cos(-2\hat{\theta}_r) \end{bmatrix} \begin{bmatrix} \hat{v}_{2d} \\ \hat{v}_{2q} \end{bmatrix} \tag{6.18}$$

Further, with (6.17) and (6.18), the actual injected three phase voltages can be expressed as:

$$\begin{aligned} v_{\mathrm{AM}} &= V_h \cos \omega_h t \sin(2\hat{\theta}_r) \\ v_{\mathrm{BM}} &= V_h \cos \omega_h t \sin\left(2\hat{\theta}_r + \frac{2}{3}\pi\right) \\ v_{\mathrm{CM}} &= V_h \cos \omega_h t \sin\left(2\hat{\theta}_r - \frac{2}{3}\pi\right) \end{aligned} \tag{6.19}$$

Similarly, with (6.1)–(6.3), (6.5)–(6.9), and (6.19), the zero sequence HF voltage can be derived as:

$$
\begin{aligned}
v_{SN} &= v_{o1} \cos \omega_h t \sin(2\Delta\theta) - v_{o2} \cos \omega_h t \sin\left(4\theta_r + 2\hat{\theta}_r\right) \\
v_{o1} &= V_h \frac{2(-M_{s0} + L_{s0})}{4{L_{s0}}^2 - 8L_{s0}M_{s0} - {L_{s2}}^2 - 4{M_{s2}}^2 - 4L_{s2}M_{s2} + 4{M_{s0}}^2}(-M_{s2} + L_{s2}) \\
v_{o2} &= V_h \frac{(-2M_2 + L_2)}{4{L_{s0}}^2 - 8L_{s0}M_{s0} - {L_{s2}}^2 - 4{M_{s2}}^2 - 4L_{s2}M_{s2} + 4{M_{s0}}^2}(-M_{s2} + L_{s2})
\end{aligned}
\tag{6.20}
$$

Then, similar to (6.10) and (6.11), the first term of (6.20) has a much larger amplitude and will be utilized for the position estimation, i.e.

$$
v_{SN} = V_h \frac{2(-M_{s0} + L_{s0})(-M_{s2} + L_{s2})}{4{L_{s0}}^2 - 8L_{s0}M_{s0} - {L_{s2}}^2 - 4{M_{s2}}^2 - 4L_{s2}M_{s2} + 4{M_{s0}}^2} \cos \omega_h t \sin(2\Delta\theta)
\tag{6.21}
$$

From (6.21), it is observed that the ZSV response is very simple. Furthermore, from (6.21), the ZSV is amplitude-modulated by machine saliency, and therefore, the phase shifts of saliency position due to signal processing delays can be approximately canceled. Also, compared to the conventional pulsating injection on the estimated synchronous reference frame in (6.16), the undesirable sixth harmonic amplitude is much smaller.

Thus, it is expected that the improved method can combine the synergies of both pulsating sinusoidal signal injection and ZSV sensing, i.e. high accuracy and bandwidths, great dynamic response and robustness, etc., as will be verified experimentally later. It should be noted that the principle of position tracking enables the ZSV to be zero in steady conditions, i.e. $v_{SN} = 0$ when $\Delta\theta = 0$.

6.4.2 Signal Demodulation

To further obtain the estimated rotor position, the ZSV response in (6.21) can be processed as:

$$
\begin{aligned}
v_{SN-sum} &= \mathrm{LPF}(v_{SN} \cdot 2\cos \omega_h t) \\
&= \frac{2V_h(-M_{s0} + L_{s0})(-M_{s2} + L_{s2})}{4{L_{s0}}^2 - 8L_{s0}M_{s0} - {L_{s2}}^2 - 4{M_{s2}}^2 - 4L_{s2}M_{s2} + 4{M_{s0}}^2} \sin(2\Delta\theta)
\end{aligned}
\tag{6.22}
$$

Accordingly, the signal demodulation for the zero sequence HF voltage is shown in Figure 6.9. Due to the simpler signal response in (6.21) and (6.22), the signal demodulation is also much simpler than that of the rotating sinusoidal signal injection method as introduced in Section 6.2.2.

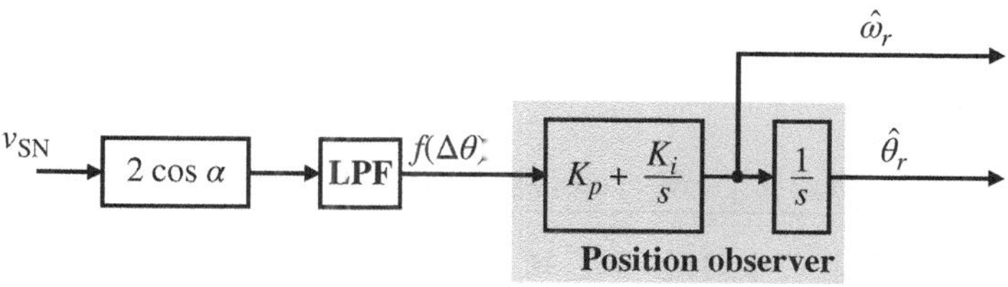

Figure 6.9 Signal demodulation process for anti-rotating pulsating sinusoidal signal injection method [5].

6.4.3 Cross-Saturation Effect

The cross-saturation effect has been systematically investigated for the HF current sensing-based HF signal injection methods in Chapter 5. The current section discusses the cross-saturation effect for the ZSV sensing-based methods.

First, by considering the saturation effects, the three phase inductances can be expressed as [7]:

$$\begin{cases} L_{AA} = L_{s0} - L_{s2}\cos(2\theta_r) - L_c\sin(2\theta_r) \\ L_{BB} = L_{s0} - L_{s2}\cos\left(2\theta_r + \frac{2}{3}\pi\right) - L_c\sin\left(2\theta_r + \frac{2}{3}\pi\right) \\ L_{CC} = L_{s0} - L_{s2}\cos\left(2\theta_r - \frac{2}{3}\pi\right) - L_c\sin\left(2\theta_r - \frac{2}{3}\pi\right) \end{cases} \tag{6.23}$$

$$\begin{cases} M_{AB} = M_{BA} = M_{s0} - M_{s2}\cos\left(2\theta_r - \frac{2}{3}\pi\right) - M_c\sin\left(2\theta_r - \frac{2}{3}\pi\right) \\ M_{BC} = M_{CB} = M_{s0} - M_{s2}\cos(2\theta_r) - M_c\sin(2\theta_r) \\ M_{CA} = M_{AC} = M_{s0} - M_{s2}\cos\left(2\theta_r + \frac{2}{3}\pi\right) - M_c\sin\left(2\theta_r + \frac{2}{3}\pi\right) \end{cases} \tag{6.24}$$

where L_c and M_c represent the second harmonic amplitude of sine inductance terms in the self- and mutual-inductances, respectively. By way of example, to introduce the phase shift angle caused by the cross-saturation effect, the self- and mutual-inductances of phase A can be rewritten as:

$$\begin{aligned} L_{AA} &= L_{s0} - L'_{s2}\cos(2\theta_r + \theta_{s1}) \\ M_{AB} &= M_{s0} - M'_{s2}\cos\left(2\theta_r - \frac{2\pi}{3} + \theta_{s2}\right) \end{aligned} \tag{6.25}$$

where $L'_{s2} = \sqrt{L_{s2}^2 + L_c^2}$, $M'_{s2} = \sqrt{M_{s2}^2 + M_c^2}$, $\theta_{s1} = \tan^{-1}(-L_c/L_{s2})$, and $\theta_{s2} = \tan^{-1}(-M_c/M_{s2})$.

The same amplitude and phase shifts can also be derived for phases B and C, respectively. Then, with rotating signal injection, the zero sequence HF voltage can be obtained as:

$$v_{SN} = \frac{V_h(L_{s0} - M_{s0})(L'_{s2} - M'_{s2})\cos(\omega_h t + 2\theta_r + \theta_{s3})}{2L_{s0}^2 + 2M_{s0}^2 - 4L_{s0}M_{s0} - 2L'_{s2}M'_{s2} - \frac{{L'_{s2}}^2}{2} - 2{M'_{s2}}^2} \tag{6.26}$$

And for the anti-rotating pulsating sinusoidal signal injection, the zero sequence HF voltage can be obtained as:

$$v_{SN} = \frac{V_h(L_{s0} - M_{s0})(L'_{s2} - M'_{s2})\cos\omega_h t \sin(2\Delta\theta + \theta_{s3})}{2L_{s0}^2 + 2M_{s0}^2 - 4L_{s0}M_{s0} - 2L'_{s2}M'_{s2} - \frac{L_s^2}{2} - 2M'_{s2}} \tag{6.27}$$

where θ_{s3} presents the phase shift of the zero sequence HF voltage due to the cross-saturation effect and can be derived as:

$$\theta_{s3} = -\arctan\frac{L'_{s2}\sin(\theta_{s1}) - M'_{s2}\sin(\theta_{s2})}{L'_{s2}\cos(\theta_{s1}) - M'_{s2}\cos(\theta_{s2})} = -\arctan\frac{L_c - M_c}{L_{s2} - M_{s2}} \tag{6.28}$$

Therefore, from (6.26) and (6.27), the position estimation errors due to the cross-saturation effect are exactly the same for the two injection methods with ZSV sensing. Accordingly, the position estimation errors can be obtained from (6.28) as:

$$\Delta\theta = -\frac{1}{2}\theta_{s3} = \frac{1}{2}\arctan\frac{L_c - M_c}{L_{s2} - M_{s2}} \tag{6.29}$$

In order to compensate for the position errors due to the cross-saturation effect, offline measurements introduced in Chapter 5, etc. are normally used for the compensation. Similarly, in this chapter, the estimation error caused by the cross-saturation effect is approximately proportional to the fundamental q-axis current, i.e. $\theta_m \approx k_c i_q$, as shown in Figure 6.10. Thus, the measured results can be directly employed for online compensation.

6.4.4 Experimental Results

Experiments are implemented to validate the anti-rotating pulsating sinusoidal signal injection method on the prototype IPMSM-II (Appendix B). The overall control block diagram for the anti-rotating injection method with ZSV sensing is shown in Figure 6.11. In order to measure the zero sequence HF voltage, one balanced resistor network, one additional A/D sampling channel, one hardware circuit for signal conditioning, and also the access to the machine neutral point are required, as shown in Figure 6.1. The inverter switching frequency is set to 10 kHz and is the same as the current/voltage sampling frequency.

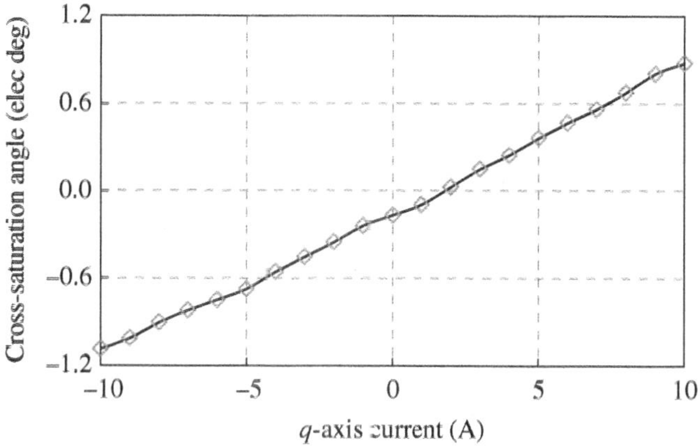

Figure 6.10 Measured phase shift due to cross-saturation effect with q-axis current. (*Source:* [5]/IEEE.)

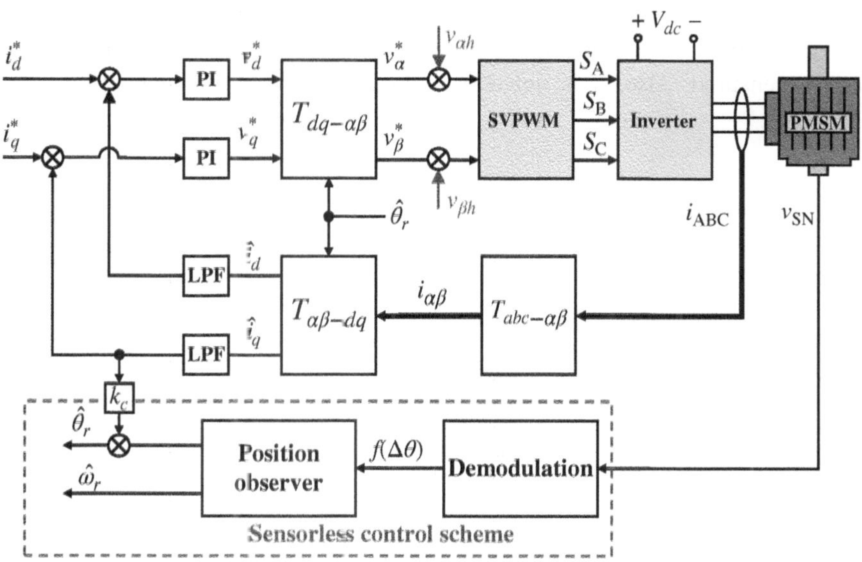

Figure 6.11 Block diagram of overall control system with HF signal injection using zero sequence voltage.

6.4.4.1 Zero Sequence Voltage Model Verification

Figure 6.12 illustrates the measured zero sequence HF voltage spectra. Specifically, for the anti-rotating method, the position error is set to $\pi/4$ with the assistance of a position sensor to maximize the zero sequence signal. The injected signal is set as 8 V and 600 Hz, and the fundamental frequency f_e is 2.5 Hz. From Figure 6.12, it is observed that the ZSV components for rotating sinusoidal signal injection are primarily at $f_h + 2f_e$ and $f_h - 4f_e$ (242nd and 236th), while the conventional pulsating sinusoidal signal injection in the estimated synchronous reference frame is mainly at $f_h \pm 3f_e$ (237th and 243rd). By comparison, for the anti-rotating pulsating sinusoidal signal injection method, the frequency components at f_h and $f_h \pm 6f_e$ (240th and 234th/246th) arise as illustrated in Figure 6.12. The measured results for all three are consistent with the theoretical analyses in (6.10), (6.15), and (6.20), respectively.

6.4.4.2 Steady- and Dynamic-State Position Estimation Performances

The steady-state sensorless control performance with the anti-rotating pulsating sinusoidal signal injection method is illustrated in Figure 6.13 with rated load at 10 A. The injection signal is set as 4 V and 800 Hz with the rotor reference speed at 40 r/min. It can be seen that under load condition, due to the cross-saturation effect, there exist significant position estimation errors. Then, by using the compensation methods as given in Figure 6.10, more steady estimation performance can be ensured. Also, it is noted from Figure 6.13 that the noticeable sixth

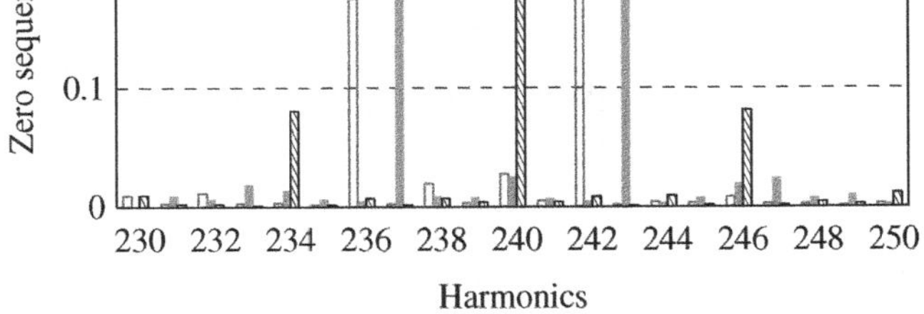

Figure 6.12 Measured zero sequence HF voltage with different injection methods. (*Source:* [5]/IEEE.)

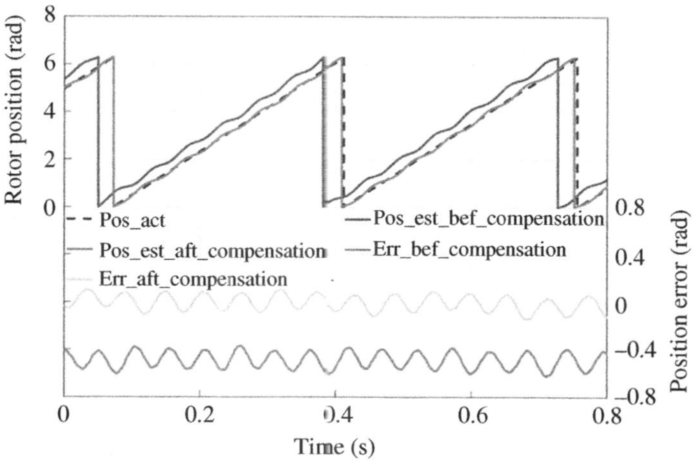

Figure 6.13 Position estimation with and without compensation of cross-saturation effect in steady state. (*Source:* [5]/IEEE.)

harmonic position error exists and may mainly arise from the second term of (6.10) and (6.20), and therefore, the decoupling of the undesirable effect is important [1], [2].

The dynamic performance with the anti-rotating pulsating sinusoidal signal injection method is shown in Figure 6.14, where the rotor speed reference is given as 0–50–100 r/min. The fast dynamic response can be observed, and the overall position errors are small with cross-saturation effect compensation.

The step-load test is also implemented on the prototype PM machine as shown in Figure 6.15. Good position tracking performances are observed when the q-axis current is rapidly increased from 0.3 to 10 A. Therefore, it can be concluded that the anti-rotating pulsating sinusoidal signal injection method presents good dynamic performance at various speed and load conditions.

6.4.4.3 Robustness and Accuracy Comparison

To further illustrate the robustness and accuracy of the anti-rotating signal injection method, the conventional injection methods (i.e. rotating and conventional pulsating sinusoidal signal injections) with ZSV sensing are compared to the anti-rotating-based method. Specifically, the sensitivity to the injection frequency and the cutoff frequency of the low-pass filter (LPF) for signal demodulation are compared for the three injection methods. The comparative conditions for the three injection methods are listed in Table 6.1.

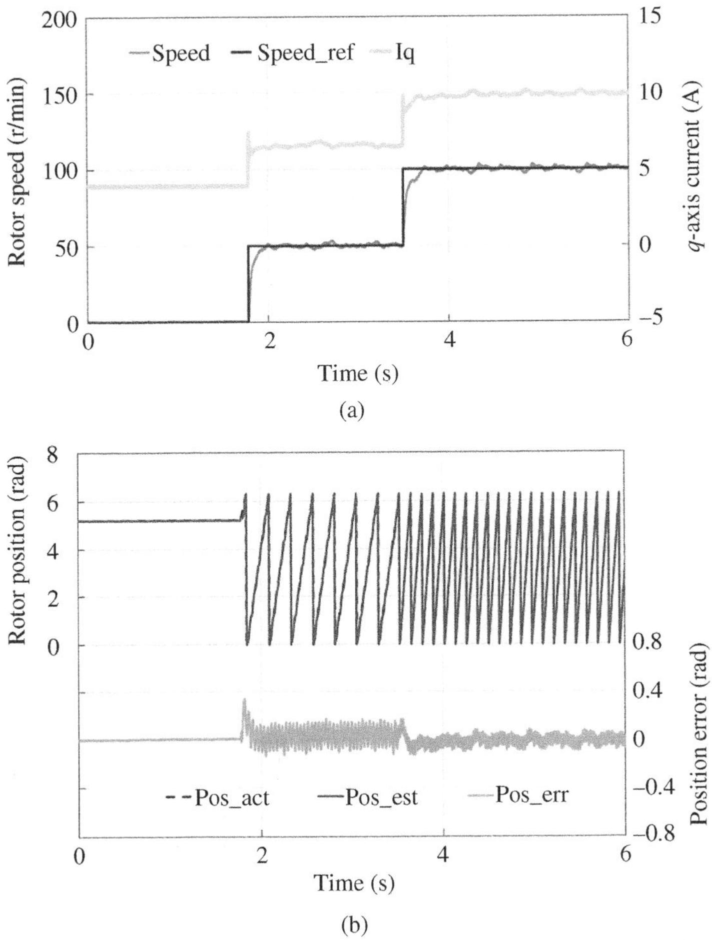

Figure 6.14 Dynamic position estimation performances under step-speed condition. (a) Rotor speed and q-axis current. (b) Rotor position estimation. (*Source:* [5]/IEEE.)

Figure 6.16 illustrates the position errors for three HFs with HF injection frequencies ranging from 300 to 1000 Hz while the cutoff frequency of LPF is kept the same at 60 Hz. It can be seen that position errors for rotating and conventional pulsating sinusoidal signal injection methods are very sensitive to the injection frequency. This is due to the signal processing delays (PWM updating delay), which introduce additional phase shifts in the HF responses as discussed earlier. While for the anti-rotating based signal injection method, the figure demonstrates good robustness due to the amplitude-modulation by saliency as analyzed above.

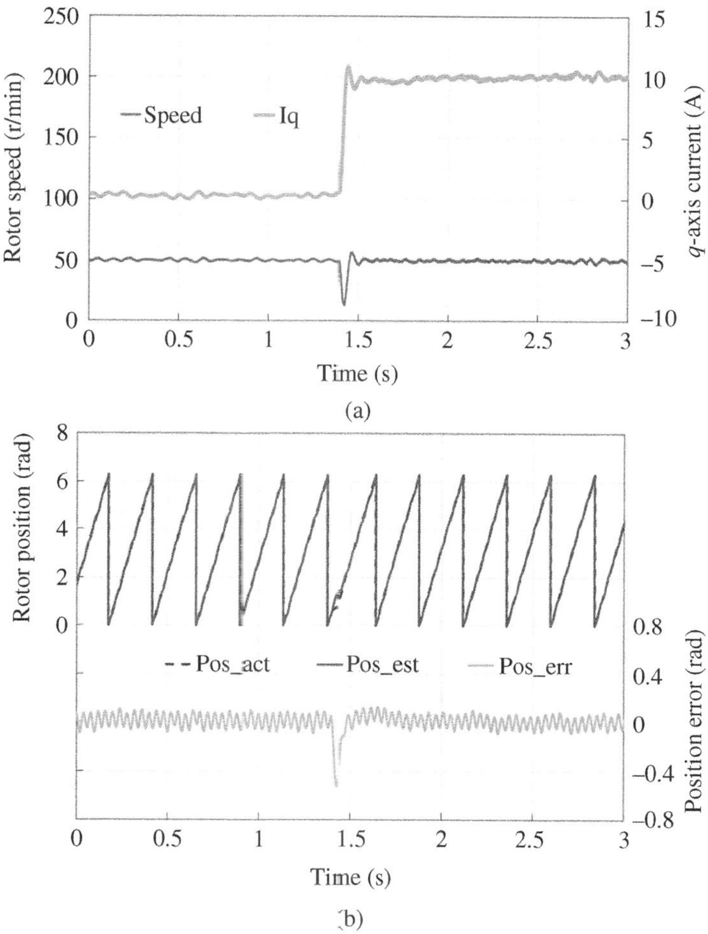

Figure 6.15 Dynamic position estimation performances under step-load condition. (a) Rotor speed and q-axis current. (b) Rotor position estimation. (*Source:* [5]/IEEE.)

Table 6.1 Comparison conditions for three injection methods.

				Position observer	
	Method	**HF signal**	***LPF* (Hz)**	K_p	K_i
Figure 6.16	Rotating	4 V (0.3–1 kHz)	60	400	1000
	Conventional pulsating		60	400	1000
	Presented		60	400	1000
Figure 6.17	Rotating	4 V/800 Hz	10–60	400	1000
	Conventional pulsating			400	1000
	Presented			400	1000

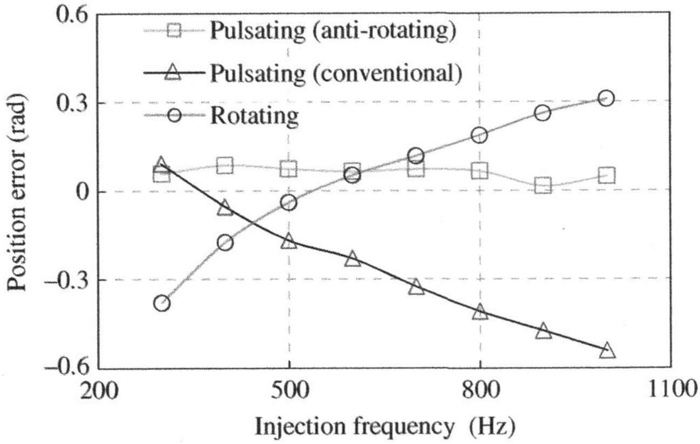

Figure 6.16 Position errors with sensitivity to injection frequency. (*Source:* [5]/IEEE.)

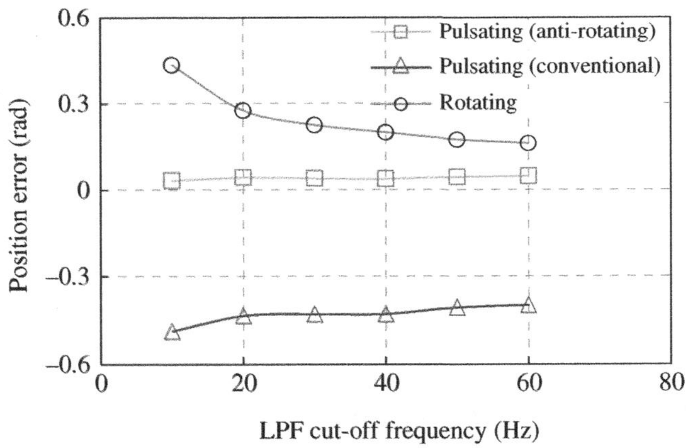

Figure 6.17 Position errors with sensitivity to cut-off frequency of LPF. (*Source:* [5]/IEEE.)

Similarly, as shown in Figure 6.17, at the same injection frequency 800 Hz, the position estimation errors of rotating sinusoidal signal injection are observed to vary with the cutoff frequency of LPF. This is due to the rotor angular frequency causing the phase shifts of the saliency position due to LPF. In contrast, the conventional and anti-rotating pulsating sinusoidal signal injection methods are not sensitive to the cutoff frequency of LPF. However, it is clearly observed that the anti-rotating injection method is more accurate. Thus, compared to conventional injection methods, the anti-rotating-based injection method utilizing ZSV is more advantageous.

6.5 Conventional Pulsating Square-Wave Signal Injection

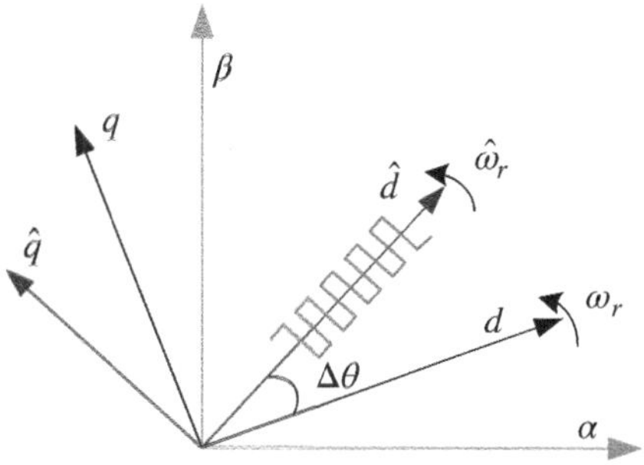

Figure 6.18 Pulsating square-wave signal injection in estimated synchronous reference frame. (*Source:* [6]/IEEE.)

In the previous section, a pulsating sinusoidal signal injection strategy is presented utilizing the zero sequence HF voltage for sensorless control of PMSMs. As introduced in Chapter 4, a pulsating square-wave signal can also be injected to increase the system bandwidth [8], [9]. However, similar to the conclusion in Section 6.3 for the conventional pulsating square-wave signal injection methods, the resultant zero sequence HF voltage responses deteriorate the position estimation performance, as will be analyzed in detail in this section.

First, for the pulsating square-wave injection in the estimated synchronous reference frame [9] in Figure 6.18, the injected signals are expressed as:

$$\begin{bmatrix} \hat{v}_{dh} \\ \hat{v}_{qh} \end{bmatrix} = \begin{bmatrix} V_h \cdot (-1)^n \\ 0 \end{bmatrix}, n = 0, 1, 2... \tag{6.30}$$

where V_h is the HF voltage, and n is an integer and the number of the injected voltages.

Considering the Fourier series of a square-wave signal, (6.30) can be further expressed as:

$$\begin{aligned} \hat{v}_{dh} &= \frac{4}{\pi} V_h \sum_{k=1} \frac{1}{2k-1} \cos k\omega_h t \\ \hat{v}_{qh} &= 0 \end{aligned} \tag{6.31}$$

Then, with the coordinate transformation, the three phase injection voltages can be described as:

$$\begin{aligned} v_{\mathrm{AM}} &= \frac{4}{\pi} V_h \cos(\hat{\theta}_r) \sum_{k=1} \frac{1}{2k-1} \cos k\omega_h t \\ v_{\mathrm{BM}} &= \frac{4}{\pi} V_h \cos\left(\hat{\theta}_r - \frac{2}{3}\pi\right) \sum_{k=1} \frac{1}{2k-1} \cos k\omega_h t \\ v_{\mathrm{CM}} &= \frac{4}{\pi} V_h \cos\left(\hat{\theta}_r + \frac{2}{3}\pi\right) \sum_{k=1} \frac{1}{2k-1} \cos k\omega_h t \end{aligned} \tag{6.32}$$

Then, based on (6.1)–(6.3), (6.5)–(6.7), (6.9), and (6.30)–(6.32), the zero sequence HF voltage v_{SN} as shown in Figure 6.1 can be obtained as:

$$v_{SN} = \frac{2}{\pi}\frac{V_h(L_{s0}-M_{s0})(L_{s2}-M_{s2})\cos\left(2\theta_r+\hat{\theta}_r\right)}{(L_{s0}-M_{s0})^2-\left(\frac{L_{s2}}{2}+M_{s2}\right)^2}\sum_{k=1}\frac{1}{2k-1}\cos k\omega_h t \tag{6.33}$$

However, as aforementioned, the ZSV in (6.33) for the conventional pulsating square-wave signal injection method deteriorates the position estimation performance. To be more specific, from (6.33), the corresponding signal demodulation is expressed as:

$$\begin{aligned} v_{SN\_d} &= v_{RN}\cdot\frac{\hat{v}_d}{V_h}\cdot\left(-\sin\left(3\,\hat{\theta}_r\right)\right) \\ &= \frac{1}{\pi}\frac{V_h(L_{s0}-M_{s0})(L_{s2}-M_{s2})}{(L_{s0}-M_{s0})^2-\left(\frac{L_{s2}}{2}+M_{s2}\right)^2}\left(\sin(2\Delta\theta)-\sin\left(2\theta_r+4\hat{\theta}_r\right)\right) \end{aligned} \tag{6.34}$$

From (6.34), it can be seen that the additional sixth harmonic (i.e. $\sin\left(2\theta_r+4\hat{\theta}_r\right)$) having the same amplitude with the term $\sin(2\Delta\theta)$ exists, which can severely affect the position estimation performance as shown in Figure 6.19, which was tested on the prototype IPMSM-II with a 4 V/2.5 kHz pulsating

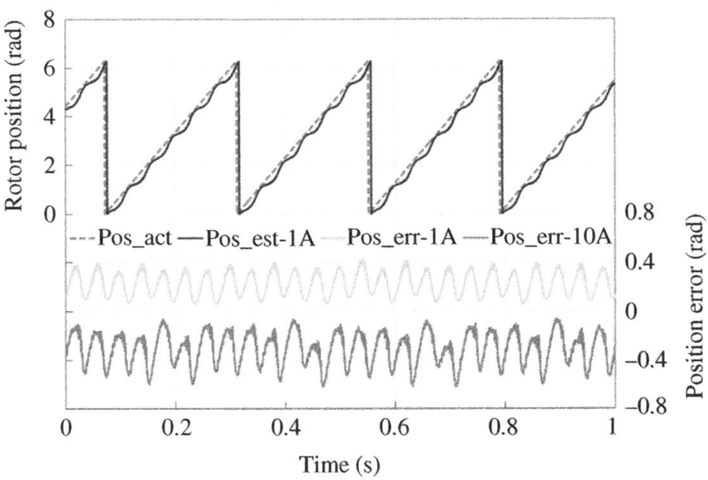

Figure 6.19 Position estimation performance conventional pulsating square-wave injection in estimated synchronous reference frame. (*Source:* [6]/IEEE.)

square-wave signal injected. Large oscillation errors can be observed. It is very difficult to suppress these errors from the disturbance of the sixth harmonics in (6.34).

From the above analysis, the resultant estimated position errors for the pulsating square-wave injections in the estimated synchronous reference frame are intolerable. Therefore, an improved pulsating square-wave signal injection method with ZSV sensing is presented in the next section.

6.6 Anti-rotating Pulsating Square-Wave Signal Injection

From the above analyses, it can be seen that the conventional pulsating square-wave signal injection using ZSV has undesirable dc and large sixth harmonic errors. In order to utilize the synergies of both ZSV sensing and pulsating signal injection, an improved pulsating square-wave signal injection method utilizing zero sequence HF voltage is introduced in this section. The square-wave signal is injected on the estimated reference frame, anti-rotating at twice the rotor's estimated electrical speed. The resultant HF response for the anti-rotating-based method is simpler for signal demodulation and has higher accuracy compared to the conventional pulsating square-wave methods.

6.6.1 Anti-rotating Signal Injection

An improved pulsating square-wave injection method with ZSV sensing is presented, as shown in Figure 6.20. It can be seen that the injection strategy is exactly the same as that presented in Section 6.4, but with different signal injection waveforms Specifically, from Figure 6.20, the pulsating square-wave injection is at the q_2-axis in the estimated reference frame, anti-rotating at twice the rotor's estimated electrical speed. Accordingly, the injected signal can be expressed as:

$$\hat{v}_{2d} = 0$$
$$\hat{v}_{2q} = \frac{4}{\pi} V_h \sum_{k=1} \frac{1}{2k-1} \cos k\omega_h t \tag{6.35}$$

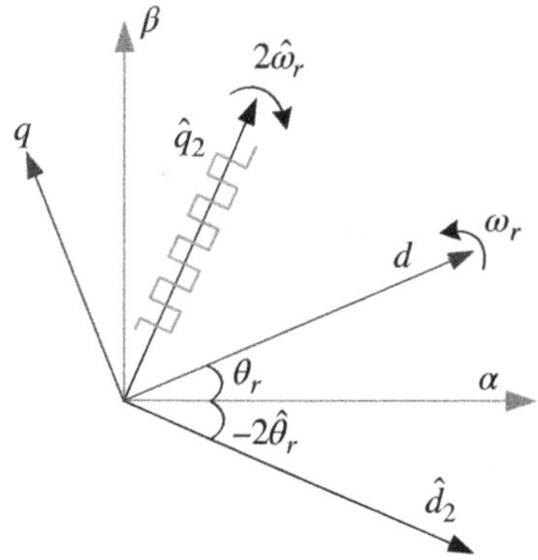

Figure 6.20 Anti-rotating pulsating square-wave signal injection. (*Source:* [6]/IEEE.)

Then, by coordinate transformation, the actual injected phase voltages can be expressed as:

$$
\begin{aligned}
v_{\mathrm{AM}} &= \frac{4}{\pi} V_h \sum_{k=1} \frac{1}{2k-1} \cos k\omega_h t \sin\left(2\hat{\theta}_r\right) \\
v_{\mathrm{BM}} &= \frac{4}{\pi} V_h \sum_{k=1} \frac{1}{2k-1} \cos k\omega_h t \sin\left(2\hat{\theta}_r + \frac{2}{3}\pi\right) \\
v_{\mathrm{CM}} &= \frac{4}{\pi} V_h \sum_{k=1} \frac{1}{2k-1} \cos k\omega_h t \sin\left(2\hat{\theta}_r - \frac{2}{3}\pi\right)
\end{aligned}
\tag{6.36}
$$

Similarly, with (6.1)–(6.3), (6.5)–(6.7), (6.9), and (6.36), the zero sequence HF voltage for the injection strategy can be derived as:

$$
\begin{aligned}
v_{\mathrm{SN}} &= \frac{4}{\pi}\left[v_{o1}\sin(2\Delta\theta) - v_{o2}\sin\left(4\theta_r + 2\hat{\theta}_r\right)\right] \sum_{k=1} \frac{1}{2k-1} \cos k\omega_h t \\
v_{o1} &= \frac{V_h}{2} \frac{(L_{s0} - M_{s0})}{(L_{s0} - M_{s0})^2 - \left(\frac{L_{s2}}{2} + M_{s2}\right)^2} (L_{s2} - M_{s2}) \\
v_{o2} &= \frac{V_h}{4} \frac{(L_{s2} - 2M_{s2})}{(L_{s0} - M_{s0})^2 - \left(\frac{L_{s2}}{2} + M_{s2}\right)^2} (L_{s2} - M_{s2})
\end{aligned}
\tag{6.37}
$$

Typically $(L_{s0} - M_{s0})$ is much larger than $(L_{s2} - 2M_{s2})$. The first term of (6.37) has much larger amplitude [1] and therefore will be utilized for the position estimation, i.e.

$$
v_{\mathrm{SN}} = \frac{2}{\pi} \frac{V_h(L_{s0} - M_{s0})(L_{s2} - M_{s2})\sin(2\Delta\theta)}{(L_{s0} - M_{s0})^2 - \left(\frac{L_{s2}}{2} + M_{s2}\right)^2} \sum_{k=1} \frac{1}{2k-1} \cos k\omega_h t
\tag{6.38}
$$

6.6.2 Signal Demodulation

By signal demodulation, (6.38) becomes:

$$
v_{\mathrm{SN\_}d} = v_{\mathrm{SN}} \cdot \frac{\hat{v}_{2q}}{V_h} = \frac{2}{\pi} \frac{V_h(L_{s0} - M_{s0})(L_{s2} - M_{s2})}{(L_{s0} - M_{s0})^2 - \left(\frac{L_{s2}}{2} + M_{s2}\right)^2} \sin(2\Delta\theta)
\tag{6.39}
$$

From (6.39), it can be seen that the zero sequence HF signal responses for the above method are amplitude-modulated by machine saliency, leading to higher estimation accuracy [4]. Also, compared to the conventional pulsating square-wave injections in the estimated synchronous reference frame in (6.34), the undesirable disturbance components do not exist for the improved

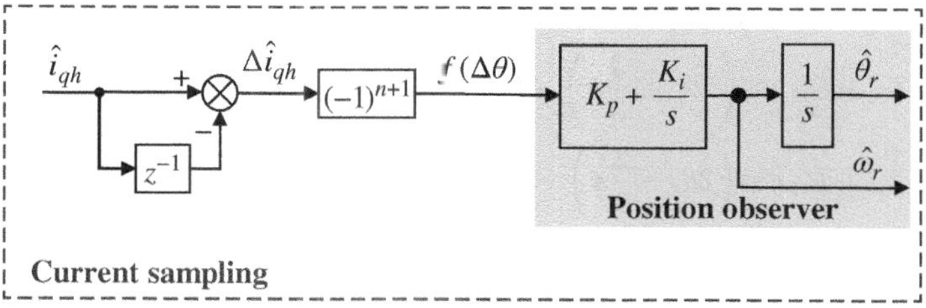

(a)

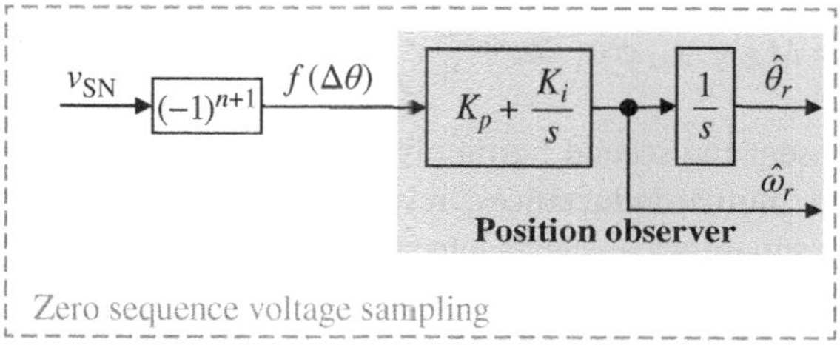

(b)

Figure 6.21 HF signal demodulation process. (a) Conventional current sampling-based method [9]. (b) Improved zero sequence voltage sampling-based method [6].

anti-rotating pulsating square-wave injection method. Thus, it is expected that the position estimation accuracy will be greatly enhanced.

Furthermore, from (6.38) and (6.39), it can be seen that the zero sequence HF voltage for the anti-rotating-based method also keeps the advantages of the conventional pulsating square-wave injection methods. First, the signal response is also modulated by the square waveform, and accordingly the signal differentiation for rotor position estimation is not required. Consequently, the signal demodulation process is also much simpler, as shown in Figure 6.21, where the conventional current sampling-based method in [8], [9] is also illustrated for comparison. Second, the ZSV for the anti-rotating-based method is also not related to the injection frequency, i.e. higher injection voltage is not required with the increased injection frequency, resulting in reduced noises and increased available voltage for control, etc.

It is expected that the improved anti-rotating pulsating square-wave signal injection method has the advantages of high accuracy and bandwidths, increased voltage margin, good dynamic response and robustness, etc.

6.6.3 Cross-Saturation Effect

The cross-saturation effect for the anti-rotating pulsating square-wave signal injection strategy using ZSV is also investigated in this section. By considering the saturation effect, phase inductances can be expressed as [7]:

$$\begin{cases} L_{AA} = L_{s0} - L_{s2}\cos(2\theta_r) - L_c \sin(2\theta_r) \\ L_{BB} = L_{s0} - L_{s2}\cos\left(2\theta_r + \frac{2}{3}\pi\right) - L_c \sin\left(2\theta_r + \frac{2}{3}\pi\right) \\ L_{CC} = L_{s0} - L_{s2}\cos\left(2\theta_r - \frac{2}{3}\pi\right) - L_c \sin\left(2\theta_r - \frac{2}{3}\pi\right) \end{cases} \tag{6.40}$$

$$\begin{cases} M_{AB} = M_{BA} = M_{s0} - M_{s2}\cos\left(2\theta_r - \frac{2}{3}\pi\right) - M_c \sin\left(2\theta_r - \frac{2}{3}\pi\right) \\ M_{BC} = M_{CB} = M_{s0} - M_{s2}\cos(2\theta_r) - M_c \sin(2\theta_r) \\ M_{CA} = M_{AC} = M_{s0} - M_{s2}\cos\left(2\theta_r + \frac{2}{3}\pi\right) - M_c \sin\left(2\theta_r + \frac{2}{3}\pi\right) \end{cases} \tag{6.41}$$

where L_c and M_c represent the second harmonic amplitudes of sine inductance terms in the self- and mutual-inductances, respectively. Accordingly, for the anti-rotating pulsating square-wave signal injection method, the zero sequence HF voltage can be obtained as:

$$v_{SN} = \frac{2}{\pi} \frac{V_h(L_{s0} - M_{s0})(L'_{s2} - M'_{s2})\sin(2\Delta\theta + \Delta_{s3})}{(L_{s0} - M_{s0})^2 - \left(\frac{L'_{s2}}{2} + M'_{s2}\right)^2} \sum_{k=1} \frac{1}{2k-1}\cos k\omega_h t \tag{6.42}$$

where θ_{s3} represents the phase shift of the zero sequence HF voltage due to the cross-saturation effect and can be obtained as:

$$\theta_{s3} = -\arctan\frac{L_c - M_c}{L_{s2} - M_{s2}} \tag{6.43}$$

Therefore, the position error due to the cross-saturation effect is obtained as

$$\Delta\theta = \frac{1}{2}\arctan\frac{L_c - M_c}{L_{s2} - M_{s2}} \tag{6.44}$$

Thus, from (6.29) in Section 6.4 and (6.44), it is seen that actually the cross-saturation effect causes exactly the same position estimation errors regardless of signal injection waveforms. Accordingly, the offline measured cross-saturation correction angle in Figure 6.10 can also be utilized for the online compensation of the above square-wave injection method.

6.6.4 Experimental Results

Experiments were carried out for the anti-rotating pulsating square-wave signal injection method on the prototype IPMSM-II (Appendix B). The overall control diagram for the anti-rotating pulsating square-wave injection method with ZSV sensing is shown in Figure 6.22.

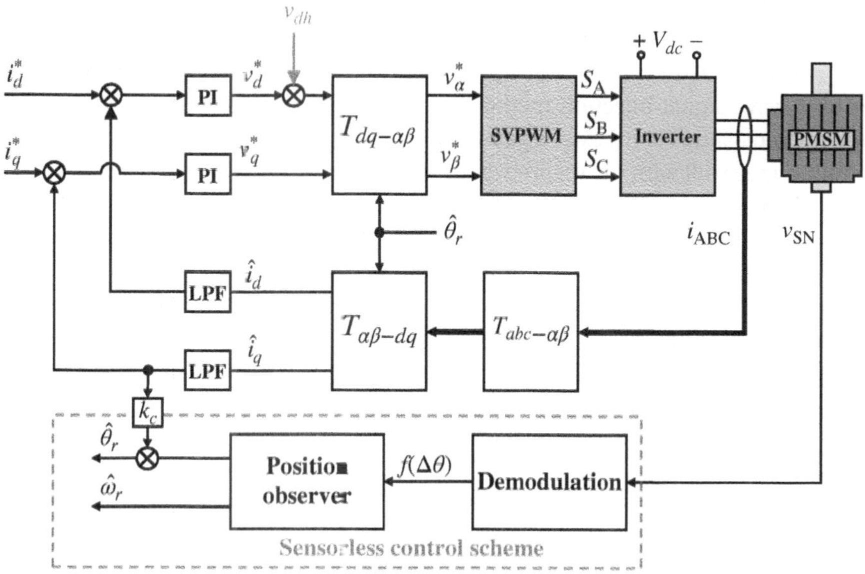

Figure 6.22 Overall block diagram of anti-rotating pulsating square-wave signal injection using zero sequence voltage.

6.6.4.1 Zero Sequence Voltage Model Verification

Figure 6.23 illustrates the measured zero sequence HF voltage spectra (the injected square-wave signal is set as 4 V/1 kHz, and the fundamental frequency f_e is 2.5 Hz, and only the main harmonics are illustrated). Note that in order to maximize the zero sequence signal in (6.37) for the described injection strategy, the position

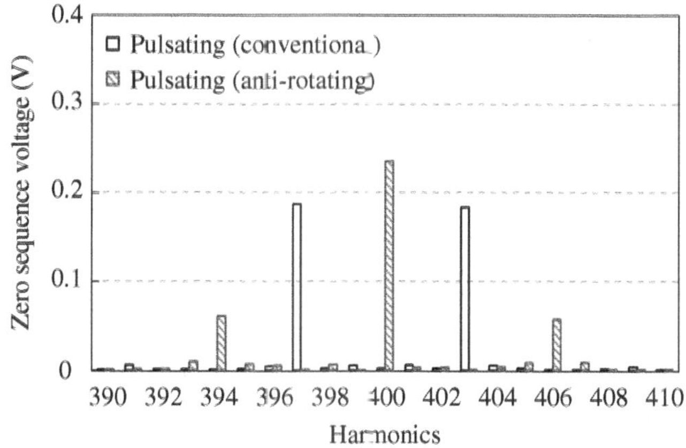

Figure 6.23 Measured zero sequence HF voltage with different square-wave injection methods [6].

error is set to $\pi/4$ with the assistance of a position sensor. From Figure 6.23, it is observed that the conventional pulsating square-wave signal injection is primarily at $f_h \pm 3fe$ (397th and 403rd). By comparison, for the anti-rotating pulsating square-wave signal injection method, the frequency components f_h and $f_h \pm 6f_e$ (400th and 394th/406th) exist as shown in Figure 6.23. Hence, all of the measured results agree well with the theoretical derivations in (6.33) and (6.37), respectively. Also, from the measured spectra in Figure 6.23, obviously the anti-rotating based method is also much easier for signal processing and has a higher signal-to-noise ratio (SNR) compared to conventional method.

The spectra for the anti-rotating pulsating square-wave signal injection method under different loads are also shown in Figure 6.24. Considering the cross-saturation effect, the position error in (6.42) is set as 0 and $\pi/4$ rad, respectively, with the assistance of a position sensor, i.e.

$$v_{SN} = \frac{2}{\pi}\frac{V_h(L_{s0}-M_{s0})(L'_{s2}-M'_{s2})\cos(\theta_{s3})}{(L_{s0}-M_{s0})^2-\left(\frac{L'_{s2}}{2}+M'_{s2}\right)^2}\sum_{k=1}\frac{1}{2k-1}\cos k\omega_h t \tag{6.45}$$

for $\Delta\theta = \pi/4$, and

$$v_{SN} = \frac{2}{\pi}\frac{V_h(L_{s0}-M_{s0})(L'_{s2}-M'_{s2})\cos(\theta_{s3})}{(L_{s0}-M_{s0})^2-\left(\frac{L'_{s2}}{2}+M'_{s2}\right)^2}\sum_{k=1}\frac{1}{2k-1}\cos k\omega_h t \tag{6.46}$$

for $\Delta\theta = 0$ rad, respectively.

Under no-load condition, since the cross-saturation angle θ_{s3} is quite small, as shown in Figure 4.10, the measured ZSV amplitude (400th) is maximized in (6.45),

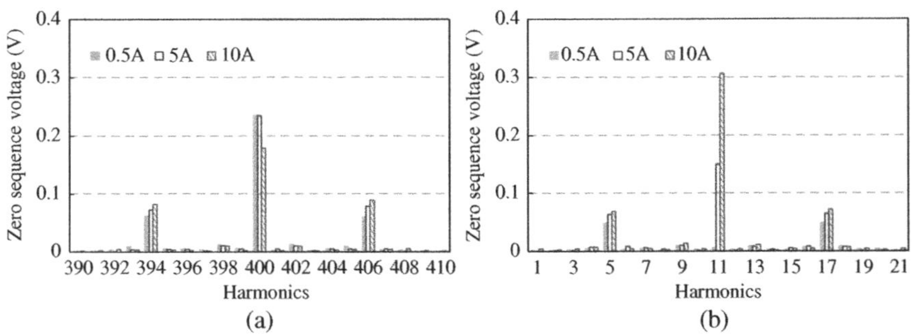

Figure 6.24 Measured zero sequence HF voltages for anti-rotating pulsating square-wave-based method under different loads. (a) $\Delta\theta = \pi/4$ rad. (b) $\Delta\theta = 0$ rad. (*Source:* [6]/IEEE.)

Figure 6.24a, while the amplitude in (6.46) is negligible, Figure 6.24b. In contrast, with the increase of load, particularly under rated load condition ($I_q = 10$ A), due to the significantly increased cross-saturation angle θ_{s3}, the amplitude in (6.46) is even larger than that in (6.45), Figures 6.24a,b. This shows that $\theta_{s3} > \pi/4$ according to (6.45) and (6.46) and also agrees with the measured results in Figure 4.10. It is also noted that under load conditions, the disturbance harmonics (e.g. 394th and 406th) in the ZSV only slightly increase, Figure 6.24.

6.6.4.2 Steady- and Dynamic-State Position Estimation Performance

The steady-state position estimation performance with the anti-rotating pulsating square-wave signal injection method is illustrated in Figure 6.25. The q-axis current is set as 5 A, and the injection signal is set as 4 V/2.5 kHz. It can be observed that under load conditions, due to the cross-saturation effect, there exist dc position estimation errors. Then, with the compensation information from Figure 4.10, the estimation accuracy is greatly improved as shown in Figure 6.25.

The dynamic position estimation performance with the anti-rotating pulsating square-wave signal injection method is also shown in Figure 6.26 in which the speed reference is given as 0–50–100 r/min. The injected HF signals are the same as those in the steady-state conditions. The quick dynamic response can be observed, and the overall position errors are small after cross-saturation effect compensation.

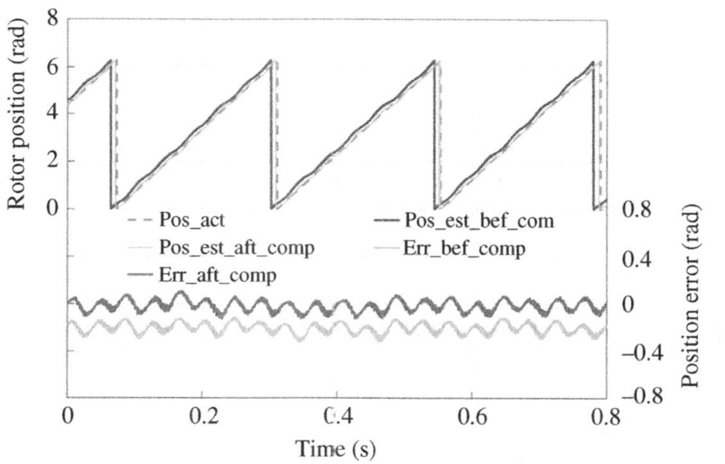

Figure 6.25 Position estimation with and without compensation of cross-saturation effect in steady state. (*Source:* [6]/IEEE.)

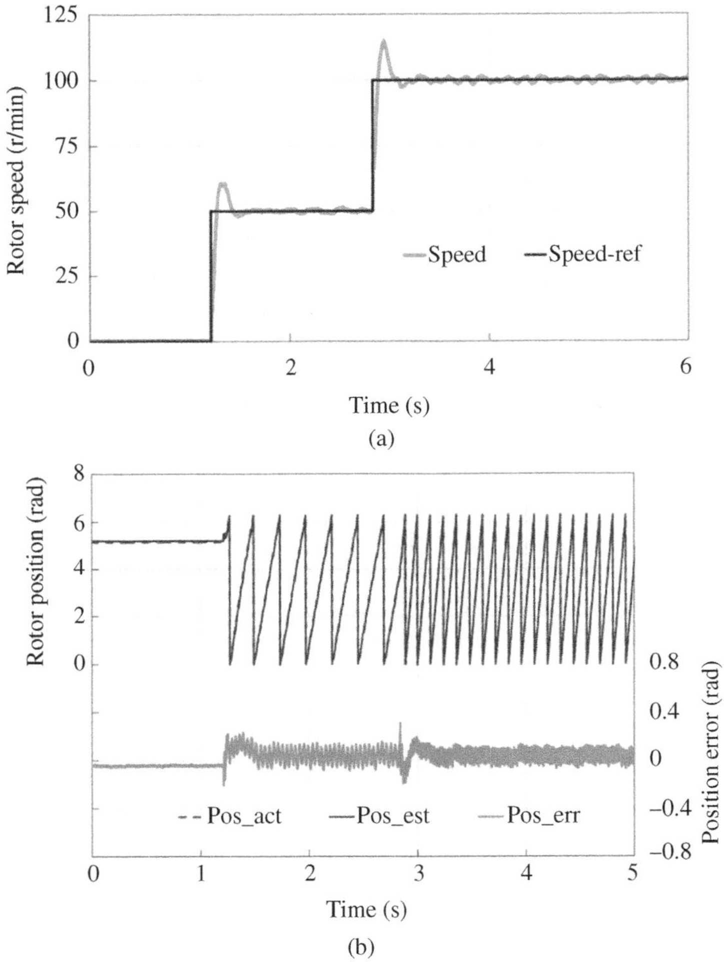

Figure 6.26 Dynamic rotor position performance under step-speed condition. (a) Rotor speed. (b) Rotor position estimation. (*Source:* [6]/IEEE.)

6.6.4.3 Comparison to Square-Wave Injection Method with HF Current Sensing

In this section, the conventional pulsating square-wave signal injection method with current sensing [8] is also implemented to compare the position estimation performance with that of the anti-rotating pulsating square-wave injection method with ZSV sensing as shown in Figure 6.27. Three injection conditions have been considered, i.e. 2.5 kHz/4 V and 2.5 kHz/8 V injection for the conventional square-wave injection method, and 2.5 kHz/4 V injection for the described method.

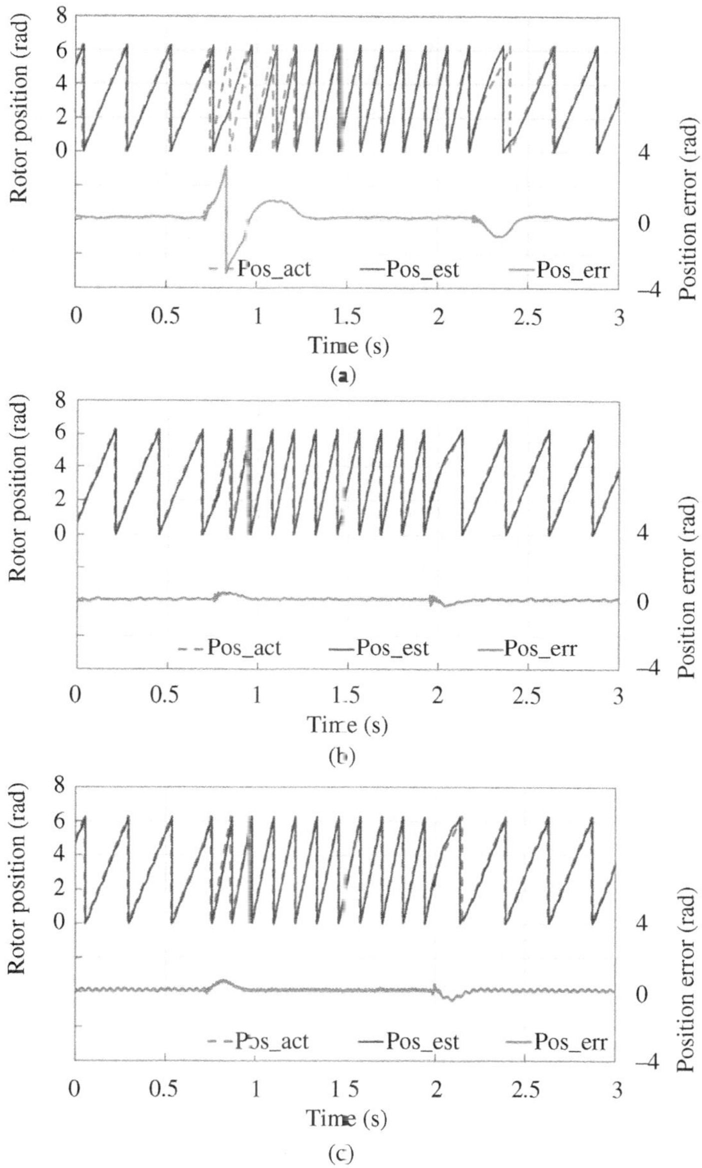

Figure 6.27 Estimated rotor position with pulsating square-wave signal injection. (a) Conventional current sensing-based method (2.5 kHz/4 V). (b) Conventional current sensing-based method (2.5 kHz/8 V). (c) Zero sequence voltage sensing-based method (2.5 kHz/4 V). (*Source:* [6]/ EEE.)

In order to clearly illustrate the position tracking capability, the estimated position is observed in the sensored mode to avoid losing synchronization. As can be seen from Figure 6.27, if the conventional current sensing-based method is performed with only 4 V injection voltage, the position estimation is apparently degraded as shown in Figure 6.27a, losing the position tracking capability during the transient process. It is found experimentally that with around twice the injection voltage, i.e. 8 V, the stable position estimation capability can then be ensured as can be seen from Figure 6.27b for the conventional current sensing-based method. In contrast, for the ZSV sensing-based method, with only a 4 V injection voltage, the same stable position estimation performances (as the 8 V injection for the conventional current sensing method, Figure 6.27b) can be obtained as illustrated in Figure 6.27c.

Further, in the sensorless mode, with 2.5 kHz/8 V injection for the conventional current sensing-based method and 2.5 kHz/4 V injection for the ZSV sensing-based method, the step-load tests for the two methods are also performed, as shown in Figures 6.28 and 6.29. It can be seen that the two injection methods both show good dynamic performance. Therefore, the available voltage margins for control, noise issues, etc. are significantly improved for the anti-rotating pulsating square-wave signal injection method with ZSV sensing.

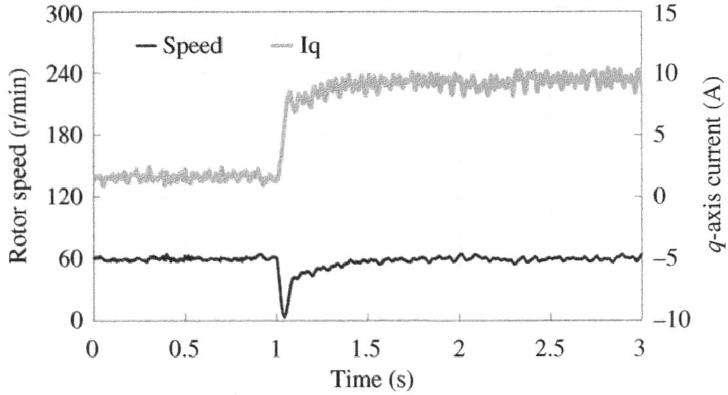

(a)

Figure 6.28 Step-load tests for conventional current sensing-based method with 2.5 kHz/8 V injection. (Pos_err_aft, position error after cross-saturation compensation; Pos_err_bef, position error before compensation). (a) Rotor speed and q-axis current. (b) Position estimation performance. (*Source:* [6]/IEEE.)

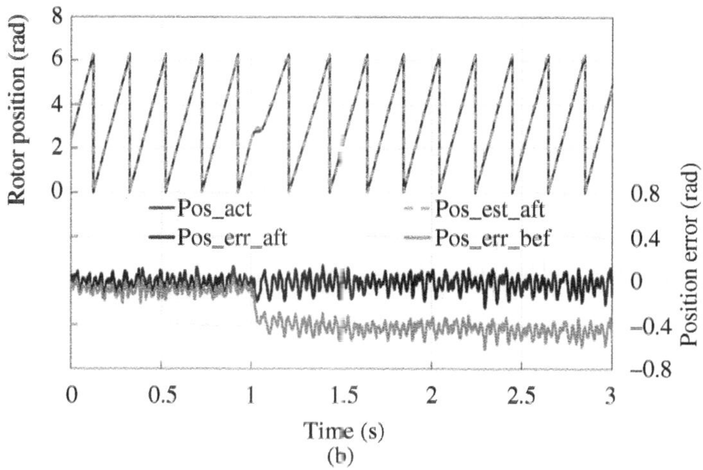

Figure 6.28 (Continued)

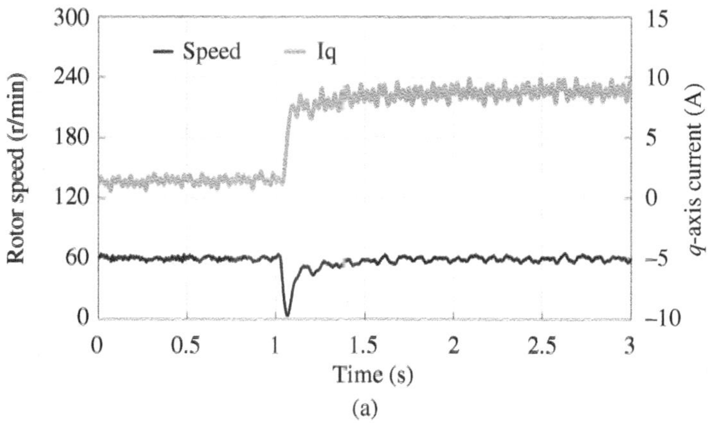

Figure 6.29 Step-load tests for zero sequence voltage sensing-based method with 2.5 kHz/4 V injection. (Pos_err_aft, position error after cross-saturation compensation; Pos_err_bef, position error before compensation). (a) Rotor speed and q-axis current. (b) Position estimation performance. (*Source:* [6]/IEEE.)

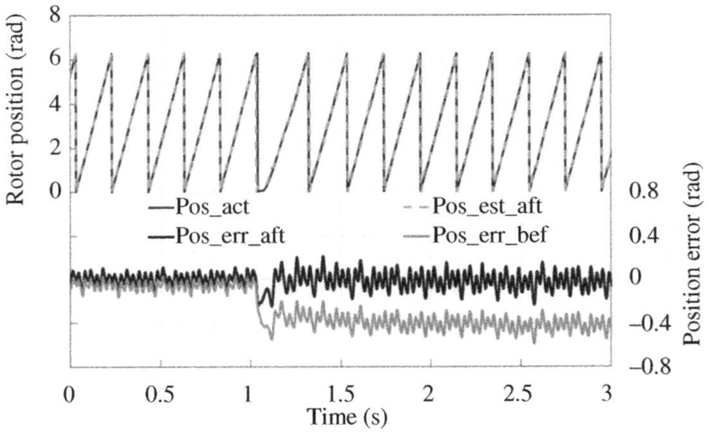

(b)

Figure 6.29 (Continued)

6.7 Summary

In this chapter, the ZSV response for HF signal injection methods was introduced, which was shown to have the advantages of less injection distortion, enhanced estimation bandwidth and stability, etc. For the conventional ZSV-based method with rotating sinusoidal signal injection, the ZSV response is phase-modulated by machine saliency, which easily results in position estimation errors. Pulsating-based signal injection methods have a good robustness due to amplitude-modulation. However, conventional pulsating signal injection methods using ZSV response will result in extra dc and sixth order harmonic position errors.

To combine the merits of the zero sequence method with the amplitude-modulation technique, an improved anti-rotating signal injection strategy was presented in this chapter. This injection strategy was performed on the estimated reference frame, anti-rotating at twice the rotor's estimated electrical speed. It was shown that compared to the conventional ZSV method, the improved strategy is easier for signal processing and less sensitive to signal processing delays because it is amplitude-modulated by saliency position and the phase shifts due to signal processing delays are intrinsically canceled. Therefore, the improved method has combined the synergies of the zero sequence method (i.e. high bandwidth and stability) and the amplitude-modulation technique (i.e. increased accuracy), and shows great rotor position estimation performance, albeit with the requirement of an additional balanced resistor network and the need to access the machine's neutral point in order to obtain ZSV. The comparison of the different signal injection methods with ZSV sensing and current sensing is summarized in Table 6.2.

Table 6.2 Comparison of high-frequency signal injection methods.

Injection signal	Rotating sinusoidal		Pulsating sinusoidal		Pulsating square wave	
Sensing signal	**Current**	**ZSV**	**Current**	**ZSV**	**Current**	**ZSV**
Reference frame	Stationary	Stationary	Estimated synchronous	Estimated anti-rotating	Estimated synchronous	Estimated anti-rotating
Coordinate	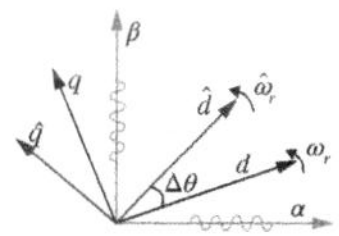		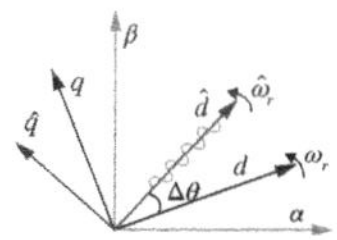	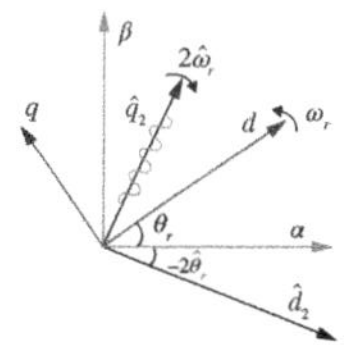	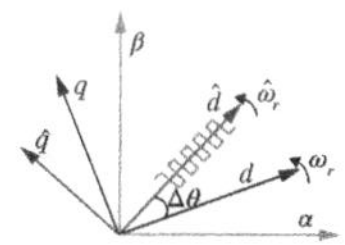	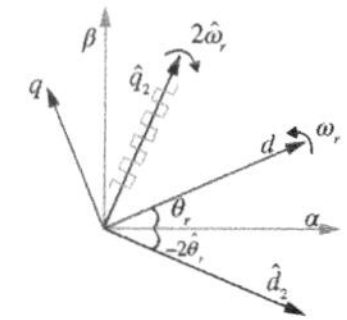
Modulation	Phase	Phase	Amplitude	Amplitude	Amplitude	Amplitude
Additional hardware	No	Yes	No	Yes	No	Yes
Cross-saturation error	$-\theta_m/2$	$-\theta_m/2$	$-\theta_m/2$	$-\theta_m/2$	$-\theta_m/2$	$-\theta_m/2$
Demodulation	Complex	Complex	Medium	Medium	Simple	Simplest
Dynamic performance	Low	Low	Medium	Medium	Good	Good
SNR	Low	High	Low	High	Low	High
Injection frequency dependency	Yes	No	Yes	No	Yes	No
Robustness	Poor	Poor	Good	Good	Good	Good

References

1 F. Briz, M. W. Degner, P. Garcia, and J. M. Guerrero, "Rotor position estimation of AC machines using the zero-sequence carrier-signal voltage," *IEEE Trans. Ind. Appl.*, vol. 41, no. 6, pp. 1637–1646, Nov. 2005.

2 P. Garcia, F. Briz, M. W. Degner, and D. Reigosa, "Accuracy, bandwidth, and stability limits of carrier-signal-injection-based sensorless control methods," *IEEE Trans. Ind. Appl.*, vol. 43, no. 4, pp. 990–1000, Jul. 2007.

3 A. Consoli, G. Scarcella, G. Bottiglieri, and A. Testa, "Harmonic analysis of voltage zero-sequence-based encoderless techniques," *IEEE Trans. Ind. Appl.*, vol. 42, no. 6, pp. 1548–1557, Nov. 2006.

4 D. Raca, P. Garcia, D. Reigosa, F. Briz, and R. D. Lorenz, "Carrier-signal selection for sensorless control of PM synchronous machines at zero and very low speeds," *IEEE Trans. Ind. Appl.*, vol. 46, no. 1, pp. 167–178, Jan. 2010.

5 P. L. Xu and Z. Q. Zhu, "Novel carrier signal injection method using zero-sequence voltage for sensorless control of PMSM drives," *IEEE Trans. Ind. Electron.*, vol. 63, no. 4, pp. 2053–2061, Apr. 2016.

6 P. L. Xu and Z. Q. Zhu, "Novel square-wave signal injection method using zero-sequence voltage for sensorless control of PMSM drives," *IEEE Trans. Ind. Electron.*, vol. 63, no. 12, pp. 7444–7454, Dec. 2016.

7 Y. Yan, J. Zhu, H. Lu, Y. Guo, and S. Wang, "Study of a PMSM model incorporating structural and saturation saliencies," in *2005 Int. Conf. Power Electron. Drives Syst.*, vol. 1, pp. 575–580, Nov. 2005.

8 Y. Yoon, S. K. Sul, S. Morimoto, and K. Ide, "High-bandwidth sensorless algorithm for ac machines based on square-wave-type voltage injection," *IEEE Trans. Ind. Appl.*, vol. 47, no. 3, pp. 1361–1370, May 2011.

9 S. Kim, J. I. Ha, and S. K. Sul, "PWM switching frequency signal injection sensorless method in IPMSM," *IEEE Trans. Ind. Appl.*, vol. 48, no. 5, pp. 1576–1587, Sep. 2012.

7

Sensorless Control of Dual Three-Phase PMSMs and Open-Winding PMSMs

7.1 Introduction

In the previous chapters, sensorless control, including fundamental model-based methods and saliency tracking-based methods, were introduced. These methods are based on Y-connected single three-phase permanent magnet synchronous machines (STP-PMSM). Apart from this type of machine drive, other topologies have also become popular in recent years, e.g. dual three-phase PMSM (DTP-PMSM) and open-winding PMSM (OW-PMSM), as shown in Figure 7.1. Compared to STP-PMSM, DTP-PMSM [1]–[13] and OW-PMSM [14]–[16] show advantages in torque control, fault tolerance control, and sensorless control. For DTP-PMSM and OW-PMSM, due to additional degrees of freedom, many novel rotor position estimation techniques have been proposed in recent years, which demonstrates the superiority of DTP-PMSMs [8]–[13] and OW-PMSMs [15], [16] in the area of sensorless control. Therefore, in this chapter, sensorless control strategies on DTP-PMSMs and OW-PMSMs are presented and discussed.

7.2 Dual Three-Phase PMSMs

The interest in multi-phase machine drives has particularly increased over the last decade, especially for high-power and fault-tolerant applications [1], [2]. This is because of their excellent features, such as improved efficiency, reduced dc-link current harmonics (and thus reduced requirement of the dc-link capacitor), reduced torque ripples, improved power sharing capability, and increased reliability of the system. The most popular and well-discussed multi-phase machine is the DTP-PMSM with two independent three-phase stator winding sets displaced by

Sensorless Control of Permanent Magnet Synchronous Machine Drives, First Edition.
Zi Qiang Zhu and Xi Meng Wu.

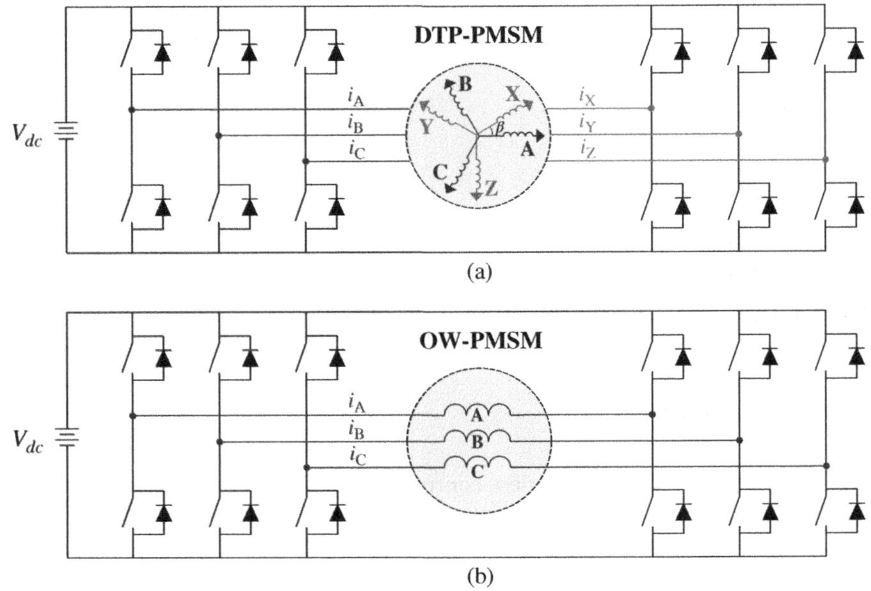

Figure 7.1 Dual inverter drive topologies. (a) DTP-PMSM. (b) OW-PMSM.

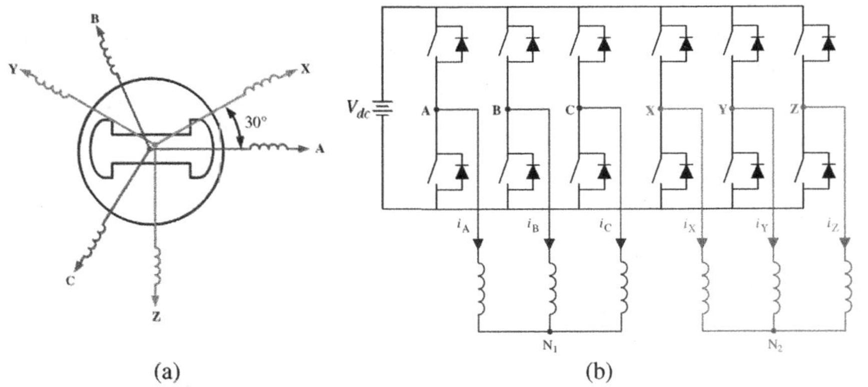

Figure 7.2 Six-phase voltage source inverter (VSI) fed DTP-PMSM. (a) Schematic. (b) Drive configuration.

30 electrical degrees [3] with isolated neutral points, as depicted in Figure 7.2. Compared to other topologies, 30° displacement has a reduced sixth order harmonic torque ripple [1]–[3]. In this section, the basic modeling and sensorless control of DTP-PMSM will be introduced.

7.2.1 Modeling of DTP-PMSM Drive

The modeling of a DTP-PMSM can be represented using two different approaches, i.e. the double dq model [2]–[6] and vector space decomposition (VSD) [7]. In both approaches, the two zero-sequence components can be neglected since the neutral points N_1 and N_2 of the two three-phase windings are isolated as shown in Figure 7.2b.

7.2.1.1 Double *dq* Model

For this approach, the machine is represented with two pairs of dq windings for the two three-phase winding sets, which could be regarded as two single three-phase machines with coupling voltage [4], [5]. By assuming the two stator windings are identical, the voltages of both stator winding sets in the dq reference frame can be derived as:

$$\begin{cases} v_{d1} = R_s i_{d1} + L_{d1}\dfrac{di_{d1}}{dt} - M_{d21}\dfrac{di_{d2}}{dt} - \omega_r L_{q1} i_{q1} - \omega_r M_{q21} i_{q2} \\ v_{q1} = R_s i_{q1} + L_{q1}\dfrac{di_{q1}}{dt} - M_{q21}\dfrac{di_{q2}}{dt} + \omega L_{d1} i_{d1} + \omega_r M_{d2} i_{d2} + \omega_r \psi_m \\ v_{d1} = R_s i_{d2} + L_{d2}\dfrac{di_{d2}}{dt} - M_{d12}\dfrac{di_{d1}}{dt} - \omega_r L_{q2} i_{q2} - \omega_r M_{q12} i_{q1} \\ v_{q1} = R_s i_{q2} + L_{q2}\dfrac{di_{q2}}{dt} + M_{q12}\dfrac{di_{q1}}{dt} + \omega_r L_{d2} i_{d2} + \omega_r M_{d12} i_{d1} + \omega_r \psi_m \end{cases} \tag{7.1}$$

where i_d, i_q, v_d, and v_q are the d- and q-axis currents and voltages, respectively, ψ_m is the PM flux-linkage, $L_{d1} = L_{d2} = L_d$ and $L_{q1} = L_{q2} = L_q$ are the d- and q-axis self-inductances of the two winding sets, $M_{d21} = M_{d12} = M_d$ and $M_{q12} = M_{q21} = M_q$ are the d- and q-axis mutual-inductances between the two winding sets. Subscripts 1 and 2 represent the two stator winding sets.

For DTP-PMSM control, the same control strategies (including the PWM module) used for the STP-PMSM, i.e. FOC and DTC, can be applied. In this section, taking FOC as an example, the control scheme based on the double dq model is shown in Figure 7.3.

The reference currents in d- and q-axes are shared by two individual current controllers. Four PI current regulators are employed, as shown in Figure 7.3 [2]. Overall, the double dq model is straightforward and each three-phase set can be controlled independently. However, there usually exists magnetic coupling between two sets, which may cause system instability.

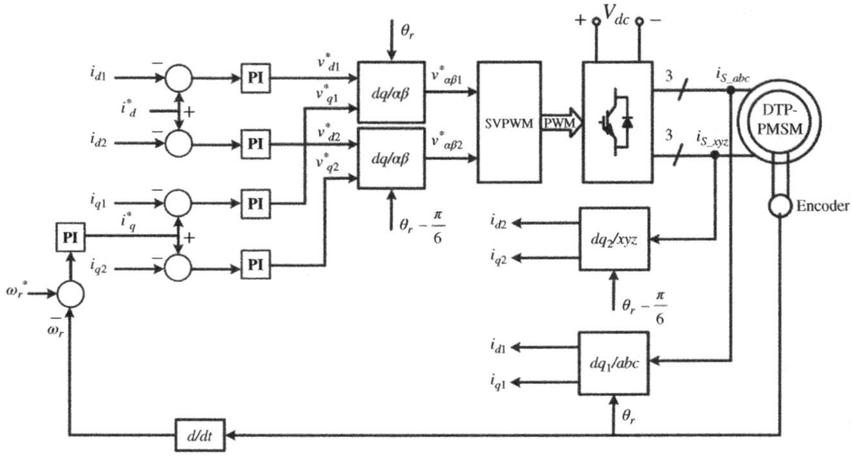

Figure 7.3 Double *dq* current control for dual three-phase PMSM.

7.2.1.2 Vector Space Decomposition

The VSD method [7] can simplify the DTP-PMSM system into three decoupled subspaces so that the current in each subspace can be controlled independently. The VSD transformation for DTP-PMSM can be expressed as:

$$\begin{bmatrix} x_\alpha \\ x_\beta \\ x_{z1} \\ x_{z2} \\ x_{o1} \\ x_{o2} \end{bmatrix} = \frac{1}{3}\begin{bmatrix} 1 & -1/2 & -1/2 & \sqrt{3}/2 & -\sqrt{3}/2 & 0 \\ 0 & -\sqrt{3}/2 & \sqrt{3}/2 & 1/2 & 1/2 & -1 \\ 1 & -1/2 & -1/2 & -1 & 1/2 & 1/2 \\ 0 & -\sqrt{3}/2 & \sqrt{3}/2 & 0 & \sqrt{3}/2 & -\sqrt{3}/2 \\ 1 & 1 & 1 & 0 & 0 & 0 \\ 0 & 0 & 0 & 1 & 1 & 1 \end{bmatrix}\begin{bmatrix} x_A \\ x_B \\ x_C \\ x_X \\ x_Y \\ x_Z \end{bmatrix} \tag{7.2}$$

where x could represent the machine variables, i.e. voltage, current, and flux-linkage. The harmonics with orders of $12k \pm 1$ ($k = 1, 2, 3, \ldots$) and the fundamental component of the machine variables are mapped into the $\alpha\beta$ subspace. The harmonics with orders of $6k \pm 1(k = 1, 3, 5, \ldots)$ are located in the z_1z_2 subspace, while the zero sequence components with $3k$ ($k = 0, 1, 2, \ldots$) are mapped into the o_1o_2 subspace. After transformation, the model of DTP-PMSM is expressed as:

$$v_{\alpha\beta} = R_s i_{\alpha\beta} + \frac{d\psi_{\alpha\beta}}{dt}, \psi_{\alpha\beta} = L_s i_{\alpha\beta} + \psi_{m\alpha\beta}, \psi_{m\alpha\beta} = \psi_m e^{j\theta_r} \tag{7.3}$$

$$v_{z1z2} = R_s i_{z1z2} + \frac{d\psi_{z1z2}}{dt}, \psi_{z1z2} = L_{ls} i_{sz1z2} \tag{7.4}$$

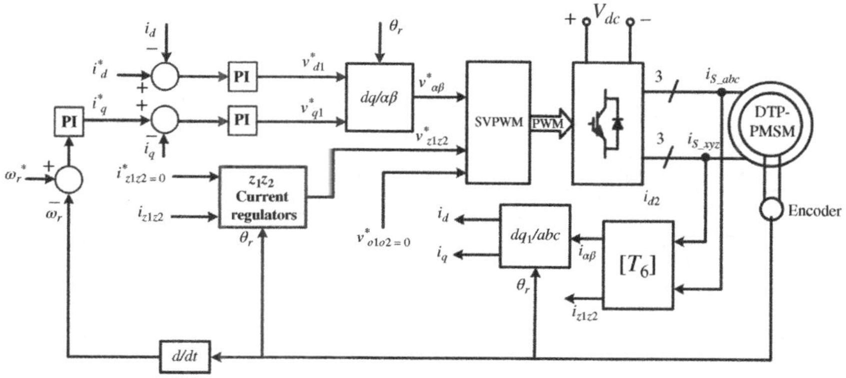

Figure 7.4 Current control based on VSD for dual three-phase PMSM.

where $v_{\alpha\beta}$, $\psi_{\alpha\beta}$, $i_{\alpha\beta}$, v_{z1z2}, ψ_{z1z2}, i_{z1z2} are stator voltage, flux, and current in $\alpha\beta$ and z_1z_2 subspaces, respectively, and L_{ls} is the leakage self-inductance.

Similarly, both FOC and DTC can be extended to DTP-PMSM based on the VSD model. Taking the FOC scheme as an example, Figure 7.4 shows the FOC current control based on VSD for DTP-PMSM. Since the impedance in the z_1z_2 subspace is relatively small, large current harmonics can be found in z_1z_2 subspace. Thus, the z_1z_2 subspace variables should be controlled as well. Moreover, dual three-phase drives have much more inverter voltage vectors than a single three-phase system due to their six inverter legs. By neglecting four zero vectors, the $\alpha\beta$ and z_1z_2 subspaces voltage vectors are illustrated in Figures 7.5a,b, respectively. Since the available vectors are different from the STP-PMSM case, specific PWM techniques should be applied for VSD-based DTP-PMSM drives [1].

7.2.2 HFSI Sensorless Control with Current Response

For DTP-PMSM, conventional high-frequency signal injection (HFSI) methods including pulsating and rotating sinusoidal signals are discussed in this section. These two sensorless control methods, introduced in Chapter 4, are extended to DTP-PMSM and their sensorless control performance is investigated experimentally for both the double dq model and the VSD model. In the experiment, the specifications of the prototype dual three-phase SPMSM-I are given in Appendix B. For the high-frequency signals, the amplitude and frequency are selected as 8 V and 550 Hz. For the double dq model, the HF signals are injected on both stator winding sets independently. For the VSD model, the HF signals are injected on the transformed fundamental subspace.

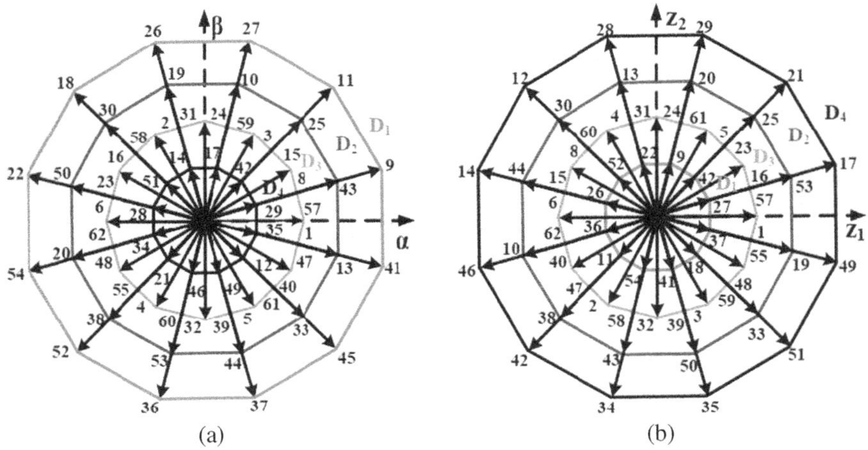

Figure 7.5 Inverter voltage vectors in two subspaces. (a) $\alpha\beta$ subspace. (b) z_1z_2 subspace. (*Source:* [1]/MDPI/CC BY 4.0.)

First, the steady-state performance is investigated for two DTP-PMSM models and two sensorless methods. Rotor speed is 30 r/min and q-axis current is 2 A. The position estimation performance is illustrated in Figures 7.6 and 7.7. Second, the dynamic-state performance is shown in Figures 7.8 and 7.9, where the rotor speed reference is given as 0–30–60 r/min. From the experimental results, it can be seen that the pulsating injection method has a smaller position oscillation error, which is the same as described in Chapter 4. For both sensorless control methods, the oscillation errors can be reduced with the VSD model due to the mutual decoupling between the two sets.

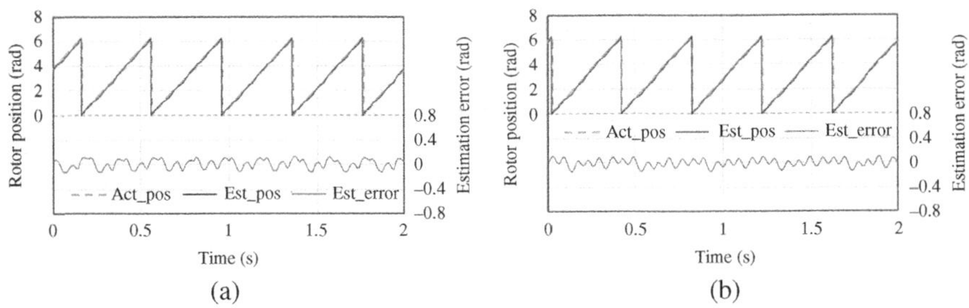

Figure 7.6 Steady-state performance with pulsating sinusoidal signal injection. (a) Double dq. (b) VSD.

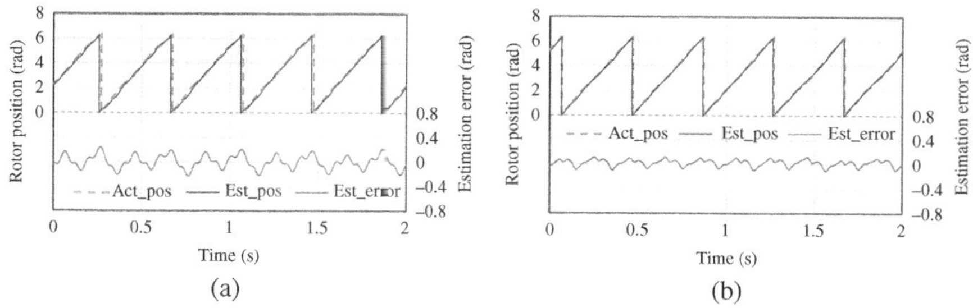

Figure 7.7 Steady-state performance with rotating sinusoidal signal injection. (a) Double *dq*. (b) VSD.

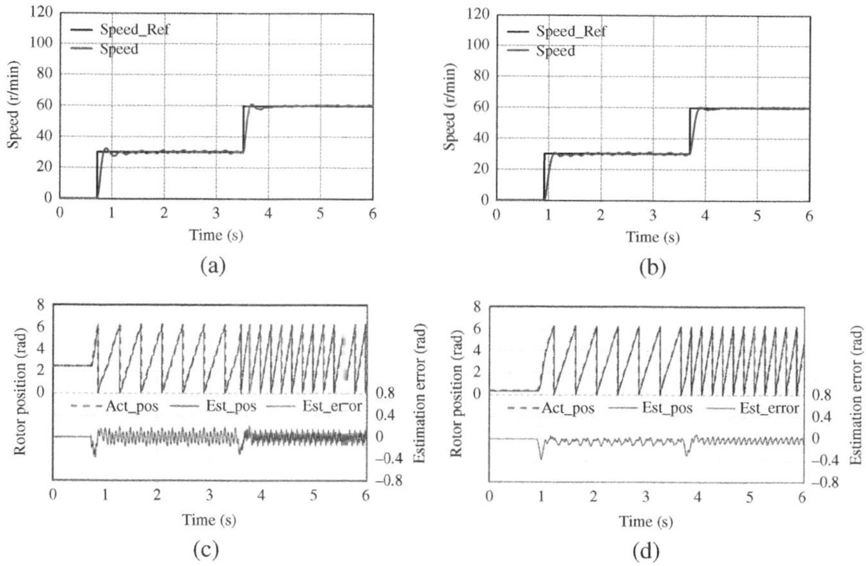

Figure 7.8 Dynamic-state performance with pulsating sinusoidal signal injection. (a) Speed reference (double *dq*). (b) Speed reference (VSD). (c) Position estimation (double *dq*). (d) Position estimation (VSD).

7.2.3 HFSI Sensorless Control with Voltage Response

After injecting HF signals, instead of the current response, the zero sequence voltage response can also be used for position estimation on DTP-PMSM [8], [9]. As mentioned in Chapter 6, the pulsating signal injection and rotating signal injection with the zero sequence voltage-based sensorless control method for single three-phase PMSM exhibits large oscillation in the position estimation error due to large

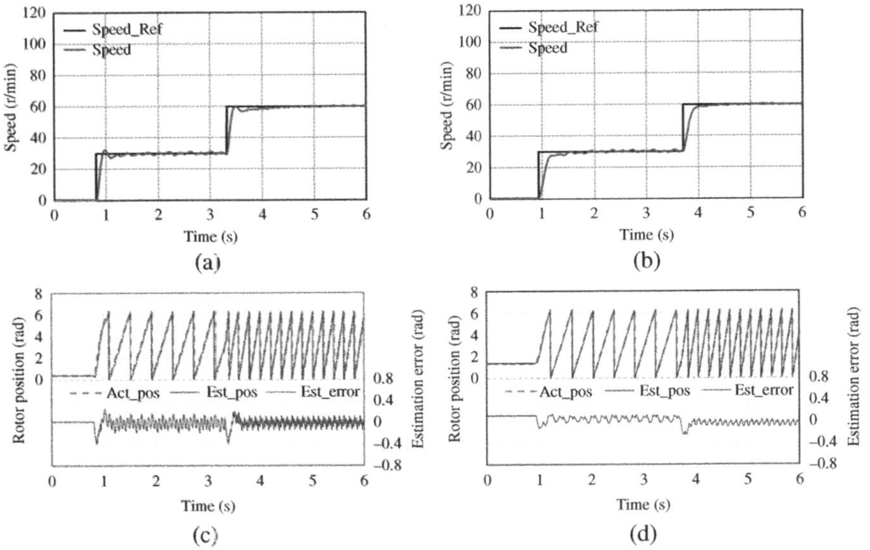

Figure 7.9 Dynamic-state performance with rotating sinusoidal signal injection. (a) Speed reference (double *dq*). (b) Speed reference (VSD). (c) Position estimation (double *dq*). (d) Position estimation (VSD).

amplitude of the undesirable sixth harmonic component in the zero sequence HF voltage signal. Moreover, the measurement of zero sequence voltage is relatively complicated. Hence, this section investigates the sensorless control method based on zero sequence voltage for DTP-PMSM drives. A simple method of zero sequence voltage measurement is presented at first by using one voltage sensor between two isolated neutral points of both winding sets. Then, based on the double *dq* model, an improved zero sequence-based method is presented where undesirable sixth harmonic components can be suppressed by applying an optimum phase shift between the two independent high-frequency injected signals. Consequently, the steady-state position estimation accuracy is improved.

7.2.3.1 Zero Sequence Voltage Measurement

After HF signal injection, the resultant zero sequence HF voltage is normally measured based on one of two methods, i.e. measurement using phase-to-neutral voltages [11] as shown in Figure 7.10, or measurement using the auxiliary resistor network [12], Figure 7.11. Both of these two methods can measure the zero

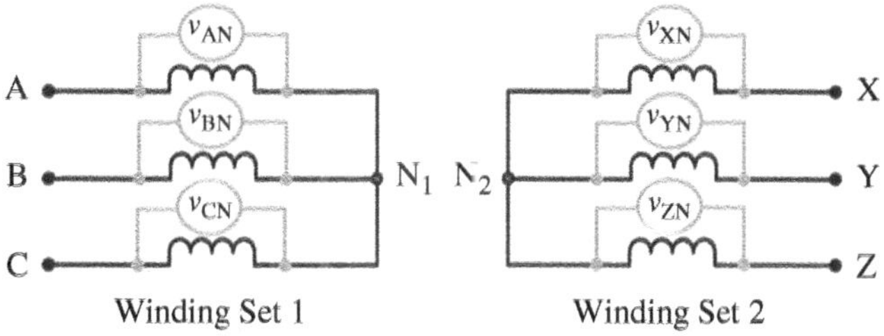

Figure 7.10 Measurement of zero sequence voltage using phase-to-neutral voltages [11].

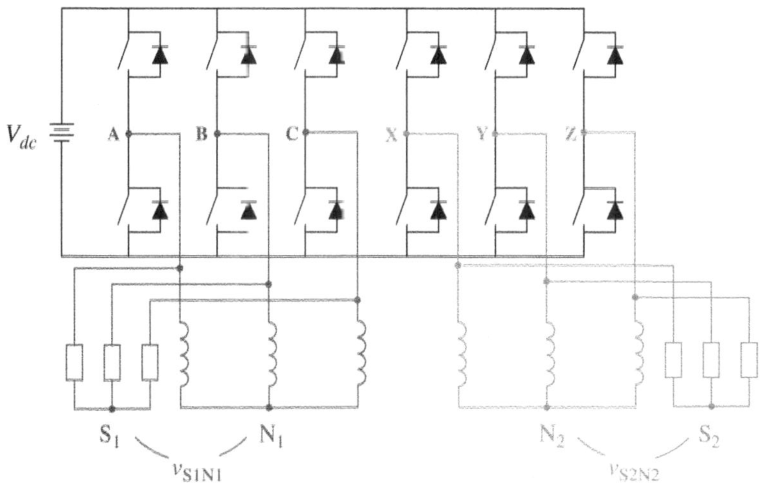

Figure 7.11 Measurement of zero sequence voltage using auxiliary resistor network [12].

sequence component effectively. However, for the first method, the six voltage sensors requirement inevitably increases the cost. The second method also requires two auxiliary resistance networks with two voltage sensors, which makes the measurement of zero sequence complicated.

Thus, this section introduces a simple zero sequence voltage measurement method [8] in which one voltage sensor is placed between the two isolated neutral points, i.e. v_{0sn1n2}, as shown in Figure 7.12. This method can effectively measure the zero sequence HF voltage response after injecting the HF signal. Compared to the aforementioned measurement methods, this method is simple in structure and easy to implement. Hence, the zero sequence HF voltage signals presented hereafter are obtained by this simple measurement method.

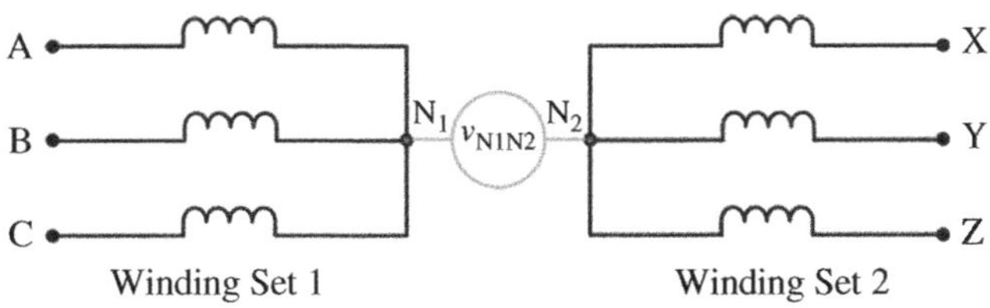

Figure 7.12 Measurement method of zero sequence voltage between two neutral points. (*Source:* [8]/IEEE.)

7.2.3.2 Modeling of Dual Three-Phase PMSM

For DTP-PMSM having two Y-connected winding sets with two isolated neutral points, the voltage equations and inductances for the first and the second winding sets are expressed as:

$$\begin{bmatrix} v_{AN} \\ v_{BN} \\ v_{CN} \end{bmatrix} = \begin{bmatrix} L_{AA} & M_{AB} & M_{AC} \\ M_{BA} & L_{BB} & M_{BC} \\ M_{CA} & M_{CB} & L_{CC} \end{bmatrix} p \begin{bmatrix} i_A \\ i_B \\ i_C \end{bmatrix}, \begin{bmatrix} v_{XN} \\ v_{YN} \\ v_{ZN} \end{bmatrix} = \begin{bmatrix} L_{XX} & M_{XY} & M_{XZ} \\ M_{YX} & L_{YY} & M_{YZ} \\ M_{ZX} & M_{ZY} & L_{ZZ} \end{bmatrix} p \begin{bmatrix} i_X \\ i_Y \\ i_Z \end{bmatrix} \tag{7.5}$$

$$\begin{aligned} &L_{AA} = L_{s0} - L_{s2}\cos(2\theta_r) && M_{AB} = M_{BA} = M_{s0} - M_{s2}\cos(2\theta_r - 2\pi/3) \\ &L_{BB} = L_{s0} - L_{s2}\cos(2\theta_r + 2\pi/3) && M_{BC} = M_{CB} = M_{s0} - M_{s2}\cos(2\theta_r) \\ &L_{CC} = L_{s0} - L_{s2}\cos(2\theta_r - 2\pi/3) && M_{CA} = M_{AC} = M_{s0} - M_{s2}\cos(2\theta_r + 2\pi/3) \\ &L_{XX} = L_{s0} - L_{s2}\cos(2\theta_r - 2\pi/6) && M_{XY} = M_{YX} = M_{s0} - M_{s2}\cos(2\theta_r + \pi) \\ &L_{YY} = L_{s0} - L_{s2}\cos(2\theta_r - 10\pi/6) && M_{YZ} = M_{ZY} = M_{s0} - M_{s2}\cos(2\theta_r - 10\pi/6) \\ &L_{ZZ} = L_{s0} - L_{s2}\cos(2\theta_r + \pi) && M_{ZX} = M_{XZ} = M_{s0} - M_{s2}\cos(2\theta_r - 2\pi/6) \end{aligned} \tag{7.6}$$

where p is the differential operator, v_{AN}, v_{BN}, v_{CN} and i_A, i_B, i_C are referred to as the phase voltages and currents of the first winding set, respectively. v_{XN}, v_{YN}, v_{ZN} and i_X, i_Y, i_Z are referred to as the phase voltages and currents of the second winding set, respectively. L_{AA}, L_{BB}, L_{CC} and M_{AB}, M_{BA}, M_{BC}, M_{CB}, M_{CA}, M_{AC} are the three-phase incremental self- and mutual-inductances of the first winding sets, respectively. L_{XX}, L_{YY}, L_{ZZ} and M_{XY}, M_{YX}, M_{YZ}, M_{ZY}, M_{ZX}, M_{XZ} are the three-phase incremental self- and mutual-inductances of the second winding sets, respectively.

7.2.3.3 Pulsating Sinusoidal Signal Injection

For the pulsating sinusoidal signal injection, two independent high-frequency signals are injected into the estimated d-axes for both winding sets, and the injected voltages of the first and the second winding sets are given by (7.7).

$$\begin{cases} v_{Ah} = V_h \cos(\omega_h t)\cos(\hat{\theta}_r) \\ v_{Bh} = V_h \cos(\omega_h t)\cos(\hat{\theta}_r - 2\pi/3) \\ v_{Ch} = V_h \cos(\omega_h t)\cos(\hat{\theta}_r + 2\pi/3) \end{cases}, \begin{cases} v_{Xh} = V_h \cos(\omega_h t)\cos(\hat{\theta}_r - \pi/6) \\ v_{Yh} = V_h \cos(\omega_h t)\cos(\hat{\theta}_r - 5\pi/6) \\ v_{Zh} = V_h \cos(\omega_h t)\cos(\hat{\theta}_r + 3\pi/6) \end{cases} \tag{7.7}$$

Hence, the resultant zero sequence HF voltages for both winding sets of DTP-PMSM can be obtained approximately as:

$$v_{S1N1} = K_v \cos(\omega_h t)\cos(2\theta_r + \hat{\theta}_r) \tag{7.8}$$

$$v_{S2N2} = K_v \cos(\omega_h t)\cos(2\theta_r + \hat{\theta}_r - 3\pi/6) \tag{7.9}$$

$$K_v = V_h \frac{(L_{s0} - M_{s0})(L_{s2} - M_{s2})}{2L_{s0}^2 + 2M_{s0}^2 - 2M_{s2}^2 - 4L_{s0}M_{s0} - 2L_{s2}M_{s2} - \frac{L_{s2}^2}{2}} \tag{7.10}$$

Thus, the resultant zero sequence HF voltage between the two isolated neutral points of DTP-PMSM can be given as:

$$v_{N1N2} = v_{S1N1} - v_{S2N2} = -\sqrt{2}K_v \cos(\omega_h t)\cos(2\theta_r + \hat{\theta}_r - 3\pi/4) \tag{7.11}$$

The corresponding demodulated zero sequence HF voltage, V_{N1N2}, is expressed as:

$$\begin{aligned} V_{N1N2} &= \mathrm{LPF}\left[4v_{N1N2}\sin(\omega_h t - 3\hat{\theta}_r)\right] \\ &= -\sqrt{2}K_v \times \left[\sin\left(2\Delta\theta - \frac{3\pi}{4}\right) + \sin\left(2\theta_r + 4\hat{\theta}_r - \frac{3\pi}{4}\right)\right] \end{aligned} \tag{7.12}$$

From (7.12), it can be seen that the sixth harmonic component with large amplitude also exists under the steady-state condition, as is the case of the single three-phase PMSM, which deteriorates the position estimation significantly. Therefore, a modified method that fully utilizes the additional degrees of freedom in DTP-PMSM drives to improve the position estimation performance is presented below.

The expression of zero sequence HF voltage between the two isolated neutral points, considering a phase shift in the injected signal of the second winding set, is given by:

$$v_{N1N2} = -K_v \times \begin{bmatrix} \sin\left(\omega_h t + 2\theta_r + \hat{\theta}_r - \frac{\pi}{4} - \frac{\varphi_{12}}{2}\right)\sin\left(\frac{\pi}{4} + \frac{\varphi_{12}}{2}\right) \\ + \sin\left(\omega_h t - 2\theta_r - \hat{\theta}_r + \frac{\pi}{4} - \frac{\varphi_{12}}{2}\right)\sin\left(-\frac{\pi}{4} + \frac{\varphi_{12}}{2}\right) \end{bmatrix} \tag{7.13}$$

where φ_{12} is the phase shift between the two injected high-frequency signals. From (7.13), it can be clearly seen that the zero sequence voltage involves two high-frequency components at $f_h \pm 3f_e$, and the high-frequency components at $f_h - 3f_e$ can deteriorate the position estimation performance where the undesirable sixth harmonic in the estimation error occurs. In order to suppress the undesirable sixth harmonic in the estimation error, an optimum phase shift is required. It is worth noting that when the phase shift between the two high-frequency signals injected into the two sets of three-phase windings is $\varphi_{12} = \pi/2$, the second term, which causes large oscillation in the estimation error, can be eliminated as undesirable components at $f_h - 3f_e$ equal to zero. Meanwhile, the first term at high frequency of $f_h + 3f_e$ that is used for position estimations can be maximized. Therefore, with the optimum phase shift, $\pi/2$, the zero sequence HF voltage becomes:

$$v_{\mathrm{N1N2}} = K_v \cos\left(\omega_h t + 2\theta_r + \hat{\theta}_r\right) \tag{7.14}$$

From (7.14), it can be clearly seen that the zero sequence HF voltage expression is simplified with one term only when applying the optimum phase shift in the injected HF signal of the second winding set. Furthermore, the undesirable sixth harmonic component does not exist for the position tracking observer. Thus, it is expected that the improved pulsating injection with zero sequence HF voltage for DTP-PMSM can achieve high accuracy of rotor position estimation.

The demodulation process for the zero sequence HF voltage, by considering the optimum phase angle shift, v_{N1N2}, is given by (7.15).

$$V_{\mathrm{N1N2}} = \mathrm{LPF}\left[4v_{\mathrm{N1N2}} \sin\left(\omega_h t - 3\hat{\theta}_r\right)\right] = K_v \sin(2\Delta\theta) \tag{7.15}$$

7.2.3.4 Rotating Signal Injection Method

For the rotating sinusoidal signal injection, the balanced rotating HF voltage is injected into two winding sets of DTP-PMSM independently, and the injected HF voltages in the two sets of three-phase windings are expressed as

$$\begin{cases} v_{\mathrm{A}h} = V_h \cos(\omega_h t) \\ v_{\mathrm{B}h} = V_h \cos(\omega_h t - 2\pi/3) \\ v_{\mathrm{C}h} = V_h \cos(\omega_h t + 2\pi/3) \end{cases}, \begin{cases} v_{\mathrm{X}h} = V_h \cos(\omega_h t) \\ v_{\mathrm{Y}h} = V_h \cos(\omega_h t - 2\pi/3) \\ v_{\mathrm{Z}h} = V_h \cos(\omega_h t + 2\pi/3) \end{cases} \tag{7.16}$$

Similarly, the resultant zero sequence HF voltage [9] between the two isolated neutral points of DTP-PMSM can be given as:

$$v_{\mathrm{N1N2}} = K_v \sin(\omega_h t + 2\theta_r + 5\pi/6) - \frac{\sqrt{3}}{2} K_{v2} \sin(\omega_h t - 4\theta_r + 2\pi/6) \tag{7.17}$$

$$K_{v2} = V_h \frac{(L_{s2} - 2M_{s2})(L_{s2} - M_{s2})}{2L_{s0}^2 + 2M_{s0}^2 - 2M_{s2}^2 - 4L_{s0}M_{s0} - 2L_{s2}M_{s2} - \dfrac{L_{s2}^2}{2}} \tag{7.18}$$

From (7.17), it can be seen that the zero sequence HF voltage includes two high-frequency components at $f_h + 2f_e$ and $f_h - 4f_e$, similar to the case of a single three-phase PMSM. The expression of zero sequence HF voltage between the two isolated neutral points, considering a phase shift in the injected signal of the second winding set, is given by

$$\begin{aligned} v_{N1N2} = & -2K_v \sin(\omega_h t + 2\theta_r + 5\pi/6 - \varphi_{12}/2)\sin(-5\pi/6 + \varphi_{12}/2) \\ & + K_{v2}\sin(\omega_h t - 4\theta_r + 2\pi/6 - \varphi_{12}/2)\sin(-2\pi/6 + \varphi_{12}/2) \end{aligned} \tag{7.19}$$

where φ_{12} is the phase shift between the two injected high-frequency signals. From (7.19), it is found that the second term, which causes oscillation in the position estimation error, can be suppressed when the phase shift between the two injected high-frequency signals is $\varphi_{12} = 2\pi/3$, as the undesirable harmonic component at $f_h - 4f_e$ equals to zero. Meanwhile, the amplitude of the high-frequency component at $f_h + 2f_e$, which is used for position estimation, is maximized. Thus, $2\pi/3$ is the optimum phase shift in order to suppress such oscillation in the estimation error. Then, the zero sequence voltage while also considering the optimum phase shift becomes:

$$v_{N1N2} = -2K_v \cos(\omega_h t + 2\theta_r) \tag{7.20}$$

From (7.20), it can be clearly seen that the zero sequence HF voltage expression is simplified with one term only when applying the optimum phase shift in the injected HF signal of the second winding set. The undesirable sixth harmonic component does not exist for the position tracking observer. Thus, at steady state, it is expected that the improved rotating signal injection with zero sequence HF voltage for DTP-PMSM can achieve a high accuracy of rotor position estimation. Then, the demodulation process for the zero sequence HF voltage, by considering the optimum phase angle shift, v_{N1N2}, is given by:

$$\begin{aligned} V_{N1N2} & = \mathrm{LPF}[v_{N1N2}\cos(\omega_h t)]\sin(2\hat{\theta}_r) - \mathrm{LPF}[v_{N1N2} \times \sin(\omega_h t)]\cos(2\hat{\theta}_r) \\ & = -K_v \sin(2\Delta\theta_e) \end{aligned} \tag{7.21}$$

7.2.3.5 Experimental Results and Analysis for DTP-PMSM

The specifications of the test rig are given in Appendix B, SPMSM-I. The overall sensorless control system is shown in Figure 7.13. For both injected pulsating and rotating signals, they are injected into two winding sets independently. For both HF signals, the amplitude is 8 V and the frequency is 500 Hz. The rotor speed during the steady-state condition is set as 30 r/min ($f_e = 2.5$ Hz).

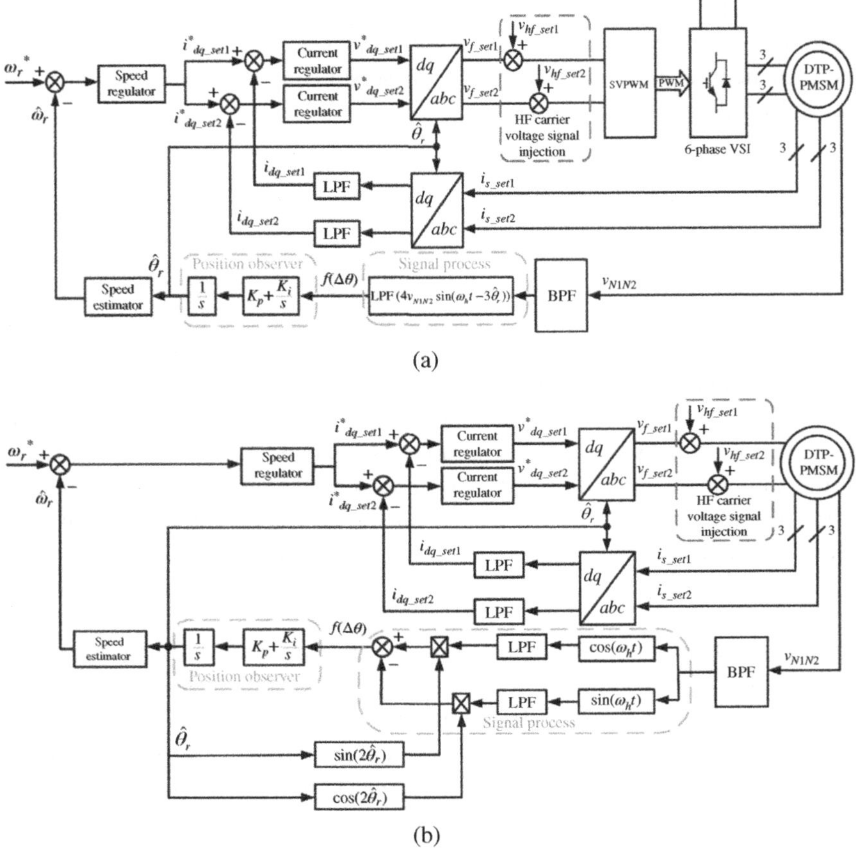

Figure 7.13 Overall control systems. (a) Pulsating signal injection. (*Source:* [8]/IEEE.) (b) Rotating signal injection. (*Source:* [9]/IEEE.)

First, the spectra of measured zero sequence HF voltages between the two isolated neutral points while employing different phase shifts are illustrated in Figure 7.14. For the pulsating signal injection method, it can be seen that the undesirable high-frequency component of the 197th order, which causes oscillation in the estimation error, is significantly suppressed when applying a 90° phase angle shift between the two injected signals among different phase shifts. For the rotating signal injection method, the undesirable component of 196th is suppressed after applying a 120° phase angle shift. Figures 7.15 and 7.16 show the position estimation results without/with an optimum phase shift, respectively. By applying an optimum phase shift between the injected high-frequency signals, the 197th

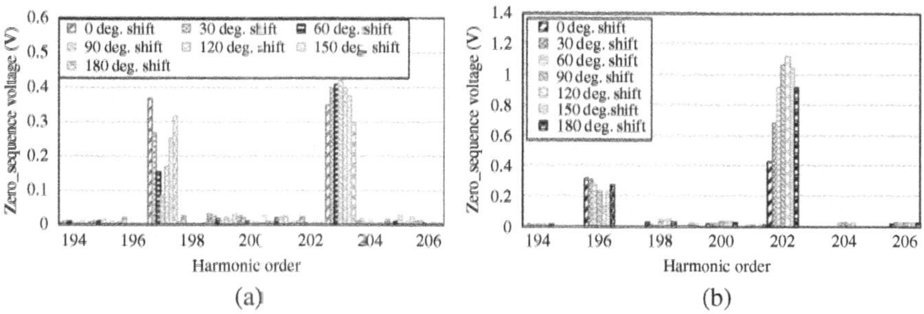

Figure 7.14 Spectra of measured zero sequence HF voltages with different phase shifts. (a) Pulsating signal injection. (*Source:* [8]/IEEE.) (b) Rotating signal injection. (*Source:* [9]/IEEE.)

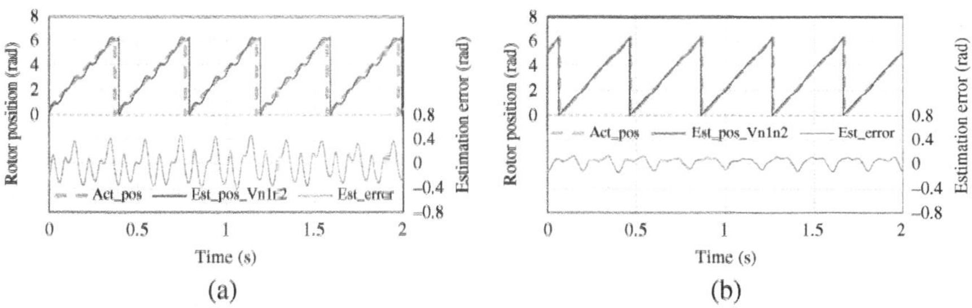

Figure 7.15 Rotor position estimation of DTP-PMSM with pulsating sinusoidal signal injection. (a) Without phase shift. (b) With optimum phase shift. (*Source:* Adapted from [8].)

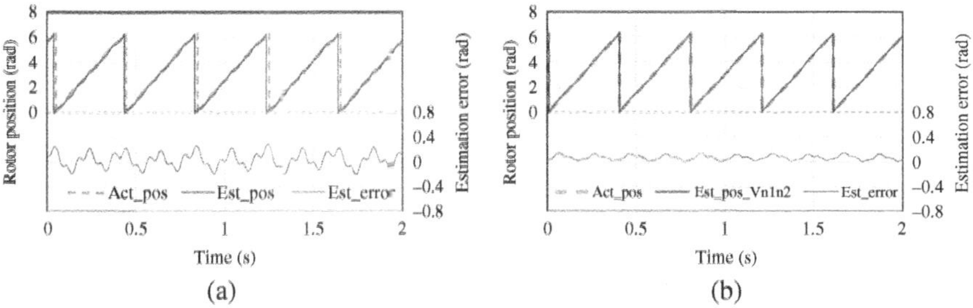

Figure 7.16 Rotor position estimation of DTP-PMSM with rotating sinusoidal signal injection. (a) Without phase shift. (b) With optimum phase shift. (*Source:* Adapted from [9].)

(pulsating) and 196th (rotating) frequency components of zero sequence voltage are significantly suppressed. Meanwhile, the 203rd (pulsating) and 202nd (rotating) frequency components, which are utilized for rotor position estimation, are maximized, Figure 7.14. As a result, the rotor position estimation is improved.

7.2.4 Fundamental Model-Based Sensorless Control

The aforementioned HFSI sensorless methods can be used in both zero- and low-speed range. For the middle- and high-speed ranges, fundamental model-based sensorless control on STP-PMSM introduced in Chapter 2 can be extended to DTP-PMSM. For the fundamental model-based sensorless control methods of DTP-PMSM, the estimation equation models can be different with the VSD model and double *dq* model. For the VSD model, conventional fundamental model-based sensorless control methods can be directly applied, while for the double *dq* model, the coupling between two winding sets should be considered [10], which will be shown in this section. Although extra coupling needs to be considered, thanks to the additional freedom of the double *dq* model, a simple correction method is developed in [10] to compensate for the dc position error caused by machine parameter mismatches. This method will be presented in this section.

7.2.4.1 Extended Back-EMF Model on DTP-PMSM

With the extended back-EMF method [17] as an example, the voltage equations for the double *dq* modeled DTP-PMSM are derived. For simplicity, only the model of Set 1 is presented here as the model for Set 2 is the same. The voltage equations can be written as [10]:

$$\begin{bmatrix} v_{d1} \\ v_{q1} \end{bmatrix} = \begin{bmatrix} R_s + pL_D & -\omega_r L_Q \\ \omega_r L_Q & R_s + pL_D \end{bmatrix} \begin{bmatrix} i_d \\ i_q \end{bmatrix} + \begin{bmatrix} 0 \\ E_{ex1} \end{bmatrix} \tag{7.22}$$

where E_{ex1} can be expressed as:

$$E_{ex1} = (L_D - L_Q)\left(\omega_r i_d - p i_q\right) + \omega_r \psi_m \tag{7.23}$$

where $L_D = L_{d1} + M_{d21} = L_{d2} + M_{d12}$ and $L_Q = L_{q1} + M_{q21} = L_{q2} + M_{q12}$ are defined as the equivalent inductances of both sets by considering the two three-phase windings being identical, as well as the phase resistances $R_s = R_{s1} = R_{s2}$. Moreover, $i_d = i_{d1} = i_{d2}$ and $i_q = i_{q1} = i_{q2}$ due to the two sets are generally operated at the same condition for double *dq*-based control.

In sensorless control operation, since the actual rotor position is unknown, (7.22) is transformed to the estimated rotating reference frame as:

$$\begin{bmatrix} \hat{v}_{d1} \\ \hat{v}_{q1} \end{bmatrix} = \begin{bmatrix} R_s + L_D p & -\omega_r L_Q \\ \omega_r L_Q & R_s + L_D p \end{bmatrix} \begin{bmatrix} \hat{i}_d \\ \hat{i}_d \end{bmatrix} + \begin{bmatrix} \hat{E}_{d1} \\ \hat{E}_{q1} \end{bmatrix} \tag{7.24}$$

$$\begin{bmatrix} \hat{E}_{d1} \\ \hat{E}_{q1} \end{bmatrix} = E_{ex1} \begin{bmatrix} -\sin \Delta\theta_1 \\ \cos \Delta\theta_1 \end{bmatrix} + (\hat{\omega}_r - \omega_r) L_D \begin{bmatrix} -\hat{i}_q \\ \hat{i}_d \end{bmatrix} \tag{7.25}$$

Based on (7.25), the rotor position can be tracked by forcing the estimated d-axis back-EMF to zero by a position observer (PO) so that the position and speed can be tracked. It is worth noting that compared to the EEMF model for STP-PMSM in Chapter 2, mutual-inductance M_{q21} is required by DTP-PMSM [10].

7.2.4.2 Parameter Mismatch Effect

According to Chapter 3, the relationship between the parameter mismatches and the estimated error at steady state can be derived as:

$$\Delta\theta_{\Delta R_s} = \tan^{-1}\left(\frac{\Delta R_s \hat{i}_d}{E_{ex} + \Delta R_s \hat{i}_q}\right), \; \Delta\theta_{\Delta L_D} = 0, \Delta\theta_{\Delta L_Q} = \tan^{-1}\left(\frac{-\Delta L_Q \hat{i}_q}{E_{ex}/\hat{\omega}_r + \Delta L_Q \hat{i}_d}\right) \tag{7.26}$$

It can be concluded that only the phase resistance and q-axis equivalent inductance deviations cause the dc estimated errors [10]. Moreover, $\Delta\theta_{\Delta R_s}$ decreases as the speed increases, whereas $\Delta\theta_{\Delta L_Q}$ has little dependency on speed levels.

7.2.4.3 Parameter Mismatch Correction

From the aforementioned analysis, the dc position error could be the result of the deviation of the parameters. In this section, a simple position error correction method [10] will be described. For a DTP-PMSM machine, two winding sets are controlled independently. Therefore, it is straightforward to apply two sensorless observers individually for each set. In the real application, the parameter nominal values set in both observers are the same. Then, the estimated position of Set 1 $\hat{\theta}_{r1}$ and Set 2 $\hat{\theta}_{r2}$ can be expressed as:

$$\hat{\theta}_{r1} = \theta_r + \Delta\theta_{par1}, \; \hat{\theta}_{r2} = \theta_r + \Delta\theta_{par2} + \Delta\theta_{12} \tag{7.27}$$

where $\Delta\theta_{par1}$, $\Delta\theta_{par2}$ are the position error caused by the parameter deviation of Set 1 and Set 2, respectively, and $\Delta\theta_{12}$ is the shifted angle between the two sets of three-phase windings. Since the dq-axis currents for both sets are controlled to the same references, the estimated position error can be expressed as:

$$\Delta\theta_{par1} = \Delta\theta_{\Delta R_{s1}} + \Delta\theta_{\Delta L_{Q1}} = \Delta\theta_{par2} = \Delta\theta_{\Delta R_{s2}} + \Delta\theta_{\Delta L_{Q2}} \tag{7.28}$$

Based on (7.26) and (7.28), the estimated error can be simplified as:

$$\Delta\theta_{par1} = \Delta\theta_{par2} \approx K_{\Delta R_s} \hat{i}_d + K_{\Delta L_Q} \hat{i}_q \tag{7.29}$$

where $K_{\Delta R_s}$ and $K_{\Delta R_Q}$ are defined as the deviation factors:

$$K_{\Delta R_s} = \frac{\Delta R_s}{E_{ex} + \Delta R_s \hat{i}_q}, \quad K_{\Delta L_Q} = -\frac{\Delta L_Q}{E_{ex}/\hat{\omega}_r + \Delta L_Q \hat{i}_d} \tag{7.30}$$

It is noted that if there is no parameter deviation ($K_{\Delta R_s} = 0$, $K_{\Delta L_Q} = 0$), the change in the current should not make any difference in terms of position error. Based on this principle, by changing the current in one set, there will be a position error difference between two observers if there is a parameter mismatch; otherwise, the difference is zero. Accordingly, a controller can be designed by forcing this difference error to zero to correct the parameter mismatch, thus eliminating the position error.

For the extra current signal, it can be expressed as:

$$i^*_{(Extra)}(t) = \begin{cases} I^*, & t_0 < t < t_1 \\ 0, & else \end{cases} \tag{7.31}$$

where I^* is the magnitude of the injected current, $T_{inj} = t_1 - t_0$ is defined as the time duration of injection, and t_0 and t_1 are the start time and stop time of the injection, respectively.

First, to correct R_s deviation, $i^*_{(Extra)}$ can be injected into the d-axis of Set 1 (or 2, take Set 1 as an example), during the injection procedure, and the estimated error of Set 1 can be expressed as:

$$\Delta\theta^{inj}_{par1} = K_{\Delta L_Q}\hat{i}_{q1} + K_{\Delta R_s}\hat{i}_{d1} + K_{\Delta R_s} i^*_{(Extra)} = \Delta\theta_{par1} + \delta_{\Delta R_s} \tag{7.32}$$

where $\delta_{\Delta R_s}$ is defined as $\delta_{\Delta R_s} = K_{\Delta R_s}\hat{i}_{(Extra)}$ and is proportional to resistance deviation. The estimated position after the injection in Set 1 $\hat{\theta}_{r1(new)}$ can be rewritten as:

$$\hat{\theta}_{r1(new)} = \theta_r + \Delta\theta^{inj}_{par} = \theta_r + \Delta\theta_{par1} + \delta_{\Delta R_s} \tag{7.33}$$

Moreover, $\delta_{\Delta R_s}$ can be calculated from the difference between the estimated positions of Sets 1 and 2, which can be expressed as:

$$\delta_{\Delta R_s} = \hat{\theta}_{r1(new)} - \hat{\theta}_{r2} - \Delta\theta_{12} \tag{7.34}$$

After $\delta_{\Delta R_s}$ is obtained, a PI controller (7.35) is used to adjust R_s in the EEMF observers of both sets, so that $\delta_{\Delta R_s}$ would approach zero.

$$R_s = \tilde{R}_s + \left(K_p + K_i/s\right)\delta_{\Delta R_s} \tag{7.35}$$

where K_p and K_i are the coefficients of the proportional and integral terms, respectively.

The configuration of the R_s correction is shown in Figure 7.17. Two adaptive observers are based on the conventional EEMF observer [17] but with adjustable parameters. The difference between the two sets of estimated positions $\delta_{\Delta R_s}$ is the input of the update controller. By forcing $\delta_{\Delta R_s}$ to zero, the phase resistance can be adjusted to the appropriate value to correct the estimated error caused by R_s deviation.

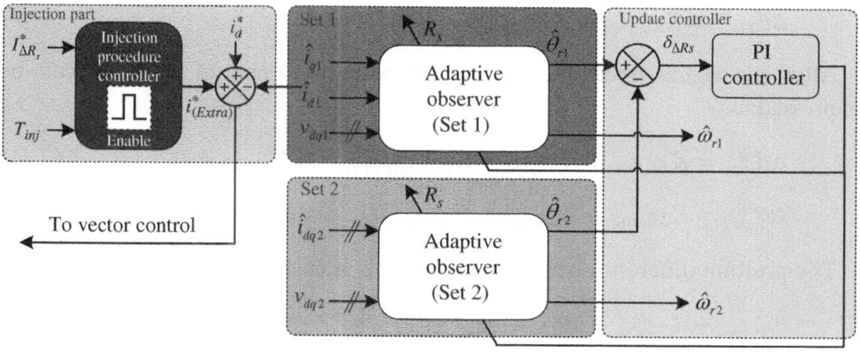

Figure 7.17 Adaptive compensation technique for estimated rotor position error due to R_s deviation. (*Source:* [10]/IEEE.)

A similar procedure as in Figure 7.17 can be applied to correct the position error caused by L_Q inaccuracy. According to (7.29), instead of d-axis current injection, q-axis current injection is used. The estimated error of Set 1 after injection can be expressed as (7.36) if the mutual-inductance deviation is neglected ($K_{\Delta L_Q} = K_{\Delta L_q}$):

$$\Delta\theta_{par1}^{inj} = K_{\Delta L_Q}\left(\hat{i}_{q1} + i^*_{(Extra)}\right) + K_{\Delta R_s}\hat{i}_{d1} = \Delta\theta_{par1} + \delta_{\Delta L_Q} \tag{7.36}$$

where $\delta_{\Delta L_Q}$ is defined as $\delta_{\Delta L_Q} = K_{\Delta L_Q} i^*_{(Extra)}$, and the updated rule can be expressed as:

$$L_Q = \tilde{L}_Q + \left(K_p + K_i/s\right)\delta_{\Delta L_Q} \tag{7.37}$$

However, in some DTP machine topologies, the mutual-inductance deviation could not be neglected, and thus, the equivalent inductance deviation ΔL_Q can be expressed as:

$$\Delta L_Q = \Delta L_{q1} + \Delta M_{q21} = \Delta L_{q2} + \Delta M_{q12} \tag{7.38}$$

where ΔL_{q1}, ΔL_{q2} are the deviation values of L_q while ΔM_{q21}, ΔM_{q12} denote the M_q deviations between the two sets. Therefore, after the injection, when the reference of q-axis current of two sets is no longer equal, i.e. $i^*_{q1} \neq i^*_{q2}$, the error caused by inductance deviation of each set can be expressed as:

$$\Delta\theta_{\Delta L_Q 1} = \left(\frac{-\Delta L_{q1}}{E_{ex}/\hat{\omega}_r + \Delta L_Q \hat{i}_d}\right)\hat{i}_{q1} + \left(\frac{-\Delta M_{q21}}{E_{ex}/\hat{\omega}_r + \Delta L_Q \hat{i}_d}\right)\hat{i}_{q2} = K_{\Delta L_{q1}}\hat{i}_{q1} + K_{\Delta M_{q21}}\hat{i}_{q2} \tag{7.39}$$

$$\Delta\theta_{\Delta L_Q 2} = K_{\Delta L_{q2}}\hat{i}_{q2} + K_{\Delta M_{q12}}\hat{i}_{q1} \tag{7.40}$$

After the signal injection, the overall estimated error of each set can be expressed as:

$$\Delta\theta_{par1}^{inj} = K_{\Delta L_Q}\hat{i}_{q1} + K_{\Delta R_s}\hat{i}_{d1} + K_{\Delta L_{q1}} i^*_{q(Extra)} = \Delta\theta_{par1} + \delta_{\Delta L_{q1}} \tag{7.41}$$

$$\Delta\theta_{par2}^{inj} = K_{\Delta L_Q}\hat{i}_{q2} + K_{\Delta R_s}\hat{i}_{d2} + K_{\Delta M_{q12}} i^*_{q(Extra)} = \Delta\theta_{par2} + \delta_{\Delta M_{q12}} \tag{7.42}$$

The position difference between the two sets is calculated as:

$$\hat{\theta}_{r1(new)} - \hat{\theta}_{r2(new)} - \Delta\theta_{12} = \delta_{\Delta L_{q1}} - \delta_{\delta M_{q12}} = \left(K_{\Delta L_{q1}} - K_{\Delta M_{q12}}\right) i^*_{(Extra)} \tag{7.43}$$

Therefore, with the same correction procedure, the system reaches a steady state when the deviation in L_q is updated to ΔM_{q12} but is not zero. However, on the practical side, for the majority of DTP machines, the deviation in M_q is usually much smaller than the deviation in L_q [10], which can be neglected.

7.2.4.4 Experimental Results

The test machine is a DTP-PMSM (Appendix B, SPMSM-III). The overall configuration is depicted in Figure 7.18. The machine is rotating at 30 r/min and operating in generating mode. The magnitude and duration of the injected signal are chosen to be 0.5 A and 2 seconds.

Based on (7.29), a higher magnitude of injection current can be beneficial to the correction procedure with an increased signal-to-noise ratio (SNR). On the other hand, the signal injection as disturbance to the system should be as low as possible. Nevertheless, in the case of q-axis current injection, the same magnitude of current with opposite direction can be applied to both sets, so that the torque disturbance can be minimized. Moreover, the magnitude of injected current and the time duration may be considered together. The higher injected signal results in faster speed of convergence, and thus, the time duration of injection can be shorter; a lower magnitude would require a longer time duration of injection for convergence. In either case, the overall effect on the system from the current injection as a disturbance may be considered similar, and there exists a trade-off depending on the characteristic of the machine.

Figure 7.19 shows the parameter correction performance. The initial nominal L_q value is set at 10 mH. The correction technique is applied at around 1.5 seconds and lasts for two seconds when $i^*_{(Extra)}$ is injected. During the procedure, L_q is adjusted to the accurate value. With the corrected value, the position error can be eliminated. However, as shown in Figure 7.19a, extra torque and pulsating position errors are introduced after q-axis current injection in one set. In order to solve this issue, the current can be injected in both sets, with the same amplitude but

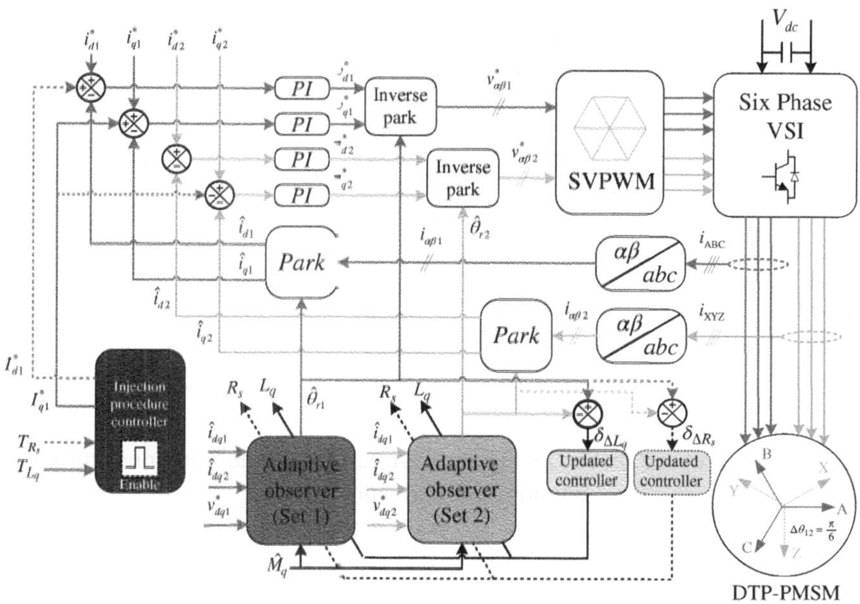

Figure 7.18 Overall block diagram of DTP-PMSMs control system with position error correction. (*Source:* [10]/IEEE.)

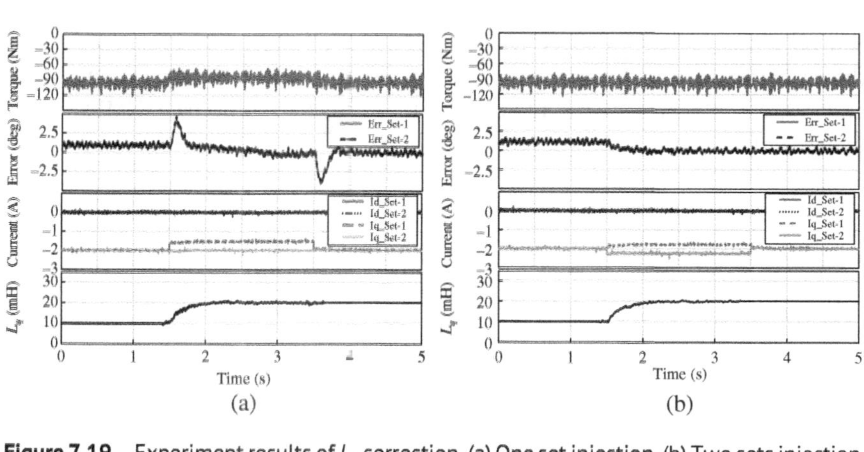

Figure 7.19 Experiment results of L_q correction. (a) One set injection. (b) Two sets injection. (*Source:* [10]/IEEE.)

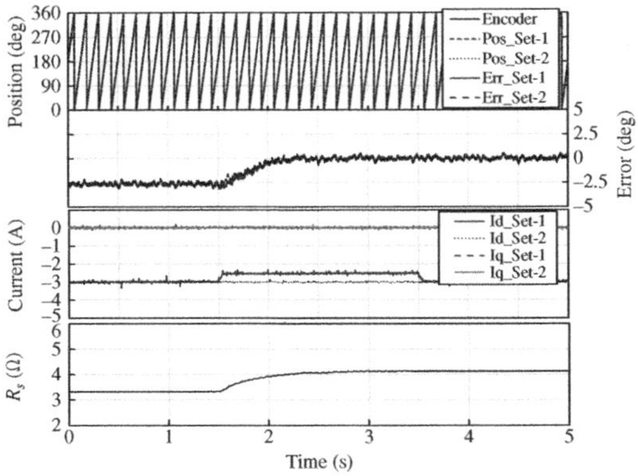

Figure 7.20 Experiment results of R_s correction. (*Source:* [10]/IEEE.)

opposite polarity as shown in Figure 7.19b, and thus, no additional torque is induced, together with a smooth error correction.

For the resistance deviation, $i^*_{(Extra)}$ is injected into the d-axis. The results are shown in Figure 7.20. The nominal R_s value set in the observers is equal to 3.3 Ω. By applying the correction technique, the position error caused by the R_s deviation can be compensated for effectively.

7.2.5 Third Harmonic Back-EMF-Based Sensorless Control

Apart from the fundamental model, the third harmonic back-EMF on DTP-PMSM can also be utilized for position estimation [13]. First, the third harmonic EMFs can be measured by auxiliary resistor networks as shown in Figure 7.11. For the DTP-PMSM, the measured voltages between S1 and N1, and between S2 and N2, could represent the two sets of the third harmonic back-EMFs of the two sets of windings as:

$$\begin{cases} v_{\text{S1N1}} \approx e_{3\_set1} \\ v_{\text{S2N2}} \approx e_{3\_set2} \end{cases} \tag{7.44}$$

Considering that the two sets of winding have a spatial 30 electrical degrees shift as shown in Figure 7.2a, e_{3_set1} and e_{3_set2} can be expressed as:

$$\begin{cases} e_{3\_set1} = E_3 \cdot \sin(3\theta_r) \\ e_{3\_set2} = E_3 \cdot \sin(3 \times (\theta_r - \pi/6)) = -E_3 \cdot \cos(3\theta_r) \end{cases} \tag{7.45}$$

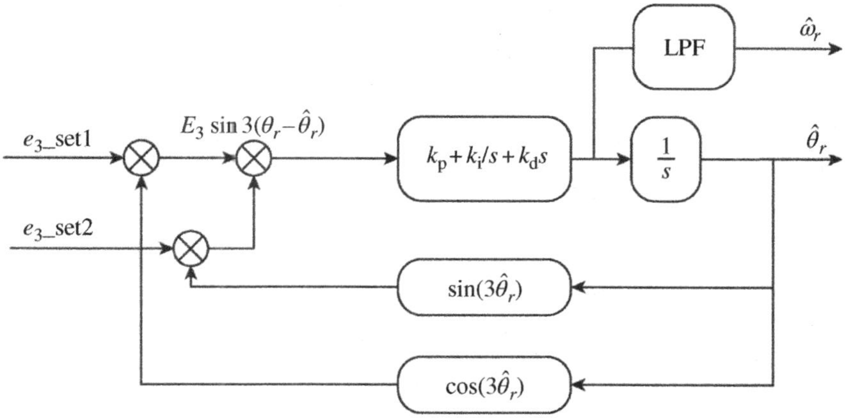

Figure 7.21 Rotor position estimator. (*Source:* [13]/CMP/CC BY.)

Moreover, the third harmonic flux-linkage can also be obtained by applying a low-pass filter (LPF) on the third harmonic back-EMF as:

$$\begin{cases} \psi_{3\_set1} = -\psi_{m3}\cos(3\theta_r) \\ \psi_{3\_set2} = -\psi_{m3}\cos(3\times(\theta_r - \pi/6)) = -\psi_{m3}\cos(3\theta_r - \pi/2) = -\psi_{m3}\sin(3\theta_r) \end{cases} \tag{7.46}$$

From (7.48) and (7.49), it can be seen that the third harmonic back-EMFs or flux-linkages in the two sets are orthogonal to each other, which can be directly used as inputs of a position observer. As shown in Figure 7.21, a simplified extended Kalman filter is used in [13] as a position observer, which has a similar structure as a phase-locked loop (PLL). The details of this observer will be given in Chapter 12.

Experimental results for the prototype DTP-PM machine (Appendix-B, SPMSM-V) by using both third harmonic back-EMFs and flux-linkages are shown here. First, as shown in Figure 7.22, the steady-state tests are carried out when the rotor speed is 200 r/min with 1 A q-axis current. Figure 7.23 shows the step-speed performance with speed reference changes from 200 to 300 r/min. The experimental results prove that both strategies have accurate estimation under steady state, while the strategy based on the third harmonic back-EMF can achieve better dynamic performance due to the elimination of filters compared to the flux-linkage strategy.

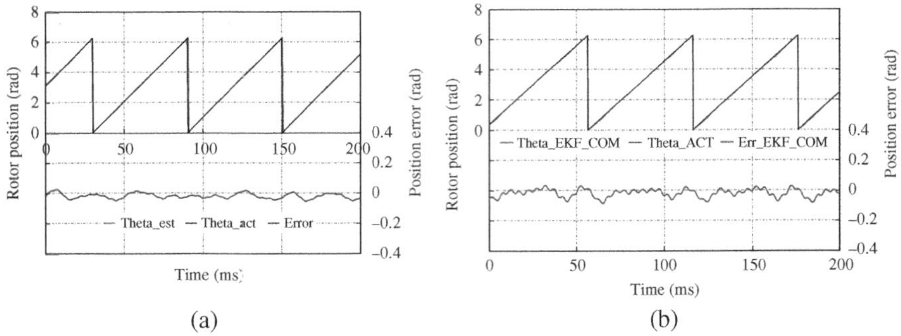

Figure 7.22 Steady-state estimation performance. (a) Third harmonic flux-linkage. (b) Third harmonic back-EMF. (*Source:* Adapted from [13].)

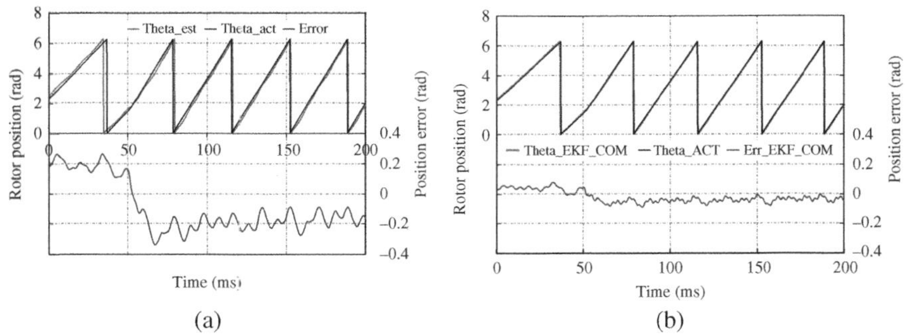

Figure 7.23 Step speed estimation performance. (a) Third harmonic flux-linkage. (b) Third harmonic back-EMF. (*Source:* Adapted from [13].)

7.3 Open Winding PMSMs

The OW-PMSM drives are investigated in [14]–[16]. Compared to the Y-connected PMSM shown in Figure 7.24a, the three phase windings of the PMSM can be disconnected and placed as three independent phase windings as shown in Figure 7.24b, which is driven by an H-bridge for each phase. Compared to the Y-connected or delta-connected PMSM topology, the number of the driven terminals increases from 3 to 6. Meanwhile, more flexible controllability can be expected. Due to the independence of the phase winding, better fault tolerant capability can also be obtained. Moreover, the zero sequence is essentially controllable for the OW-PMSM; as a result, more stable torque performance can be achieved under fault mode. Since the dc bus utilization ratio of the OW-PMSM is much higher, the capability of the power delivery for the OW-PMSM is higher as well.

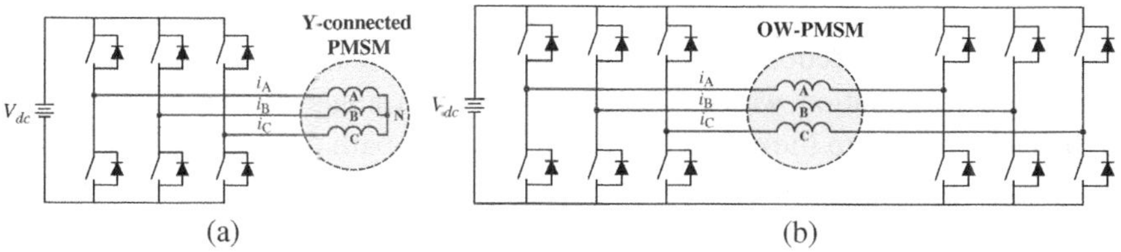

Figure 7.24 PMSM topologies. (a) Y-connected PMSM. (b) OW-PMSM.

Regarding sensorless control, the conventional sensorless control methods in Chapters 2–5 can be also extended to OW-PMSM. Nevertheless, due to the specific configuration, the OW-PMSM also has specific superiority on sensorless control [15], [16], which will be described in this section.

7.3.1 Modeling of OW-PMSM Drive

There are many different topologies of OW-PMSM drives [14]; two main types are introduced next.

The most common drive topology for OW-PMSM is the common dc bus one, which is shown in Figure 7.25a [14]. Two voltage source inverters (VSIs) are connected to the six terminals of the OW-PMSM. The dc bus for these two inverters is the same. In this topology, the equivalent zero sequence is no longer open circuit. The third harmonic back-EMF contained in the OW-PMSM, and the zero sequence modulated voltage from the inverter will consequently induce a circulating current, which can be utilized for position estimation.

Another drive topology for OW-PMSM is the isolated dc bus one that is shown in Figure 7.25b. The dual inverter drive is supplied by two isolated dc sources and hence the zero sequence is open circuit. Consequently, the zero sequence circulating current can be avoided. Compared to the common dc bus, the isolated dc bus supplies increase the system complexity, cost, and size. For this topology, since zero sequence is open circuit, conventional sensorless control can be directly applied.

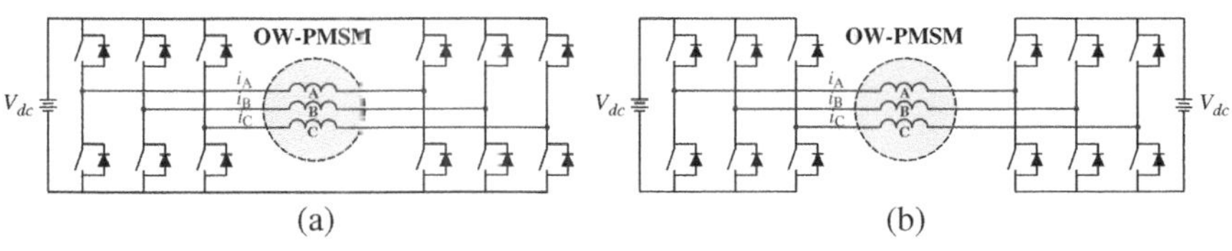

Figure 7.25 OW-PMSM drive topologies. (a) Common dc bus. (b) Isolated dc bus.

The OW-PMSM model is similar to the conventional single inverter-controlled PMSM with Y-connection. However, for an OW-PMSM with common dc bus, the zero sequence equation, which is usually ignored in a single inverter system, needs to be included due to the existence of zero sequence path. Hence, the OW-PMSM model considering the zero sequence equation can be described as:

$$\mathbf{v} = \begin{bmatrix} R_s & -\omega_r L_q & 0 \\ \omega_r L_d & R_s & 0 \\ 0 & 0 & R_s \end{bmatrix} \cdot \mathbf{i} + \begin{bmatrix} L_d & 0 & 0 \\ 0 & L_q & 0 \\ 0 & 0 & L_0 \end{bmatrix} \cdot \dot{\mathbf{i}} + \begin{bmatrix} 0 \\ \omega_r \psi_m \\ 3\omega_r \psi_0 \sin 3\theta_r \end{bmatrix} \tag{7.47}$$

where $\mathbf{v} = [v_d, v_q, v_0]^T$, $\mathbf{i} = [i_d, i_q, i_0]^T$, v_d is the d-axis voltage, v_q is the q-axis voltage, v_0 is the zero sequence voltage, i_d is the d-axis current, i_q is the q-axis current, i_0 is the zero sequence current, R_s is the winding resistance, L_d is the d-axis inductance, L_q is the q-axis inductance, L_0 is the zero sequence inductance, ω_r represents the electrical angular speed, "·" denotes the differential operator, ψ_m is the fundamental flux-linkage, ψ_0 is the zero sequence flux-linkage (the third harmonic flux-linkage), and θ_r is the electrical rotor position.

Hence, the equivalent zero sequence circuit of the OW-PMSM is shown in Figure 7.26.

The zero sequence voltage v_0 from the inverter side is generated from the specific PWM strategy [14]. The zero sequence inductance is described as:

$$L_0 = L_{s0} - 2M_{s0} \tag{7.48}$$

where represents the average value of the arbitrary phase self-inductance and M_{s0} represents the average value of mutual-inductance between any two phases.

The winding phase resistance is equivalent to zero sequence resistance. The zero sequence back-EMF is expressed as:

$$e_0 = 3\omega_r \psi_0 \sin 3\theta_r \tag{7.49}$$

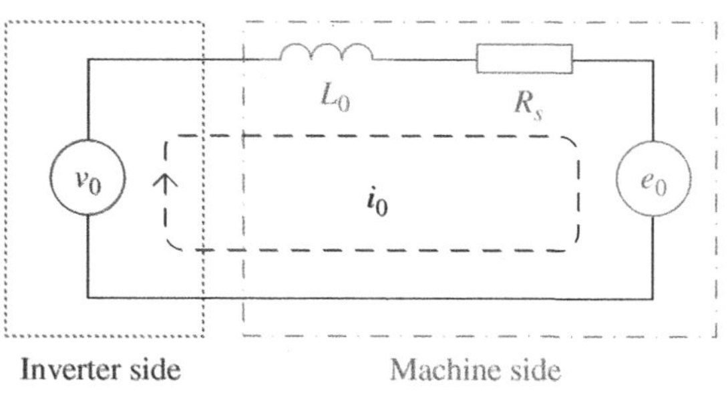

Figure 7.26 Equivalent zero sequence circuit of OW-PMSM [14].

Hence, the zero sequence equivalent equation is:

$$v_0 = (L_{s0} - 2M_{s0})\frac{di_0}{dt} + i_c R_s + 3\omega_r \psi_0 \sin 3\theta_r \tag{7.50}$$

It can be found that the rotor position information is contained in the zero sequence back-EMF e_0. According to (7.50), the rotor position can be extracted. Therefore, in this section, based on the zero sequence, two specific sensorless control methods for OW-PMSM with common dc bus are introduced with experimental results.

7.3.2 Phase Shift-Based SVPWM for OW-PMSM

Before introducing sensorless control, specific SVPWM for OW-PMSM should be explained [15]. In an OW-PMSM drive system, the conventional SVPWM strategy is not suitable for operation since the existing zero sequence path allows the voltage disturbance in the zero axis to induce the disturbance current. The disturbance voltage includes the low-frequency common mode voltage (CMV) produced by SVPWM and also the back-EMF harmonics such as 3rd, 9th, 15th, ... components sharing the same phase angle in three-phase back-EMFs. Hence, in order to eliminate the interference from the SVPWM, a phase shift-based SVPWM [15] is introduced and adopted for the whole control system in this section, which is shown in Figure 7.27.

The conventional SVPWM strategy contains quasi-triangle waves in modulation signals in three phases to improve the dc bus voltage utilization. It imposes almost no influence on the single inverter drive system, but it will induce current in an OW configuration with common dc bus. The basic countermeasure is to shift the voltage reference signals for two sets of the inverters and obtain a counteraction between zero sequence voltages of the two modules. The actual d- and q-axis

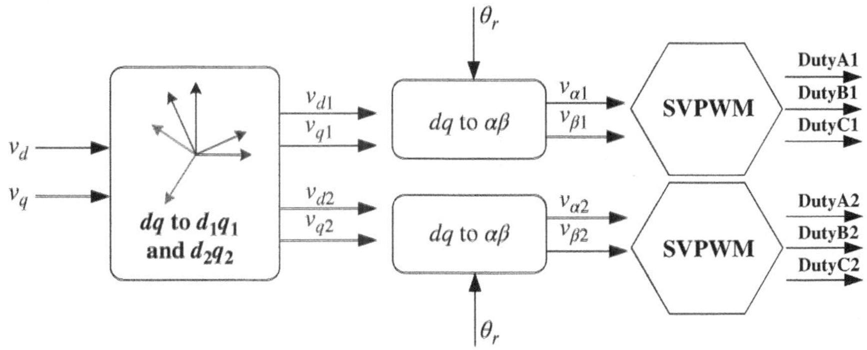

Figure 7.27 Phase shift-based SVPWM for OW-PMSM. (*Source:* [15]/IEEE.)

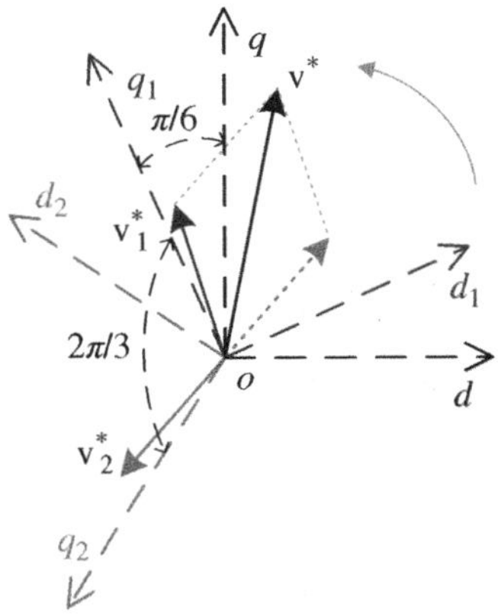

Figure 7.28 Synthetic voltage vector of two sets of inverters, (*Source:* [15]/IEEE.)

voltage reference signals can be decomposed into two sub-coordinates, d_1q_1 and d_2q_2, which are synchronized with the dq-coordinate but have $\pi/6$ and $5\pi/6$ shifted angles to the dq-coordinate, respectively, as shown in Figure 7.28.

Meanwhile, the decomposition transformation can be deduced as:

$$\begin{bmatrix} v_{d1} \\ v_{q1} \\ v_{d2} \\ v_{q2} \end{bmatrix} = \frac{1}{2} \cdot \begin{bmatrix} 1 & \tan\frac{\pi}{6} \\ -\tan\frac{\pi}{6} & 1 \\ -1 & \tan\frac{\pi}{6} \\ -\tan\frac{\pi}{6} & -1 \end{bmatrix} \cdot \begin{bmatrix} v_d \\ v_q \end{bmatrix} \tag{7.51}$$

This phase shift-based SVPWM achieves the phase shifting by reconstructing the reference signals for each SVPWM module. Moreover, the zero sequence voltage output from the inverter side is eliminated and the equivalent zero sequence circuit can be simplified as $v_0 = 0$, which is necessary for the sensorless method described in the next sections.

7.3.3 Zero Sequence Current-Based Sensorless Control

With the phase shift-based SVPWM introduced in Section 7.3.2 adopted, the inverter side only shows the short circuit characteristic ($v_0 = 0$ V) and the machine side stays the same. The equivalent zero sequence circuit can be simplified as Figure 7.29. Hence, the third harmonic back-EMF e_0 induces a corresponding zero sequence current, which in turn can be used to reconstruct the third harmonic back-EMF and finally used for position estimation [15].

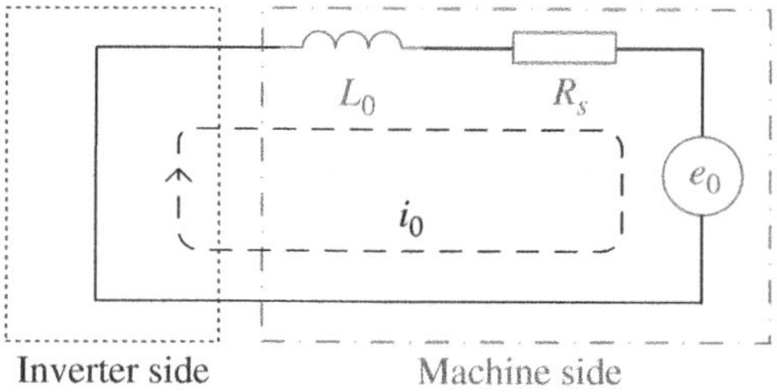

Figure 7.29 Simplified equivalent zero sequence circuit of the OW-PMSM by using phase shift-based SVPWM algorithm. (*Source:* [15]/IEEE.)

According to Figure 7.29, the zero sequence equation is written as:

$$0 = L_0 \frac{di_0}{dt} + i_0 R_s + 3\omega_r \psi_0 \sin 3\theta_r \tag{7.52}$$

Meanwhile, in (7.52), i_0 can be obtained via

$$i_0 = \frac{i_a + i_b + i_c}{3} \tag{7.53}$$

where i_a, i_b, and i_c are the phase A, phase B, and phase C currents, respectively.

According to this principle, the zero sequence model-based sensorless strategy is shown in Figure 7.30 The synchronous PLL-based quadrature signal generator (QSG) in Figure 6.8 is used to generate the equivalent zero sequence current signals, which satisfy:

$$\begin{cases} i'_{\alpha 0} = i_0 \\ i'_{\beta 0} = \dfrac{di_0}{dt} \end{cases} \tag{7.54}$$

Meanwhile, the outputs of the synchronous PLL-based QSG are used to reconstruct the zero sequence back-EMF $\hat{e}_0$ according to (7.52). The estimated $\hat{e}_0$ then serves as the input signal of the synchronous PLL-based PO to estimate the rotor position signal.

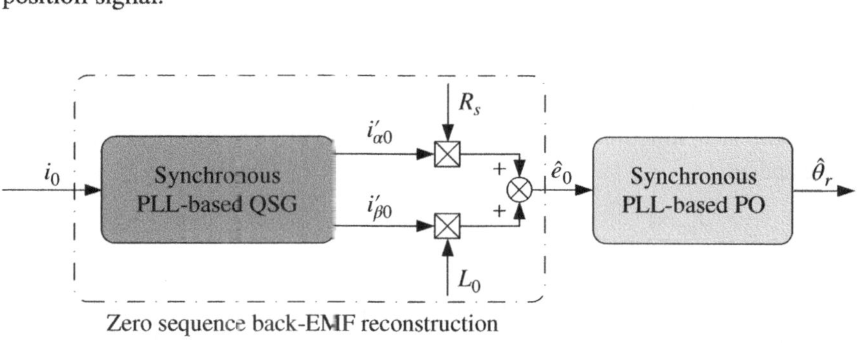

Figure 7.30 Zero sequence current-based sensorless strategy. (*Source:* [15]/IEEE.)

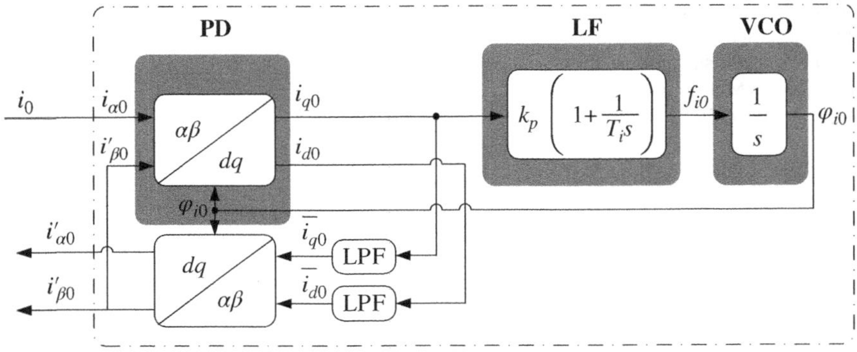

Figure 7.31 Synchronous PLL-based QSG. (*Source:* [15]/IEEE.)

In (7.52), differential information of i_0 is required. It would not be a good choice since the pure digital differentiator suffers from noise amplification and phase error. Hence, the synchronous PLL-based QSG is used. The structure of the QSG is shown in Figure 7.31.

The transfer function of the LPFs in Figure 7.31 on the synchronous reference coordinate are both first order and expressed as:

$$H(s) = \frac{\bar{I}_{d0}(s)}{I_{d0}(s)} = \frac{\bar{I}_{q0}(s)}{I_{q0}(s)} = \frac{\omega_f}{s + \omega_f} \tag{7.55}$$

where ω_f is the cutoff frequency.

Meanwhile, when they are transformed into the dq-axis, the transfer functions are

$$D(s) = \frac{I'_{\alpha 0}(s)}{I_{\alpha 0}(s)} = \frac{k\omega_0 s}{s^2 + k\omega_0 s + \omega_0^2} \tag{7.56}$$

$$Q(s) = \frac{I'_{\beta 0}(s)}{I_{\alpha 0}(s)} = \frac{k\omega_0^2}{s^2 + k\omega_0 s + \omega_0^2} \tag{7.57}$$

where $k = \omega_f / \omega_0$ and $\omega_0 = 3\omega_r$.

The transfer function describing the relationship between $i'_{\alpha 0}$ and $i'_{\beta 0}$ is expressed as

$$G(s) = \frac{I'_{\alpha 0}(s)}{I'_{\beta 0}(s)} = \frac{s}{\omega_0} \tag{7.58}$$

The dual output signals from the dq to $\alpha\beta$ transformation can be transformed into a time-domain differential equation as (6.14), and hence the two signals are orthogonal if they are both sinusoidal.

$$i'_{\alpha0} = \frac{1}{\omega_0}\frac{di'_{\beta0}}{dt} \tag{7.59}$$

Similarly, the synchronous PLL-based PO is shown in Figure 7.32, whereby, the difference is that a frequency multiplication index h is introduced to make the estimated position signal match the third harmonic back-EMF. In the inner loop of the PO, the phase angle signal θ_{i0} is triple the observed rotor position angle $\hat{\theta}_r$. The input signal is the reconstructed zero sequence back-EMF.

The above algorithm is tested on a prototype 3 kW OW-PMSM drive (Appendix B, SPMSM-VI). Figure 7.33 shows the steady-state position estimation performance under different load conditions. Figure 7.34a shows the closed loop transient estimation performance under rated load disturbance. Figure 7.34b shows the corresponding d-axis, q-axis, and the phase A currents. The rated load disturbance is applied at 6 seconds and released at 20 seconds. Figure 7.35a shows the transient estimation performance with step speed reference from 15 to 40 r/min. Figure 7.35b shows the corresponding d-axis, q-axis, and the phase A currents under the same condition. Clearly, good estimation performance is obtained.

7.3.4 Nonparametric Zero Sequence Voltage-Based Sensorless Control

For the method introduced in Section 7.3.3, zero sequence current exists that causes extra losses and torque ripples [15]. Hence, a novel zero sequence back-EMF-based sensorless method together with circulating current suppression in [16] is presented in this section. The method extracts the rotor position information from the output signal of the zero sequence current controller (ZSC). The method is nonparametric when compared with conventional fundamental back-EMFbased methods.

As illustrated in Section 7.3.1, the zero sequence differential equation can be expressed as:

$$v_0 = (L_{s0} - 2M_{s0})\frac{di_0}{dt} + i_0R_s + 3\omega_r\psi_0 \sin 3\theta_r \tag{7.60}$$

where $e_0 = 3\omega_r\psi_0 \sin 3\theta_r$, L_{s0} represents the average phase self-inductance and M_{s0} represents the average mutual-inductance between any two phases.

The zero sequence current can be suppressed to zero by a PI controller as shown in Figure 7.36. If the zero sequence current is completely suppressed to zero, the output zero sequence voltage v_0 produced by the dual inverters satisfies:

$$v_0 = 3\omega_r\psi_0 \sin 3\vartheta_r = e_0 \tag{7.61}$$

It means that the regulation output signal of the ZSC is equivalent to zero sequence back-EMF e_0. Hence, this signal can be consequently used for position estimation. Note also that no voltage measurement is required here to obtain the zero sequence back-EMF information. Moreover, when compared to the conventional fundamental back-EMF-based methods using d- and q-axis equations as:

$$\begin{bmatrix} v_d \\ v_q \end{bmatrix} = \underbrace{\begin{bmatrix} R_s + pL_d & -\omega_r L_q \\ \omega_r L_d & R_s + pL_q \end{bmatrix} \begin{bmatrix} i_d \\ i_q \end{bmatrix}}_{\text{disturbance voltage drops}} + \begin{bmatrix} 0 \\ \omega_r \psi_m \end{bmatrix} \tag{7.62}$$

The output signal of ZSC is simply the zero sequence back-EMF without any other disturbance voltage drops. Once the equivalent zero sequence back-EMF signal is obtained, it can be used as the input signal of a synchronous PLL-based PO, which is the same as the one in Figure 7.32. The position estimation process is shown in Figure 7.36.

The experiment is conducted on the prototype 3 kW OW-PMSM drive (Appendix B, SPMSM-IV). Figure 7.37 shows the suppression performance for the circulating

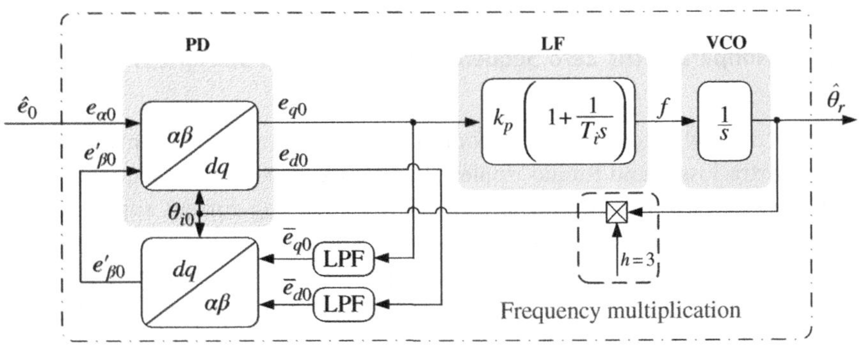

Figure 7.32 Synchronous PLL-based PO. (*Source:* [15]/IEEE.)

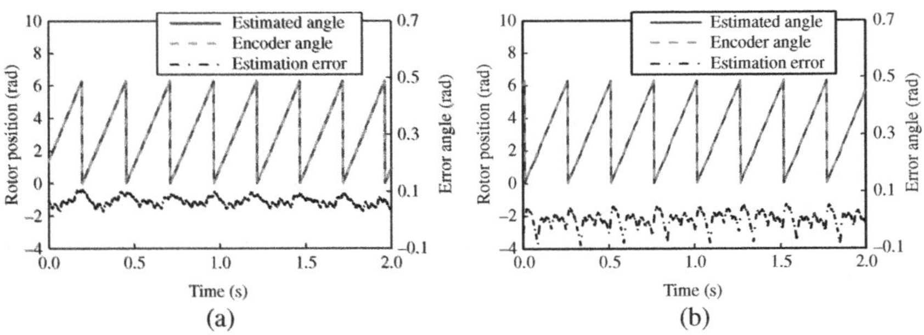

Figure 7.33 Steady-state estimation performance. (a) No load. (b) Rated load. (*Source:* [15]/IEEE.)

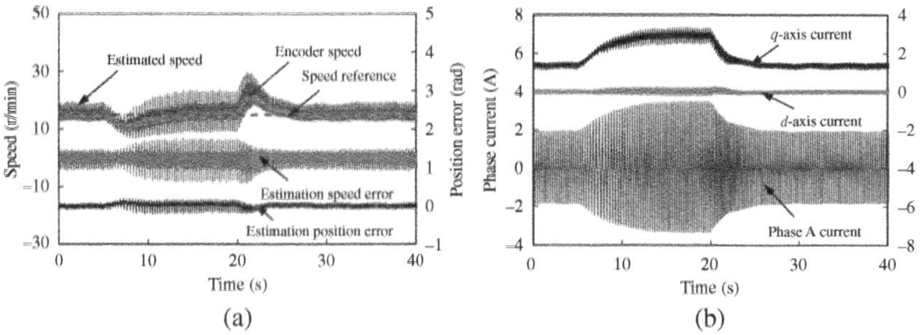

Figure 7.34 Transient estimation performance and current signals under rated load disturbance (which is applied at 6 seconds and released at 20 seconds). (a) Position estimation. (b) Stator current. (*Source:* [15]/IEEE.)

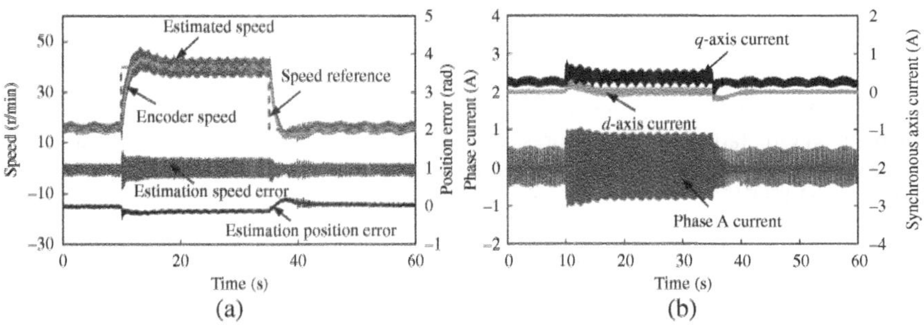

Figure 7.35 Transient estimation performance and current signals with step speed reference from 15 to 40 r/min at 10 seconds and from 40 to 15 r/min at 35 seconds. (a) Position estimation. (b) Stator current. (*Source:* [15]/IEEE.)

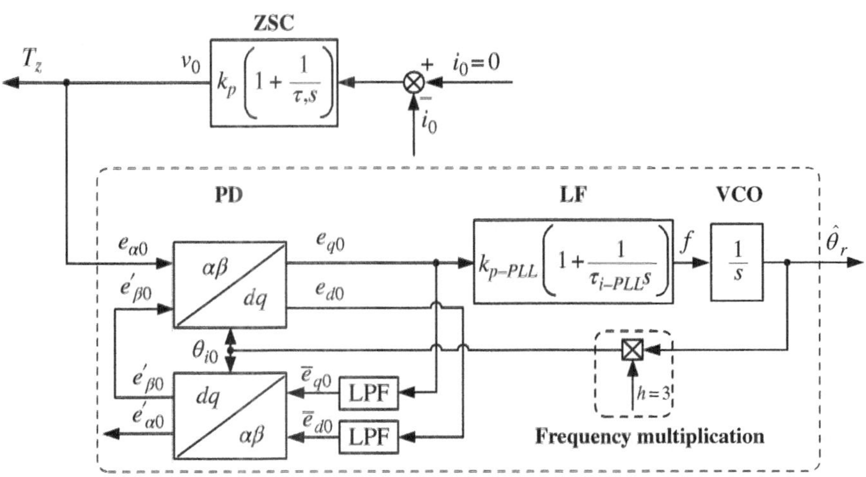

Figure 7.36 Position estimation process using synchronous PLL-based PO. (*Source:* [16]/IEEE.)

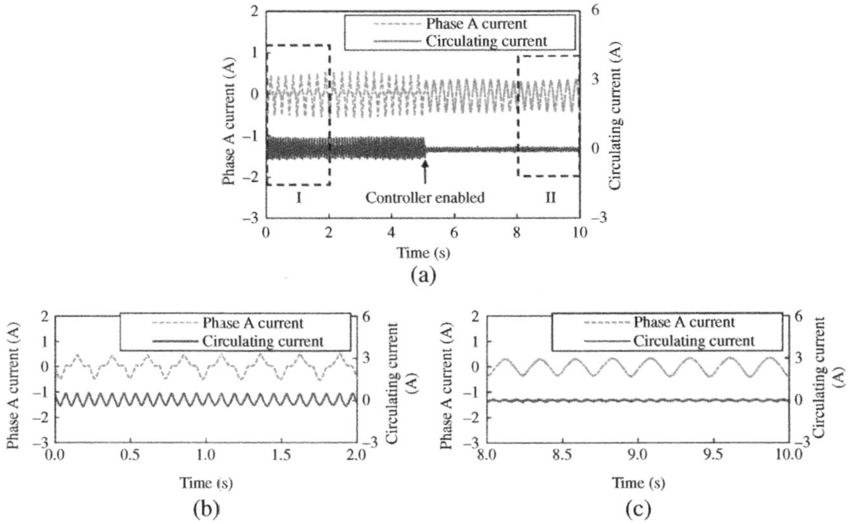

Figure 7.37 Circulating current suppression performance before and after controller enabled. (a) Circulating current suppression performance with controller enabled at 5 seconds. (b) Before the ZSC enabled, region I. (c) After the ZSC enabled, region II. (*Source:* [16]/IEEE.)

current at 15 r/min including the phase A current and circulating current waveforms. The ZSC is enabled at 5 seconds in Figure 7.37a and the circulating current is suppressed after the ZSC is enabled. Figures 7.37b,c show the magnified regions marked as I and II in Figure 7.37a, respectively. There exists dominant third harmonic current in the circulating current before suppression.

Figure 7.38 shows the steady-state estimation performance under no load and rated load conditions. Figure 7.39 shows the transient-state estimation

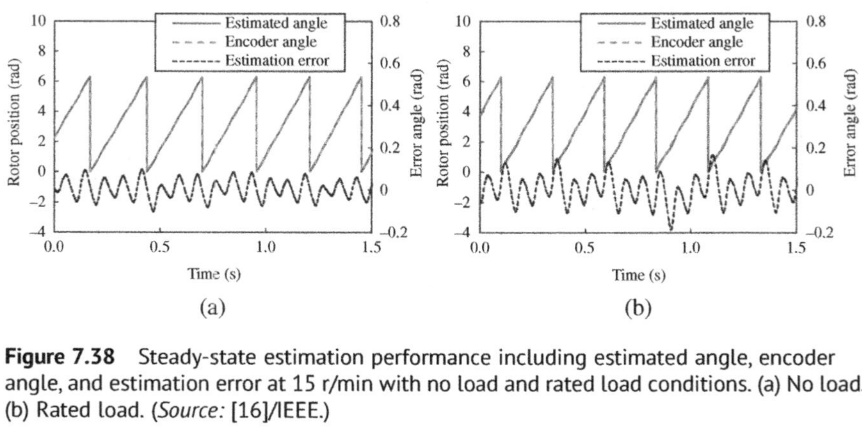

Figure 7.38 Steady-state estimation performance including estimated angle, encoder angle, and estimation error at 15 r/min with no load and rated load conditions. (a) No load. (b) Rated load. (*Source:* [16]/IEEE.)

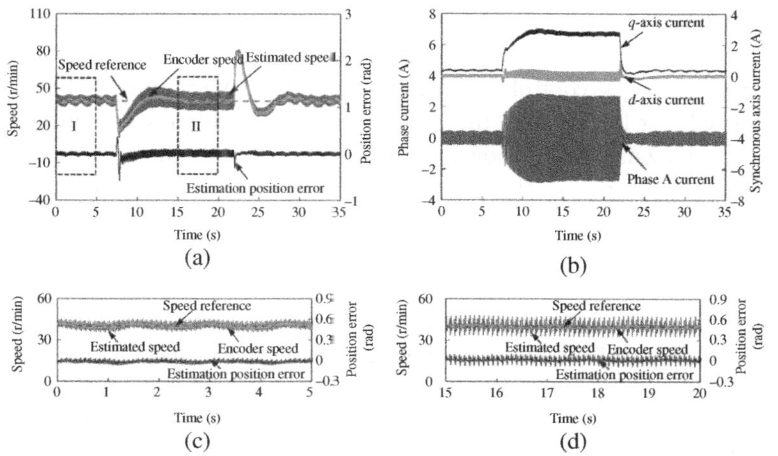

Figure 7.39 Transient-state estimation performance and current signals under step-load. (a) Speed and position estimation. (b) d-axis, q-axis, and phase A currents. (c) Magnified region I (40 r/min, no load). (d) Magnified region II (40 r/min, rated load). (*Source:* [16]/IEEE.)

performance and current signals under rated step-load disturbance. The step-load is imposed at 7 seconds and released at 22 seconds. Figure 7.40 shows the transient state estimation performance and current signals with step speed reference from 40 to 80 r/min. The step speed reference is given at 10 seconds and steps back at 25 seconds.

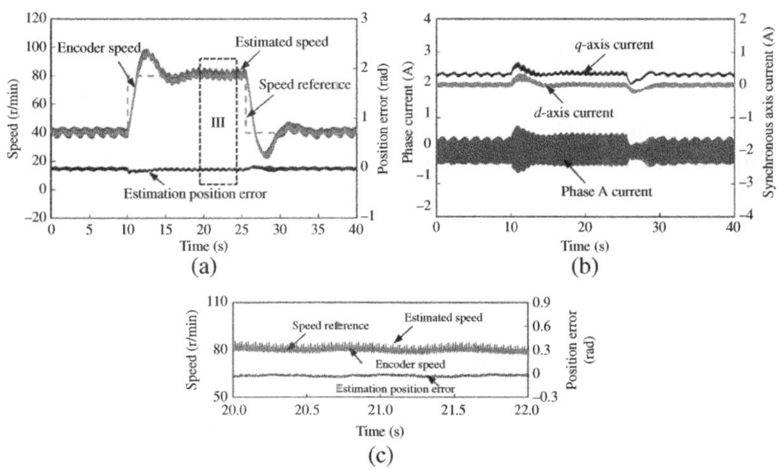

Figure 7.40 Transient-state estimation performance and current signals under step speed. (a) Speed and position estimation. (b) d-axis, q-axis, and phase A currents. (c) Magnified region III. (*Source:* [16]/IEEE.)

7.4 Summary

In this chapter, sensorless control strategies based on DTP-PMSM and OW-PMSM were introduced. For DTP-PMSM, the conventional HFSI methods were investigated first. The VSD model was shown to have a better steady-state performance due to decoupled subspaces. Then, the HF zero-sequence voltage response is utilized by a simple voltage measurement method [8], [9]. Based on the double-*dq* model, the injection phase shift between two sets can be optimized [8], [9] to obtain the improved steady-state estimation performance with reduced sixth order harmonic. Then, the fundamental model-based [10] and third harmonic model-based methods [13] were described. Based on the double *dq* model, a simple correction method [10] was introduced that corrects the resistance and *q*-axis inductance mismatch and thus minimizes the position error. Moreover, by utilizing the 30 electrical degrees shift between the two sets, the two orthogonal zero sequence voltages of the two sets can be used for position estimation [13]. For an OW-PMSM with common dc bus, due to the existence of zero sequence current, specific sensorless control methods [15], [16] utilizing zero sequence were presented and show better robustness than the conventional methods.

References

1 Z. Q. Zhu, S. Wang, B. Shao, L. Yan, P. Xu, and Y. Ren, "Advances in dual-three-phase permanent magnet synchronous machines and control techniques," *Energies*, vol. 14, no. 22, paper 7508, pp. 1–46, Nov. 2021.

2 R. Bojoi, M. Lazzari, F. Profumo, and A. Tenconi, "Digital field-oriented control for dual three-phase induction motor drives," *IEEE Trans. Ind. Appl.*, vol. 39, no. 3, pp. 752–760, May/Jun. 2003.

3 M. Barcaro, N. Bianchi, and F. Magnussen, "Analysis and tests of a dual three-phase 12-slot 10-pole permanent-magnet motor," *IEEE Trans. Ind. Appl.*, vol. 46, no. 6, pp. 2355–2362, Nov. 2010.

4 R. H. Nelson and P. C. Krause, "Induction machine analysis for arbitrary displacement between multiple winding sets," *IEEE Trans. Power App. Syst.*, vol. 93, no. 3, pp. 841–848, May 1974.

5 T. A. Lipo, "A d-q model for six-phase induction machines," in *Proc. Int. Conf. Elect. Mach. ICEM*, Athens, Greece, pp. 860–867, 1980.

6 Y. Hu, Z. Q. Zhu, and M. Odavic, "Comparison of two-individual current control and vector space decomposition control for dual three-phase PMSM," *IEEE Trans. Ind. Appl.*, vol. 53, no. 5, pp. 4483–4492, Sep. 2017.

7 Y. Zhao and T. A. Lipo, "Space vector PWM control of dual three-phase induction machine using vector space decomposition," *IEEE Trans. Ind. Appl.*, vol. 31, no. 5, pp. 1100–1109, Sep./Oct. 1995.

8 A. H. Almarhoon, Z. Q. Zhu, and P. Xu, "Improved pulsating signal injection using zero-sequence carrier voltage for sensorless control of dual three-phase PMSM," *IEEE Trans. Energy Convers.*, vol. 32, no. 2, pp. 436–446, Jun. 2017.

9 A. H. Almarhoon, Z. Q. Zhu, and P. L. Xu, "Improved rotor position estimation accuracy by rotating carrier signal injection utilizing zero-sequence carrier voltage for dual three-phase PMSM," *IEEE Trans. Ind. Electron.*, vol. 64, no. 6, pp. 4454–4462, Jun. 2017.

10 T. Liu, Z. Q. Zhu, Z. Y. Wu, D. Stone, and M. Foster, "A simple sensorless position error correction method for dual three-phase permanent magnet synchronous machines," *IEEE Trans. Energy Convers.*, vol. 36, no. 2, pp. 895–906, Jun. 2021.

11 F. Briz, M. W. Degner, P. Garcia, and R. D. Lorenz, "Comparison of saliency-based sensorless control techniques for AC machines," *IEEE Trans. Ind. Appl.*, vol. 40, no. 4, pp. 1107–1115, Jul./Aug. 2004.

12 F. Briz, M. W. Degner, P. Garcia, and J. M. Guerrero, "Rotor position estimation of AC machines using the zero sequence carrier signal voltage," *IEEE Trans. Ind. Appl.*, vol. 41, no. 6, pp. 1637–1646, Nov./Dec. 2005.

13 J. Liu and Z. Q. Zhu, "Rotor position estimation for single- and dual-three-phase permanent magnet synchronous machines based on third harmonic back-EMF under imbalanced situation," *Chin. J. Elect. Eng.*, vol. 3, no. 1, pp. 63–72, 2017.

14 H. Zhan, Z. Q. Zhu, and M. Odavic, "Analysis and suppression of zero sequence circulating current in open winding PMSM drives with common dc bus," *IEEE Trans. Ind. Appl.*, vol. 53, no. 4, pp. 3609–3620, Jul. 2017.

15 H. Zhan, Z. Q. Zhu, M. Odavic, and Y. Li, "A novel zero-sequence model-based sensorless method for open-winding PMSM with common dc bus," *IEEE Trans. Ind. Electron.*, vol. 63, no. 11, pp. 6777–6789, 2016.

16 H. Zhan, Z. Q. Zhu, and M. Odavic, "Nonparametric sensorless drive method for open-winding PMSM based on zero-sequence back emf with circulating current suppression," *IEEE Trans. Power Electron.*, vol. 32, no. 5, pp. 3808–3817, 2017.

17 S. Morimoto, K. Kawamoto, M. Sanada, and Y. Takeda, "Sensorless control strategy for salient-pole PMSM based on extended EMF in rotating reference frame," *IEEE Trans. Ind. Appl.*, vol. 38, no. 4, pp. 1054–1061, Jul. 2002.

8

Magnetic Polarity Identification

8.1 Introduction

As shown in Figure 8.1, the rotor position θ_r is defined as the angle between the winding axis of phase A (A-axis) and the north magnetic pole (d-axis) of the rotor. As described in Chapter 4, the geometric inductance saliency in IPMSM and the magnetic inductance saliency in SPMSM both vary periodically as a second order spatial harmonic. Thus, the estimated position (or estimated d-axis $\hat{d}$) obtained by saliency tracking-based sensorless control is the rotor position without the magnetic polarity information, i.e. north or south magnetic pole, which may cause ambiguity of π in position estimation error $\Delta\theta$. As shown in Figure 8.1, the position estimation error is zero at the north magnetic pole position and π at the south magnetic pole position. If the estimated rotor position is assumed to be aligned at the south magnetic pole position but the rotor position is actually at the north magnetic pole position, the sign of the output torque will be negative and the system will be unstable. Therefore, it is important to identify the magnetic polarity before operation.

The basic principle of detecting polarity is based on the magnetic saturation effect, which can be utilized on both IPMSM and SPMSM. As shown in Figure 8.2, a positive d-axis current will increase the saturation of the stator iron core and therefore decrease the d-axis inductance, whereas a negative d-axis current decreases the saturation and increases the d-axis inductance.

A general frame of the magnetic polarity identification is depicted in Figure 8.3. Based on the magnetic saturation effect, different signals can be injected into the d-axis of the machine and the corresponding responses are used to identify the

Sensorless Control of Permanent Magnet Synchronous Machine Drives, First Edition.
Zi Qiang Zhu and Xi Meng Wu.

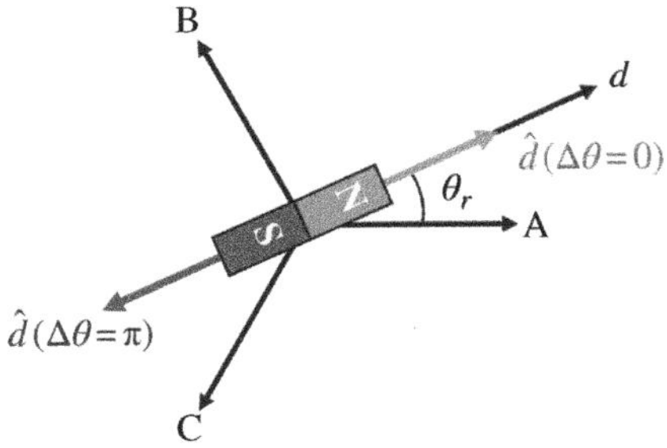

Figure 8.1 Illustration of estimated reference frame with magnetic polarity ambiguity.

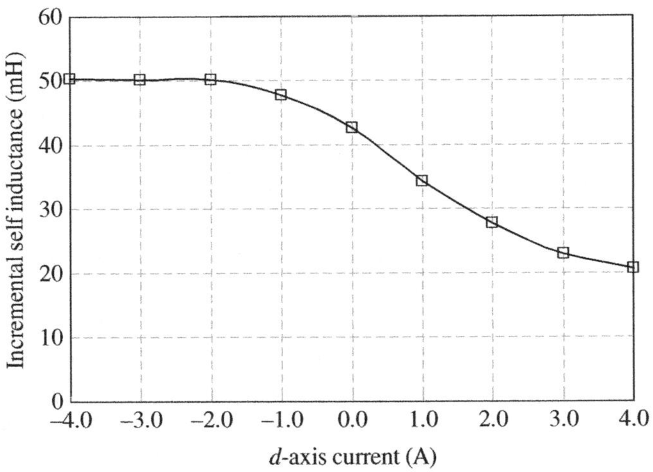

Figure 8.2 Measured *d*-axis incremental self-inductance against *d*-axis current. (*Source:* [1]/IEEE.)

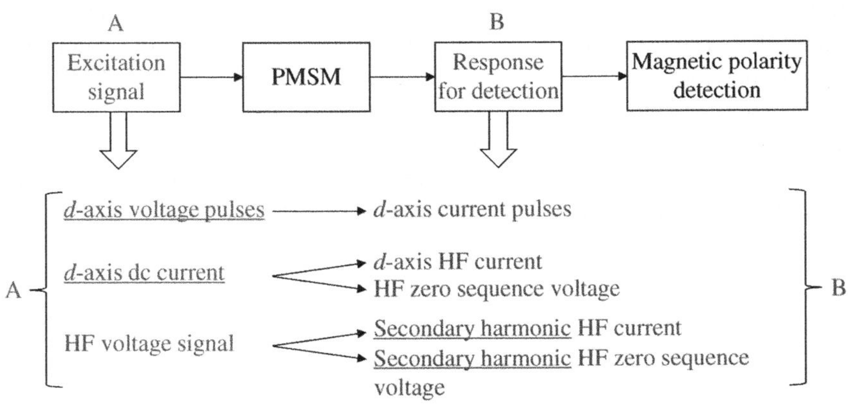

Figure 8.3 General frame of magnetic polarity identification.

polarity. Depending on the injection signal or response used for polarity identification, which are underlined in Figure 8.3, the magnetic polarity can be basically classified by

- Dual voltage pulses injection-based method [2]
- *d*-axis current injection-based method [1], [3]
- Secondary harmonics of the high-frequency (HF) response-based method [3]–[5]

8.2 Dual Voltage Pulses Injection-Based Method

After the rotor position is estimated by the saliency tracking-based sensorless control, the dual voltage pulses injection method [2] can be applied at standstill to identify the magnetic polarity. Two voltage pulses with opposite amplitudes are injected into the estimated *d*-axis (or $\hat{d}$-axis) and the difference between their current response peaks indicates the magnetic polarity. The injection is demonstrated in Figure 8.4. Based on the magnetic saturation effect, the pulse with a higher current response peak value indicates the positive direction of the actual *d*-axis, while the pulse with a lower current response peak value indicates the negative direction of the actual *d*-axis. The overall identification process is demonstrated in

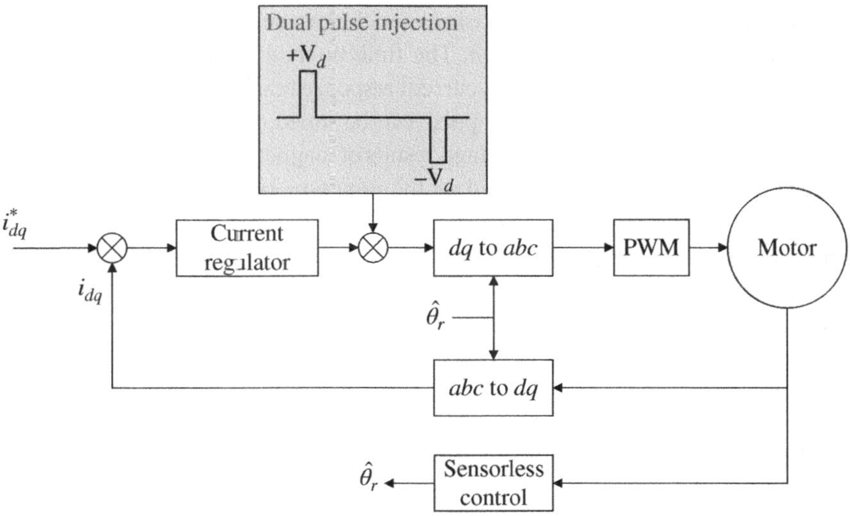

Figure 8.4 Block diagram of sensorless drive system with dual pulse injection.

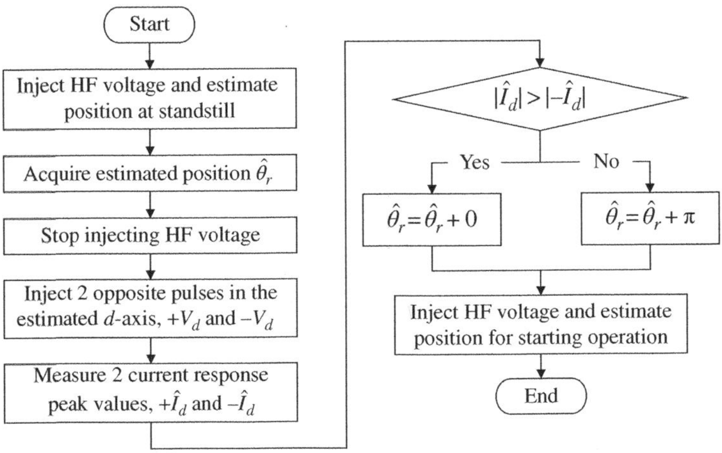

Figure 8.5 Flow chart of dual voltage pulses injection method.

Figure 8.5. Moreover, since the pulse is injected into the d-axis, the electromagnetic torque that occurs with the pulse voltage injected into the machine is negligible and takes a short time, which hardly moves the rotor [2].

Regarding the voltage pulse, the width of each pulse and the time interval between the two pulses should be short enough for a fast identification process. The width and amplitude of the voltage pulses should be selected large enough to guarantee an observable difference between the two current responses for a reliable magnetic polarity identification. The time interval between the two pulses should be long enough until the first current response reaches zero. Based on these design rules, the settings of voltage pulse can be tuned offline.

Figures 8.6 and 8.7 show the measured results of magnetic polarity detection with the dual voltage pulses method based on the prototype IPMSM-I (Appendix B).

If the estimated rotor position is correct, i.e. $\Delta\theta = 0°$, the applied positive d-axis voltage pulse will result in a higher d-axis current pulse, Figure 8.6. On the other hand, a lower amplitude current response to the positive d-axis voltage pulse suggests that the estimated position has an error of π, Figure 8.7.

Although the dual voltage pulses injection method has robust and reliable identification performance due to a good signal-to-noise ratio (SNR), it suffers from a slower convergence time [6]. Moreover, as shown in Figure 8.5, the HF voltage signal injection of saliency tracking-based sensorless control should be disabled during the pulse injection to ensure that (i) the injected pulse is not affected by the HF signal injection, and (ii) the current response of the pulse is not affected by the HF current response of HFSI. Thus, the dual voltage pulses-based magnetic

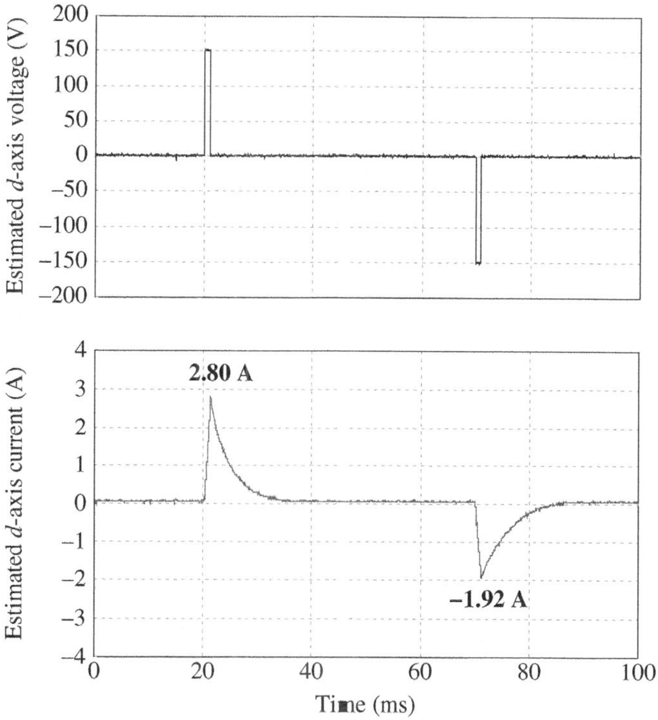

Figure 8.6 Magnetic polarity detection with dual voltage pulses ($\Delta\theta = 0°$). (*Source:* [1]/IEEE.)

polarity identification is performed as an independent process to the saliency tracking-based sensorless control, which is difficult to integrate with the sensorless algorithm, thus giving rise to computational complexity.

8.3 *d*-Axis Current Injection-Based Method

8.3.1 HF Current Response

In [1], based on the *d*-axis HF pulsating voltage signal injection sensorless control, the *d*-axis HF current response is used to identify the magnetic polarity by injecting a dc current at the *d*-axis.

As described in Chapter 4, when an HF pulsating voltage signal injection is injected at the estimated *d*-axis, i.e. $\hat{v}_{dh} = V_h \cos(\omega_h t)$, the resultant *d*-axis HF current response can be derived as:

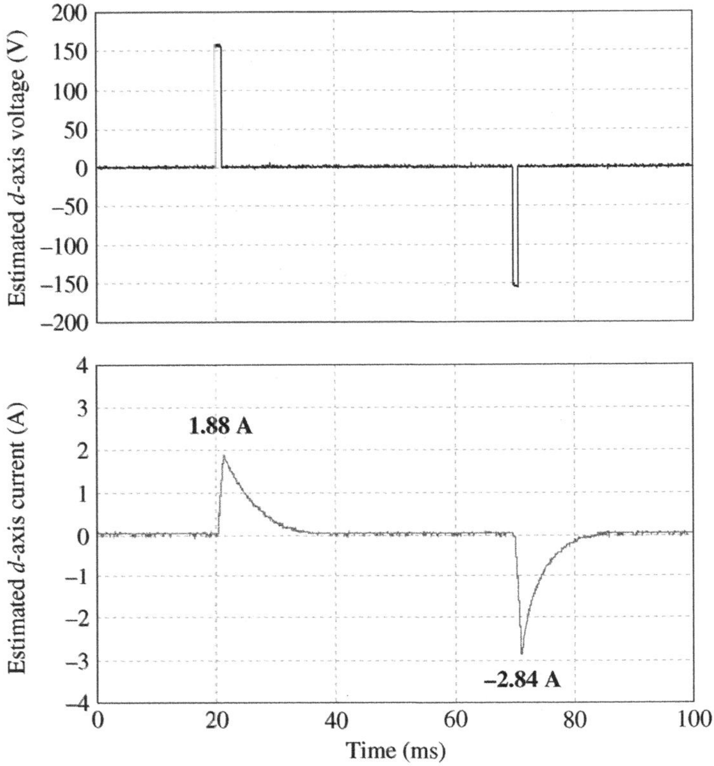

Figure 8.7 Magnetic polarity detection with dual voltage pulses ($\Delta\theta = 180°$). (*Source:* [1]/IEEE.)

$$\hat{i}_{dh} = \left[I_p + I_n\cos(2\Delta\theta + \theta_m)\right]\sin(\omega_h t) \tag{8.1}$$

where $I_p = \dfrac{V_h}{\omega_h L_p}$, $I_n = \dfrac{V_h}{\omega_h L_n}$, $L_p = \dfrac{L_{dh}L_{qh} - L_{dqh}^2}{L_{sa}}$, $L_n = \dfrac{L_{dh}L_{qh} - L_{dqh}^2}{\sqrt{L_{sd}^2 + L_{dqh}^2}}$, and $\theta_m = \tan^{-1}\left(\dfrac{-L_{dqh}}{L_{sd}}\right)$. The amplitude of the d-axis HF current response can be extracted by:

$$\left|\hat{i}_{dh}\right| = \mathrm{LPF}\left(\hat{i}_{dh}\cdot 2\sin(\omega_h t)\right) = I_p + I_n\cos(2\Delta\theta + \theta_m) \tag{8.2}$$

Considering that the position estimation error is sufficiently small, the amplitude of d-axis HF current in the estimated synchronous reference frame can be simplified as:

$$\left|\hat{i}_{dh}\right| = I_p + I_n\cos(2\Delta\theta + \theta_m) \approx I_p + I_n\cos\theta_m = \frac{V_h}{\omega_h}\cdot\frac{L_{qh}}{L_{dh}L_{qh} - L_{dqh}^2} \tag{8.3}$$

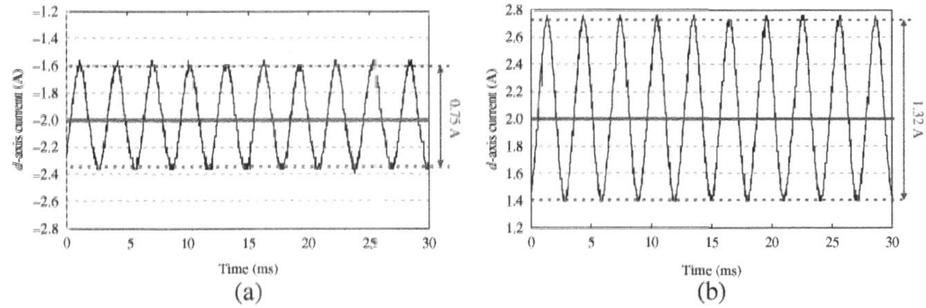

Figure 8.8 Measured d-axis HF current with different d-axis dc currents. (a) $i_{df} = -2$ A, $i_{qf} = 0$ A. (b) $i_{df} = 2$ A, $i_{qf} = 0$ A. (*Source:* [1]/IEEE.)

Considering that the mutual-inductance is very small, i.e. $L_{dqh}^2 \ll L_{dh}L_{qh}$, therefore (8.3) can be simplified as

$$\left|\hat{i}_{dh}\right| \approx \frac{V_h}{\omega_h} \cdot \frac{L_{qh}}{L_{dh}L_{qh} - 0} \approx \frac{V_h}{\omega_h} \cdot \frac{1}{L_{dh}} \tag{8.4}$$

For a given injected HF voltage signal (V_h/ω_h), (8.4) indicates that the amplitude of the d-axis HF current is only determined by L_{dh}. Due to magnetic saturation, L_{dh} is significantly dependent on the d-axis dc current. Consequently, it is possible to identify the magnetic polarity from the variation of $\left|\hat{i}_{dh}\right|$ against the d-axis dc current.

Based on the IPMSM-I, one experimental result [1] is provided in Figure 8.8 showing the d-axis HF current with different d-axis dc current. Clearly, the magnetic polarity can be detected based on the comparison of $\left|\hat{i}_{dh}\right|$ at different d-axis dc current levels, i.e. 2 and −2 A. The dc current level can be selected in the pre-tests. To achieve a reliable identification, the dc current level should be high enough to create a sufficient current difference between the north pole and the south pole.

The block diagram of the d-axis current injection-based method is shown in Figure 8.9. In Figure 8.9, based on a FOC drive system, the d-axis current injection is combined with a conventional d-axis pulsating HF voltage signal injection sensorless control introduced in Chapter 4. The HF current responses are obtained by a band-pass filter (BPF) and used to estimate the rotor position and polarity identification. As depicted in Figure 8.9, this method can be seamlessly integrated into the d-axis pulsating HF signal injection-based sensorless control since it utilizes the HF current response to identify the polarity. Compared to the dual voltage pulse injection method, the d-axis current injection method can be executed during the HF voltage signal injection, and hence there is no need to stop the HF voltage signal injection to identify the polarity.

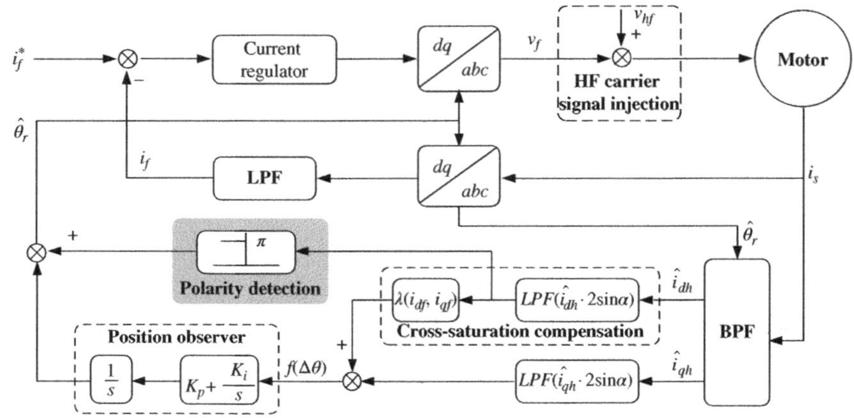

Figure 8.9 Block diagram of *d*-axis current injection-based method for pulsating HF signal injection. (*Source:* [1]/IEEE.)

The detailed description of the magnetic polarity detection is shown in Figure 8.10. It should be noted that torque must not be applied to the machine during the detection process ($i_{qf} = 0$ A). Since the polarity is not identified, doing so may cause a negative torque. Before the drive is initially started, the rotor direction information can be obtained from a conventional sensorless algorithm without dc excitation ($i_{df} = 0$ A, $i_{qf} = 0$ A). Meanwhile, the amplitude of the *d*-axis HF

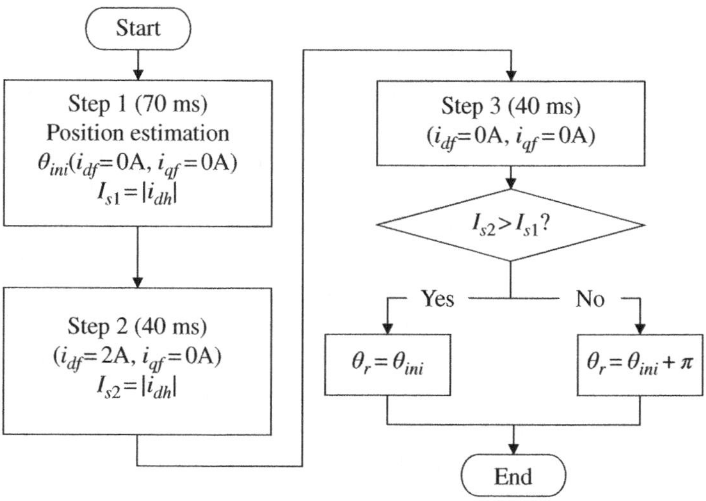

Figure 8.10 Flow chart of magnetic polarity detection. (*Source:* [1]/IEEE.)

current at this condition is recorded (Step 1). The estimated rotor direction either indicates the correct rotor position or is shifted by π from the correct rotor position. Referring to the estimated rotor direction, a given d-axis dc current reference is applied to the machine. The dc current reference of ($i_{df} = 2$ A, $i_{qf} = 0$ A) is selected for the prototype machine. At the same time, the amplitude of the d-axis HF current at this load condition is also stored into processor memory (Step 2). After that, the dc current reference is set back to ($i_{df} = 0$ A, $i_{qf} = 0$ A) (Step 3). Finally, the amplitude of the d-axis HF current at a different load condition is compared. The increase of amplitude indicates the estimated rotor direction is at the correct rotor position. Otherwise, the phase shift of π should be added.

The experimental results shown in Figures 8.11–8.13 confirm the effectiveness of polarity detection. It can be found that the initial rotor position including magnetic polarity detection can be successfully estimated within 150 ms.

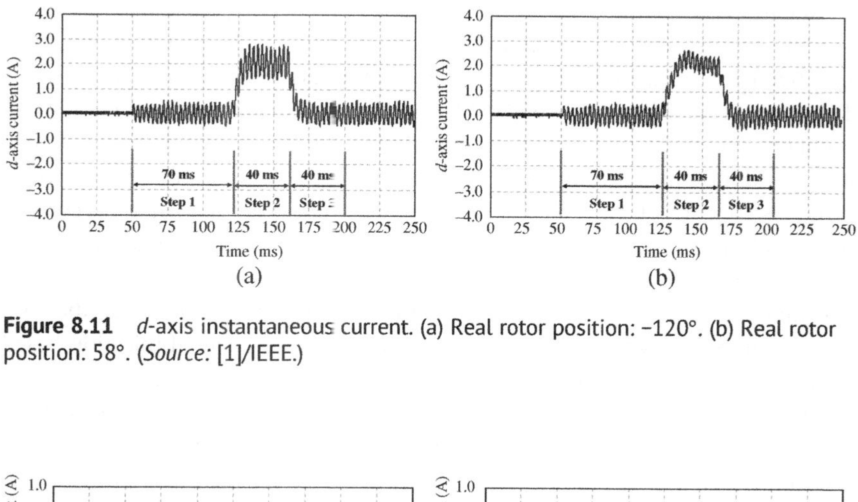

Figure 8.11 *d*-axis instantaneous current. (a) Real rotor position: −120°. (b) Real rotor position: 58°. (*Source:* [1]/IEEE.)

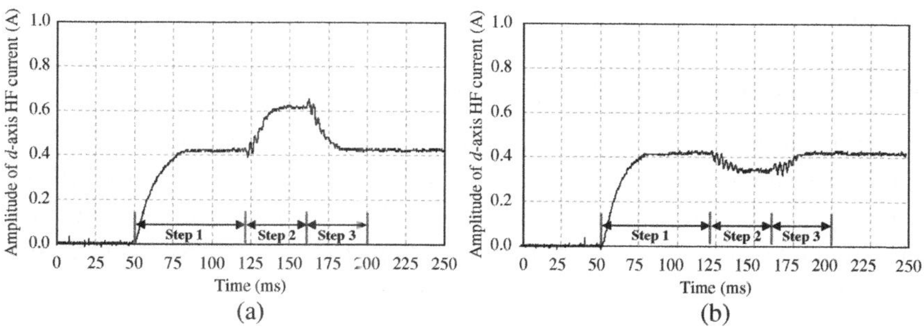

Figure 8.12 Magnitude of *d*-axis HF current. (a) Real rotor position: −120°. (b) Real rotor position: 58°. (*Source:* [1]/IEEE.)

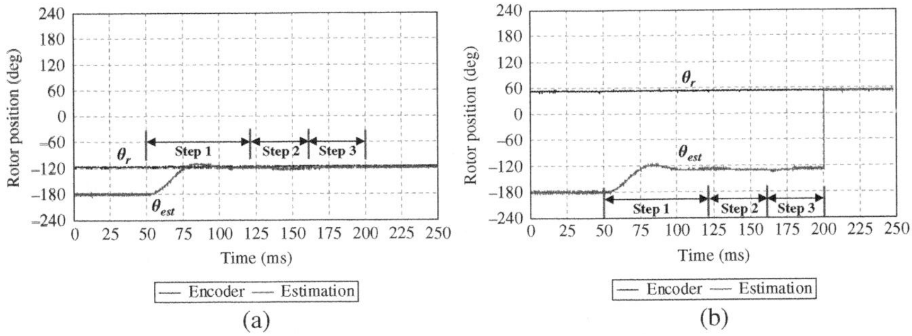

Figure 8.13 Estimated rotor position information. (a) Real rotor position: −120°. (b) Real rotor position: 58°. (*Source:* [1]/IEEE.)

In Figure 8.13, the estimation process is activated at the time of 50 ms. During Step 1, the rotor position without magnetic polarity information is obtained at the dc excitation of ($i_{df} = 0$ A, $i_{qf} = 0$ A). At Step 2, the prototype machine is excited with the dc reference of ($i_{df} = 2$ A, $i_{qf} = 0$ A). Finally, the dc excitation is reset to ($i_{df} = 0$ A, $i_{qf} = 0$ A) at Step 3. During the whole process, the amplitude of the d-axis HF current is recorded. At the end of Step 3, the magnetic polarity information can be obtained from the amplitude variation of the d-axis HF current with the change of the d-axis dc current from 0 A (Step 1) to 2 A (Step 2). The amplitude increase of the d-axis HF current suggests that the estimated rotor position at Step 1 is correct, as shown in Figure 8.13a. Otherwise, a phase shift of π should be added, as shown in Figure 8.13b.

In summary, the d-axis current injection-based magnetic polarity identification scheme can be easily integrated with the conventional HF signal injection-based sensorless technique with a slight increase of computational effort. Compared with secondary harmonics-based methods, the d-axis current injection-based method has the advantage of robust detection due to higher SNR and less computational intensity.

8.3.2 HF Zero Sequence Voltage Response

The principle of the d-axis dc current injection-based method is based on the d-axis inductance variation with d-axis magnetic saturation [1]. Apart from the HF current response, the HF zero sequence voltage (ZSV) response can also be used to identify the polarity [3]. As introduced in Chapter 6, the HF ZSV response of a rotating signal injection can be expressed as:

$$v_{SN} = \frac{V_h}{4}\left(\frac{L_{qh}}{L_{dh}} - \frac{L_{dh}}{L_{qh}}\right)\cos(\omega_h t + 2\theta_r) \tag{8.5}$$

And the amplitude of the HF ZSV response can be extracted as:

$$|v_{SN}| = \text{LPF}\left(v_{SN} \cdot 2\cos\left(\omega_h t + 2\hat{\theta}_r\right)\right) = \frac{V_h}{4}\left(\frac{L_{qh}}{L_{dh}} - \frac{L_{dh}}{L_{qh}}\right)\cos(2\Delta\theta) \tag{8.6}$$

The effectiveness of using the amplitude of the HF ZSV response for polarity identification is verified based on the IPMSM-II (Appendix B); its finite element (FE) calculated inductance values are given in Table 8.1.

For the prototype machine, the incremental d- and q-axis inductances calculated by the FE simulation are illustrated in Figure 8.14, from which it can be seen that L_{dh} is significantly changed with the d-axis current (i.e. d-axis magnetic saturation) while L_{qh} is approximately unchanged. Then, based on the simulated inductances in Figure 8.14, the value of $(L_{qh}/L_{dh} - L_{dh}/L_{qh})$ related to the ZSV amplitude in (8.6) can be further calculated and is shown in Figure 8.15. Since the ZSV amplitude, which is proportional to $(L_{qh}/L_{dh} - L_{dh}/L_{qh})$, changes significantly with the magnetic saturation from Figure 8.15, the magnetic polarity can then be identified with the d-axis dc excitation variations.

Table 8.1 FE calculated inductances of prototype IPMSM-II.

Variable	Value (mH)
Average phase inductance	2.4
Second harmonic phase inductance	0.6
$(f_h/f_e \pm 1)$th HF phase inductance	0.08/0.09

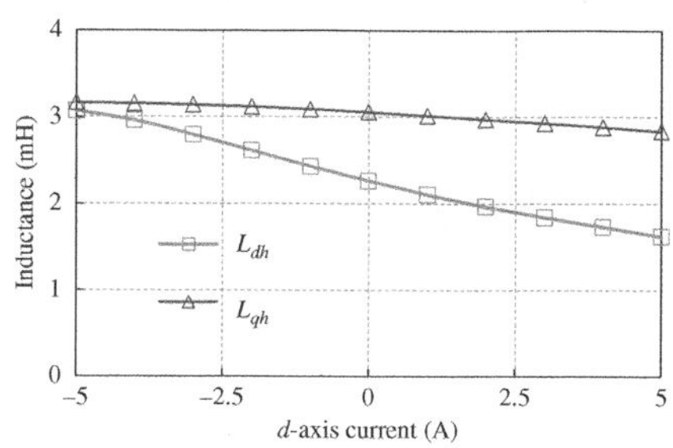

Figure 8.14 Incremental inductance with d-axis current based on FEA. (*Source:* Adapted from [3].)

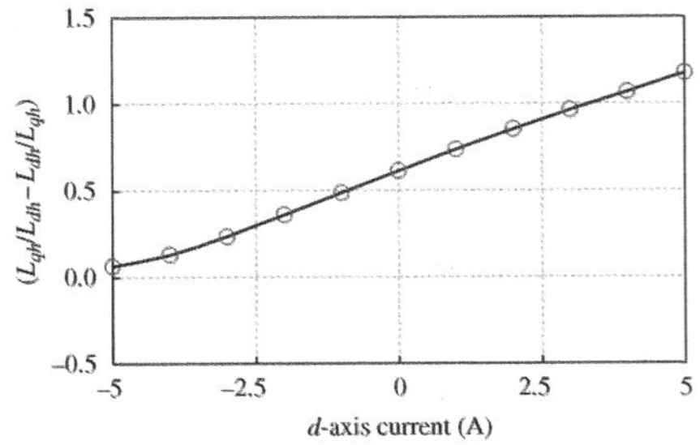

Figure 8.15 Calculated (L_{qh}/L_{dh} − L_{dh}/L_{qh}) with d-axis current. (*Source:* [3]/IEEE.)

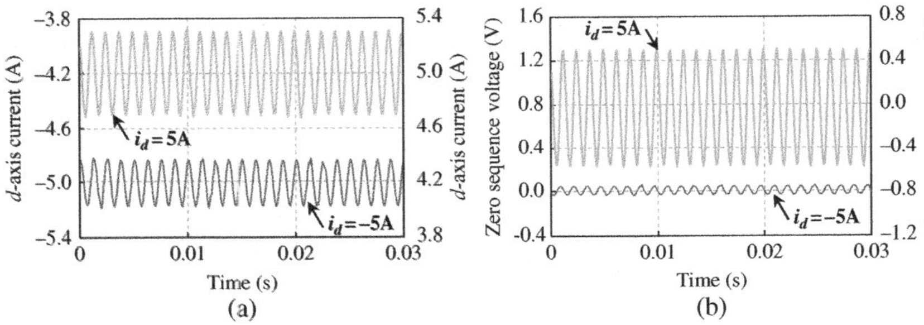

Figure 8.16 Measured HF response amplitude variations with d-axis currents. (a) d-axis current. (b) Zero sequence voltage. (*Source:* [3]/IEEE.)

With the same estimation procedure described in Figure 8.10, the amplitude variations of HF d-axis current and HF ZSV with d-axis dc current at −5 and 5 A are compared and shown in Figure 8.16. Clearly, the ZSV has much more significant variations than the HF d-axis current. Therefore, the detection sensitivity for the magnetic polarity using ZSV can be significantly enhanced.

8.4 Secondary Harmonic-Based Method

The dual voltage pulse injection method and the d-axis dc current injection method require additional excitation signals apart from the HF signal and therefore have slow convergence. The secondary harmonic-based methods have faster convergence since they do not need to change the magnetic saturation with additional signals as a further step [4], [5]. The secondary harmonics in the HF current response [4], [5] and the HF ZSV response [3], [7] can both be used to identify the polarity, which will be introduced and compared in this section.

8.4.1 Modeling of Secondary Harmonics

The modeling of the secondary harmonics is described in [4], [5] using the Taylor series, which is based on the nonlinear magnetic saturation relationship between the current and the flux-linkage in the d-axis. In [7], a more insightful explanation is given for the secondary harmonics. Considering a rotating signal injection, two dominant HF inductance harmonics exist, i.e. $f_h + f_e$ and $f_h - f_e$ (f_h and f_e are the HF frequency and fundamental frequency), which is due to the saturation modulation between the main flux field and the HF fields [8], [9]. The detailed modeling for saturation modulation effects is introduced in Chapter 6. Accordingly, the equivalent three-phase inductances can be described as

$$\begin{aligned} L_{AA} &= L_{s0} - L_{s2}\cos(2\theta_r) + L_{f_h-f_e}\sin(\theta_h-\theta_r) + L_{f_h+f_e}\sin(\theta_h+\theta_r) \\ L_{BB} &= L_{s0} - L_{s2}\cos(2\theta_r + 2\pi/3) + L_{f_h-f_e}\sin(\theta_h-\theta_r) + L_{f_h+f_e}\sin(\theta_h+\theta_r+2\pi/3) \\ L_{CC} &= L_{s0} - L_{s2}\cos(2\theta_r - 2\pi/3) + L_{f_h-f_e}\sin(\theta_h-\theta_r) + L_{f_h+f_e}\sin(\theta_h+\theta_r-2\pi/3) \end{aligned} \tag{8.7}$$

where $\theta_h = \omega_h t$, and L_{fh-fe} and L_{fh+fe} are the (f_h/f_e − 1)th and (f_h/f_e + 1)th harmonic inductance amplitudes, respectively. It can be seen that in the derived inductances, two HF inductance harmonics exist, i.e. $f_h + f_e$ and $f_h - f_e$, which contain the magnetic polarity information. Moreover, from (8.8), by coordinate transformation, the d-axis and q-axis inductances can be derived as

$$\begin{aligned} L_d &= L_{s0} - L_{s2}/2 + L_{f_h-f_e}\sin(\omega_h t-\theta_r)/2 + L_{f_h-f_e}\sin(\omega_h t-\theta_r)/2 \\ L_q &= L_{s0} + L_{s2}/2 + L_{f_h-f_e}\sin(\omega_h t-\theta_r)/2 - L_{f_h-f_e}\sin(\omega_h t-\theta_r)/2 \\ L_{dq} &= L_{f_h+f_e}\cos(\omega_h t-\theta_r)/2 \end{aligned} \tag{8.8}$$

8.4.2 HF Current Response

In this section, the analysis is based on the conventional rotating voltage signal injection sensorless control, the basic principles of which were introduced in Chapter 4.

Consider a rotating voltage signal injection in the stationary reference frame, which is derived as:

$$v_\alpha + jv_\beta = V_h(\cos\omega_h t + j\sin\omega_h t) \tag{8.9}$$

Accordingly, by including the HF inductance harmonics in (8.8), the HF current responses can be expressed as:

$$i_{\alpha\beta h} \approx \frac{V_h}{\omega_h \left(L_{s0}^2 - L_{s2}^2/4\right)} \left[2L_{s0}e^{j\left(\omega_h t - \frac{\pi}{2}\right)} + L_{s2}e^{j\left(-\omega_h t + 2\theta_r + \frac{\pi}{2}\right)} + \frac{L_{f_h - f_e}}{2}e^{j(2\omega_h t - \theta_r)} - \frac{L_{f_h + f_e}}{2}e^{j(-2\omega_h t + 3\theta_r)}\right] \tag{8.10}$$

For the HF current response, additional secondary positive sequence harmonics at $2f_h - f_e$ and a negative sequence HF current at $-2f_h + 3f_e$ arise due to the HF inductance harmonics in (8.8). Since the positive sequence HF current $2f_h - f_e$ contains the actual rotor position information, it can be used for magnetic polarity identification. The extraction of polarity information from secondary harmonics in the HF current response is expressed as:

$$i_{2nd} = \mathrm{Re}\left[\mathrm{LPF}\left(i_{\alpha\beta h}e^{j\left(-2\omega_h t + \hat{\theta}_r\right)}\right)\right] = \frac{V_h L_{f_h - f_e}}{2\omega_h \left(L_{s0}^2 - L_{s2}^2/4\right)} \cos \Delta\theta \tag{8.11}$$

Moreover, the polarity identification process is shown in Figure 8.17.

8.4.3 HF Zero Sequence Voltage Response

Apart from the HF current response, the HF ZSV response can also be used to extract the secondary harmonics component for polarity identification. The basic principles of the ZSV-based sensorless control were introduced in Chapter 6.

The HF ZSV response by including the HF inductance harmonics in (8.8) can be derived as:

$$v_{\mathrm{SN}} \approx V_h \frac{\left[L_{s0}L_{s2}\cos(\omega_h t + 2\theta_r) - L_{s2}{}^2\cos(\omega_h t - 4\theta_r) - \left(L_{s0}L_{f_h + f_e} + \frac{L_{s2}L_{f_h - f_e}}{2}\right)\sin(2\omega_h t + \theta_r)\right]}{\left(2L_{s0}{}^2 - L_{s2}{}^2/2\right)} \tag{8.12}$$

For the HF ZSV response, only one secondary positive sequence harmonic appears at $2f_h + f_e$ that can be used for magnetic polarity identification.

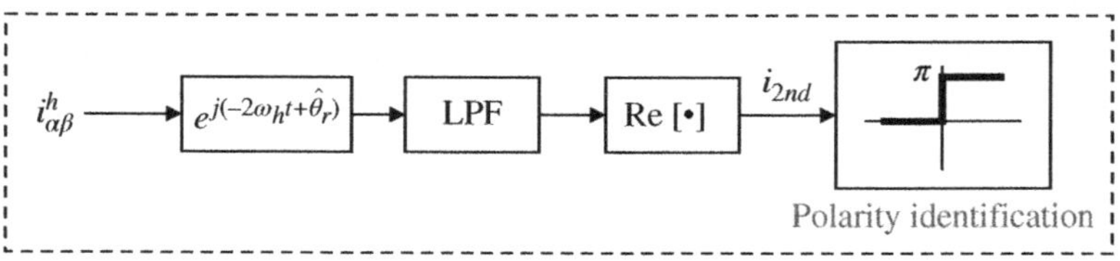

Figure 8.17 Demodulation process of secondary harmonics HF current responses.

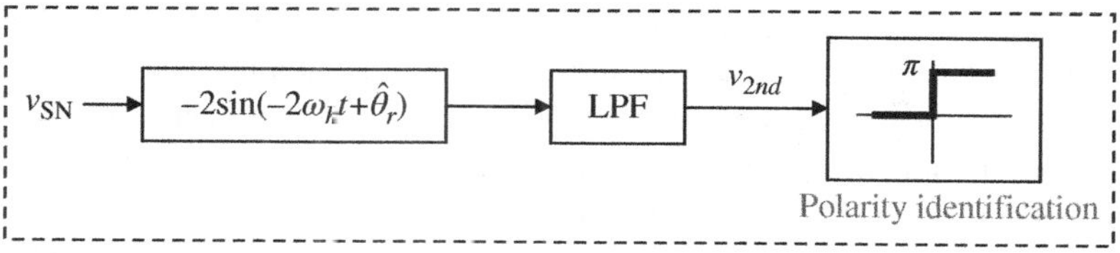

Figure 8.18 Demodulation process of secondary harmonics HF zero sequence voltage response.

Furthermore, the secondary positive sequence harmonic HF ZSV responses can be extracted by:

$$v_{2nd} = \mathrm{LPF}\left[v_{SN} \cdot \left(-2\sin\left(2\omega_h t + \hat{\theta}_r\right)\right)\right] = k\cos(\Delta\theta), (k > 0) \tag{8.13}$$

where k is the related secondary harmonic amplitude. Clearly, only by judging that the value in (8.13) is positive or negative, the ambiguity of magnetic polarity can be cleared, and the polarity identification process is demonstrated in Figure 8.18.

8.4.4 Experiment Results

Experimental results of the secondary harmonic-based method are provided based on the prototype IPMSM-II (Appendix B). Figure 8.19 illustrates the spectra of measured secondary harmonics with an 8 V/300 Hz rotating signal injection with the fundamental frequency at 2.5 Hz. The selection of amplitude and frequency of the HF rotating signal is based on the method introduced in Chapter 5.

It can be seen that the 241st ($2f_h + f_e$) harmonic arises for zero sequence HF voltage while the 239th ($2f_h - f_e$) and 237th ($2f_h - 3f_e$) harmonics exist for the HF current, which are all consistent with the analyses in (8.10) and (8.12). Moreover, it is observed that the secondary zero sequence HF voltage harmonics in Figure 8.19a

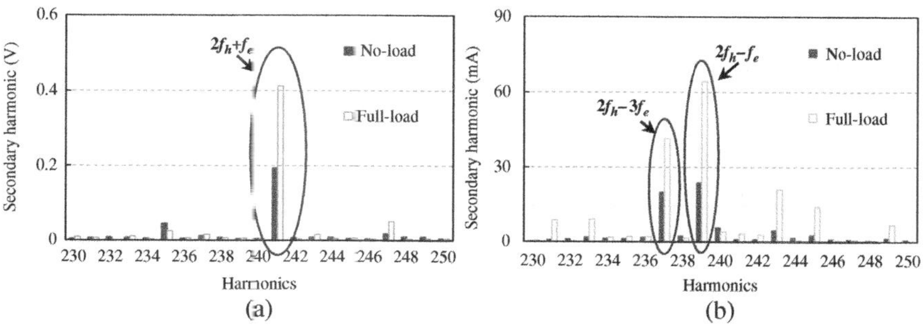

Figure 8.19 Spectra of measured secondary harmonic responses for rotating signal injection. (a) Zero sequence HF voltage. (b) HF current. (*Source:* [3]/IEEE.)

have considerably larger signal amplitude and less distortion compared to the HF current harmonics in Figure 8.19b, i.e. larger SNR can be obtained for the ZSV. Therefore, the secondary harmonic-based detection method using zero sequence HF voltage is superior since both the fast response and the increased SNR can be obtained. Besides, since only one dominant frequency component ($2f_h + f_e$) exists for the ZSV, the signal demodulation for the polarity identification is also simpler compared to that of the HF current having two dominant harmonics ($2f_h - f_e$ and $2f_h - 3f_e$). However, as described in Chapter 6, the major issues of the ZSV-based method are the requirement of an additional voltage measurement sensor and the resistor network to create an artificial neutral point.

8.5 Summary

In this chapter, the magnetic polarity identifications for HF signal injection-based sensorless methods were described. Magnetic polarity identification is always required for saliency-based sensorless techniques since the machine saliency undergoes two cycles in a single electrical period. Based on the *d*-axis magnetic saturation effect, the polarity can be identified with different signal injections and the corresponding responses. Three typical methods were illustrated in this chapter with experimental results, and the overall comparison between them is shown in Figure 8.20.

In summary:

- The dual voltage pulses injection-based method is very simple and robust, but it is a separate process against the position estimation that stops the HF signal injection and hence slows down the overall estimation.

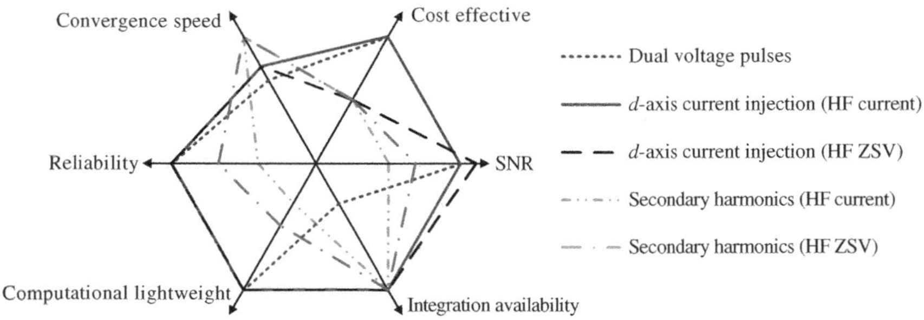

Figure 8.20 Comparison between different magnetic polarity detection methods.

- For the d-axis dc current injection-based method, by either HF current response or HF ZSV response, a reliable and robust performance can be obtained. The HF ZSV response offers a larger difference against the polarity than the HF current response but requires additional hardware, which brings with it increased costs. Compared to dual voltage pulses injection, this method can be integrated with conventional HF signal injection-based sensorless control. Since a dc current is injected to detect the polarity the convergence needs to be improved as well.
- The secondary harmonic-based method has the fastest convergence [10] without using an additional excitation signal for polarity identification, but the low secondary component limits its robustness. Although a larger magnitude is obtained by using the HF ZSV response, additional hardware is required, which adds to the cost of the system.

References

1 L. M. Gong and Z. Q. Zhu, "Robust initial rotor position estimation of permanent-magnet brushless ac machines with carrier-signal-injection-based sensorless control," *IEEE Trans. Ind. Appl.*, vol. 49, no. 6, pp. 2602–2609, Nov. 2013.

2 T. Aihara, A. Toba, T. Yanase, A. Mashimo, and K. Endo, "Sensorless torque control of salient-pole synchronous motor at zero-speed operation," *IEEE Trans. Power Electron.*, vol. 14, no. 1, pp. 202–208, Jan. 1999.

3 P. L. Xu and Z. Q. Zhu, "Initial rotor position estimation using zero-sequence carrier voltage for permanent-magnet synchronous machines," *IEEE Trans. Ind. Electron.*, vol. 64, no. 1, pp. 149–158, Jan. 2017.

4 H. Kim, K. K. Huh, and R. D. Lorenz, "A novel method for initial rotor position estimation for IPM synchronous machine drives," *IEEE Trans. Ind. Appl.*, vol. 40, no. 5, pp. 1369–1378, 2004.

5 Y. Jeong, R. D. Lorenz, T. M. Jahns, and S. K. Sul, "Initial rotor position estimation of an interior permanent-magnet synchronous machine using carrier-frequency injection methods," *IEEE Trans. Ind. Appl.*, vol. 41, no. 1, pp. 38–45, Jan. 2005.

6 J. Holtz, "Acquisition of position error and magnet polarity for sensorless control of PM synchronous machines," *IEEE Trans. Ind. Appl.*, vol. 44, no. 4, pp. 1172–1180, 2008.

7 P. L. Xu, Z. Q. Zhu, and D. Wu, "Carrier signal injection-based sensorless control of permanent magnet synchronous machines without the need of magnetic polarity identification," *IEEE Trans. Ind. Appl.*, vol. 52, no. 5, pp. 3916–3926, Sep. 2016.

8 A. Consoli, G. Scarcella, G. Bottiglieri, and A. Testa, "Harmonic analysis of voltage zero-sequence-based encoderless techniques," *IEEE Trans. Ind. Appl.*, vol. 42, no. 6, pp. 1548–1557, Nov./Dec. 2006.

9 A. Consoli, G. Bottiglieri, G. Scarcella, and G. Scelba, "Flux and voltage calculations of induction motors supplied by low- and high-frequency currents," *IEEE Trans. Ind. Appl.*, vol. 45, no. 2, pp. 737–746, Mar. 2009.

10 M. C. Harke, D. Raca, and R. D. Lorenz, "Implementation issues for fast initial position and magnet polarity identification of PM synchronous machines with near zero saliency," *Conf. Power Electron. Appl.*, pp. 1–10, 2005.

9

Rotor Initial Position Estimation

9.1 Introduction

For PM machines including both BLAC machines (or PMSMs) and BLDC machines starting from standstill, rotor initial position information is crucial to guarantee a stable and efficient starting performance [1]. Without the knowledge of the rotor initial position, undesired reverse rotation or oscillation during startup may happen, which could lead to failure of a sensorless drive system. Rotor initial position can be estimated in two ways. The first is a continuous position, which varies from 0 to 2π. The second is based on discrete sectors; i.e. the whole position range from 0 to 2π is divided into several equal sectors as shown in Figure 9.1.

As classified in Figure 9.2, at standstill, the continuous position can be estimated by high-frequency signal injection (HFSI)-based sensorless control introduced in Chapter 4. Then, the discrete sector can be estimated by short pulse injection-based methods [2]–[6] and also by the high-frequency rotating sinusoidal signal injection method [7]. Both methods can be applied to BLDC and BLAC (PMSM) machine drives with satisfactory estimation performance.

For HFSI-based sensorless control methods, a continuous rotor initial position can be estimated. However, for extra magnetic polarity detection as described in Chapter 8, filters and signal demodulation processes are necessary, which are relatively complicated to implement and result in a higher cost requirement for the overall drive system [8]. Furthermore, this type of method requires sufficient inductance saliency, which is more suitable for IPMs with geometric inductance saliency.

On the other hand, for discrete sectors estimation with short pulse injection-based methods, several voltage pulses with a short duration are injected into the stator windings and rotor initial position can be obtained from the different responses, including three-phase currents [2], [3], dc-link current [4], dc-link

Sensorless Control of Permanent Magnet Synchronous Machine Drives, First Edition.
Zi Qiang Zhu and Xi Meng Wu.

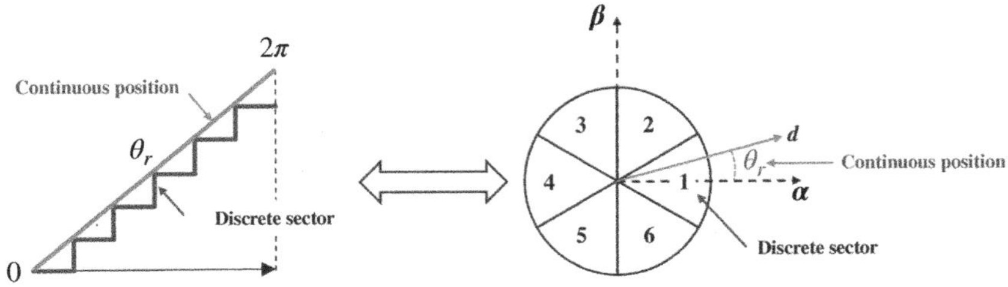

Figure 9.1 Illustration of rotor initial position classification.

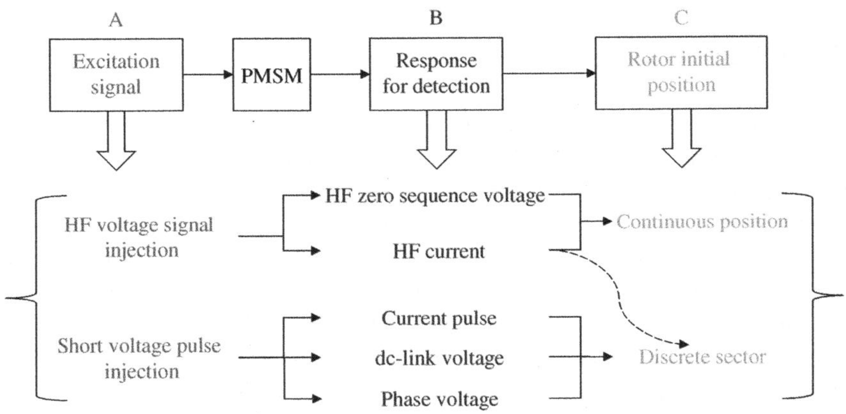

Figure 9.2 Classification of rotor initial position estimation.

voltage [5], and phase voltage [6]. For these methods, magnetic polarity is intrinsically identified during the estimation. Although estimation resolution may be lower than for HFSI-based methods, short pulse injection-based methods can still provide a reliable estimation with a satisfying signal-to-noise ratio (SNR) [9]. Since filters and signal demodulation processes are not necessary, short pulse injection-based methods are more robust and simpler for implementation. Therefore, they can be a desirable option for low-cost systems. Apart from IPMs, this type of method can also be an appropriate option for SPMs by utilizing the magnetic saturation effect.

Moreover, by adopting the advantages of short pulse injection-based methods, a hybrid method is presented in [7] to simplify the conventional HFSI-based rotor initial position estimation. A HF rotating signal is injected, but the rotor sector, instead of the continuous position, is estimated by comparing the amplitude of the HF current response. In this way, the estimation is simplified and the execution time is reduced compared to continuous position estimation with conventional HF signal injection sensorless control.

In this chapter, the discrete sector-based rotor initial position estimation is introduced first. The magnetic saturation effect and the basic estimation principle [2] using the short pulse injection-based method are demonstrated. Furthermore, two improved estimation methods [3], [5] are described in detail with experiment results. In addition, it is necessary to determine the magnitude and duration of the pulse for injection in order to obtain a reliable estimation performance, and therefore a selection guidance of the pulse is given in this chapter. Finally, the hybrid method [7] is introduced and compared to short pulse injection-based methods.

9.2 Magnetic Saturation Effect

The majority of the short pulse injection-based rotor initial position estimation methods rely on the magnetic saturation effect, which is illustrated in this section. Stator iron core is a non-ideal magnetic material. As the magnetic flux in the iron core increases, it begins to saturate, resulting in a decrease of inductance.

In Figure 9.3, a simplified schematic of an SPM is shown. The total magnetic flux that saturates the stator iron core consists of two parts: PM flux ψ_{magnet} and

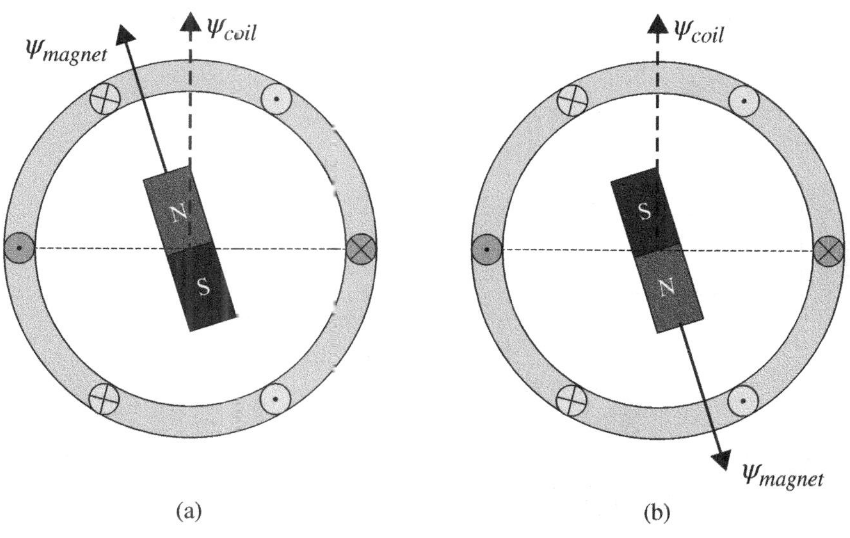

Figure 9.3 Simplified SPM schematic for polarity identification. (a) Alignment with north pole. (b) Alignment with south pole.

the flux produced by coil current ψ_{coil}. In Chapter 8, the magnetic saturation caused by coil current alone is illustrated, which is utilized for magnetic polarity identification. In the present chapter, the magnetic saturation effect caused by the PM flux and the coil flux together are described, which is utilized for both magnetic pole direction detection and polarity identification in methods [2]–[6].

First, for the magnetic circuit due to PM only, the stator iron core becomes more saturated when it is close to the rotor magnetic poles of PM, causing inductance to decrease. As the rotor magnetic poles move away, the inductance will increase. Therefore, inductance will vary against rotor position. However, PM has a north pole and a south pole that introduce the same saturation level, and inductance will vary twice over one electrical period, which causes an ambiguity of π in the initial position estimation. Then, ψ_{coil} produced by the coil current can be used for polarity identification. As demonstrated in Figure 9.3a, when the north pole is aligned with the coil, i.e. flux produced by the coil current has the same direction with PM flux, the total flux increases and saturation level of the stator iron core is increased, and consequently the inductance is slightly decreased. Then, as shown in Figure 9.3b, when the south pole is aligned with the coil, i.e. the coil flux is opposite to the PM flux, the total flux is reduced, resulting in a lower saturation level and the inductance is slightly increased. In this way, the polarity of the magnet can be identified. Overall, due to the magnetic saturation effect, the inductance varies with the rotor position as shown in Figure 9.4. Thus, rotor initial position can be estimated based on this phenomenon.

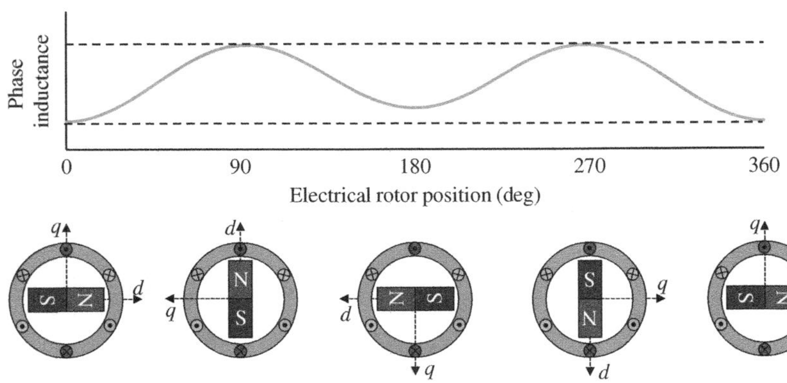

Figure 9.4 Inductance variation against rotor position.

9.3 Basic Pulse Injection Method Using Three Phase Currents

Based on the magnetic saturation effect, rotor initial position can be easily detected by looking at the current responses by injecting several short voltage pulses. The basic rotor initial position estimation procedure [2] with the short pulse injection-based method is described in this section.

9.3.1 Pulse Excitation Configuration

A two-level three-phase voltage source inverter has eight switching states in total, and six of them are used to energize the coil for initial position estimation as summarized in Table 9.1.

As summarized in Table 9.1, with six switching states, six voltage pulses can be injected into six positions, at six selected voltage vectors ($\mathbf{v}_1$~$\mathbf{v}_6$). Moreover, each injection position also corresponds to an injection phase and the resultant phase current is used for position estimation, as also given in Table 9.1. Finally, each injection position is associated with a sector numbered from 1 to 6, which can be used for identifying the location of the rotor north pole (Figure 9.5).

Table 9.1 Summary of excitation configuration.

Switching action	100	110	010	011	001	101
Voltage vector	$\mathbf{v}_1$	$\mathbf{v}_2$	$\mathbf{v}_3$	$\mathbf{v}_4$	$\mathbf{v}_5$	$\mathbf{v}_6$
Injection position (degree)	0	60	120	180	240	300
Injection phase	A+	C−	B+	A−	C+	B−
Phase current for position estimation	i_A	i_C	i_B	i_A	i_C	i_B
Sector	1	2	3	4	5	6

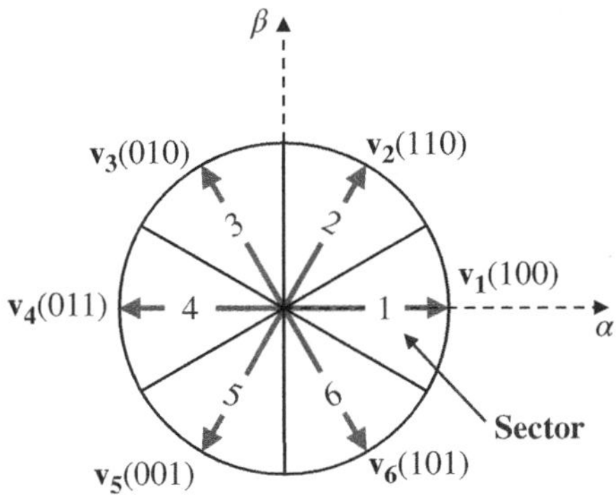

Figure 9.5 Voltage vectors associated with sectors.

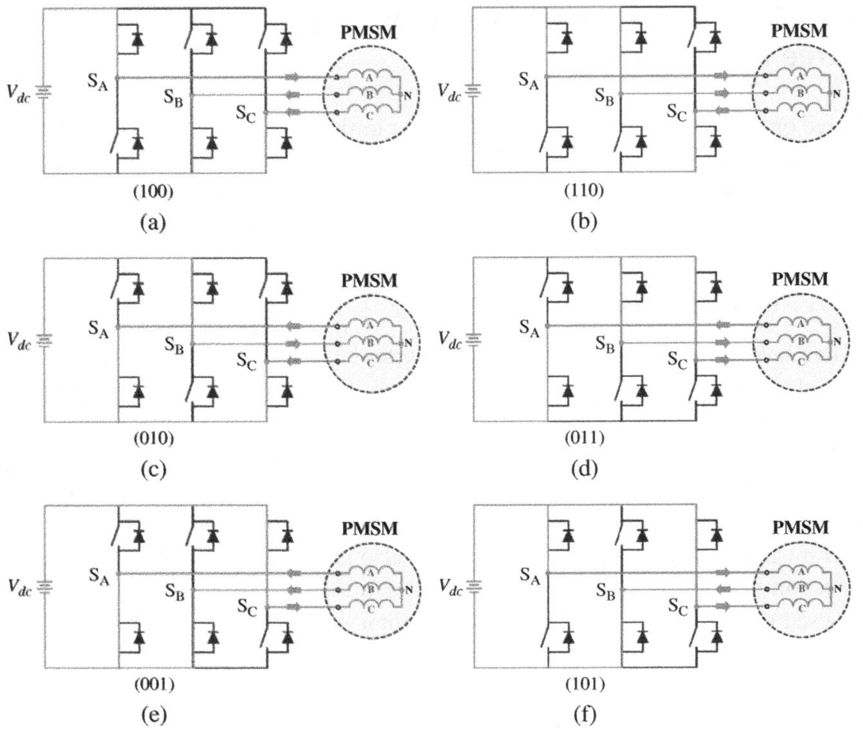

Figure 9.6 Excitation configuration associated with voltage vectors. (a) $\mathbf{v}_1$. (b) $\mathbf{v}_2$. (c) $\mathbf{v}_3$. (d) $\mathbf{v}_4$. (e) $\mathbf{v}_5$. (f) $\mathbf{v}_6$.

Moreover, for different voltage vectors, the excitation configurations are shown in Figure 9.6.

9.3.2 Current Response Model

At a random rotor position, the current response after injecting a voltage pulse can be regarded as a step response, which is expressed as:

$$i(t) = \frac{V_P}{R_{eq}}\left(1 - e^{-\frac{R_{eq}}{L_{eq}}t}\right) \tag{9.1}$$

where i is the current response, t is the time, V_P is the voltage magnitude, and R_{eq} and L_{eq} are the equivalent resistance and inductance, respectively, in the excitation circuit. Then, the peak value of the current response is measured. Since the inductance varies against the rotor position, the measured peak value of the current response is rotor position-dependent and can be used for position estimation.

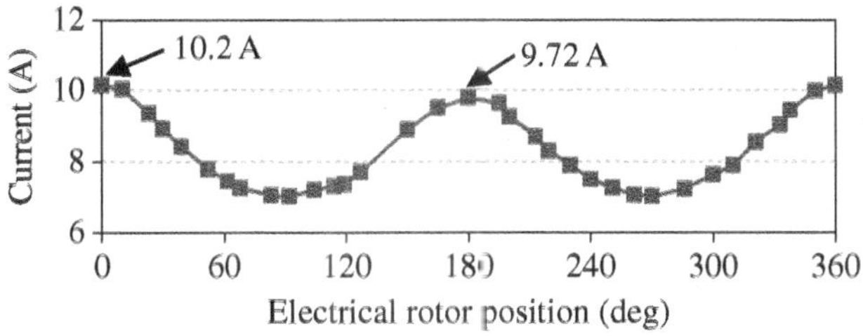

Figure 9.7 Measured peak values of phase A current response against rotor position. (*Source:* [3]/IEEE.)

Taking the phase A current response as an example, by injecting voltage pulse ($\mathbf{v}_1$) into phase A at different rotor positions, the measured peak values of the phase A current response against the rotor position are shown in Figure 9.7.

The measurement is taken from the prototype SPMSM-III (Appendix B). It can be seen that the measured current responses vary with rotor position. Also, it is clear that there is a 0.71 A difference between the current responses corresponding to the north and the south poles of the rotor, which is large enough for polarity identification.

9.3.3 Initial Position Estimation

A commonly used traditional rotor initial position estimation method [2] is introduced in this section and the overall estimation flow chart is shown in Figure 9.8. The principle is based on the magnetic saturation effect that a larger peak value indicates that the voltage pulse injection position in that sector, as shown in Table 9.1, is closer to the rotor magnetic poles.

At standstill, after injecting six different voltage pulses with vectors $\mathbf{v}_1$~$\mathbf{v}_6$, the peak values of current responses are measured and compared. By comparing the largest peak values among the six current responses, the rotor initial position can be determined as one of the sectors in Figure 9.5. An experimental example of the estimation is shown in Figure 9.9 based on the prototype SPMSM-III (Appendix B). The magnitude and duration of the voltage pulse are 60 V and 5 ms, respectively. The real rotor

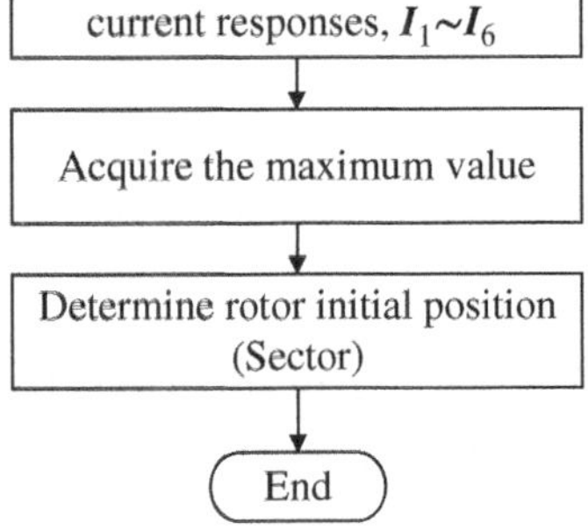

Figure 9.8 Flow chart of rotor initial position estimation.

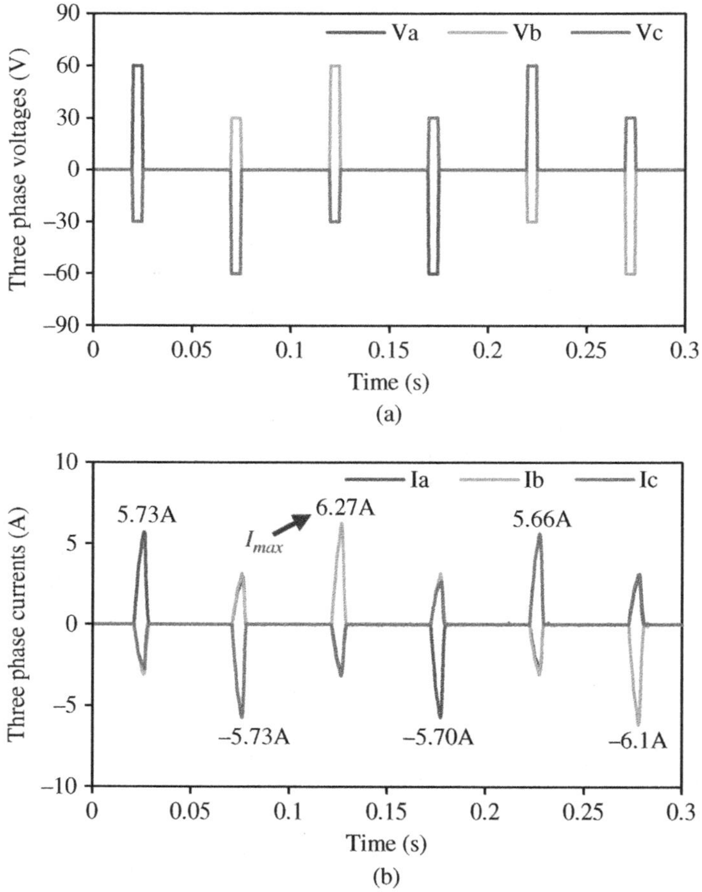

Figure 9.9 Example of rotor initial position estimation. (a) Three phase voltage pulses. (b) Three phase currents.

position is 120°. Six voltage pulses are injected. It is found that the peak value of the third current response of vector $\mathbf{v}_3$ is the largest. Thus, the rotor initial position is estimated as 120° (or in Sector 3).

9.4 Improved Pulse Injection Method Using Three Phase Currents

The rotor initial position estimation method described in Section 9.3.3 works effectively, but the estimation resolution is only 60°, i.e. the maximum estimation error can be 30°, which can reduce the maximum starting torque. Moreover, six pulses are required, which slows down the startup procedure. Hence, in [3], an improved method is developed and introduced in this section. The required number of pulses is reduced from six to three for a faster execution and the estimation resolution can be increased from 60° to 30°, i.e. the maximum estimation error is reduced from 30° to 15°. Moreover, by using a so-called "boundary detection strategy," the impact of current sampling noise on position estimation is reduced and the estimation accuracy is enhanced with minimized influence of current sampling noises.

9.4.1 Utilization of Three Phase Current Responses

First, for the estimation method introduced in Section 9.3, after injecting a voltage pulse in one phase, there are currents induced in all three phase windings, but only the phase current in the phase with the injected voltage pulse is utilized for estimation [2]. However, apart from the current of the injection phase, the other two phase currents also contain the position information, which can be used to estimate the initial position. Based on this phenomenon, the required number of pulses can be reduced.

Assuming one voltage pulse is injected in phase A at different rotor positions, the peak values of three-phase current responses against rotor position can be represented as:

$$
\begin{aligned}
I_A^P &= I_0 + I_2\cos(2\theta_r) \\
I_B^S &= -\frac{1}{2}I_0 - I_2\cos\left(2\theta_r + \frac{\pi}{3}\right) \\
I_C^S &= -\frac{1}{2}I_0 - I_2\cos\left(2\theta_r - \frac{\pi}{3}\right)
\end{aligned}
\tag{9.2}
$$

where I_0 and I_2 are the amplitudes of the dc and second order harmonic components of three-phase current responses. The superscript P indicates the primary current response and S indicates the secondary current response. For the three-phase current responses, the phase A current is called "primary current response" since the voltage pulse is injected in phase A with the largest amplitude of current response. The phase B and C currents are called "secondary current responses."

Figure 9.10 shows the measured results of three-phase current responses based on the prototype SPMSM-III (Appendix B). Voltage pulses are injected in phase A and the three-phase current responses at different rotor positions are measured and shown in Figure 9.10. Clearly, apart from the injection phase, both phase B and phase C currents vary with the rotor position. Therefore, by utilizing all three-phase current responses, more information can be obtained within one pulse injection, which helps to reduce the number of required voltage pulses, and thus, save the estimation time.

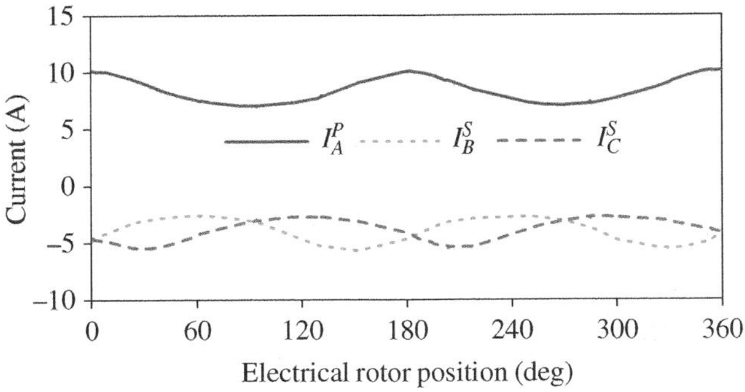

Figure 9.10 Measured peak values of three-phase current responses against rotor position. (*Source:* [3]/IEEE.)

9.4.2 Pulse Injection Sequence

Based on utilization of both the primary and secondary current responses, a specific voltage pulse injection sequence is presented in this subsection. During the injection process, the injection position is selected based on the results from the previous pulse. Consequently, only three voltage pulses in total are required.

After injection of voltage pulses, the peak values of the current responses are recorded and are used for rotor initial position estimation. In this section, for simplicity, $|i_A|$, $|i_B|$, $|i_C|$ represent the absolute peak values of current responses in three phases. $|i_{A+}|$ and $|i_{A-}|$ are the absolute peak values of positive and negative current responses in phase A, respectively, and they are the same for phases B and C. The subscript number of the current response indicates the number of the injected pulse, e.g. $|i_{A1+}|$ is the phase A current response of the first voltage pulse. The whole process can be divided into five steps, which are demonstrated in the following. The overall process of rotor initial position estimation is demonstrated by flow chart in Figure 9.11.

1) *First Voltage Pulse Injection:* In the first step, a positive voltage pulse is injected at phase A. After injection, current responses of three phases, i_{A1+}, i_{B1-}, i_{C1-}, are recorded. Then, i_{B1-} and i_{C1-} are compared to determine whether the rotor magnetic pole is closer to phase B or phase C. This is demonstrated by Figure 9.12a, where only the phases with the solid black line are compared and the dashed gray lines are not. If $|i_{B1-}| > |i_{C1-}|$, the rotor magnetic pole is closer to phase B, then a positive voltage pulse will be injected at phase B in the next step. Otherwise, $|i_{B1-}| < |i_{C1-}|$, the rotor magnetic pole is closer to phase C, and a positive voltage pulse will be injected at phase C in the next step.

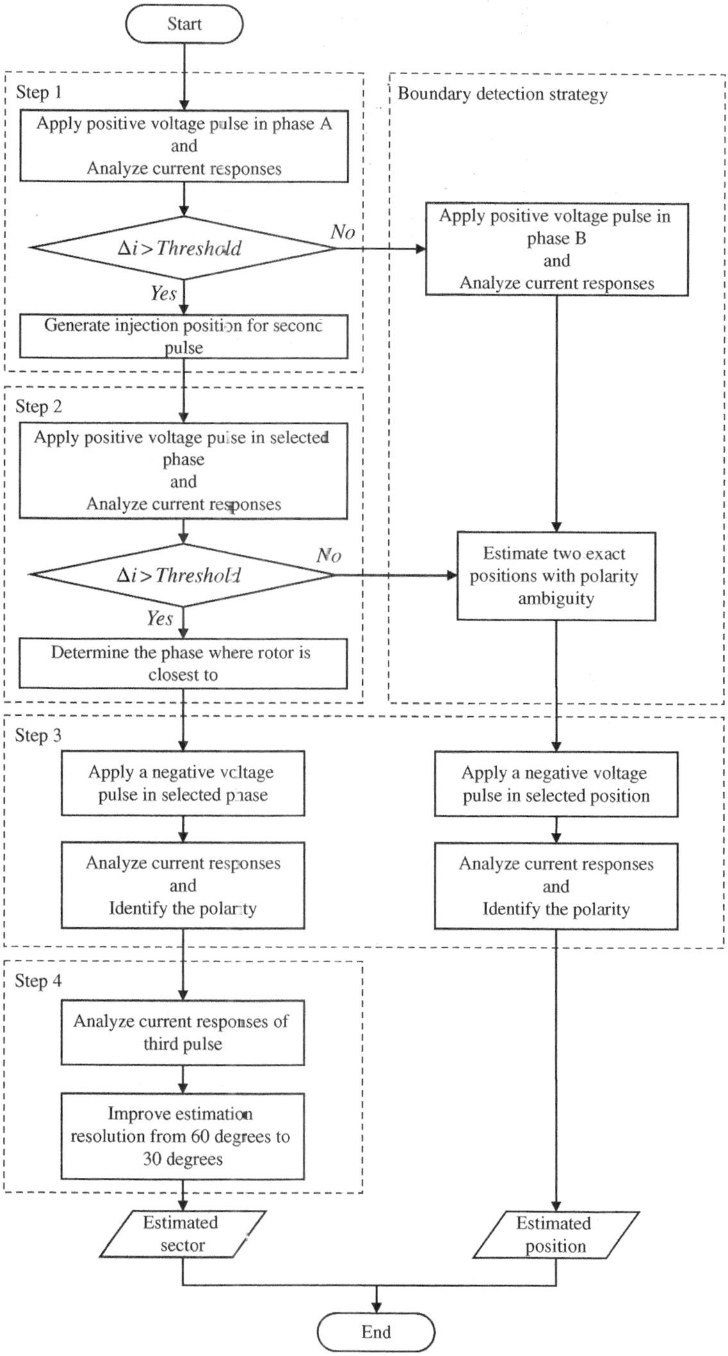

Figure 9.11 Flow chart of the whole estimation process. (*Source:* [3]/IEEE.)

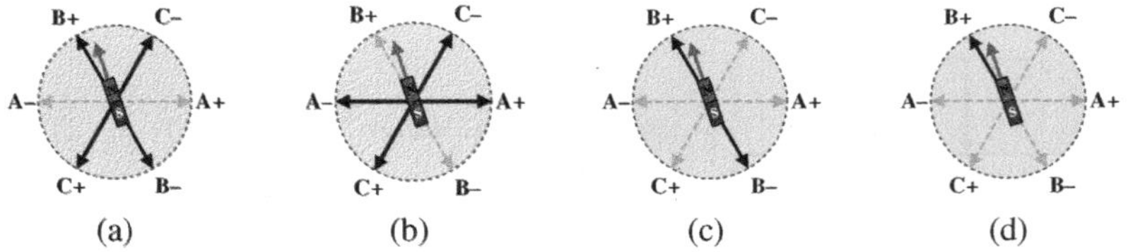

Figure 9.12 Illustration of current responses between different phases. (a) Phase B and phase C. (b) Phase C and phase A. (c) Positive phase B and negative phase B. (d) Positive phase B determined. (*Source:* [3]/IEEE.)

2) *Second Voltage Pulse Injection:* For the second voltage pulse, the injection position is based on the results from Step 1. For example, if it is known that $|i_{B1-}| > |i_{C1-}|$ from the first pulse results, then a positive pulse is injected at phase B. After injection, there are three current responses, i_{A2-}, i_{B2+}, i_{C2-}. Then, as shown in Figure 9.12b, current responses in phase A and phase C are compared, which are represented by solid black lines, i.e. i_{A2-} and i_{C2-}. An example is shown in Figure 9.12b. If $|i_{A2-}| < |i_{C2-}|$, it means that the rotor magnetic pole is closer to phase C than to phase A. Combining the results from Step 1, i.e. $|i_{B1-}| > |i_{C1-}|$, the rotor magnetic pole is closer to phase B than to phase C. It can be determined at last that the magnetic pole of rotor is closest to phase B, as shown in Figure 9.12c. Therefore, the phase that the rotor magnetic pole is closest to can be identified in this step. Then, a corresponding negative voltage pulse will be injected at the selected phase in the next step to identify the polarity of the rotor magnetic pole.
3) *Polarity Determination:* After the identification of which phase the rotor magnetic pole is closest to in Step 2, the north pole will be determined in this step. Based on the results from Step 2, a corresponding negative voltage pulse is injected at the selected phase in this step. If the rotor magnetic pole is the closest to phase B, then a negative voltage pulse is injected at phase B to determine the north pole. By way of example, Figure 9.12d shows that it is determined that the rotor north pole locates at the positive phase B, by only showing phase B with solid black lines. In this step, as shown in Figure 9.13a, the sector associated with the phase that the rotor's north pole is closest to is determined. By this step, a 60° estimation resolution is achieved, which is the same as the method in [2] but only three pulses are injected.
4) *Improving Estimation Resolution:* In this step, the estimation resolution is improved from 60° to 30° as shown in Figures 9.13a,b. After injection of the third voltage pulse, there are three phase current responses. One of them is used for polarity determination in the previous step and another two will be subsequently used for improving estimation resolution in this step. By way of example, if the rotor magnetic pole locates at phase B, after injecting a

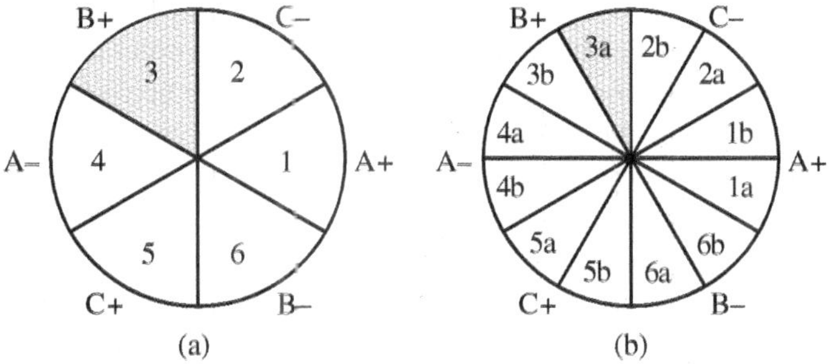

Figure 9.13 Improvement of estimation resolution. (a) 60° resolution estimation result. (b) 30° resolution estimation result. (*Source:* [3]/IEEE.)

negative voltage pulse at phase B, three current responses are recorded, i.e. i_{A3+}, i_{B3-}, i_{C3+}. After determination of polarity in the previous step, the rotor magnetic pole is in sector 3 as shown in Figure 9.13a. If $|i_{A3+}| < |i_{C3+}|$, then the rotor magnetic pole locates at sector 3a as depicted in Figure 9.13b. Hence, a 30° resolution estimation is obtained in this step. The conventional 6 sectors are subsequently increased to 12 sectors.

9.4.3 Boundary Detection Strategy

In some cases the rotor magnetic pole is located at the boundary of two adjacent sectors as highlighted by shaded areas in Figure 9.14. The difference between two current responses may not be easily observed and become more sensitive to noise. Consequently, an estimation error will be introduced and will affect the estimation accuracy. Therefore, a simple boundary detection strategy (BDS) is presented in this section. With BDS, the estimation performance can be enhanced. In total, there are 12 boundary cases as shown in Figure 9.14.

In real systems, due to the existence of noises, the current recorded in the controller may not be the same as the real current. Therefore, the recorded current I_{record} may be represented as:

$$I_{record} = I_{real} + \Delta I_{error} \tag{9.3}$$

where I_{real} is the real current, and ΔI_{error} is the error (difference) between real and recorded currents.

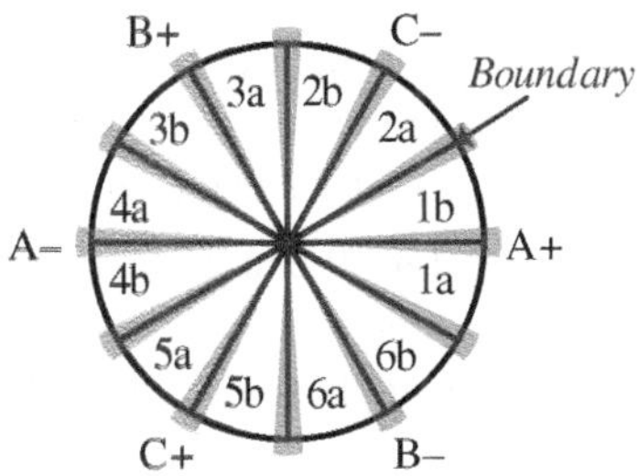

Figure 9.14 Illustration of boundary cases. (*Source:* [3]/IEEE.)

If the rotor magnetic pole is close to the boundary, the small current difference becomes more sensitive to noise, and the sign of the difference may be opposite from the actual value, which affects the estimation performance. In order to solve this, a threshold value of the difference ΔI_{th} is set; if the difference between two current responses is lower than the threshold value, the rotor magnetic pole is judged to be located at the boundary and an exact initial position can be estimated to improve the estimation accuracy. The threshold ΔI_{th} considering noise impact is determined through pre-test. The measurement error ΔI_{error} is obtained at first and the threshold value is defined as $\Delta I_{th} = \Delta I_{error}$. A simple way to determine the value of ΔI_{error} is introduced here. Several voltage pulses with the same magnitude and duration are injected at the same position, the difference between the maximum and average values of current response peak values are recorded, and ΔI_{error} can be calculated by (9.4).

$$\Delta I_{error} = 2(I_{\max} - I_{\text{mean}}) \tag{9.4}$$

9.4.4 Experiment Results

Experiment results are presented in this section. The experiment is based on the prototype SPMSM-III (Appendix B). In this section, the rotor initial position estimation process is demonstrated via one estimation example, followed by the overall estimation against different rotor positions. Finally, estimation performance improvement using BDS is also given.

9.4.4.1 Estimation Example

An estimation example is provided to illustrate the whole estimation process. The actual electrical rotor position is at 41°. The experimental results are demonstrated in Figure 9.15. In Figure 9.15, there are a total of three current responses corresponding to three applied voltage pulses.

In Figures 9.15a–c, the voltage pulse and current responses of each pulse are shown. In addition, each sample point is marked. In Figure 9.15a, the current responses of the first voltage pulse show that $|i_{B1-}| < |i_{C1-}|$, indicating that the rotor magnetic pole is closer to phase C than to phase B. A positive voltage pulse is injected at phase C in Step 2, and the current responses are shown in Figure 9.15b. It can be seen that $|i_{B2-}| < |i_{A2-}|$ and $|i_{A1+}| < |i_{C2+}|$, and hence, the rotor magnetic pole is closest to phase C. In Step 3, a negative voltage pulse is injected at phase C to identify the polarity, as shown in Figure 9.15c. The current responses in Figure 9.15c show that $|i_{C3-}| < |i_{C2+}|$, and therefore, the rotor north pole is in sector 2. In the last step, the current responses from Figure 9.15c are utilized to upgrade the estimation resolution from 60° to 30°. It is shown that

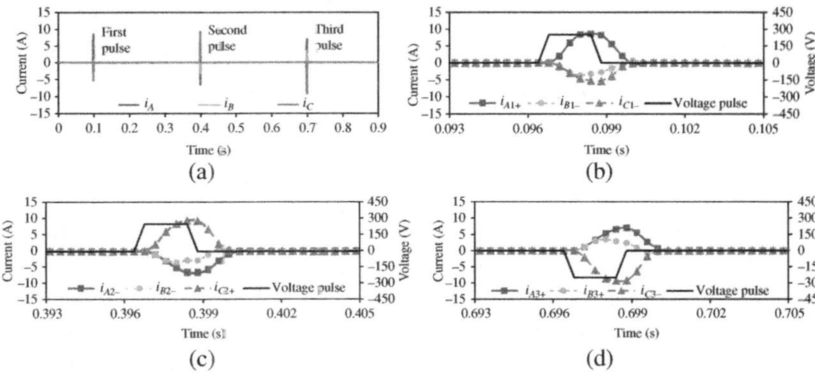

Figure 9.15 Current responses during estimation. (a) Overall. (b) First pulse. (c) Second pulse. (d) Third pulse. (*Source:* [3]/IEEE.)

$\left|i_{B3+}\right| < \left|i_{A3+}\right|$. Therefore, the rotor magnetic pole is closer to phase A, indicating that the rotor north pole is in sector 2a, which is the same as the actual rotor position.

9.4.4.2 Overall Rotor Initial Position Estimation Performance

For further illustration, the described method is tested over the whole range of rotor positions. During the tests, some random initial positions within the whole rotor position range are selected for estimation at standstill. The estimation results and estimation errors are shown in Figure 9.16a. The rotor initial position can be detected over the whole range of rotor positions. The majority of results are within 15°.

9.4.4.3 Boundary Detection Performance

For the overall range of rotor initial position estimation shown in Figure 9.16a, the majority of the results are within 15°. However, there are some cases in which the errors are larger than 15° due to a smaller difference in the current responses

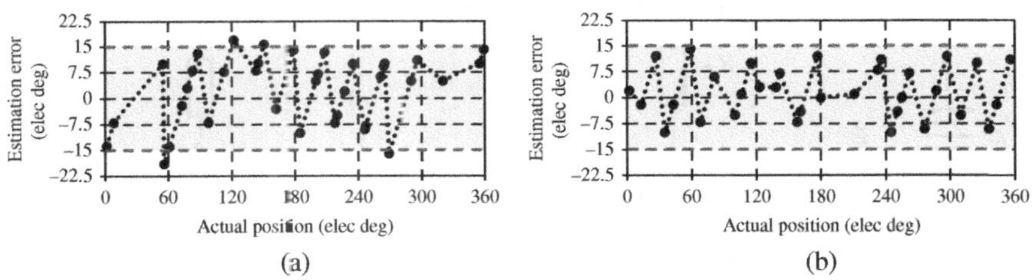

Figure 9.16 Initial rotor position estimation error with/without applying BDS. (a) Without. (b) With. (*Source:* [3]/IEEE.)

at the boundary. After applying BDS during estimation, the improved results are shown in Figure 9.16b, and it can be seen that the estimation errors are all within 15°.

9.5 Pulse Injection Method Using DC-Link Voltage

Apart from estimation methods by using current responses [2], [3], the variation of dc-link voltage variation due to pulse injection is measured and used to estimate the rotor initial position in [5]. This method, which is presented in this section, provides a 30° estimation resolution with maximum error of ±15°. It shows a better performance with a larger dc-link resistance compared with conventional current response-based methods.

9.5.1 Utilization of DC-Link Voltage Variation

First, the dc-link voltage source can be approximately regarded as one internal dc-link resistor R_{dc} and a controlled voltage source V_c in series as shown in Figure 9.17. Next, an equivalent circuit during the voltage pulse injection is also shown in Figure 9.17, where R_{eq} and L_{eq} are the equivalent resistance and inductance in the circuit, respectively, during excitation.

Based on Figure 9.17, after injection of the voltage pulse, there is a current I_{dc} flowing through the circuit and the measured dc-link voltage V_{dc} can be represented by:

$$V_{dc} = V_c - R_{dc}I_{dc} \tag{9.5}$$

During the excitation, it can be seen that there is a voltage drop on the dc-link resistor, i.e. $R_{dc}I_{dc}$ causing a variation on the dc-link voltage, $\Delta V_{dc} = R_{dc}I_{dc}$. As

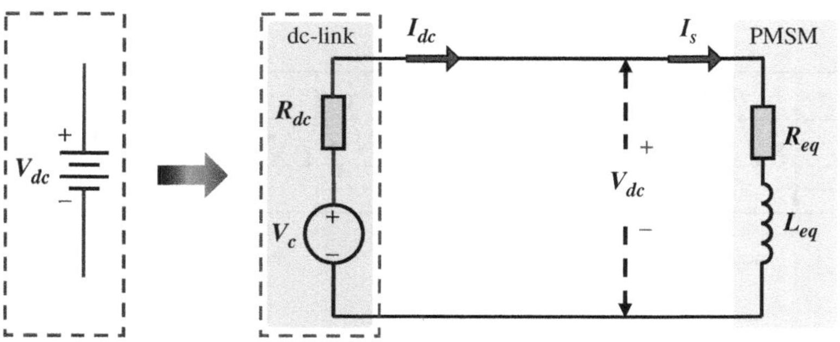

Figure 9.17 Equivalent circuit during excitation. (*Source:* [5]/IEEE.)

described in Section 9.3, the induced current response is position dependent, and the dc-link voltage variation also contains the rotor initial position information.

9.5.2 Pulse Injection Process

Similarly, six voltage pulses are injected as introduced in Section 9.3. Then, the dc-link voltage variations are recorded. Based on the principle described in the above section, the maximum dc-link voltage variation indicates which sector the north pole of the rotor is closest to. The magnetic pole can be located in the sectors shown in Figure 9.18a.

Up to now, a 60° estimation resolution has been acquired. Next, the estimation resolution is improved to 30°. An example is given here for a clear illustration, which is shown as the highlighted areas in Figure 9.18. If the rotor position is estimated at 30°, i.e. sector 1, then by comparing two adjacent dc-link voltage variations of voltage pulses injected in 90° and 330°, if the variation of 90° is larger than that of 330°, the rotor position is then estimated at 45° (sector 1b); otherwise it is 15° (sector 1a), and other cases are similar. By doing this, a 30° estimation resolution can be obtained at last. The overall estimation process is shown in Figure 9.19.

9.5.3 Experiment Results

Experimental results with the dc-link voltage-based method are presented in this section. The experiment is based on the prototype SPMSM-III (Appendix B). In the test, the voltage pulse is selected as 100 V, 3 ms, the rotor movement or vibration is minimized, and a large enough dc-link voltage response is also obtained for rotor initial position estimation. In this section, the rotor initial position estimation

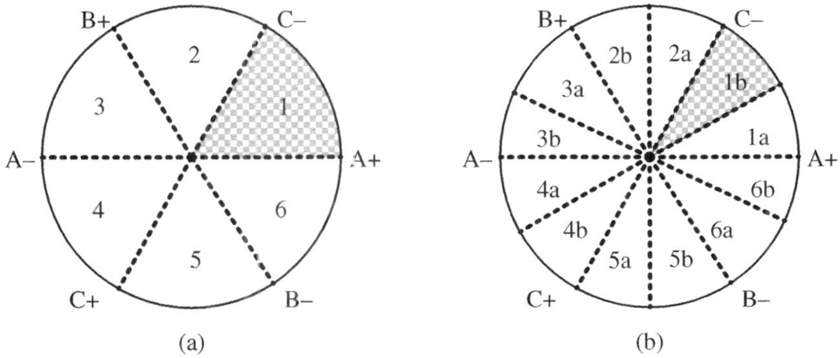

Figure 9.18 Sector demonstration. (a) 60° resolution estimation. (b) 30° resolution estimation. (*Source:* [5]/IEEE.)

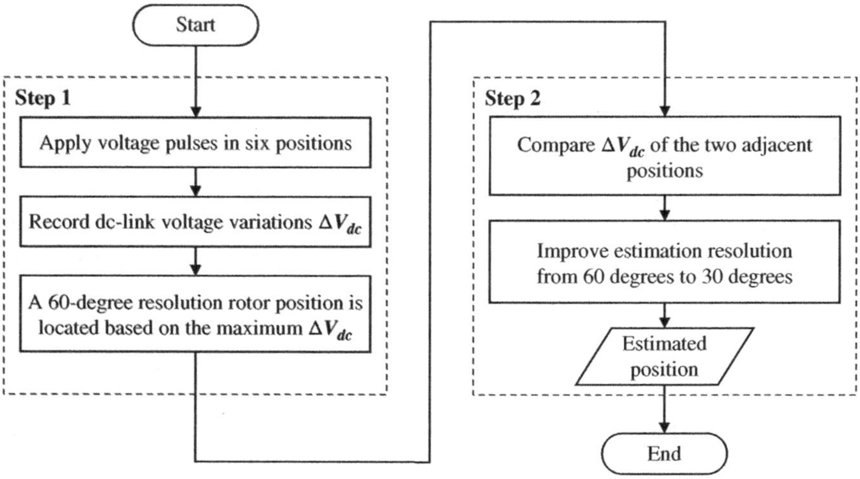

Figure 9.19 Overall estimation process flow chart. (*Source:* [5]/IEEE.)

process is demonstrated via one example, followed by the overall estimation against different rotor positions.

9.5.3.1 Estimation Example

An example of rotor initial position estimation with the described method is shown in Figure 9.20.

The values of the voltage variations after injection pulses are given in blue in Figure 9.20b. The actual rotor position is at 103°. Six voltage pulses are injected at first, followed by an injection sequence of 30°, 90°, 150°, 210°, 270°, and 330°. The dc-link voltage variations are recorded and compared. By comparing six dc-link voltage variations, it is found that the largest one is the second pulse, which

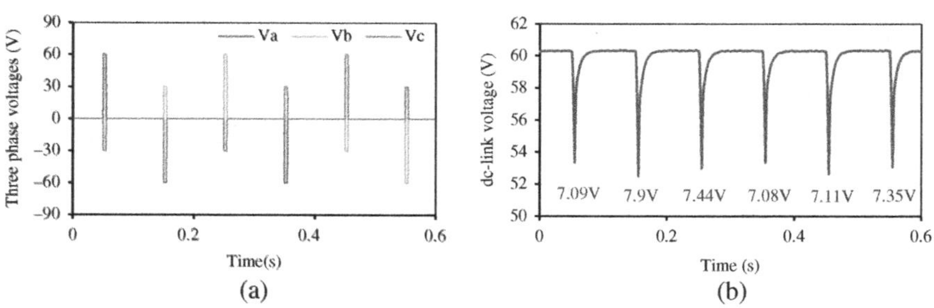

Figure 9.20 Example of estimation. (a) Three phase voltage pulses. (b) dc-link voltage marked with voltage variations. (*Source:* [5]/IEEE.)

is injected at 90°. The largest response indicates that the rotor north pole is closest to 90°. By now, a 60° resolution is obtained and the next step is to increase the estimation resolution by comparing the two neighbor voltage variations, i.e. the first and third pulses; it is obvious that the third one is larger than the first one, and therefore the rotor is closer to the position of the third one. Hence, the new updated estimation position is 105°. Compared with the actual position, the error is smaller than 15°, and therefore, a 30° resolution is obtained at last.

9.5.3.2 Overall Estimation Performance

The overall estimation performance over different rotor positions is illustrated in Figure 9.21. In Figure 9.21b, all the estimation errors of different positions are within 15°, and therefore, a 30° resolution is achieved.

9.5.3.3 Comparison with Estimation Using Current Responses

In this section, the conventional estimation using current response and the proposed estimation using dc-link voltage variation are compared considering different dc-link resistances. Six voltage pulses are injected into six positions, and the peak values of phase current responses and dc-link voltage variations are both recorded. In the test, a 100 V and 5 ms voltage pulse is selected and the real rotor position is fixed at 90°. The peak values of each case are presented in Table 9.2, where Δ (values shown in bold) represents the difference between the maximum and minimum values among the six responses in each case. A larger Δ indicates a better performance in terms of measurement noise or sensor resolution. In other words, a smaller Δ is more sensitive to measurement errors including noises and resolution.

From Table 9.2, it can be concluded that as dc-link resistance increases, a larger difference is obtained with the proposed method and thus a better estimation performance is obtained compared with the conventional current response-based methods.

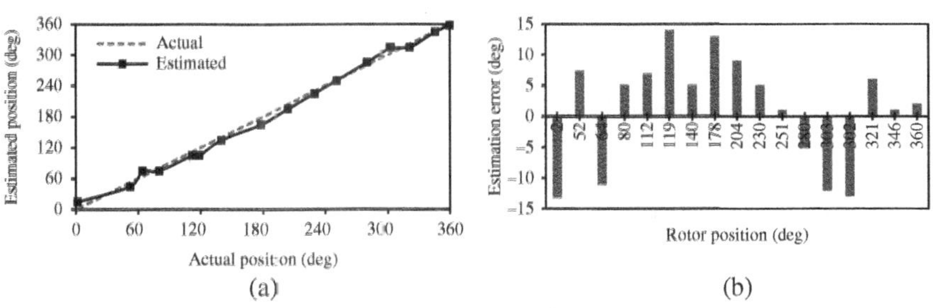

Figure 9.21 Overall estimation performance [5]. (a) Position. (b) Error. (*Source:* [5]/IEEE.)

Table 9.2 Peak value comparison between phase current and dc-link voltage variation.

Injection position (degree)		30	90	150	210	270	330	Δ
0.5 Ω	I (A)	8.027	8.771	8.104	8.043	8.464	8.036	**0.75**
	ΔV_{dc} (V)	3.58	3.94	3.6	3.59	3.81	3.58	**0.36**
1 Ω	I (A)	7.895	8.614	7.899	7.906	8.291	7.9	**0.72**
	ΔV_{dc} (V)	7.06	7.85	7.09	7.10	7.53	7.06	**0.79**
2 Ω	I (A)	7.811	8.509	7.887	7.841	8.235	7.888	**0.682**
	ΔV_{dc} (V)	9.49	10.58	9.56	9.60	10.26	9.6	**1.09**
5 Ω	I (A)	7.725	8.381	7.777	7.736	8.152	7.799	**0.656**
	ΔV_{dc} (V)	12.2	13.32	12.25	12.04	13.12	12.13	**1.12**

(*Source:* [5]/IEEE.)

9.6 Voltage Pulse Selection

For the short pulse injection-based rotor initial position estimation method, the magnitude and duration of the pulse should be determined before injection to obtain a good estimation performance. Hence, based on the estimation method introduced in Section 9.3, a selection guidance in [10] is introduced in this section.

The current response after injecting the voltage pulse is given as:

$$i(t) = \frac{V_P}{R_{eq}}\left(1 - e^{-\frac{R_{eq}}{L_{eq}}T_P}\right) \tag{9.6}$$

where the magnitude and duration of the voltage pulse are defined as V_P and T_P. From (9.6), it can be seen that the magnitude and duration of the voltage pulse can affect the current response and therefore the estimation performance. During the selection of magnitude and duration of the pulse, there are two main rules:

- The torque produced by voltage injection should be limited to avoid rotor movement.
- Make sure the current responses against rotor position are observable. The observable current responses guarantee a reliable estimation.

9.6.1 Selection of Duration

Considering the fixed voltage pulse magnitude, the current response varies with duration. The optimal duration T_{opt} of voltage pulses can be selected in order to generate the maximum difference between current responses [4].

Considering two voltage pulses injected at two different positions, the corresponding current responses I_1 and I_2 are given in (9.7) and the difference ΔI between two current responses is calculated by (9.8).

$$I_1(t) = \frac{V_P}{R_{eq}}\left(1 - e^{-\frac{t}{\tau_{e1}}}\right)$$
$$I_2(t) = \frac{V_P}{R_{eq}}\left(1 - e^{-\frac{t}{\tau_{e2}}}\right) \tag{9.7}$$

$$\Delta I(t) = I_1(t) - I_2(t) = \frac{V_P}{R_{eq}}\left(e^{-\frac{t}{\tau_{e2}}} - e^{-\frac{t}{\tau_{e1}}}\right) \tag{9.8}$$

where τ_{e1} and τ_{e2} are the electrical time constants of two injection cases. By the differential of $\Delta I(t)$ by time t, the theoretical duration that produces the maximum difference can be obtained by (9.10).

$$T_{opt} = \frac{\tau_{e1}\tau_{e2}\log\frac{\tau_{e2}}{\tau_{e1}}}{\tau_{e2} - \tau_{e1}} \tag{9.9}$$

By way of example, Figure 9.22 illustrates that the maximum difference of two current responses occurs at T_{opt} according to (9.9).

However, T_{opt} in (9.9) is too complicated for practical use. Therefore, an approximation is used instead. Defining $\tau_{e2} = \tau_{e1} + \Delta\tau$, for a small $\Delta\tau$, T_{opt} can be approximated as:

$$T_{opt} = \frac{\tau_{e1} + \tau_{e2}}{2} \approx \tau_{eN} \tag{9.10}$$

where τ_{eN} is the nominal electrical time constant that can be selected as a start value for T_P. Moreover, considering the variation of the electrical time constant

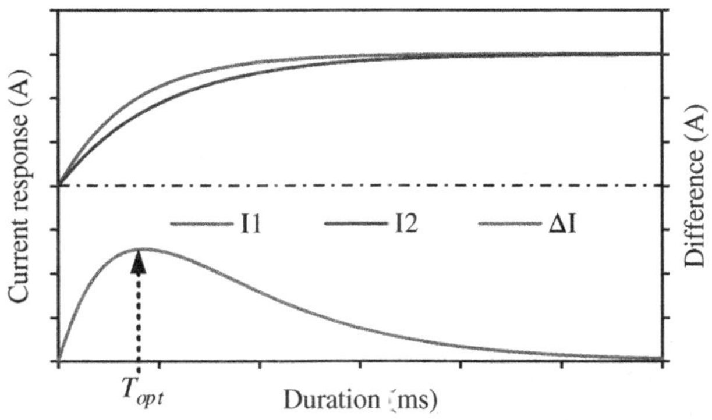

Figure 9.22 Current responses and variation of current difference against duration. (*Source:* [10]/with permission of IET.)

against rotor position, i.e. inductance changes, rather than to select a single duration value it is better to select a duration range between the maximum and minimum electrical time constants given by:

$$\begin{aligned} T_{P\_MAX} &= \tau_{e\_MAX} = \frac{L_{MAX}}{R_{eq}} \\ T_{P\_MIN} &= \tau_{e\_MIN} = \frac{L_{MIN}}{R_{eq}} \end{aligned} \tag{9.11}$$

where L_{MAX} and L_{MIN} are the maximum and minimum inductances, respectively, over all rotor positions. These two time constants can be obtained based on FE analysis or experiments.

9.6.2 Selection of Magnitude

For the magnitude of voltage pulses, there are two limitations, upper limitation V_{P_MAX} and lower limitation V_{P_MIN}, which should be determined for magnitude selection.

First, the upper limitation V_{P_MAX} can be calculated based on the torque or q-axis current limitation and the rotor movement limitation.

A maximum q-axis current I_{q_MAX} can be determined according to the requirement of real application. Based on the selection of duration T_{opt}, which is chosen around the electrical time constant, the maximum magnitude can be chosen as:

$$V_{P\_MAX\_1} \leq \frac{R_{eq} I_{q\_MAX}}{0.63} \tag{9.12}$$

Apart from the q-axis current limitation, the rotor movement should also be limited by $\theta_{\max}$, which is the maximum moving angle depending on the applications. By neglecting load torque and friction, the maximum magnitude is:

$$V_{P\_MAX\_2} \leq 5 \frac{J R_{eq}}{P^2 \psi_m} \left(\frac{R_{eq}}{L_{eq}} \right)^2 \theta_{\max} \tag{9.13}$$

where J is the inertia, P is the number of pole pairs, and ψ_m is the permanent magnet flux-linkage. Finally, the upper limitation of voltage pulse magnitude is given as:

$$V_{P\_MAX\_2} = \min(V_{P\_MAX\_1}, V_{P\_MAX\_2}) \tag{9.14}$$

Second, the lower limitation V_{P_MIN} is defined to guarantee that the rotor initial position estimation is reliable. According to the magnetic saturation effect, in order to identify the polarity, enough voltage pulse should be given to produce sufficient armature reaction to produce an observable current difference between the current responses associated with north and south poles during polarity identifications. Otherwise, a reverse starting can happen with 180° position error. Therefore, it is necessary to determine an appropriate V_{P_MIN}. Meanwhile, in real-world applications, during estimation, the current responses sampled in the controllers could be different from the actual values as shown by (9.15). This current recording error can affect the estimation performance.

$$I_{record} = I_{real} + \varepsilon_{sample} \tag{9.15}$$

The current sampling error ε_{sample} can be caused by measurement errors, analog-to-digital converter (ADC) quantization, noise, and dc-link voltage variation. In order to determine the maximum value of ε_{sample}, several voltage pulses with the same magnitude and duration are injected at the same position, the difference between the maximum and average values, i.e. $I_{\max}$ and I_{mean}, of current responses are recorded, and ε_{sample} can be calculated by:

$$\varepsilon_{sample} \geq 2(I_{\max} - I_{\text{mean}}) \tag{9.16}$$

Then, as shown in (9.17), two opposite voltage pulses are injected into one position, and the difference between the two current responses is recorded and compared with ε_{sample}.

$$||I(V_{P\_MIN})| - |I(-V_{P\_MIN})|| = \Delta I > \varepsilon_{sample} \tag{9.17}$$

According to (9.16) and (9.17), the minimum limitation of voltage pulse magnitude V_{P_MIN} should be selected by ensuring that $\Delta I > \varepsilon_{sample}$.

9.6.3 Experiment Results

Experimental results with the voltage pulse selection method are presented in this section. The experiment is based on the prototype SPMSM-III (Appendix B). From the determined duration and magnitude of the voltage pulse, a reliable selection area (RSA) can be obtained. Selecting a voltage pulse within this RSA, a more reliable and effective estimation performance can be guaranteed. In the experiment, V_{P_MAX}, V_{P_MIN}, V_{P_MAX}, and T_{P_MIN} are obtained and the RSA is depicted in Figure 9.23. Based on the estimation method described in Section 9.3, several tests are carried out to investigate the rotor initial position estimation performance. Different voltage pulse magnitude and duration combinations are considered, both inside and outside RSA. In Figure 9.23, the circles indicate the acceptable estimation performance with the correct estimation result and allowable q-axis current produced. Otherwise, the estimation performance is unacceptable and is marked with a red cross.

From Figure 9.23, four different cases are chosen to illustrate the measured results when the selected voltage pulse is inside and outside RSA, Figures 9.24a–d. In Figure 9.24, according to the real rotor initial position (117 electrical degrees),

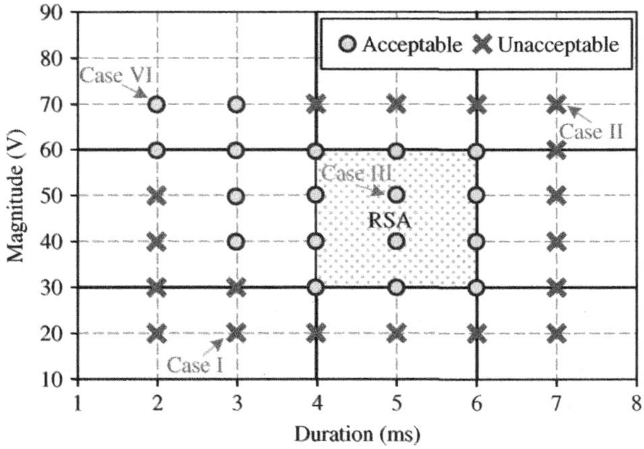

Figure 9.23 Reliable selection area. (*Source:* [10]/with permission of IET.)

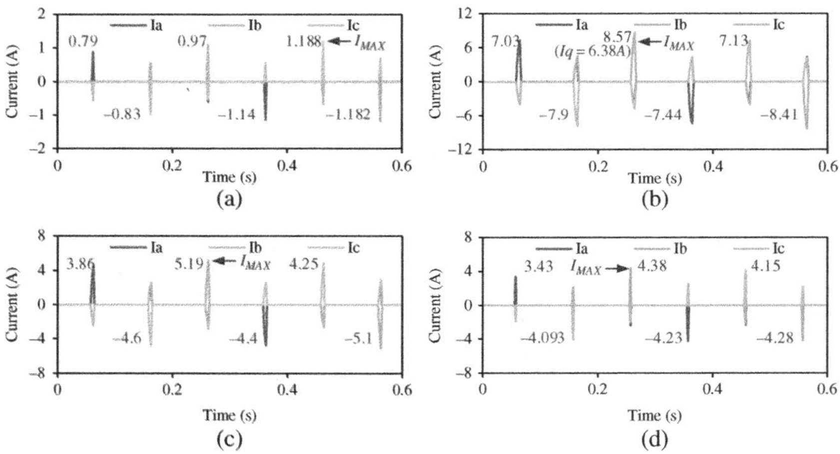

Figure 9.24 Measured current responses of typical injections. (a) Case I with wrong estimation outside RSA. (b) Case II with q-axis current over limit outside RSA. (c) Case III with acceptable estimation inside RSA. (d) Case IV with acceptable estimation outside RSA. (*Source:* [10]/with permission of IET.)

the phase B current response of the third pulse should be the largest, otherwise, the estimation will be incorrect. Also, the produced q-axis current should not exceed the 6 A limitation.

First, selected voltage pulses are outside RSA at a lower level, which will result in a wrong estimation. An example is given in Figure 9.24a, where the current response of the fifth pulse is the largest, which is wrong. Then, selected voltage pulses are outside RSA at a higher level; although the estimation could be correct, the produced q-axis current will exceed the limit. For case II in Figure 9.24b, by calculation, the produced q-axis current (6.38 A) is over the limit. Next, from case III in Figure 9.24c, selected voltage pulses are inside RSA, and estimation performances are all acceptable. Finally, it is worth mentioning that the described RSA is a simple and approximate selection strategy. From case VI in Figure 9.24d, although the voltage pulse is selected outside RSA, an acceptable estimation performance can also be obtained. Nevertheless, it can be concluded that by selecting the voltage pulse within RSA, reliable estimation performance is guaranteed.

9.7 High-Frequency Signal Injection-Based Method

Compared to HFSI methods, the short pulse-based method is simple and fast with a satisfactory estimation accuracy since it only locates rotor initial position in one of the sectors instead of an exact continuous position without using complex algorithms and observers. Therefore, a hybrid method is proposed in [7] by injecting an HF rotating signal and comparing the amplitudes of three-phase current responses to detect the sector where the north pole is located, which is similar to the short pulse injection-based method. It is worth noting that this method is suitable for IPMSM with enough saliency. For SPMSM with low saliency, it is suggested that short pulse injection-based methods be used that can utilize the magnetic saturation effect.

9.7.1 Three-Phase HF Current Amplitude

After injecting a HF rotating sinusoidal voltage signal, the three-phase HF currents can be seen as amplitude-modulated signals [8]. Three-phase HF currents can be expressed as:

$$\begin{bmatrix} i_{Ah} \\ i_{Bh} \\ i_{Ch} \end{bmatrix} = \begin{bmatrix} 1 & 0 \\ -\dfrac{1}{2} & \dfrac{\sqrt{3}}{2} \\ -\dfrac{1}{2} & -\dfrac{\sqrt{3}}{2} \end{bmatrix} \begin{bmatrix} i_{\alpha h} \\ i_{\beta h} \end{bmatrix} = \begin{bmatrix} I_p \sin(\theta_h) + I_n \sin(\theta_h - 2\theta_r) \\ I_p \sin(\theta_h) + I_n \sin(\theta_h - 2\theta_r + 2\pi/3) \\ I_p \sin(\theta_h) + I_n \sin(\theta_h - 2\theta_r - 2\pi/3) \end{bmatrix} \tag{9.18}$$

When the machine is at standstill, the three-phase HF currents can be regarded as:

$$\begin{bmatrix} i_{\mathrm{A}h} \\ i_{\mathrm{B}h} \\ i_{\mathrm{C}h} \end{bmatrix} = \begin{bmatrix} Amp\_i_{\mathrm{A}h}\sin(\theta_h + \varphi_{\mathrm{A}h}) \\ Amp\_i_{\mathrm{B}h}\sin(\theta_h + \varphi_{\mathrm{B}h}) \\ Amp\_i_{\mathrm{C}h}\sin(\theta_h + \varphi_{\mathrm{C}h}) \end{bmatrix} \tag{9.19}$$

$$\begin{bmatrix} Amp\_i_{\mathrm{A}h} \\ Amp\_i_{\mathrm{B}h} \\ Amp\_i_{\mathrm{C}h} \end{bmatrix} = \begin{bmatrix} \sqrt{I_p^2 + I_n^2 + 2I_pI_n\cos(2\theta_r)} \\ \sqrt{I_p^2 + I_n^2 + 2I_pI_n\cos(2\theta_r + 2\pi/3)} \\ \sqrt{I_p^2 + I_n^2 + 2I_pI_n\cos(2\theta_r - 2\pi/3)} \end{bmatrix} \tag{9.20}$$

where $\varphi_{\mathrm{A}h}$, $\varphi_{\mathrm{B}h}$, $\varphi_{\mathrm{C}h}$, and $Amp\_i_{\mathrm{A}h}$, $Amp\_i_{\mathrm{B}h}$, $Amp\_i_{\mathrm{C}h}$ are the phases and amplitudes of the three-phase HF currents, respectively. The amplitudes can be calculated by discrete Fourier transform (DFT) from stator currents as

$$Amp\_i_{\mathrm{ABC}h} = \frac{2}{W_s}\sqrt{\left[\Sigma_{k=0}^{W_s-1} i_{\mathrm{ABC}}(k)\cos\left(\frac{2\pi kT_s}{T_h}\right)\right]^2 + \left[\Sigma_{k=0}^{W_s-1} i_{\mathrm{ABC}}(k)\sin\left(\frac{2\pi kT_s}{T_h}\right)\right]^2} \tag{9.21}$$

where W_s is the number of sampling points in a DFT window and T_s, T_h are the periods of sampling and injection signal.

9.7.2 Sector Detection

The traces of $Amp\_i_{\mathrm{A}h}$, $Amp\_i_{\mathrm{B}h}$, and $Amp\_i_{\mathrm{C}h}$ with rotor position are plotted in Figure 9.25a according to (9.20). The whole electrical period is evenly divided into 12 sectors by amplitude crossings shown by the black dots. The angles θ_{bound} and θ_{middle} are the boundary angle between adjacent sectors and the middle angle in each sector, respectively, as shown in Figure 9.25b. The amplitude waveforms in the range of (180°, 360°), i.e. the south pole, are identical to those in the range (0°, 180°), i.e. the north pole.

Table 9.3 describes the relationship between the HF current amplitudes and the sector number, which can be used for initial position detection. The rotor position, which refers to the north pole of the magnet, may be located in one sector or on a boundary according to the three current amplitudes. If the rotor is on the boundary, the boundary angle θ_{bound} is treated as the initial position and if the rotor is in one sector, the middle angle θ_{middle} is seen as the initial position. The maximum initial position estimation error is ±15°, which is half of the sector length.

The initial position can be identified on the sector boundary or in one specific sector according to Table 9.3. However, there still exists an ambiguity of π as the saliency undergoes two cycles in one electric period. The dual voltage pulses injection method introduced in Chapter 8 can be used to identify the polarity.

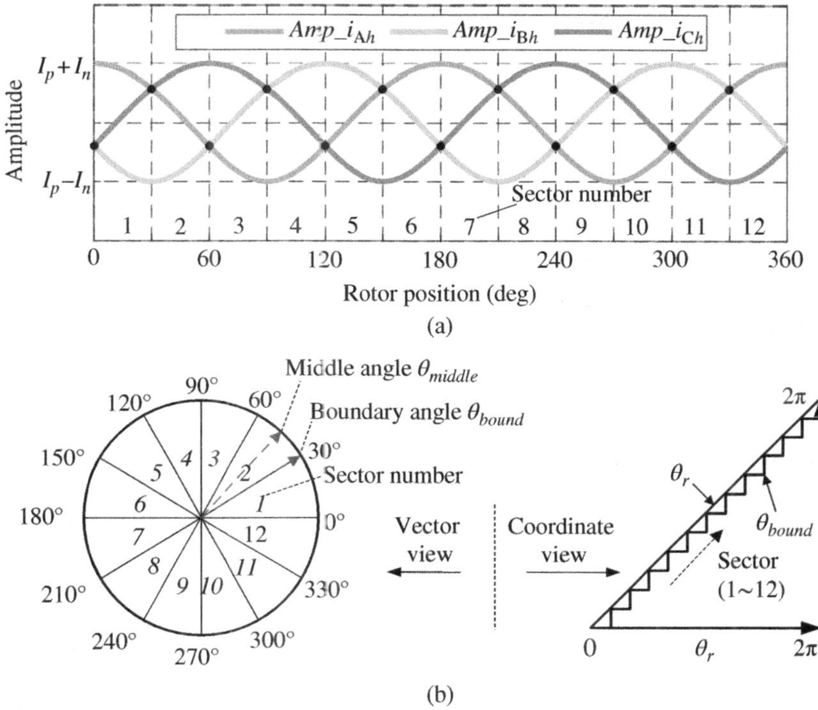

Figure 9.25 Variation of three phase amplitudes with rotor position. (a) Variation of phase current amplitudes with position. (b) 12 sectors separated by crossings. (*Source:* [7]/IEEE.)

Table 9.3 Sector or boundary judgment according to HF current amplitudes.

Relationship	Sector (θ_{middle})	Boundary (θ_{bound})
$Amp\_i_{Ah} > Amp\_i_{Ch} > Amp\_i_{Bh}$	1 or 7 ($\theta_{middle} = \pi/12$)	
$Amp\_i_{Ch} > Amp\_i_{Ah} > Amp\_i_{Bh}$	2 or 8 ($\theta_{middle} = \pi/4$)	
$Amp\_i_{Ch} > Amp\_i_{Bh} > Amp\_i_{Ah}$	3 or 9 ($\theta_{middle} = 5\pi/12$)	
$Amp\_i_{Bh} > Amp\_i_{Ch} > Amp\_i_{Ah}$	4 or 10($\theta_{middle} = 7\pi/12$)	
$Amp\_i_{Bh} > Amp\_i_{Ah} > Amp\_i_{Ch}$	5 or 11($\theta_{middle} = 3\pi/4$)	
$Amp\_i_{Ah} > Amp\_i_{Bh} > Amp\_i_{Ch}$	6 or 12 ($\theta_{middle} = \pi/12$)	
$(Amp\_i_{Bh} = Amp\_i_{Ch}) < Amp\_i_{Ah}$		0 or π
$(Amp\_i_{Ah} = Amp\_i_{Ch}) < Amp\_i_{Bh}$		$\pi/6$ or $7\pi/6$
$(Amp\_i_{Ah} = Amp\_i_{Bh}) < Amp\_i_{Ch}$		$\pi/3$ or $4\pi/3$
$(Amp\_i_{Bh} = Amp\_i_{Ch}) < Amp\_i_{Ah}$		$\pi/2$ or $3\pi/2$
$(Amp\_i_{Ah} = Amp\_i_{Ch}) > Amp\_i_{Bh}$		$2\pi/3$ or $5\pi/3$
$(Amp\_i_{Ah} = Amp\_i_{Bh}) > Amp\_i_{Ch}$		$5\pi/6$ or $11\pi/6$

(*Source:* [7]/IEEE.)

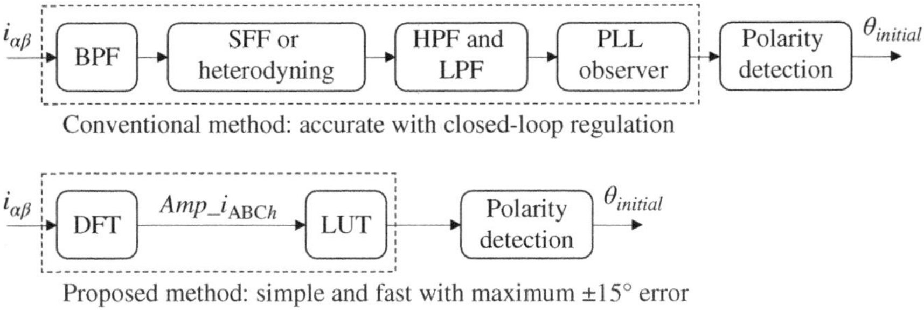

Figure 9.26 Comparison of conventional and described initial position estimation methods. (*Source:* [7]/IEEE.)

The overall block diagrams of the described method and also the conventional rotating signal injection-based method are compared and shown in Figure 9.26. Clearly, compared to the conventional method, this HF current amplitude modulation-based method is simpler and faster.

9.7.3 Experiment Results

Experimental results of the initial position detection method based on the HF signal injection-based method are shown in this section based on the prototype IPMSM-I (Appendix B). The real rotor position is 50°. A rotating HF voltage with magnitude 15 V and frequency 750 Hz is selected considering the saliency level. Both conventional rotating signal injection method and the described methods are executed in the experiment.

Figure 9.27 shows the comparison of initial position estimation between the conventional method with different PI parameters of the observer shown in Figure 9.26 and the HF signal injection-based method. The relationship between calculated current amplitudes is $Amp\_i_{Ch} > Amp\_i_{Ah} > Amp\_i_{Bh}$. According to Table 9.3, the rotor position is located in sector 2 or sector 8, and the middle angle 45° is treated as the initial rotor position.

For the HF current amplitude modulation method, it is observed that the amplitudes of i_{ABCh} are stable at point P, which is about only three injection cycles after the initial injection. The convergence time for the conventional method varies with different PI parameters. For PI parameters with overshoot and oscillation, the convergence time can be as long as one second as shown at point C3 for PI3. The convergence time can also be very short with well-tuned PI, such as point

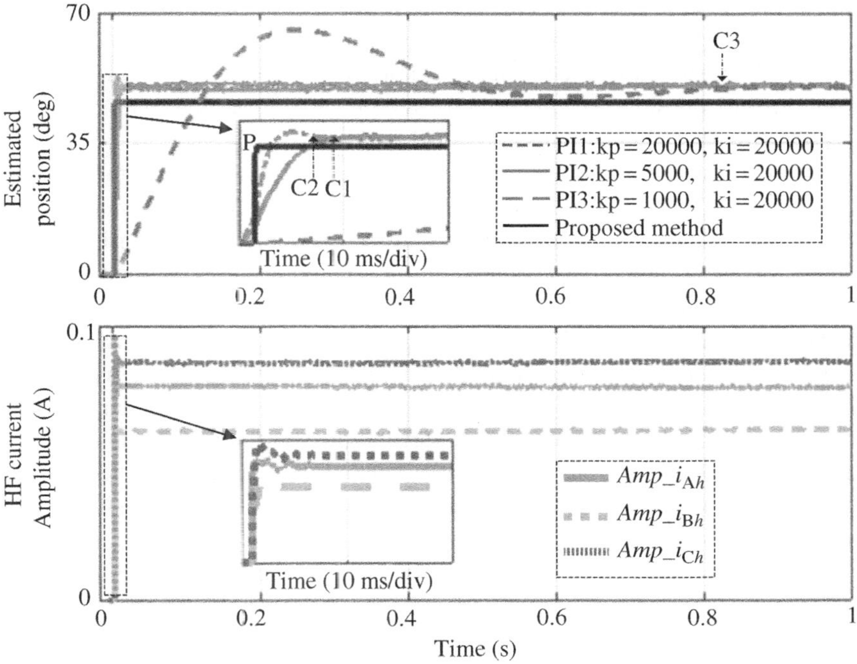

Figure 9.27 Comparison of HF signal injection-based and conventional initial position estimation methods with different PI parameters in heterodyning observer.
(*Source:* [7]/IEEE.)

C1 for PI1 and point C2 for PI2. Whereas the parameter tuning is not only limited by filters but also dependent on machine parameters, it is difficult to predetermine optimal PI parameters. Thus, enough time should be reserved between initial injection and polarity detection to guarantee reliable convergence. This time gap further increases the execution time of the conventional method.

9.8 Summary

In this chapter, four methods for rotor initial position estimation were introduced. First, the short pulse injection-based rotor initial position estimation was explained including the basic principle and two improved estimation methods with experiment results. Second, the selection of voltage pulse was discussed and a voltage pulse selection guidance was introduced to obtain a reliable estimation performance. Third, a unique HF rotating signal injection-based method was presented; by estimating the sector of the north pole instead of a continuous position, the initial position estimation process becomes simpler and faster compared to conventional HF signal injection-based methods.

An overall comparison of these methods is given in Table 9.4. First, for the short pulse injection methods, method 2 in [3] requires the least number of pulses while having the highest estimation accuracy. Also, method 2 in [3] considered the boundary cases to enhance the estimation with sampling noises. Method 3 [5] only requires one dc-link voltage sensor, which also has the same estimation accuracy as method 2 in [3]. For method 4 [7] with an HF rotating voltage signal injection, the estimation accuracy is the same as method 2 in [3] and method 4 in [5] with a maximum error of 15°. Similarly, method 4 also considers the boundary cases to improve the estimation accuracy. Compared to the short pulse injection method, method 4 in [7] is more suitable for IPMs since it requires a high saliency level, while short pulse-based methods 2–4 in [3], [5], [7] can be applied to both SPMs and IPMs.

Table 9.4 Overall comparison.

Method	Method 1 [2]	Method 2 [3]	Method 3 [5]	Method 4 [7]
Excitation signal	Short voltage pulses	Short voltage pulse	Short voltage pulse	HF rotating voltage signal
Response for position estimation	Three-phase currents	Three-phase currents	DC-link voltage	Three-phase currents
Maximum estimation error (degree)	30	15	15	15
Required pulses number	6	**3**	6	N/A
Current sensor	3	3	0	3
Voltage sensor	0	0	**1**	**0**
Consideration of sampling noise	No	Yes	Yes	No
Suitable machine type	IPM and SPM	IPM and SPM	IPM and SPM	IPM

References

1 M. Jansson, L. Harnefors, O. Wallmark, and M. Leksell, "Synchronization at startup and stable rotation reversal of sensorless nonsalient PMSM drives," *IEEE Trans. Ind. Electron.*, vol. 53, no. 2, pp. 379–387, Apr. 2006.

2 "The smart start technique for BLDC motors—Application brief for ML4428," Micro Linear, San Jose, CA, USA, Sep. 1996.

3 X. Wu, Z. Q. Zhu, and Z. Wu, "A rotor initial position estimation method for surface-mounted permanent magnet synchronous machine," *IEEE Trans. Energy Convers.*, vol. 36, no. 3, pp. 2012–2024, Sep. 2021.

4 W. J. Lee and S. K. Sul, "A new starting method of BLDC motors without position sensor," *IEEE Trans. Ind. Appl.*, vol. 42, no. 6, pp. 1532–1538, Nov. 2006.

5 X. Wu, Z. Q. Zhu, and Z. Wu, "A novel rotor initial position detection method utilizing dc-link voltage sensor," *IEEE Trans. Ind. Appl.*, vol. 56, no. 6, pp. 6486–6495, 2020.

6 Y. S. Lai, F. S. Shyu, and S. S. Tseng, "New initial position detection technique for three-phase brushless dc motor without position and current sensors," *IEEE Trans. Ind. Appl.*, vol. 39, no. 2, pp. 485–491, Mar. 2003.

7 B. Shuang and Z. Q. Zhu, "A novel sensorless initial position estimation and startup method," *IEEE Trans. Ind. Appl.*, vol. 68, no. 4, pp. 2964–2975, Apr. 2021.

8 H. Li, X. Zhang, S. Yang, F. Li, and M. Ma, "Improved initial rotor position estimation of IPMSM using amplitude demodulation method based on HF carrier signal injection," in *Conf. IECON IEEE-IES Annu. Meeting*, pp. 1996–2001, 2017.

9 L. M. Gong and Z. Q. Zhu, "Robust initial rotor position estimation of permanent-magnet brushless AC machines with carrier-signal-injection-based sensorless control," *IEEE Trans. Ind. Appl.*, vol. 49, no. 6, pp. 2602–2609, Nov. 2013.

10 X. Wu and Z. Q. Zhu, "A simple voltage pulse selection strategy for rotor initial position estimation,' in *10th Int. Conf. Power Electron. Mach. Drives (PEMD 2020)*, vol. 2020, pp. 688–693, 2020.

10

Sensorless Control of PM Brushless DC Drives Based on Detection of Zero-Crossing of Back-EMF Waveform

10.1 Introduction

For BLDC drives, as introduced in Chapter 1, the rotor position information is necessary and can be obtained from either the sensored or sensorless scheme. The Hall-effect sensors are often applied for BLDC operation, and it is quite difficult to achieve high installation accuracy, especially in small-size motors, and the inaccurate and non-uniform distributions of the Hall-effect sensors will affect the control performance [1], [2]. Also, the Hall-effect sensors are sensitive to the environment, including high temperature and electromagnetic interference. Hence, sensorless schemes of BLDC drives have attracted much attention in the last three decades. Sensorless BLDC schemes can be classified as:

- Back-EMF zero-crossing point (ZCP) detection [3]–[10]
- Back-EMF integration [11], [12]
- Freewheeling diode conduction method [13]
- Fundamental model-based method [14]

Of these, the back-EMF ZCP detection method is simpler to implement and requires minimal knowledge of machine parameters. Moreover, it does not require a complex digital signal processor (DSP). As a consequence, ZCP detection of back-EMF-based sensorless control has been widely used for BLDC drives. For the ZCP detection sensorless control method, it is usually applied based on the two-phase 120° conduction six-step control of BLDC. The basics of six-step control were introduced in Chapter 1. In this chapter, the basics of the ZCP detection method are first introduced, and several common practical issues including ZCP deviation and the long freewheeling angle are demonstrated, together with solutions.

Sensorless Control of Permanent Magnet Synchronous Machine Drives, First Edition.
Zi Qiang Zhu and Xi Meng Wu.

10.2 Basics of ZCP Detection

For sensorless BLDC drives, by measuring the back-EMF voltage, which contains position information, the commutation points can be found. As shown in Figure 10.1, the back-EMF can be sinusoidal (solid line) and trapezoidal (dashed line) in BLDC motors. For both types of back-EMF waveforms, it is clear in Figure 10.1 that the ZCPs of back-EMFs are located in the middle between two consecutive commutation points. The zero-crossing detection (ZCD) method is to search the ZCPs of the back-EMF and then identify the commutation points.

10.2.1 Mathematic Model

Figure 10.2 shows a diagram of the inverter and the PM BLDC motor.

The BLDC model can be expressed as:

$$\begin{cases} v_{AN} = v_{AG} - v_{NG} = Ri_A + (L - M)\dfrac{d}{dt}i_A + e_A \\ v_{BN} = v_{BG} - v_{NG} = Ri_B + (L - M)\dfrac{d}{dt}i_B + e_B \\ v_{CN} = v_{CG} - v_{NG} = Ri_C + (L - M)\dfrac{d}{dt}i_C + e_C \end{cases} \tag{10.1}$$

where v_{XN}, i_X, and e_X are the phase voltage, current, and back-EMF, and R, L, and M are the phase resistance, the self-inductance, and the mutual-inductance,

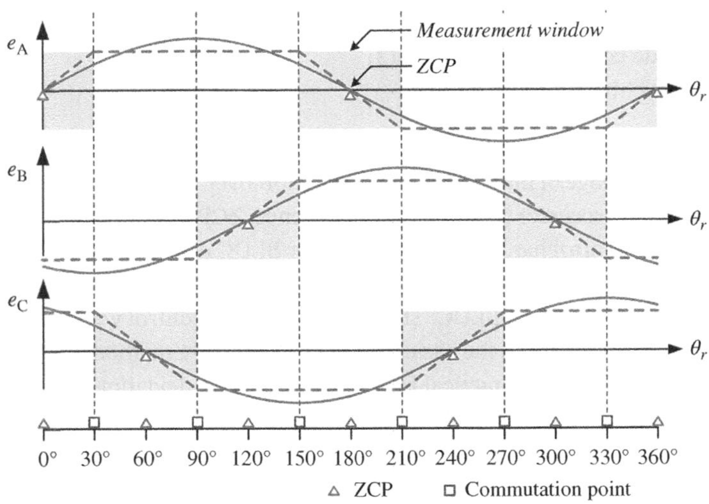

Figure 10.1 Illustration of back-EMFs and zero-crossing points.

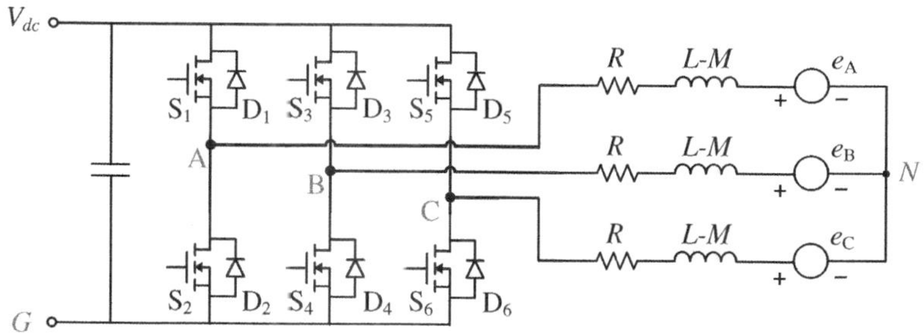

Figure 10.2 Diagram of inverter and motor.

respectively. Here, the subscript X denotes phase A, B, or C. Moreover, N is the neutral and G is the ground.

10.2.2 Typical Current Waveforms

Figure 10.3 shows a typical phase winding current, voltage, and back-EMF waveform. For simplicity, a sinusoidal back-EMF waveform has been assumed for illustrative purposes. A sinusoidal back-EMF waveform is very common for two-pole high-speed BLDC motor drives, although a trapezoidal back-EMF waveform is more common for most BLDC motor drives.

If the commutation timing on all three phases is correctly phased by 60 electrical degrees, the phase current waveform is symmetrical and consists of three 60 electrical degree intervals per half-cycle, each interval being subdivided into two periods whose relative length is determined by the conduction time of the associated freewheeling diodes, Figure 10.3b. For example, during period 1 when S_1 and S_4 are on, freewheeling diode D_6 conducts in period 1a (i.e. time interval Δ_1). All three phases then carry current, since there are two distinct current paths, the main one being from the supply through phase windings A and B, via S_1 and S_4, and the other being through one of these windings and phase winding C, via S_5 and D_6. At the end of period 1a the current through D_6 has reduced to zero, and diode D_6 does not conduct during period 1b.

10.2.3 Sensorless Operation of BLDC Drives

For the sensorless operation of detection of the zero-crossing of the back-EMF, a three-stage starting technique is generally employed [3]:

- **Step 1:** Initial rotor alignment—by exciting one or two phase windings so that the initial rotor position is known. Another method is to detect the rotor positions by the methods introduced in Chapter 9.

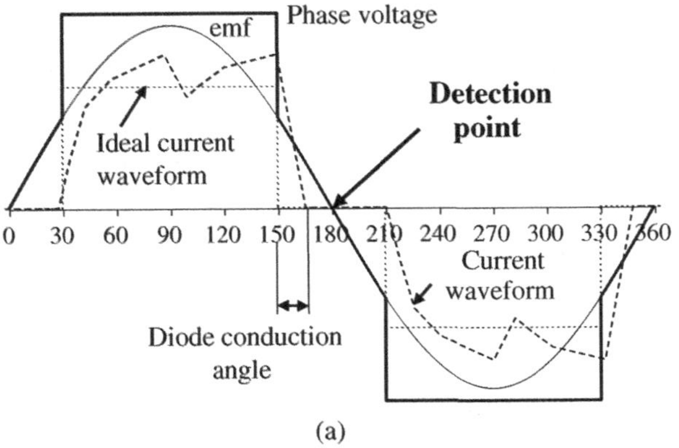

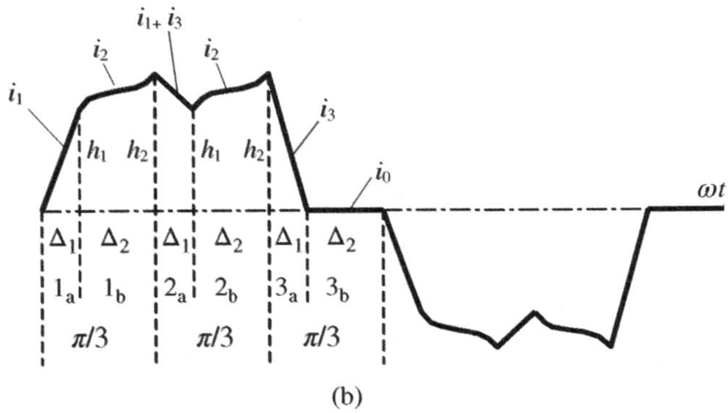

Figure 10.3 Basis of back-EMF sensorless commutation and typical phase current waveforms. (a) Back-EMF sensorless commutation. (b) Typical phase current waveforms. (*Source:* [3]/IOS Press.)

- **Step 2:** Synchronous open-loop run-up—during which the phase windings are excited using the appropriate switching logic, until the back-EMF is sufficiently high to enable the motor to be operated in Step 3.
- **Step 3:** Sensorless closed-loop mode—when the zero-crossings of the back-EMF waveforms of the three phases are detected during periods when they do not carry current.

The basic principle of back-EMF measurement is illustrated in Figure 10.3b, assuming ideal current and voltage waveforms. As can be seen, the appropriate switching devices are commutated 30 electrical degrees after detection of the zero-crossing of the associated back-EMF. Each phase will conduct for 120 electrical degrees before being switched off. When the current is zero the detected terminal voltage will be the induced back-EMF. For instance, when phases A and

B are conducted, the current of phase C is zero $i_C = 0$, which generates a measurement window. During this interval, the back-EMF of the floating phase C can be directly measured with $v_{CN} = e_C$.

Moreover, the back-EMF measurement can be implemented by detecting terminal voltage [4]–[7], phase voltage [8], [9], line-to-line voltage [10], and third harmonic voltage [15]. Among them, the terminal voltage measurement method is the simplest and most effective one and is widely used. Therefore, in this chapter, the ZCP sensorless control based on measuring terminal voltage will be described. Third harmonic voltage will be introduced in Chapter 11.

10.3 ZCP Detection with PWM

In Section 10.2, the basic principle of the ZCD is introduced. Furthermore, in order to regulate the motor speed and current, a PWM strategy should be employed as shown in Figure 10.4.

As shown in Figure 10.4, the output of the speed controller is the reference input of the current controller i_{dc}^*. By adjusting the PWM duty ratio, the feedback current i_{dc} can track the reference input i_{dc}^* and the feedback speed ω_r can track the reference input ω_r^*. In this case, the PWM effect should be considered during the ZCP detection of back-EMF. Hence, the ZCP detection with PWM is introduced in this section.

10.3.1 PWM Patterns

As introduced in Chapter 1, in a typical six-step control scheme, a drive cycle is divided into six sectors. In each 60° sector, two of the three phases are conducting.

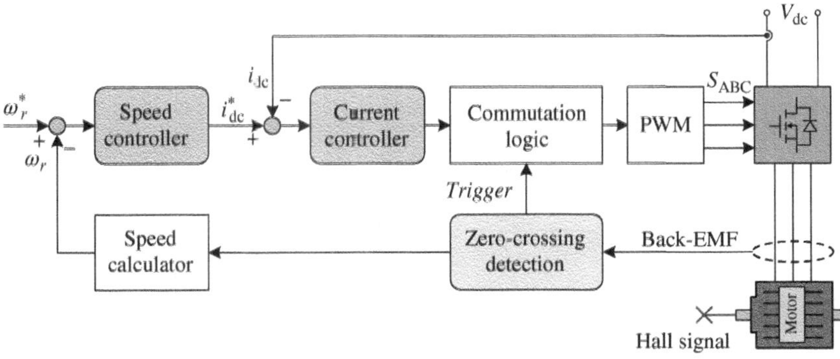

Figure 10.4 Block diagram of BLDC drive with PWM.

Usually, one of the conducting phase is controlled by the PWM signal and the other phase keeps constant on. Then, as shown in Figure 10.5, four PWM switching patterns can be used in a BLDC drive. The combination sequence of PWM and constant ON signals are defined as H-PWM-L-ON, H-ON-L-PWM, PWM-ON, and ON-PWM. Details are shown as follows.

1) For the H-PWM-L-ON pattern in Figure 10.5a, the gate signal for the upper switch S1 is controlled by the PWM signal in sectors I and II; the gate signal for the lower switch S2 keeps constant on in sectors IV and V.
2) For the H-ON-L-PWM pattern in Figure 10.5b, the PWM signal and constant-on signal are reversed, i.e. the gate signal for the upper switch S1 keeps on in sectors I and II; the gate signal for the lower switch S2 is controlled by the PWM signal in sectors IV and V.
3) For the PWM-ON pattern in Figure 10.5c, the former 60° is controlled by the PWM signal and then the latter 60° keeps constant on.
4) On the contrary, for the ON-PWM pattern in Figure 10.5d, the former 60° keeps constant on and then the latter 60° is controlled by the PWM signal.

Of these, the H-PWM-L-ON pattern is most commonly used in many applications; hence, the analysis in this section will be based on it.

However, PWM will introduce high-frequency noises in the measured voltage signals. An analog low-pass filter (LPF) circuit is therefore required to remove these noises [7]. LPF circuits will inevitably lead to commutation delay that

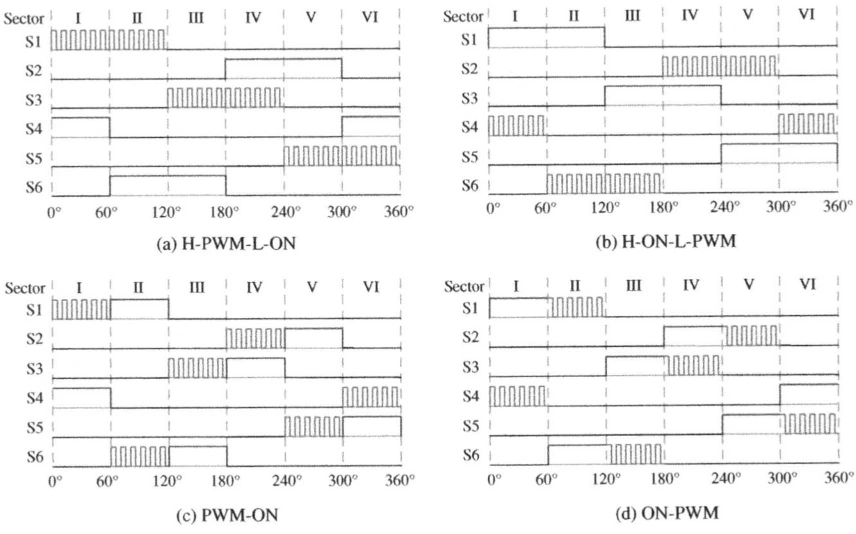

Figure 10.5 Schematic of PWM switching patterns.

increases with the motor speed. Although some improved LPF design methods and compensation methods for commutation errors have been investigated in the existing literature [16], [17], the system complexity and flexibility will be influenced. To suppress the PWM noise, the sampling of back-EMF can operate in synchronization with the PWM generation [7]. In this way, the back-EMF can be extracted without LPFs, and the adverse influences of the LPF can be avoided.

10.3.2 Back-EMF Sensing

In this section, based on the H-PWM-L-ON pattern, the relationship between back-EMF and the terminal voltage is analyzed to demonstrate the principle of back-EMF sensing with PWM control. First, consider that phase A and phase B are conducting and phase C is floating. According to Figure 10.5a, the upper switch of phase A is pulse-width modulated. The lower switch of phase B is kept on before commutation. The terminal voltage v_{CN} is sampled when the upper switch is off.

10.3.2.1 Sinusoidal Back-EMF

For sinusoidal back-EMF-type BLDC motors, the sum of three-phase back-EMFs equals zero, i.e. $e_A + e_B + e_C = 0$. In [4], the back-EMF is synchronously sampled on the floating phase during the PWM-off state as shown in Figure 10.6.

The voltage equation of phase A can be derived as:

$$v_{NG} = 0 - (i_A R + p i_A L + e_A) \tag{10.2}$$

For phase B, the voltage equation can be derived as:

$$v_{NG} = (i_B R + p i_B L - e_B) \tag{10.3}$$

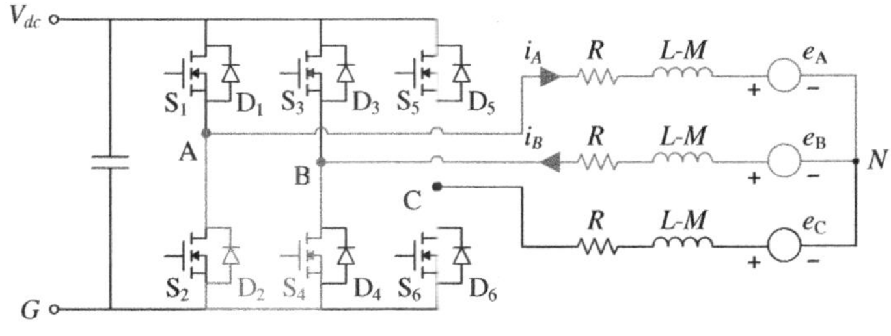

Figure 10.6 Back-EMF sensing at PWM-off state.

By adding (10.2) and (10.3), there is:

$$v_{NG} = -(e_A + e_B)/2 \tag{10.4}$$

Assuming a balanced three-phase system, i.e. $e_A + e_B + e_C = 0$, the neutral voltage can be simplified as:

$$v_N = \frac{e_C}{2} \tag{10.5}$$

Then, the terminal voltage of phase C is given by:

$$v_{CG} = \frac{3}{2}e_C \tag{10.6}$$

Clearly, it can be seen that during the PWM-off time, the terminal voltage of the floating phase is proportional to the back-EMF without high-frequency switching noises. However, this method requires a minimum PWM-off time of a few microseconds to measure the back-EMF. This requirement limits the maximum PWM duty cycle, resulting in a lower utilization of the dc-bus voltage. In this case, an improved method is proposed in [5] to sense the back-EMF at the PWM-on state, which removes the limitation of a maximum duty cycle. The back-EMF sensing technique at the PWM-on state is shown in Figure 10.7.

For phase A, the voltage equation expressed as:

$$v_{NG} = V_{dc} - (i_A R + pi_A L + e_A) \tag{10.7}$$

For phase B, the voltage equation expressed as:

$$v_{NG} = (i_B R + pi_B L - e_B) \tag{10.8}$$

Combining (10.7) and (10.8) gives:

$$v_N = \frac{V_{dc}}{2} - \frac{e_A + e_B}{2} \tag{10.9}$$

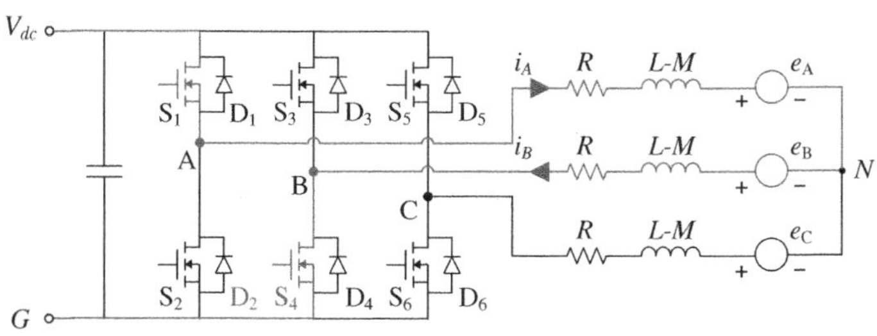

Figure 10.7 Back-EMF sensing at PWM-on state.

Assuming a balanced three-phase system, $e_A + e_B + e_C = 0$, from (10.9), there is:

$$v_N = \frac{V_{dc}}{2} + \frac{e_C}{2} \tag{10.10}$$

Hence, the terminal voltage is derived as:

$$v_{CG} = \frac{V_{dc}}{2} + \frac{3}{2}e_C \tag{10.11}$$

10.3.2.2 Trapezoidal Back-EMF

For trapezoidal back-EMFs, the sum of the back-EMFs in both upper and lower side phases equals zero, i.e. $e_B + e_C = 0$. As in Section 10.3.2.1, the phase C terminal voltage can also be sensed at the PWM-on and PWM-off states [6], [7].

At the PWM-off state, the terminal voltage is:

$$v_{CG} = e_C \tag{10.12}$$

At the PWM-on state, the terminal voltage is:

$$v_{CG} = \frac{V_{dc}}{2} + e_C \tag{10.13}$$

The relationship between back-EMF and terminal voltage for the phase C floating case is summarized in Table 10.1. From Table 10.1, although there is a different gain on the sensed back-EMFs between sinusoidal and trapezoidal cases, the detection condition for both cases is the same. At low speed and light load conditions, the duty cycle is low so that the duration of the PWM-off state is long, and it is preferable to detect the back-EMF at the PWM-off state [4]. While at high speed or high load, the PWM duty cycle increases, the PWM-on duration increases, and the PWM-off duration decreases. It is more effective to detect the back-EMF at the PWM-on state [5].

Table 10.1 Summary of back-EMF sensing at phase C terminal voltage.

	PWM on	PWM off
Sinusoidal	$V_{dc}/2 + 1.5e_C$	$1.5e_C$
Trapezoidal	$V_{dc}/2 + e_C$	e_C
ZCP detection condition	$V_{dc}/2$	0

10.4 ZCP Deviation and Solution

ZCP deviation is one of the non-ideal factors that occurs in sensorless operation. Ideally, the ZCP is located in the middle of each 60° sector, and the commutation instant is located 30° behind the ZCP. When the ZCP position deviates, it will lead to commutation errors. The most common reason of ZCP deviation is the phase delay caused by LPF circuits as shown in Figure 10.8a. In [16], [17], the method to select a suitable LPF cut-off frequency and the phase delay compensation were introduced. Apart from the phase delay, resistance tolerance of back-EMF measurement circuits [18], Figure 10.8b, and the machine parameter asymmetry [18], Figure 10.8c, will also lead to unbalanced ZCPs. Therefore, their influence on ZCP deviation will be discussed in this section.

10.4.1 Horizontal Deviation Caused by Machine Parameter Asymmetry

The drive topology of a high-speed motor is shown in Figure 10.9a. The BLDC motor can be modeled as:

$$\begin{cases} v_{AG} - v_{NG} = R_A i_A + L_A \dfrac{d}{dt} i_A + e_A \\ v_{BG} - v_{NG} = R_B i_B + L_B \dfrac{d}{dt} i_B + e_B \\ v_{CG} - v_{NG} = R_C i_C + L_C \dfrac{d}{dt} i_C + e_C \end{cases} \tag{10.14}$$

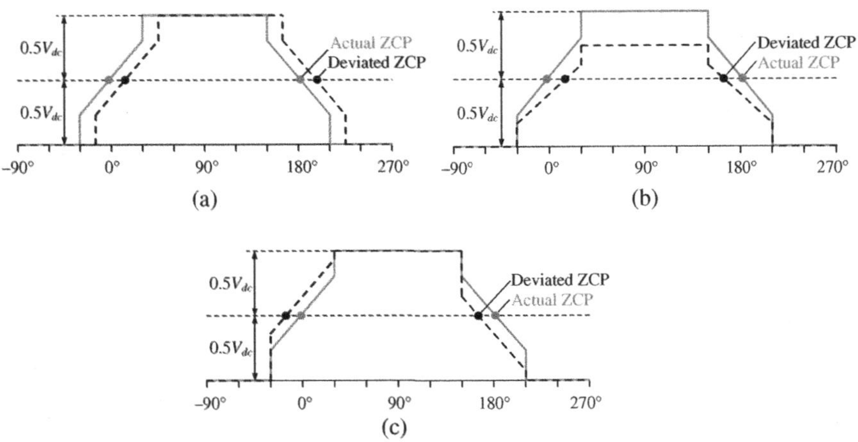

Figure 10.8 Illustration of ZCP deviation in terminal voltage measurement. (a) ZCP deviation caused by LPF circuits. (b) ZCP deviation caused by dc-link voltage drop or resistance tolerance. (c) ZCP deviation caused by asymmetric machine parameters. (*Source:* Adapted from [18].)

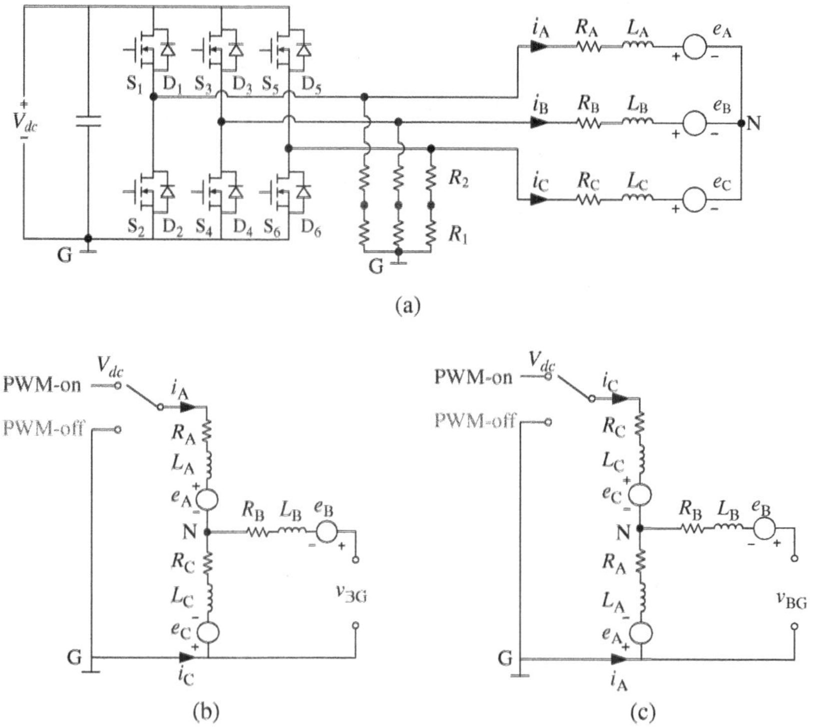

Figure 10.9 Drive topology of BLDC motor. (a) Overall drive circuit. (b) Equivalent circuit in sector II. (c) Equivalent circuit in sector V. (*Source:* [18]/IEEE.)

where v_{XG} is the terminal voltage, v_{NG} is the machine neutral voltage, i_X is the phase current, e_X is the phase back-EMF, and R_X is the stator resistance. L_X is the stator inductance combining self-inductance L'_X and mutual-inductance M, i.e. $L_X = L'_X = M$. The subscript X denotes phase A, B, or C. Here, the resistances and self-inductances in three phases are assumed to be unbalanced. Considering a high-speed BLDC motor with sinusoidal phase back-EMFs, the back-EMFs can be expressed as:

$$\begin{bmatrix} e_A \\ e_B \\ e_C \end{bmatrix} = \omega_r \psi_m \begin{bmatrix} \sin(\theta_r) \\ \sin\left(\theta_r - \frac{2}{3}\pi\right) \\ \sin\left(\theta_r + \frac{2}{3}\pi\right) \end{bmatrix} \tag{10.15}$$

where ψ_m, ω_r, and θ_r are the flux-linkage, electrical speed, and rotor position, respectively.

For the six-step commutation scheme, an electrical cycle is divided into six sectors. In each 60° sector, only two of the three phases are energized. The floating phase contains the back-EMF information, which can be used for commutation detection. In position sensorless BLDC motor drives, the commutation signal can be obtained by lagging 30° after the ZCP of the back-EMF.

10.4.1.1 Horizontal Deviation of EMF Waveforms

In this section, the influence of machine parameter asymmetry on ZCP detection will be analyzed. Two opposite sectors, II and V as shown in Figures 10.9b,c and 10.10, are illustrated first and other sectors can be derived similarly.

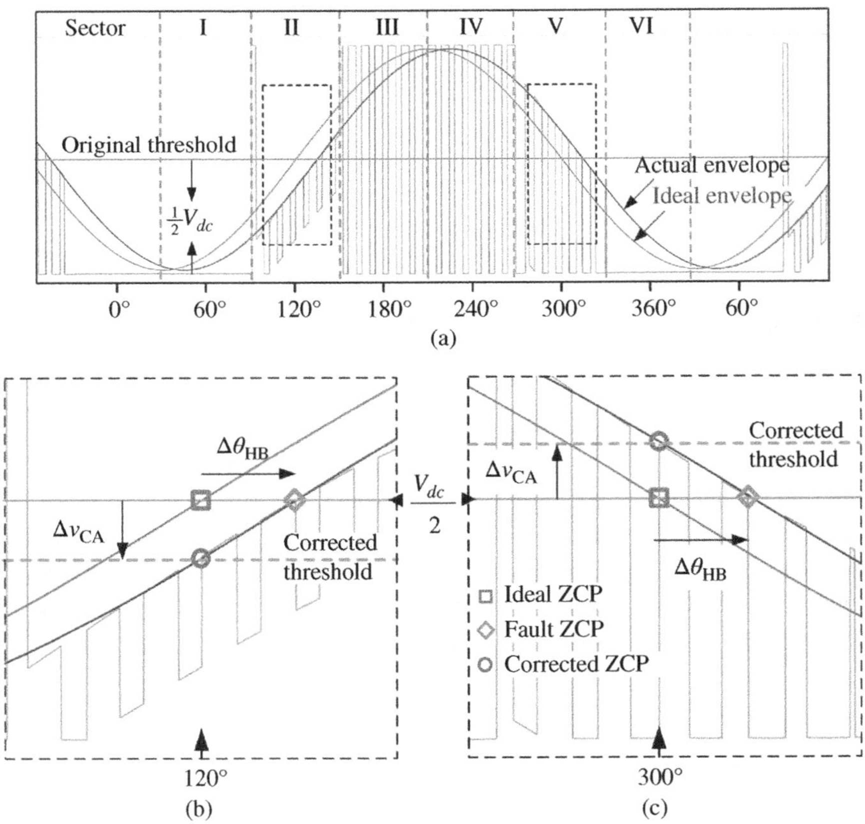

Figure 10.10 Schematic diagram of terminal voltage in phase B considering machine parameter asymmetry. (a) Terminal voltage, ideal and actual envelopes. (b) Zoomed window for sector II. (c) Zoomed window for sector V. (*Source:* [18]/IEEE.)

The BLDC model can be transformed into the Laplace domain as:

$$\begin{cases} V_{AG}(s) - V_{NG}(s) = (R_A + sL_A) \cdot I_A(s) + E_A(s) \\ V_{BG}(s) - V_{NG}(s) = (R_B + sL_B) \cdot I_B(s) + E_B(s) \\ V_{CG}(s) - V_{NG}(s) = (R_C + sL_C) \cdot I_C(s) + E_C(s) \end{cases} \tag{10.16}$$

where s denotes the Laplace operator. In sectors II and V, phases A and C are conducting and phase B is floating, and it yields $I_A(s) + I_C(s) = 0$ and $I_B(s) = 0$. The terminal voltage $V_{BG}(s)$ can be calculated as:

$$V_{BG}(s) = V_{NG}(s) + E_B(s) = \frac{1}{2}(V_{AG}(s) + V_{CG}(s)) + \frac{3}{2}E_B(s) - \frac{1}{2}\Delta V_H(s) \tag{10.17}$$

$$\Delta V_H(s) = [V_{AG}(s) - V_{CG}(s) - (E_A(s) - E_C(s))]\frac{\Delta R_{AC} + s\Delta L_{AC}}{\Sigma R_{AC} + s\Sigma L_{AC}} \tag{10.18}$$

where $\Delta V_H(s)$ is related to asymmetric parameters, and $\Delta R_{AC} = R_A - R_C$, $\Delta L_{AC} = L_A - L_C$, $\Sigma R_{AC} = R_A + R_C$, and $\Sigma L_{AC} = L_A + L_C$. Then, transforming $V_{BG}(s)$ back to the time domain, it yields:

$$v_{BG}(t) = \mathcal{L}^{-1}[V_{BG}(s)] = \frac{1}{2}(v_{AG}(t) + v_{CG}(t)) + \frac{3}{2}e_B(t) - \frac{1}{2}\mathcal{L}^{-1}[\Delta V_H(s)] \tag{10.19}$$

where $\mathcal{L}^{-1}$ indicates the inverse Laplace transform. The additional term $-\frac{1}{2}\mathcal{L}^{-1}[\Delta V_H(s)]$ is the voltage shift caused by the asymmetric parameters. The terminal voltage $v_{BG}(t)$ will be calculated in sectors II and V, respectively.

Terminal Voltage in Sector II

During sector II as shown in Figure 10.9b, phases A and C are energized and phase B is floating. This floating phase B containing the back-EMF signal information can be used for commutation detection. Power device S_1 is controlled by a chopper signal and S_6 keeps conducting in this sector. As the motor speed and PWM duty ratio increase, the PWM-on duration increases and the PWM-off duration decreases. To guarantee the detection of ZCP at the high-speed region, this described analysis is based on back-EMF sampling at PWM-on time. Then, $v_{AG} = V_{dc}$, $v_{CG} = 0$, and $\mathcal{L}^{-1}[\Delta V_H(s)]$ can be calculated accordingly. The terminal voltage envelope of the floating phase $v_{BG}(t)$ in sector II can be expressed as:

$$v_{BG}(t) = \frac{1}{2}V_{dc} + \frac{3}{2}e_B(t) + \Delta v_{AC} \tag{10.20}$$

where the voltage shift Δv_{AC} is:

$$\Delta v_{\mathrm{AC}} = -\frac{1}{2}\left(V_{dc} - \sqrt{3}\omega_r \psi_m\right)\frac{L_{\mathrm{A}} - L_{\mathrm{C}}}{L_{\mathrm{A}} + L_{\mathrm{C}}} \tag{10.21}$$

As shown in Figure 10.10b, with symmetrical parameters, the dc offset of terminal voltage is $\frac{1}{2}V_{dc}$. However, when the machine parameter asymmetry is taken into account, the additional term Δv_{AC} moves the terminal voltage as shown in Figure 10.10b. If $\frac{1}{2}V_{dc}$ is still selected as the zero-crossing threshold, an angle error will be caused.

Terminal Voltage in Sector V

As opposed to sector II, as shown in Figure 10.9c, S_5 is controlled by a chopper signal and S_2 keeps conducting in sector V. During the PWM-on period, $v_{\mathrm{AG}} = 0$, $v_{\mathrm{CG}} = V_{dc}$, and $\mathcal{L}^{-1}[\Delta V_{\mathrm{H}}(s)]$ can be calculated accordingly. Then, the terminal voltage envelope of floating phase $v_{\mathrm{BG}}(t)$ in sector V can be expressed as:

$$v_{\mathrm{BG}}(t) = \frac{1}{2}V_{dc} + \frac{3}{2}e_{\mathrm{B}}(t) + \Delta v_{\mathrm{CA}} \tag{10.22}$$

where the voltage shift Δv_{CA} is:

$$\Delta v_{\mathrm{CA}} = -\frac{1}{2}\left(V_{dc} - \sqrt{3}\omega_r \psi_m\right)\frac{L_{\mathrm{C}} - L_{\mathrm{A}}}{L_{\mathrm{C}} + L_{\mathrm{A}}} \tag{10.23}$$

As shown in Figure 10.10c, the terminal voltage envelope moves by Δv_{CA}. From (10.20) and (10.22), it can be found that the voltage errors Δv_{AC} and Δv_{CA} exist in the terminal voltage of the floating phase during sectors II and V, respectively. Also, it yields $\Delta v_{\mathrm{AC}} + \Delta v_{\mathrm{CA}} = 0$, since the motor speed will not change dramatically in steady-state operation.

10.4.1.2 Influence of Machine Parameter Asymmetry on ZCP Detection

The terminal voltage envelopes in sectors II and V have been analyzed. The results of other sectors can also be obtained in a similar way. Table 10.2 summarizes the voltage shifts, position errors, and the back-EMF direction of the floating phase, i.e. rising (R) or falling (F). The symbols +, −, and ~ denote the positive driven, negative driven, and floating phases, respectively. The voltage shift caused by parameter asymmetry is defined as:

$$\Delta v_{\mathrm{XY}} = -\frac{1}{2}\left(V_{\mathrm{dc}} - \sqrt{3}\omega_r \psi_m\right)\cdot\frac{L_{\mathrm{X}} - L_{\mathrm{Y}}}{L_{\mathrm{X}} + L_{\mathrm{Y}}} \tag{10.24}$$

From the above analysis, asymmetric machine parameters result in the movement of the terminal voltage envelope in the floating phase. As shown in Figure 10.10, the envelope moves down by Δv_{CA} in sector II and moves up by Δv_{CA}

Table 10.2 Terminal voltage shifts and ZCP position errors.

Sector	Mode	EMF (F/R)	Δv_H	Δv_V	$\Delta\theta$
I	A+B−C~	F	Δv_{AB}	Δv_{VC}	$\Delta\theta_{HC}+\Delta\theta_{VC}$
II	A+B~C−	R	$-\Delta v_{CA}$	Δv_{VB}	$\Delta\theta_{HB}-\Delta\theta_{VB}$
III	A~B+C−	F	Δv_{BC}	Δv_{VA}	$\Delta\theta_{HA}+\Delta\theta_{VA}$
IV	A−B+C~	R	$-\Delta v_{AB}$	Δv_{VC}	$\Delta\theta_{HC}-\Delta\theta_{VC}$
V	A−B~C+	F	Δv_{CA}	Δv_{VB}	$\Delta\theta_{HB}+\Delta\theta_{VB}$
VI	A~B−C+	R	$-\Delta v_{BC}$	Δv_{VA}	$\Delta\theta_{HA}-\Delta\theta_{VA}$

in sector V. In the vertical direction, the envelope moves oppositely in sectors II and V. Alternatively, both the vertical movements in sectors II and V can be unified into a horizontal movement of $\Delta\theta_{HB}$. Hence, influenced by the machine parameter asymmetry, the terminal voltage envelopes move horizontally, leading to ZCP deviations.

The envelope around each ZCP can be regarded as a straight line with a slope of $\frac{3}{2}\omega_r\psi_m$. The voltage shifts have a linear relationship with ZCP angle errors as:

$$[\Delta v_{AB}\ \Delta v_{BC}\ \Delta v_{CA}] = \frac{3}{2}\omega_r\psi_m \cdot [\Delta\theta_{HC}\ \Delta\theta_{HA}\ \Delta\theta_{HB}] \tag{10.25}$$

As shown in Table 10 2, it should be noted that Δv_{AB}, Δv_{BC}, and Δv_{CA} denote the voltage errors for sectors I, III, and V, respectively, and are opposite to the voltage errors in sectors II, IV, and VI. Also, as shown in (10.24), the sign of the voltage error depends on the relation of L_X and L_Y. According to (10.25), it will further determine the sign of the angle error (lag error is defined positive and lead error is defined as negative). The sum of the three voltage shifts can be calculated as:

$$\Delta v_{AB} + \Delta v_{BC} + \Delta v_{CA} = \frac{1}{2}\left(V_{dc} - \sqrt{3}\omega_r\psi_m\right) \cdot \underbrace{\frac{L_A - L_B}{L_A + L_B}}_{\Delta L_{AB}} \cdot \underbrace{\frac{L_B - L_C}{L_B + L_C}}_{\Delta L_{BC}} \cdot \underbrace{\frac{L_C - L_A}{L_C + L_A}}_{\Delta L_{CA}} \tag{10.26}$$

It should be noted that the three inductance-related terms, ΔL_{AB}, ΔL_{BC}, and ΔL_{CA}, are all very small. Their product is actually close to zero. Thus, (10.26) can be written as:

$$\Delta v_{AB} + \Delta v_{BC} + \Delta v_{CA} \approx 0 \tag{10.27}$$

Additionally, due to the linear relation of (10.25), the sum of three ZCP angle errors is also approximately zero.

$$\Delta\theta_{HA} + \Delta\theta_{HB} + \Delta\theta_{HC} \approx 0 \tag{10.28}$$

The commutation instant is obtained by lagging 30° after every ZCP. Without an accurate ZCP, the commutation instant will not be identified. Therefore, machine parameter asymmetry has a negative impact on position sensorless BLDC drives.

10.4.2 Vertical Deviation Caused by Resistance Tolerance of RVD

10.4.2.1 Resistance Tolerance of Measurement Circuit

The resistance tolerance of the measurement circuit is another cause of ZCP deviation. The maximum terminal voltage reaches the dc-link level. However, the on-chip analog-to-digital (AD) converter can only sample a low-voltage signal, e.g. 0–3 V. Thus, the resistance voltage divider (RVD) is usually employed as the transducer for terminal voltage sampling. The resistance precision will definitely affect the commutation accuracy. In this section, the commutation error caused by resistance tolerance is analyzed. As shown in Figure 10.9, the AD converter samples the terminal voltage signal on the low side resistor of RVD. The scale factor of the resistance can be designed as:

$$\frac{R_1}{R_1 + R_2} = \frac{V_{\mathrm{AD}}}{V_{dc}} \tag{10.29}$$

where V_{AD} denotes the maximum sampling voltage, and R_1 and R_2 are the low and high side resistance of RVD, respectively. In practice, the actual resistance has a slight difference from the nominal value. Hence, the equivalent gain K_{R} is:

$$K_{\mathrm{R}} = \frac{R_1'}{R_1' + R_2'} \cdot \frac{R_1 + R_2}{R_1} \tag{10.30}$$

where superscript $'$ denotes the value taking resistance tolerance into account. Then, the acquired phase B terminal voltage can be expressed as:

$$v_{\mathrm{BG}}' = v_{\mathrm{BG}} \cdot K_{\mathrm{R}} = \frac{1}{2} K_{\mathrm{R}} V_{\mathrm{dc}} + \frac{3}{2} K_{\mathrm{R}} e_{\mathrm{B}} \tag{10.31}$$

10.4.2.2 Voltage Shift and ZCP Angle Errors

As shown in (10.32), the original dc offset $\frac{1}{2} V_{dc}$ is shifted to $\frac{1}{2} K_{\mathrm{R}} V_{dc}$ due to the resistance tolerance of RVD, which can be expressed as:

$$\Delta v_{\mathrm{VB}} = \frac{1}{2}(K_{\mathrm{R}} - 1) V_{dc} \tag{10.32}$$

Similar to the analysis method in (10.25), the ZCP angle errors and voltage shifts shown in Table 10.2 have a linear relation as:

$$[\Delta v_{\mathrm{VA}}\ \Delta v_{\mathrm{VB}}\ \Delta v_{\mathrm{VC}}] = \frac{3}{2} \omega_r \psi_m [\Delta\theta_{\mathrm{VA}}\ \Delta\theta_{\mathrm{VB}}\ \Delta\theta_{\mathrm{VC}}] \tag{10.33}$$

Therefore, as the resistance value deviates from the nominal one, the voltage error increases. Let us consider an extreme situation: R_1' and R_2' have upper and lower limit values, i.e. $R_1' = (1+\sigma)R_1$ and $R_2' = (1-\sigma)R_2$, where σ is the tolerance ratio. Finally, the ZCP angle error can be calculated as:

$$\Delta\theta_{VB} = \frac{1}{\sqrt{3}\omega_{pu}} \cdot \left(\frac{1+\sigma}{(1+\sigma) + (1-\sigma)\left(\frac{V_{dc}}{V_{AD}} - 1\right)} \cdot \frac{V_{dc}}{V_{AD}} - 1 \right) \tag{10.34}$$

where $\omega_{pu} = \omega_r/\omega_{base}$ denotes the per-unit value of speed, the base-speed ω_{base} can be defined as $\sqrt{3}\omega_{base}\psi_m = V_{dc}$.

Obviously, the voltage acquisition accuracy is limited by the resistance tolerance. The resistance will change with the temperature. Especially for those applications in which the motor and the drive boards are integrated within a compact space, the temperature will change significantly during operation. Besides, as the resistor becomes old the resistance will change. For high-voltage application, due to voltage limit, more than two resistors are cascaded to reduce the voltage drop on each resistor. This situation is even more complicated. If the parameter design in the software algorithm is solely based on the nominal resistance, it will definitely lead to commutation errors.

Figure 10.11 shows analysis results of the ZCP angle error. Some popular resistance tolerances from 0.1% to 10% are analyzed. As can be seen, the ZCP angle error increases as the resistance precision decreases.

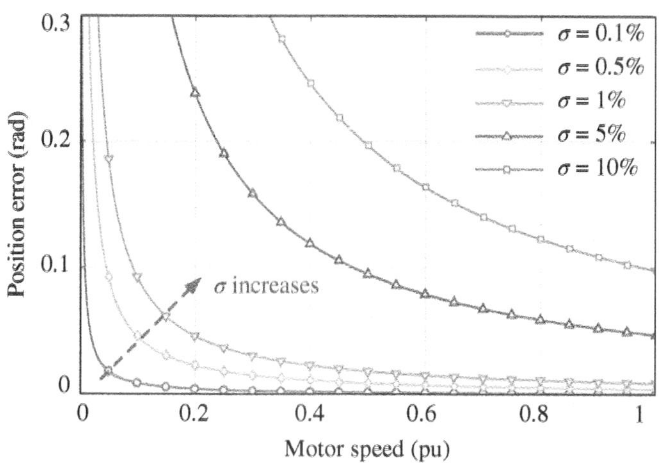

Figure 10.11 Analysis result of ZCP angle error caused by vertical shift of terminal voltage, V_{AD} = 3 V, V_{dc} = 15 V. (*Source:* [18]/IEEE.)

10.4.3 Adaptive Threshold Correction

In this section, an adaptive threshold correction (ADC) scheme [18] is presented. Figure 10.12 illustrates the implementation of the overall scheme. The terminal voltage signals are sampled with RVD by the on-chip ADC. Then, an adaptive threshold correction loop is utilized. In this algorithm, six independent zero-crossing thresholds are used for ZCP detection. These six thresholds are adjusted during operation in real time. Two CPU timers are utilized. Timer A is used for time interval measurement of ZCPs. Six latest measurement results of ZCP time intervals, denoted as $t_{12}...t_{61}$, are stored in a buffer. Timer B is used to trigger commutation 30° after each ZCP.

10.4.3.1 Vertical Correction

Figure 10.13 shows the vertical shift of the back-EMF envelope caused by the resistance tolerance of measurement circuit. Figure 10.14 shows the magnified area of the triangles in Figure 10.13. The dashed and solid lines denote the back-EMF envelopes before and after each shift, respectively. It is assumed that $K_R > 1$ and the back-EMF envelope moves upward. Ideally, the phase B ZCPs should locate at the intersection of the phase B envelope e_B and the half dc-link voltage $\frac{1}{2}V_{dc}$, i.e. 120° in sector II and 300° in sector V. However, when the resistance ratio changes, the envelope moves upward by Δv_{VB} to $\hat{e}_B$, and a fault ZCP will be detected with the original threshold $\frac{1}{2}V_{dc}$.

As shown in Figure 10.14, the purpose of the vertical correction is to superpose a common mode bias u_{VB} to $\frac{1}{2}V_{dc}$, i.e. $\frac{1}{2}V_{dc} + v_{VB}$. This new threshold is adopted for both sectors II and V. The angle error ε_{VB} can be extracted in real time. By controlling ε_{VB} to zero, the common mode bias u_{VB} will converge to Δv_{VB}. Then, even

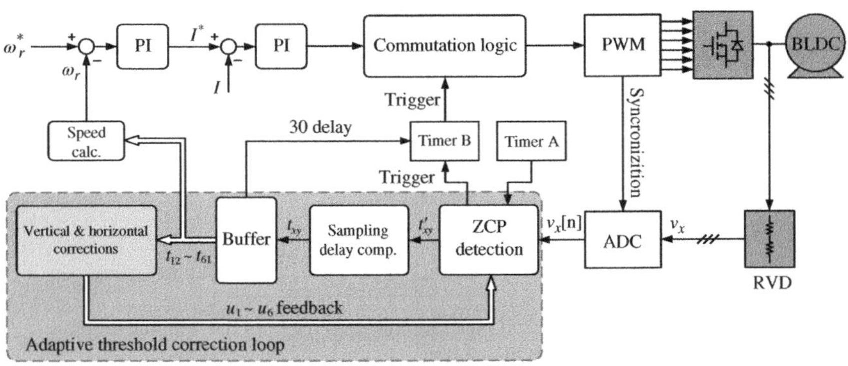

Figure 10.12 Schematic diagram of sensorless control system with adaptive threshold correction strategy. (*Source:* [18]/IEEE.)

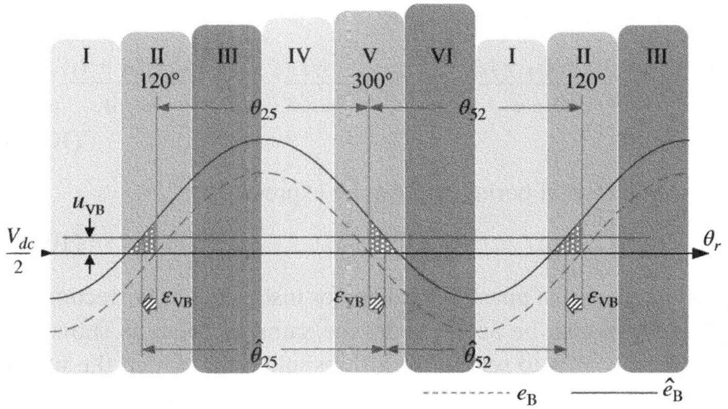

Figure 10.13 Schematic diagram of vertical deviation of back-EMF envelopes caused by resistance tolerance of measurement circuit. (*Source:* [18]/IEEE.)

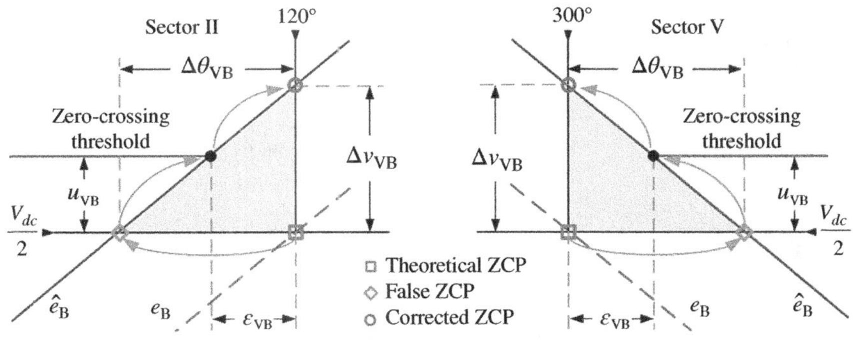

Figure 10.14 Vertical correction process of zero-crossing threshold. (*Source:* [18]/IEEE.)

if the vertical shift exists, the ZCP will be corrected back to the ideal position, i.e. 120° and 300°.

As shown in Figure 10.13, the phase ZCP interval $\hat{\theta}_{25}$ (II to V) and $\hat{\theta}_{52}$ (V to II) can be expressed as:

$$\hat{\theta}_{25} = \theta_{25} + 2\varepsilon_{VB},\ \hat{\theta}_{52} = \vartheta_{52} - 2\varepsilon_{VB} \tag{10.35}$$

where the superscript ^ denotes measurement values. It is obvious that $\theta_{25} = \theta_{52} = \pi$ in (10.35); then the angle errors can be deduced as:

$$\varepsilon_{VB} = \frac{1}{4}\left(\hat{\theta}_{25} - \hat{\vartheta}_{52}\right) \tag{10.36}$$

In practice, $\hat{\theta}_{25}$ and $\hat{\theta}_{52}$ can be calculated by:

$$\hat{\theta}_{25} = \frac{t_{25}}{t_{\text{period}}} \cdot 2\pi = \frac{t_{23} + t_{34} + t_{45}}{t_{\text{period}}} \cdot 2\pi, \hat{\theta}_{52} = \frac{t_{52}}{t_{\text{period}}} \cdot 2\pi = \frac{t_{56} + t_{61} + t_{12}}{t_{\text{period}}} \cdot 2\pi \tag{10.37}$$

where t_{period} is the fundamental period and can be expressed as:

$$t_{\text{period}} = t_{12} + t_{23} + \ldots + t_{61} \tag{10.38}$$

Here, the correction process of phase B is taken, for instance. The correction for the other two phases can also be achieved by the same method, as shown in Figure 10.15. The main goal is to search for a threshold, with which the upper and lower half-cycles have the same time duration. With higher gains, the correction controller can achieve higher convergence speed.

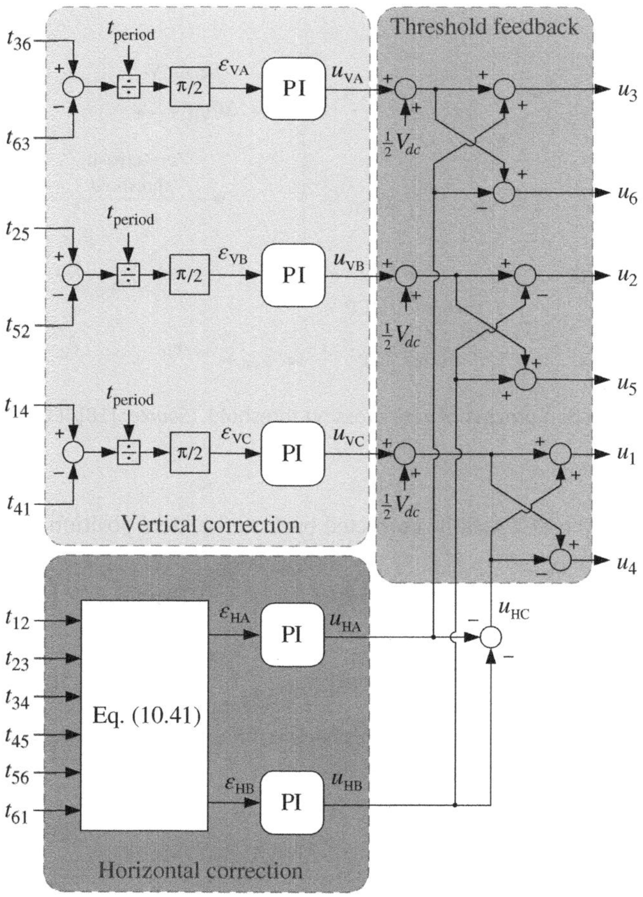

Figure 10.15 Schematic of adaptive threshold correction including vertical and horizontal corrections. (*Source:* [18]/IEEE.)

10.4.3.2 Horizontal Correction

Figure 10.16 shows the horizontal shift of the back-EMF envelope caused by machine parameter asymmetry. Figure 10.17 shows the magnified area of the triangles of phase B in Figure 10.16. Influenced by asymmetric parameters, the envelope moves downward in sector II and moves upward in sector V, which can be unified into a right movement of $\Delta\theta_{HB}$. Thus, faulty ZCPs will be detected within the original threshold $\frac{1}{2}V_{dc}$.

As shown in Figure 10.17, the horizontal correction process is similar to the vertical correction process. The difference is that differential-mode biases are used, which means $\frac{1}{2}V_{dc} - u_{HB}$ is used for sector II and $\frac{1}{2}V_{dc} + u_{HB}$ is used for sector V as new thresholds. Then, by controlling ε_{HB} to zero, the differential-mode bias u_{HB} will converge to Δv_{CA} and the ZCP will be corrected. For all three phases, only

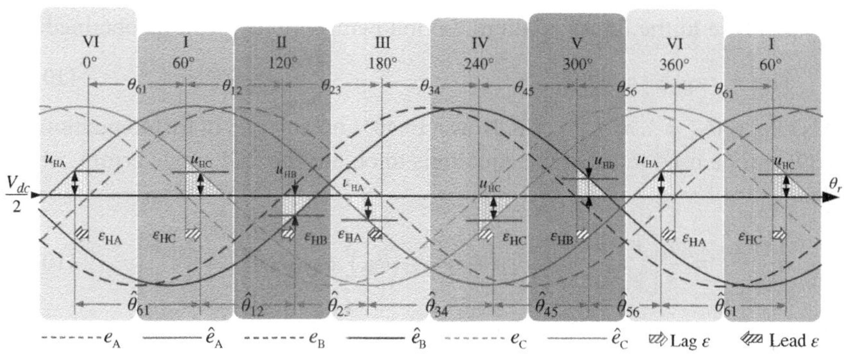

Figure 10.16 Schematic diagram of horizontal deviation of back-EMF due to machine parameter asymmetry. (*Source:* [18]/IEEE.)

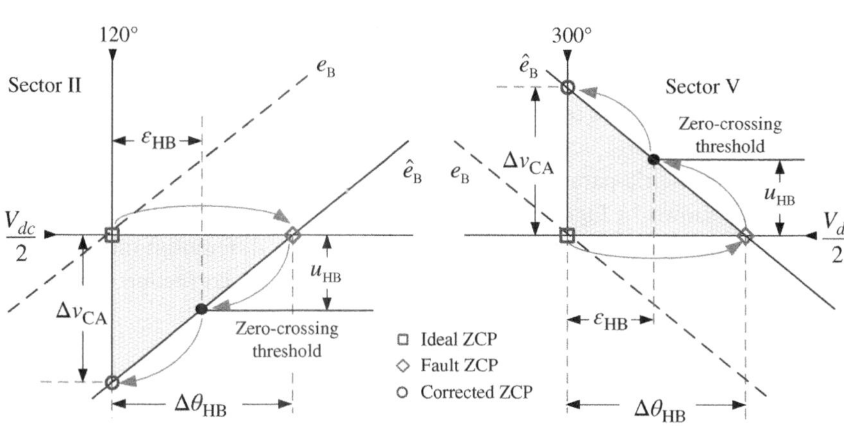

Figure 10.17 Horizontal correction process of zero-crossing threshold. (*Source:* [18]/IEEE.)

two of the three terms u_{HA} and u_{HB} are required, and the third one u_{HC} can be calculated directly from $u_{\mathrm{HA}} + u_{\mathrm{HB}} + u_{\mathrm{HC}} = 0$ as a result of (10.27).

According to (10.28), the sum of three ZCP angle errors is approximately zero. Hence, there are two advanced phases and one lagged phase, or one advanced phase and two lagged phases. In Figure 10.16, phase A is assumed advanced and the other two phases are assumed lagged. The sector ZCP intervals $\hat{\theta}_{12}...\hat{\theta}_{61}$ can be modeled as:

$$\begin{cases} \hat{\theta}_{12}\left(\hat{\theta}_{45}\right) = \theta_{12}(\theta_{45}) + \varepsilon_{\mathrm{HB}} - \varepsilon_{\mathrm{HC}} \\ \hat{\theta}_{23}\left(\hat{\theta}_{56}\right) = \theta_{23}(\theta_{56}) + \varepsilon_{\mathrm{HA}} - \varepsilon_{\mathrm{HB}} \\ \hat{\theta}_{34}\left(\hat{\theta}_{61}\right) = \theta_{34}(\theta_{61}) + \varepsilon_{\mathrm{HC}} - \varepsilon_{\mathrm{HA}} \end{cases} \tag{10.39}$$

The term in the parentheses indicates they have the same value, for instance $\hat{\theta}_{12} = \hat{\theta}_{45}$. Due to the linear relation, the following equation can be obtained:

$$\varepsilon_{\mathrm{HA}} + \varepsilon_{\mathrm{HB}} + \varepsilon_{\mathrm{HC}} = 0 \tag{10.40}$$

Also $\theta_{12} = \theta_{23} = ... = \theta_{61} = \pi/3$. Now, (10.40) and any two of three equations in (10.39) can be used to solve the equations to obtain the ZCP angle errors, which yields:

$$\begin{bmatrix} \varepsilon_{\mathrm{HA}} \\ \varepsilon_{\mathrm{HB}} \end{bmatrix} = \frac{1}{3}\begin{bmatrix} 1 & 2 & 0 \\ 2 & 0 & 1 \end{bmatrix}\left[\hat{\theta}_{12}\left(\hat{\theta}_{45}\right) \ \hat{\theta}_{23}\left(\hat{\theta}_{56}\right) \ \hat{\theta}_{34}\left(\hat{\theta}_{61}\right)\right]^{\mathrm{T}} - \frac{\pi}{3}\begin{bmatrix} 1 \\ 1 \end{bmatrix} \tag{10.41}$$

In practice, all the angle terms can be calculated by time interval signals obtained from the CPU timer as:

$$\begin{bmatrix} \varepsilon_{\mathrm{HA}} \\ \varepsilon_{\mathrm{HB}} \end{bmatrix} = \frac{1}{3} \cdot \frac{2\pi}{t_{\mathrm{period}}} \cdot \begin{bmatrix} 1 & 2 & 0 \\ 2 & 0 & 1 \end{bmatrix}\left[t_{12}(t_{45}) \quad t_{23}(t_{56}) \quad t_{34}(t_{61})\right]^{\mathrm{T}} - \frac{\pi}{3}\begin{bmatrix} 1 \\ 1 \end{bmatrix} \tag{10.42}$$

Here, (10.42) gives the method to extract the angle errors $\varepsilon_{\mathrm{HA}}$ and $\varepsilon_{\mathrm{HB}}$. By controlling these angle errors to zero, differential-mode biases u_{HA}, u_{HB}, and u_{HC} will converge to Δv_{BC}, Δv_{AB}, and Δv_{CA}, respectively. Then, the horizontal deviation caused by asymmetric parameters can be eliminated.

Finally, as shown in Figure 10.15, three common-mode biases, u_{VA}, u_{VB}, and u_{VC}, can be obtained in the vertical correction; three differential-mode biases, u_{HA}, u_{HB}, and u_{HC}, can be obtained in the horizontal correction. Based on Table 10.2, six zero-crossing thresholds can be calculated as

$$\begin{cases} u_3 = \frac{1}{2}V_{dc} + u_{\mathrm{VA}} + u_{\mathrm{HA}} \\ u_6 = \frac{1}{2}V_{dc} + u_{\mathrm{VA}} - u_{\mathrm{HA}} \end{cases}, \begin{cases} u_2 = \frac{1}{2}V_{dc} + u_{\mathrm{VB}} - u_{\mathrm{HB}} \\ u_5 = \frac{1}{2}V_{dc} + u_{\mathrm{VB}} + u_{\mathrm{HB}} \end{cases}, \begin{cases} u_1 = \frac{1}{2}V_{dc} + u_{\mathrm{VC}} + u_{\mathrm{HC}} \\ u_4 = \frac{1}{2}V_{dc} + u_{\mathrm{VC}} - u_{\mathrm{HC}} \end{cases} \tag{10.43}$$

It should be noted that during the whole process, any parameters of the machine or the measurement circuit of back-EMF are not required. Besides, the described adaptive correction strategy does not rely on a specified back-EMF. No matter whether the back-EMFs are trapezoidal or sinusoidal, the ZCP positions are the same and the mechanism is similar. Accordingly, this scheme has a good adaptive capacity and can overcome the parameter change.

10.4.4 Experimental Results

The above adaptive threshold correction strategy is tested with the BLDC-I (Appendix B). The resistance and inductance values are asymmetric as shown in Table 10.3. The PWM switching frequency and sampling frequency are both 40 kHz and the dc-link voltage is 15 V. In addition, to simulate the resistance tolerance of RVD, three different resistance ratios are used in three phases as shown in Table 10.4. The equivalent gains K_R are 110%, 100%, and 90% in phases A, B, and C, respectively. It should be noted that no capacitors are utilized in the back-EMF measurement circuits to avoid the phase delay of LPFs.

Figure 10.18 shows the comparison results with and without the adaptive threshold correction scheme at 20 000 r/min. Without correction as shown in Figure 10.18a, the sector ZCP interval ($\hat{\theta}_{12}...\hat{\theta}_{61}$) varies from 50° to 75°. Due to the machine parameter asymmetry and the resistance tolerance of RVD, the sector ZCP interval cannot remain at 60° but fluctuates within a wide range. When the vertical correction is enabled in Figure 10.18b, the sector ZCP interval ($\hat{\theta}_{12}...\hat{\theta}_{61}$) becomes closer to 60°, ranging from 55° to 65°. The vertical shift effect

Table 10.3 Parameters of BLDC-I.

Parameter	Phase A	Phase B	Phase C
Phase resistance (mΩ)	38.1	37.5	43.5
Self-inductance (μH)	27.5	27.2	28.6
Mutual-inductance (μH)		4 μH	

Table 10.4 Resistances of RVD in three phases.

Phase	R1 (kΩ)	R2 (kΩ)	Resistance ratio	K_R (%)
A	3.24	11.5	0.22	110
B	3	12	0.2	100
C	2.8	12.7	0.18	90

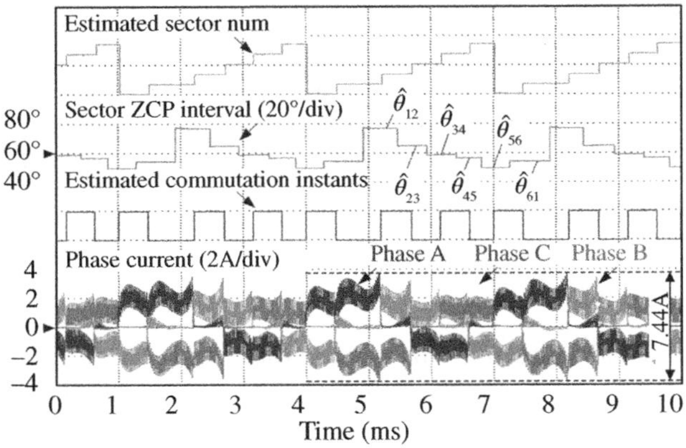

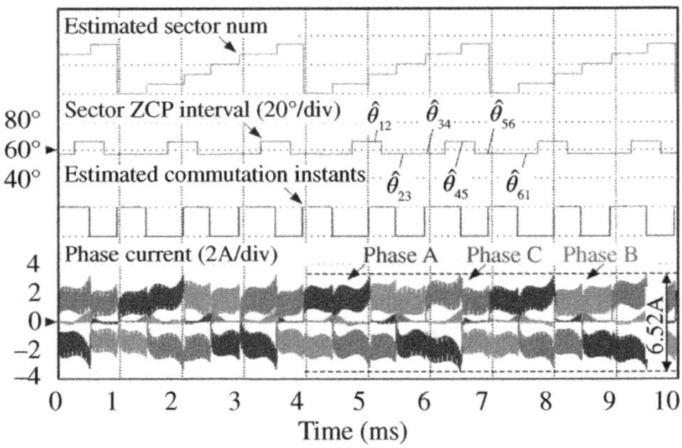

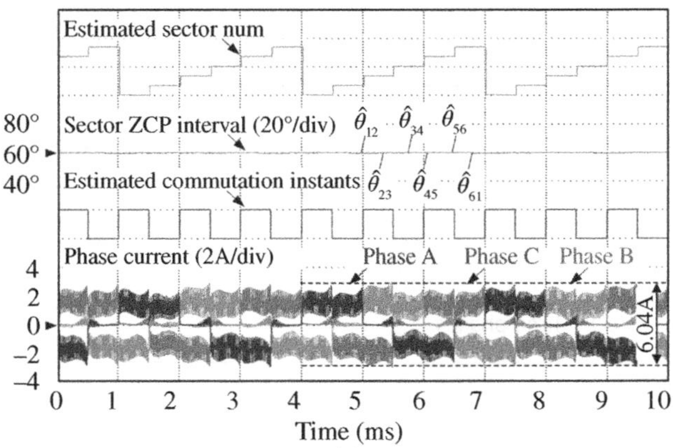

Figure 10.18 Comparison results with and without adaptive threshold correction scheme at 20 000 r/min (333 Hz). (a) Without correction. (b) With only vertical correction. (c) With both vertical and horizontal corrections. (*Source:* [18]/IEEE.)

caused by the tolerance of RVD has been eliminated, so the remaining inconsistence is mainly caused by machine parameter asymmetry. When both vertical and - horizontal corrections are enabled in Figure 10.18c, the sector ZCP interval ($\hat{\theta}_{12}...\hat{\theta}_{61}$) converges to 60°. The phase currents remain almost identical in each sector. Accordingly, both the vertical and horizontal envelope shift effects are eliminated with the correction method.

Figure 10.19 shows the comparison results with and without the adaptive threshold correction strategy at 54 000 r/min (900 Hz). The PWM duty reaches 100% and thus it is at the maximum speed. Without correction as shown in Figure 10.19a, the sector ZCP interval cannot stay at 60°. With only vertical correction enabled in Figure 10.19b, the sector ZCP interval becomes closer to 60°, but still cannot remain at 60°. When both vertical and horizontal corrections are enabled in Figure 10.19c, the sector ZCP interval converges to 60°.

10.5 Freewheeling Angle

The ZCP detection-based sensorless method suffers from a long freewheeling angle (diode conduction angle) issue. As shown in Figure 10.20, when two of the three phases are conducted, the back-EMF of the floating phase can be measured and used to identify the commutation instant. However, in practice, the current of the outgoing phase cannot vanish to zero immediately after commutation due to energy stored in the phase winding, but it flows through the freewheeling diode. This clamps the floating phase to the dc-link and increases the difficulty of detecting the ZCPs. When the freewheeling angle is larger than 30°, the ZCP will disappear and be undetectable [3]. Thus, for ZCD-based sensorless drives, a safe operating area (SOA) can be defined as the available torque and speed areas in which the freewheeling angle is within 30°.

The freewheeling angle is related to many factors, including:

- PWM switching patterns [19]
- Motor speed and current [16]
- Machine parameters [3]
- DC-link voltage [20]
- PWM duty ratio [21]

In this section, based on four PWM switching patterns described in Section 10.3.1, the freewheeling angle is first analyzed considering different speed and current conditions. Then, a simple SOA prediction method is presented that can be used to predict the maximum freewheeling angle. Based on the

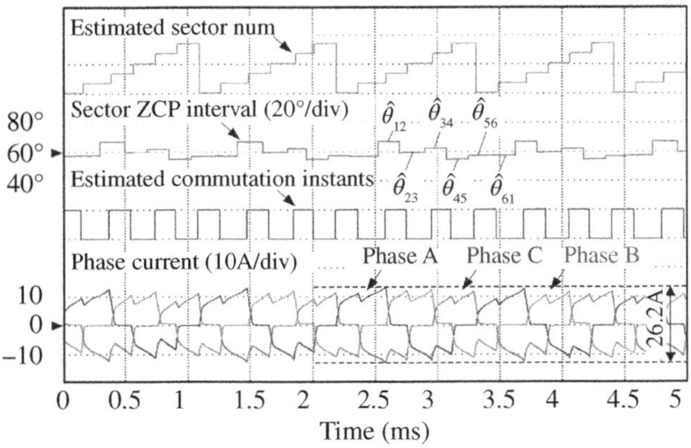

(a)

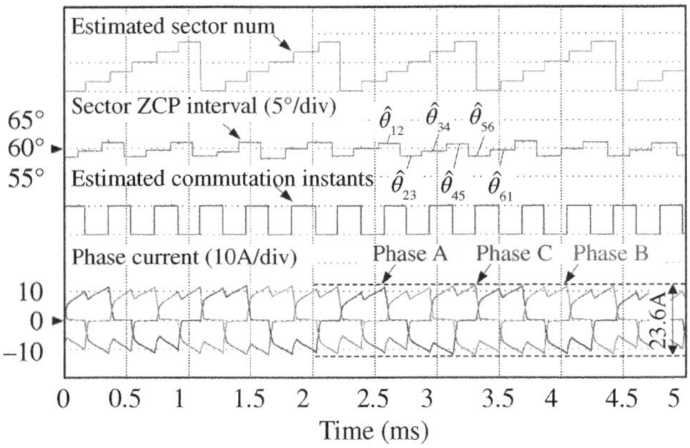

(b)

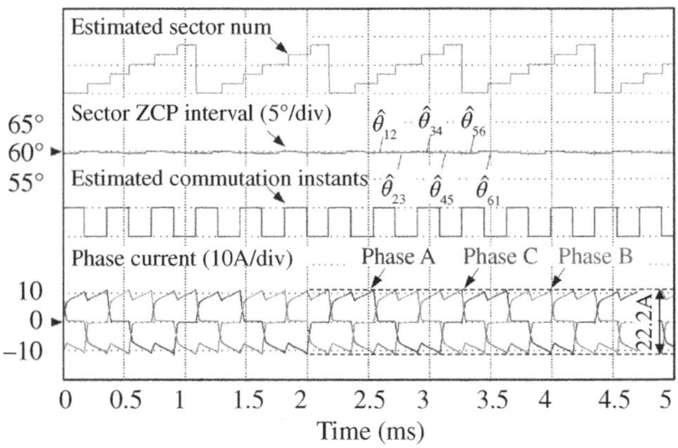

(c)

Figure 10.19 Comparison results with and without adaptive threshold correction scheme at 54 000 r/min (900 Hz). (a) Without correction. (b) With only vertical correction. (c) With both vertical and horizontal corrections. (*Source:* [18]/IEEE.)

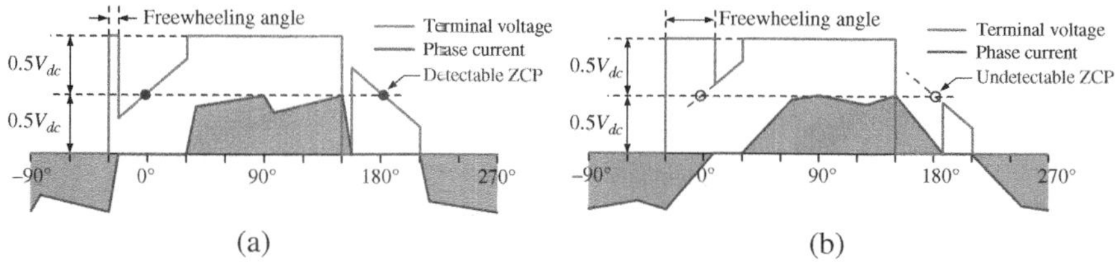

Figure 10.20 Schematics of terminal voltage and phase current. (a) Detectable ZCP. (b) Undetectable ZCP. (*Source:* Adapted from [19].)

freewheeling angle prediction method, freewheeling angles under different parameters, dc-link voltage, and PWM duty-ratio are calculated and analyzed.

10.5.1 PWM Pattern

As shown in Table 10.5, six sectors are defined according to the rotor position, which determines the conduction modes of the three phases.

According to whether the floating phase is clamped to the positive or negative dc-link, the freewheeling processes in sectors II, IV, and VI are defined as upper-diode freewheeling (UDF); those in sectors I, III, and V are defined as lower-diode freewheeling (LDF). Also, in the two conducting phases, either can be selected to perform PWM while the other one keeps on during the whole sector. Thus, according to the selection of the PWM performing phase, the commutation processes can be classified into two modes, which can be defined as follows:

Mode 1: The switch in the parallel side of the freewheeling diode is controlled by the PWM signal, and the one in the opposite side keeps conducting (Figures 10.21 and 10.22). For instance, when D1 is freewheeling in Figure 10.21, the switch S5

Table 10.5 Six-step commutation sequence.

Sector	Rotor position	Positive phase	Negative phase	Floating phase	Freewheeling diode
I	30–90°	A	B	C	Lower
II	90–150°	A	C	B	Upper
III	150–210°	B	C	A	Lower
IV	210–270°	B	A	C	Upper
V	270–330°	C	A	B	Lower
VI	330–30°	C	B	A	Upper

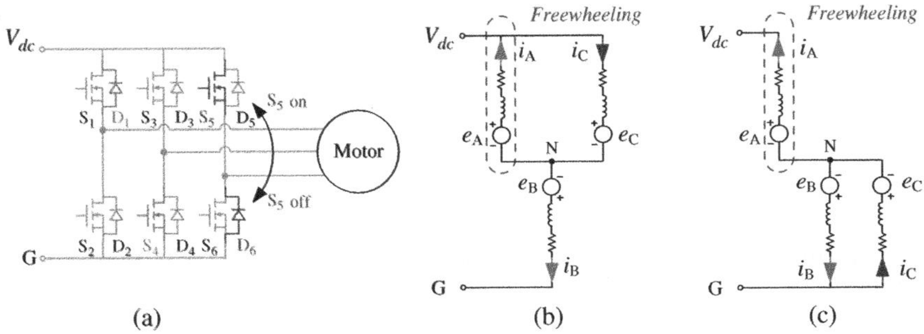

Figure 10.21 Freewheeling paths of UDF process in mode 1, sector VI. (a) Freewheeling paths. (b) Equivalent circuits in PWM chop-on period. (c) Equivalent circuits in PWM chop-off period. (*Source:* [19]/IEEE.)

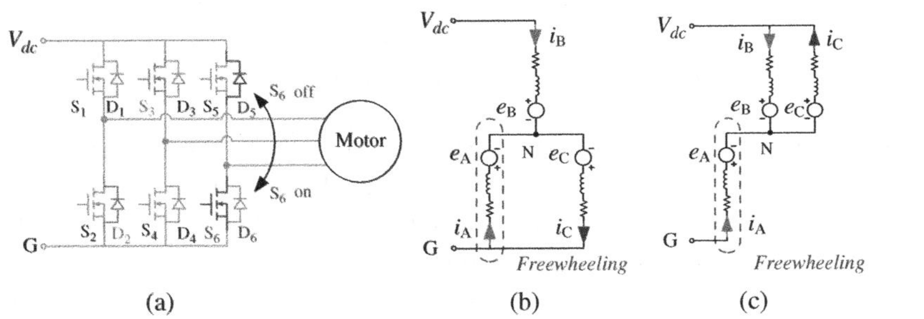

Figure 10.22 Freewheeling paths of LDF process in mode 1, sector III. (a) Freewheeling paths. (b) Equivalent circuits in PWM chop-on period. (c) Equivalent circuits in PWM chop-off period. (*Source:* [19]/IEEE.)

that is on the parallel side of D1 is controlled by the PWM signal, and the switch S4 on the opposite side keeps conducting.

Mode 2: The switch on the opposite side of the freewheeling diode is controlled by the PWM signal, and the one on the parallel side keeps conducting (Figures 10.23 and 10.24). For instance, when D1 is freewheeling in Figure 4.6, the switch S4 that is on the parallel side of D1 is controlled by the PWM signal, and the switch S5 on the opposite side keeps conducting.

In the four basic PWM switching patterns, i.e. H-PWM-L-ON, H-ON-L-PWM, PWM-ON, and ON-PWM, the commutation modes across six sectors are summarized in Table 10.6. Note that H-PWM-L-ON and H-ON-L-PWM patterns contain both modes 1 and 2, PWM-ON only contains mode 2, and ON-PWM only contains mode 1.

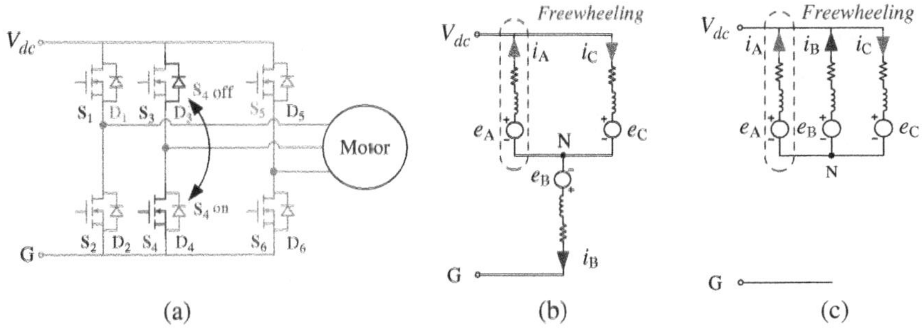

Figure 10.23 Freewheeling paths of UDF process in mode 2, sector VI. (a) Freewheeling paths. (b) Equivalent circuits in PWM chop-on period. (c) Equivalent circuits in PWM chop-off period. (*Source:* [19]/IEEE.)

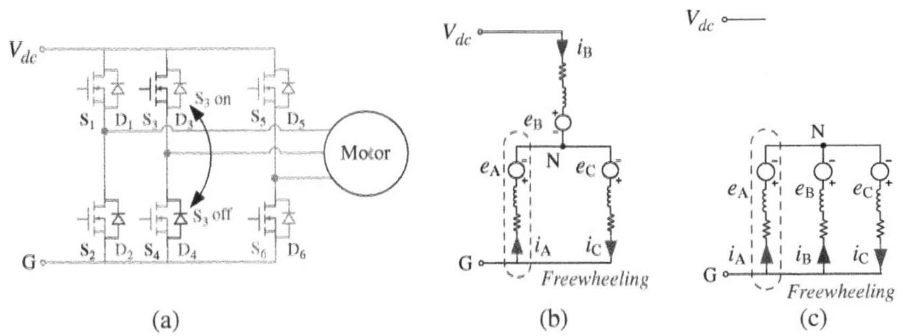

Figure 10.24 Freewheeling paths of LDF process in mode 2, sector III. (a) Freewheeling paths. (b) Equivalent circuits in PWM chop-on period. (c) Equivalent circuits in PWM chop-off period. (*Source:* [19]/IEEE.)

Table 10.6 Commutation modes and switching patterns.

Switching patterns	UDF (sectors II, IV, VI)	LDF (sectors I, III, V)
H-PWM-L-ON	Mode 1	Mode 2
H-ON-L-PWM	Mode 2	Mode 1
PWM-ON	Mode 2	Mode 2
ON-PWM	Mode 1	Mode 1

10.5.1.1 Commutation Mode 1

UDF in Mode 1

Figure 10.21 shows the freewheeling paths in UDF of mode 1. Taking sector VI for instance, the upper diode D1 is freewheeling, the upper switch S5 is controlled by the PWM signal, and the lower switch S4 keeps conducting. Figures 10.21b,c show the equivalent circuits of chop-on and chop-off periods, respectively.

Chop-on Period As shown in Figure 10.21b, during the chop-on period, phases A and C are both connected to the positive dc-link while phase B is connected to the negative one. Then, the voltage can be calculated as:

$$\begin{bmatrix} v_{AN} \\ v_{BN} \\ v_{CN} \end{bmatrix} = \begin{bmatrix} V_{dc} - v_{NG} \\ -v_{NG} \\ V_{dc} - v_{NG} \end{bmatrix} = \begin{bmatrix} i_A R + p i_A L + e_A \\ i_B R + p i_B L + e_B \\ i_C R + p i_C L + e_C \end{bmatrix} \tag{10.44}$$

For sinusoidal back-EMF-type motors, the sum of three phase back-EMFs equals zero, i.e. $e_A + e_B + e_C = 0$; while for trapezoidal ones, the sum of back-EMFs in upper and lower side phases equals zero, i.e. $e_B + e_C = 0$ in sector VI. Then, the neutral voltage v_{NG} can be calculated as:

$$v_{NG} = \begin{cases} 2V_{dc}/3 & \text{sinusoidal EMF} \\ 2V_{dc}/3 - e_A/3 & \text{trapezoidal EMF} \end{cases} \tag{10.45}$$

The current response can be calculated by substituting (10.45) into (10.44). During the chop-on period, the differential equation of the freewheeling current i_a can be expressed as:

$$i_A(t)R + p i_A(t)L = V_{dc}/3 + e_{UDF}(t) \tag{10.46}$$

where

$$e_{UDF}(t) = \begin{cases} -\omega_r \psi_m \sin(11\pi/6 + \omega_r t) & \text{sinusoidal EMF} \\ 2\omega_r \psi_m/3 & \text{trapezoidal EMF} \end{cases} \tag{10.47}$$

Chop-off Period As shown in Figure 10.21c, when S_5 switches off, the current in phase C will flow through D_6 and it will clamp phase C to the negative dc-link. Thus, the three-phase voltages can be expressed as:

$$\begin{bmatrix} v_{AN} \\ v_{BN} \\ v_{CN} \end{bmatrix} = \begin{bmatrix} V_{dc} - v_{NG} \\ -v_{NG} \\ -v_{NG} \end{bmatrix} = \begin{bmatrix} i_A R + p i_A L + e_A \\ i_B R + p i_B L + e_B \\ i_C R + p i_C L + e_C \end{bmatrix} \tag{10.48}$$

Then, the neutral voltage v_{NG} can be calculated as:

$$v_{\mathrm{NG}} = \begin{cases} 2V_{dc}/3 & \text{sinusoidal EMF} \\ 2V_{dc}/3 - e_{\mathrm{A}}/3 & \text{trapezoidal EMF} \end{cases} \tag{10.49}$$

With the same calculation method above, the freewheeling current response can be derived as:

$$i_{\mathrm{A}}(t)R + pi_{\mathrm{A}}(t)L = 2V_{dc}/3 + e_{\mathrm{UDF}}(t) \tag{10.50}$$

LDF in Mode 1

Similarly, as shown in Figure 10.22, consider the LDF event in sector III. The lower diode D_2 is freewheeling, the lower switch S_6 is controlled by the PWM signal, and the upper switch S_3 keeps conducting. The freewheeling current in both chop-on and chop-off periods is analyzed as follows.

Chop-on Period In Figure 10.22b, during the chop-on period, phases A and C are connected to the negative dc-link with D_2 and S_6, while phase B is connected to the positive dc-link with S_3. The voltage can be expressed as:

$$\begin{bmatrix} -v_{\mathrm{NG}} \\ V_{dc} - v_{\mathrm{NG}} \\ -v_{\mathrm{NG}} \end{bmatrix} = \begin{bmatrix} i_{\mathrm{A}}R + pi_{\mathrm{A}}L + e_{\mathrm{A}} \\ i_{\mathrm{B}}R + pi_{\mathrm{B}}L + e_{\mathrm{B}} \\ i_{\mathrm{C}}R + pi_{\mathrm{C}}L + e_{\mathrm{C}} \end{bmatrix} \tag{10.51}$$

By solving (10.51), the freewheeling current can be described as:

$$i_{\mathrm{A}}(t)R + pi_{\mathrm{A}}(t)L = -V_{dc}/3 + e_{\mathrm{LDF}}(t) \tag{10.52}$$

where

$$e_{\mathrm{LDF}}(t) = \begin{cases} -\omega_r\psi_m \sin(5\pi/6 + \omega_r t) & \text{sinusoidal EMF} \\ -2\omega_r\psi_m/3 & \text{trapezoidal EMF} \end{cases} \tag{10.53}$$

Chop-off Period In Figure 10.22c, during the chop-off period, D_5 will conduct after S_6 turns off, clamping phase C to the positive dc-link. Thus, the voltage can be expressed as:

$$\begin{bmatrix} -v_{\mathrm{NG}} \\ V_{dc} - v_{\mathrm{NG}} \\ V_{dc} - v_{\mathrm{NG}} \end{bmatrix} = \begin{bmatrix} i_{\mathrm{A}}R + pi_{\mathrm{A}}L + e_{\mathrm{A}} \\ i_{\mathrm{B}}R + pi_{\mathrm{B}}L + e_{\mathrm{B}} \\ i_{\mathrm{C}}R + pi_{\mathrm{C}}L + e_{\mathrm{C}} \end{bmatrix} \tag{10.54}$$

Then, the freewheeling current can be described as

$$i_{\mathrm{A}}(t)R + pi_{\mathrm{A}}(t)L = -2V_{dc}/3 + e_{\mathrm{LDF}}(t) \tag{10.55}$$

The current descent rate η depends on the voltage excitations. The current descent rates for chop-on and chop-off periods can be derived as:

$$\eta_{m1}^{on} = |V_{dc}/3 + e_{\mathrm{UDF}}(t)| = |-V_{dc}/3 + e_{\mathrm{LDF}}| \tag{10.56}$$

$$\eta_{m1}^{off} = |2V_{dc}/3 + e_{\mathrm{UDF}}(t)| = |-2V_{dc}/3 + e_{\mathrm{LDF}}(t)| \tag{10.57}$$

Thus, the conclusion is that for commutation mode 1, the current descent rate just depends on the chop-on and chop-off periods, irrespective of the UDF and LDF processes. In addition, in mode 1, the current descent rate of the chop-on period η_{m1}^{on} is lower than that of the chop-off period η_{m1}^{off} since

$$\eta_{m1}^{on} = |V_{dc}/3 + e_{\mathrm{UDF}}(t)| < |2V_{dc}/3 + e_{\mathrm{UDF}}(t)| = \eta_{m1}^{off} \tag{10.58}$$

In other words, the chop-off period will accelerate the freewheeling process in mode 1.

10.5.1.2 Commutation Mode 2

UDF in Mode 2

Figure 10.23 shows the freewheeling paths in UDF of mode 2. In sector VI, the upper diode D1 is freewheeling. The main difference from mode 1 is that the lower-switch S4 is controlled by the PWM signal and the upper-switch S5 keeps conducting.

Chop-on Period As shown in Figure 10.23a,b, the equivalent circuit in the chop-on period of mode 2 is the same as that in mode 1. Thus, the freewheeling current response can also be expressed as (10.46).

Chop-off Period As shown in Figure 10.23c, during the chop-off period of commutation mode 2, D_3 will conduct after S_4 turns off, clamping all three phases to the positive dc-link. In this situation, the three-phase voltages can be described as:

$$\begin{bmatrix} V_{dc} - v_{\mathrm{NG}} \\ V_{dc} - v_{\mathrm{NG}} \\ V_{dc} - v_{\mathrm{NG}} \end{bmatrix} = \begin{bmatrix} i_{\mathrm{A}}R + pi_{\mathrm{A}}L + e_{\mathrm{A}} \\ i_{\mathrm{B}}R + pi_{\mathrm{B}}L + e_{\mathrm{B}} \\ i_{\mathrm{C}}R + pi_{\mathrm{C}}L + e_{\mathrm{C}} \end{bmatrix} \tag{10.59}$$

Because all three phases are connected to the same point, the current response only depends on the back-EMF, as shown in:

$$i_{\mathrm{A}}(t)R + pi_{\mathrm{A}}(t)L = e_{\mathrm{UDF}}(t) \tag{10.60}$$

LDF in Mode 2

Figure 10.24 shows the freewheeling paths in LDF of mode 2. The lower diode D_2 is in freewheeling. The main difference from mode 1 is that the upper switch S_3 is controlled by the PWM signal and the lower switch S_6 keeps conducting.

Chop-on Period As shown in Figures 10.22b and 10.24b, the equivalent circuits in the chop-on period of mode 2 are the same as those in mode 1. Thus, the freewheeling current response can also be expressed as (10.52).

Chop-off Period As shown in Figure 10.24c, D_4 will conduct after S_3 turns off. Then, all three phases are connected to the negative dc-link. In this situation, the voltage can be expressed as:

$$\begin{bmatrix} -v_{\mathrm{NG}} \\ -v_{\mathrm{NG}} \\ -v_{\mathrm{NG}} \end{bmatrix} = \begin{bmatrix} i_{\mathrm{A}}R + pi_{\mathrm{A}}L + e_{\mathrm{A}} \\ i_{\mathrm{B}}R + pi_{\mathrm{B}}L + e_{\mathrm{B}} \\ i_{\mathrm{C}}R + pi_{\mathrm{C}}L + e_{\mathrm{C}} \end{bmatrix} \tag{10.61}$$

The current response can be calculated as:

$$i_{\mathrm{A}}(t)R + pi_{\mathrm{A}}(t)L = e_{\mathrm{LDF}}(t) \tag{10.62}$$

Likewise, current descent rates for chop-on and chop-off periods are expressed as:

$$\eta_{m1}^{on} = \eta_{m2}^{on} = |V_{dc}/3 + e_{\mathrm{UDF}}(t)| = |-V_{dc}/3 + e_{\mathrm{LDF}}| \tag{10.63}$$

$$\eta_{m2}^{off} = |e_{\mathrm{UDF}}(t)| = |e_{\mathrm{LDF}}| \tag{10.64}$$

In addition, the conclusion of mode 2 is opposite to that of mode 1, i.e. the current descent rate of the chop-on period η_{m2}^{on} is higher than that of the chop-off period η_{m2}^{off} since

$$\eta_{m2}^{on} = |V_{dc}/3 + e_{\mathrm{UDF}}(t)| > |e_{\mathrm{UDF}}(t)| = \eta_{m2}^{off} \tag{10.65}$$

Thus, the chop-off period slows down the freewheeling process compared with the chop-on period.

10.5.1.3 Comparison of Current Descent Rates

The above analyses illustrate the freewheeling paths and the behaviors of the freewheeling current in the two commutation modes. The relation of the current descent rates can be concluded as

$$\eta_{m1}^{off} > \eta_{m1}^{on} = \eta_{m2}^{on} > \eta_{m2}^{off} \tag{10.66}$$

Note that the overall current descent rates of mode 1 are higher than or equal to those of mode 2. Accordingly, for the same initial current value, the freewheeling process in mode 1 is always faster than that in mode 2. In this sense, the switching pattern that only contains mode 1, i.e. the ON-PWM pattern, can minimize the freewheeling angle and achieve broader sensorless SOA.

Figure 10.25a compares the phase currents in four PWM switching patterns in simulation. The speed is 60 000 r/min and the current is about 20 A. A sinusoidal back-EMF-type motor (BLDC-II, Appendix B) is simulated here. The magnified

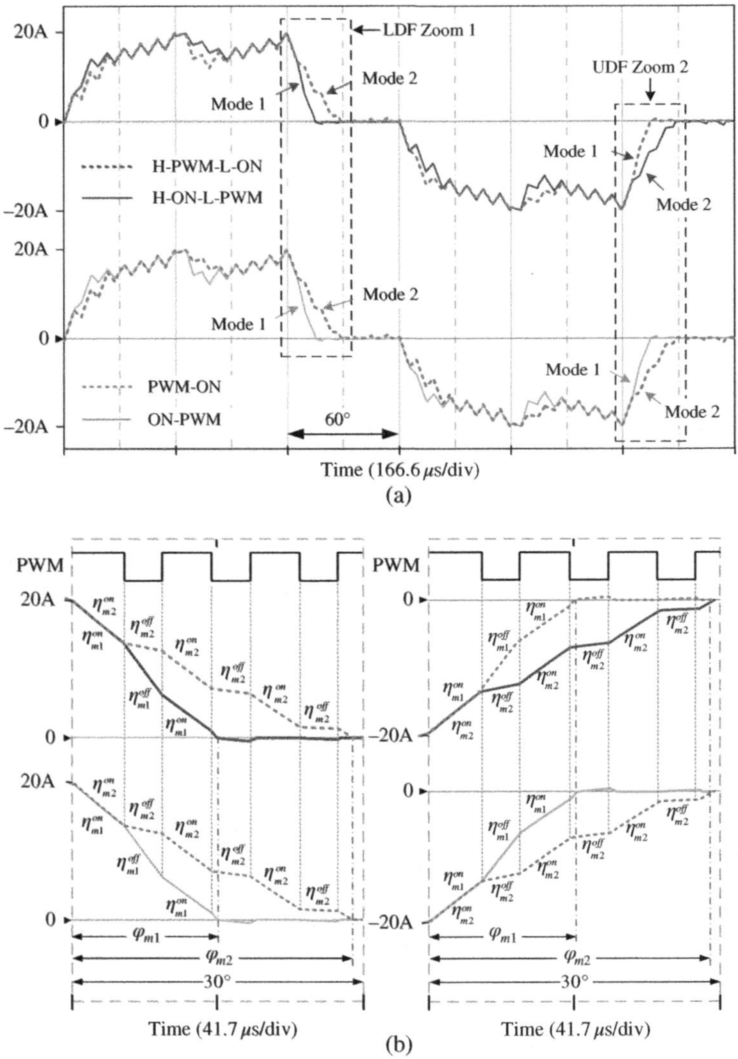

Figure 10.25 Comparison of phase currents in four PWM switching patterns at 60 000 r/min. (a) Phase currents in a fundamental period. (b) Zoomed waveforms for LDF and UDF processes. (*Source:* [19]/IEEE.)

current waveforms for LDF and UDF processes are shown in Figure 10.25b, in which the difference of current descent rates is illustrated.

For the LDF process in the left-side dashed block, the freewheeling angle φ_{m2} (H-PWM-L-ON and PWM-ON) almost reaches 30°, which means the ZCP is nearly undetectable. In contrast, the freewheeling angle φ_{m1} (H-ON-L-PWM and ON-PWM) is much smaller. On the other hand, for the UDF process in the right-side dashed block, the freewheeling angles φ_{m1} (H-PWM-L-ON and ON-PWM) are less than φ_{m2} (H-ON-L-PWM and PWM-ON). Note that among the four patterns, the ON-PWM always has less freewheeling angle irrespective of UDF or LDF process.

10.5.1.4 Freewheeling Angle

Figure 10.26 shows the analysis results of freewheeling duration and angle when the two modes of BLDC-II (Appendix B) are used. The speed range is from 10 000 to 120 000 r/min and the current range is from 10 to 30 A.

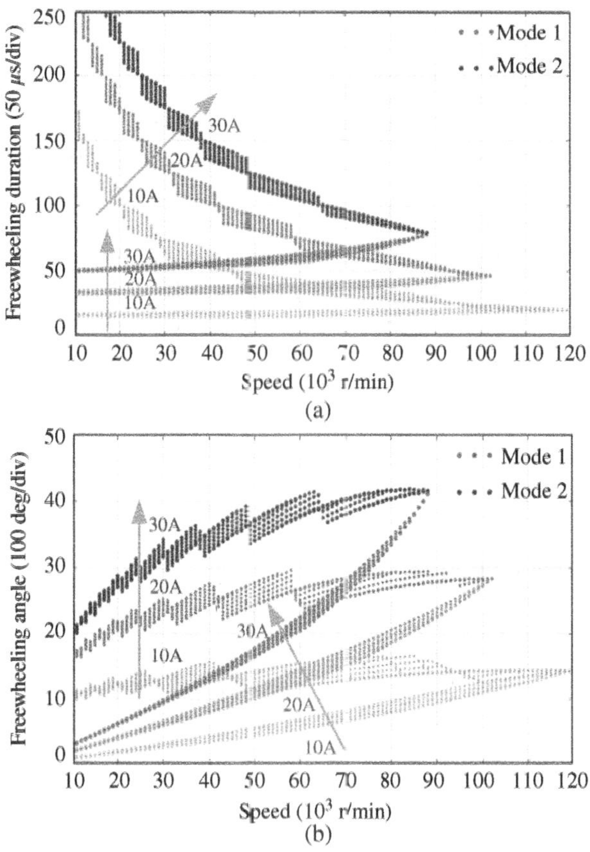

Figure 10.26 Freewheeling duration and angle for two modes with various speeds and initial currents. (a) Freewheeling duration. (b) Freewheeling angle. (*Source:* [19]/IEEE.)

For a given speed and initial current, the freewheeling duration is not fixed but varies in a certain range. This is because the commutation may occur at any instant in a PWM carrier wave, and various initial PWM states are taken into consideration. Besides, the freewheeling duration of mode 1 changes slightly with speed and mainly depends on the initial current. For mode 2, the freewheeling duration increases with the initial current and decreases with the speed. This is because the duty ratio increases with the speed. Then the chop-off period, which slows down the freewheeling process of mode 2, reduces its proportion as the duty ratio increases. When the duty ratio reaches 100%, modes 1 and 2 converge to the same freewheeling duration. Figure 10.26b shows the results in freewheeling angle. When the freewheeling angle exceeds 30°, the ZCP is undetectable. It is clear that for both modes, the freewheeling angle increases with speed and current. However, mode 1 has a relatively small freewheeling angle and thereby can achieve a broader sensorless operation range. Thus, the ON-PWM switching pattern, which only contains commutation mode 1, can achieve broader sensorless operation range in theory.

10.5.1.5 Experiment Results

The experiment results of BLDC-II (Appendix B) are shown here as an example. The PWM switching frequency and the sampling frequency are both 40 kHz, and the dc-link voltage is 25 V. Figure 10.27a–d shows the sensorless operation at 60 000 r/min using four basic PWM switching patterns. The initial current at commutation instant is about 14 A. It can be seen that the freewheeling duration for commutation mode 1 is less than 25 μs (9°). For mode 2, however, the freewheeling duration reaches over 50 μs (about 20°). Besides, as shown in the magnified figures, the ZCP of commutation mode 2 is not clear. Although it is not overridden by the freewheeling duration, it is difficult to identify the ZCP accurately. Therefore, for mode 2, the ZCP has been close to the sensorless safe operation boundary. For mode 1, the freewheeling duration is acceptable, and the ZCP can be recognized clearly. It should be noted that the ON-PWM only contains commutation mode 1, thus both the ZCPs can be clearly identified.

10.5.2 Sensorless Safe Operation Area

A long freewheeling duration increases the difficulty of ZCD. In the extreme situation of high-speed and high-load operation, the freewheeling angle perhaps exceeds 30 electrical degrees. In this circumstance, the ZCP will be overridden and the sensorless drive will fail. Thus, for ZCD-based sensorless drives, the safe operation area (SOA) can be defined as the available torque and speed areas in which the freewheeling angle is within 30°. Hence, it is desirable to predict the SOA in the design stage of the drive.

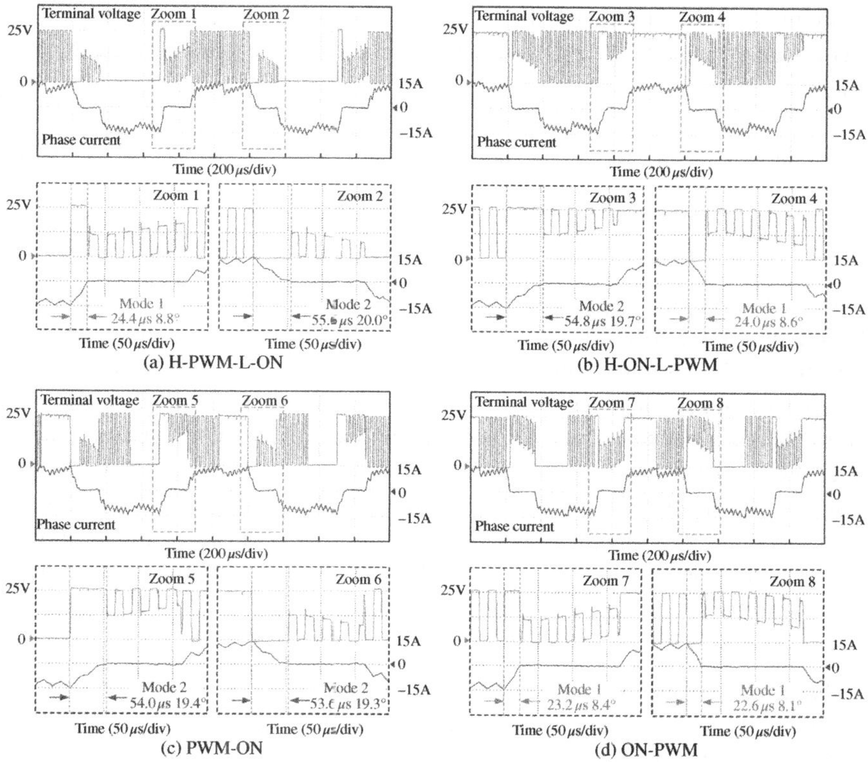

Figure 10.27 Sensorless operation at 60 000 r/min using various PWM switching patterns. (*Source:* [19]/IEEE.)

10.5.2.1 Prediction of SOA

In Section 10.5.1, it has been proved that the ON-PWM pattern can accelerate the freewheeling process and achieve a broader sensorless SOA. In addition, in the stage of motor design, a method that can predict the SOA of an ON-PWM pattern is required. This will help verify whether the predesigned torque and speed area can be achieved with the given parameters (flux density, resistance, and inductance). In this subsection, a simple numerical iterative method is illustrated.

For the ON-PWM switching pattern, since the current descent rates in chop-on period η_{m1}^{on} and chop-off period η_{m1}^{off} are different, the freewheeling angle φ_{m1} will be dependent on the PWM duty ratio. Hence, an accurate freewheeling angle of the ON-PWM pattern is difficult to obtain. Instead, it is easy to determine the upper bound of the freewheeling angle, i.e. the maximum freewheeling angle φ_{m1}^{sup}. Since the current descent rate in the chop-on period is lower than that of the chop-off period, i.e. $\eta_{m1}^{on} < \eta_{m1}^{off}$, the maximum freewheeling angle happens when the current descends with the rate η_{m1}^{on} all the way. Thus, according to (10.52), the maximum freewheeling angle $\varphi_{m1}^{\text{sup}}$ can be obtained by solving the nonlinear differential equation.

$$i[n+1] = i[n] + \frac{\Delta t}{L}\left(-\frac{1}{3}V_{dc} + e[n] - i[n]R\right) \tag{10.67}$$

$$e[n] = \begin{cases} -\omega_r \psi_m \sin(5\pi/6 + \omega_r t[n]) & \text{sinusoidal EMF} \\ -2\omega_r \psi_m/3 & \text{trapezoidal EMF} \end{cases} \tag{10.68}$$

where Δt is the time step, the initial current value $i[0]$ can be determined by the load torque. The maximum freewheeling duration (angle) is the span that the current in (10.67) descends from initial value $i[0]$ to zero. The calculation is much easier, since it does not need the PWM duty ratio information. Note that $\varphi_{m1}^{\text{sup}}$ indicates the upper bound of the freewheeling angle in the ON-PWM pattern. In practice, the freewheeling angle φ_{m1} is definitely less than the upper bound $\varphi_{m1}^{\text{sup}}$. Usually, the freewheeling angle increases with the motor speed and current. The sensorless SOA refers to the speed and current area in which the maximum freewheeling angle is less than 30° i.e. $\varphi_{m1}^{\text{sup}} \leq 30°$. By sweeping the initial current value and motor speed, the SOA can be identified.

10.5.2.2 Experiment Results

Experiments are carried out to compare the predicted and measured SOA results. Figure 10.28 compares the measured freewheeling angles φ_{m1}, φ_{m2}, and the

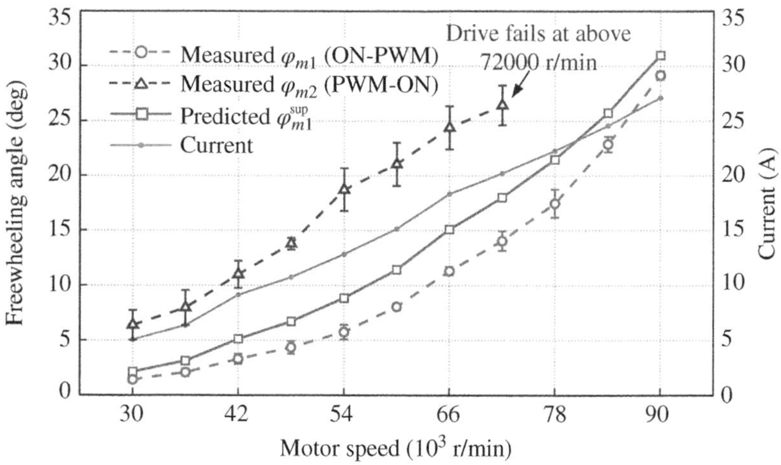

Figure 10.28 Comparison of measured freewheeling angles φ_{m1}, φ_{m2}, and predicted freewheeling angle $\varphi_{m1}^{\text{sup}}$. (*Source:* [19]/IEEE.)

predicted maximum freewheeling angle $\varphi_{m1}^{\text{sup}}$ in mode 1. The speed range is from 30 000 r/min (500 Hz) to 90 000 r/min (1500 Hz). The measured freewheeling angles φ_{m1}, φ_{m2} are obtained from ON-PWM and PWM-ON patterns, respectively. The freewheeling angles are not fixed values for a given speed but vary within a certain range due to random PWM initial states. For this reason, vertical bars are used to describe the measured range. In addition, the current increases with the motor speed, which is shown by the gray curve. As shown in both modes, the freewheeling angles rise with the motor speed. Besides, φ_{m1} is always smaller than φ_{m2}, which agrees with the theoretical analysis.

Note that the freewheeling angle of 30° is a critical point for sensorless BLDC drives. When the motor speed exceeds 72 000 r/min, the sensorless drive using the PWM-ON pattern fails because φ_{m1} reaches 30°. In this case, the SOA is limited by the long freewheeling angle. In contrast, with the pattern ON-PWM, the sensorless SOA is extended to 90 000 r/min. In addition, the curve of the predicted maximum freewheeling angle $\varphi_{m1}^{\text{sup}}$ has a similar trend with the measured φ_{m1}. Therefore, it demonstrates that the ON-PWM pattern can effectively minimize the freewheeling angle and broaden the sensorless SOA.

10.5.3 Resistance and Inductance

Figures 10.29a,b illustrate the freewheeling durations and angles for various resistances. Figures 10.30a,b illustrate the ones for various inductances. Other parameters are fixed and listed under the corresponding figure. It shows that the freewheeling durations (angle) rise as the resistances decrease and the inductances increase. This is because a larger resistance-inductance time constant (L/R) will lead to a larger freewheeling duration (angle). Besides, the inductance has a relatively significant influence on the freewheeling duration (angle) compared to the resistance. This is due to the fact that the freewheeling duration is usually much less than the resistance-inductance time constant. Within this short duration, the inductance plays a dominant role in the current descent rate. As a result, the resistance has a very small influence on the freewheeling duration (angle).

10.5.4 DC-Link Voltage

Figures 10.31a,b show the freewheeling durations and angles for various dc-link voltages ranging from 15 to 30 V. They illustrate that the freewheeling duration (angle) decreases as the dc-link voltage rises. This is because higher dc-link voltage can achieve a higher current descent rate, leading to less freewheeling duration (angle).

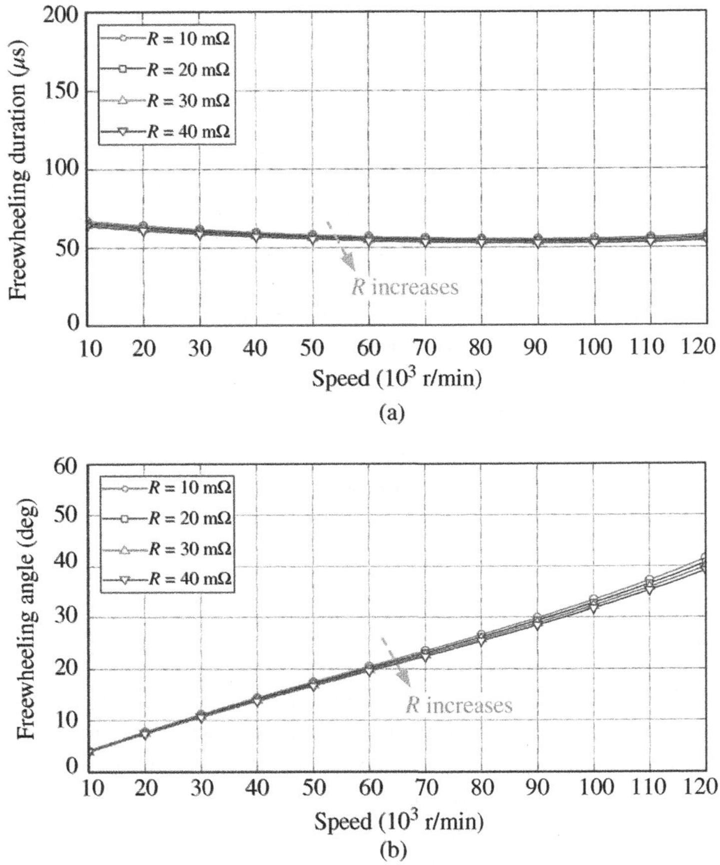

Figure 10.29 Freewheeling durations and angles for various resistances. (L = 0 μH, ψ_m = 1 mWb, I_0 = 30 A, V_{dc} = 25 V, D = 1.0). (a) Freewheeling durations. (b) Freewheeling angles. (*Source:* Adapted from [19].)

10.5.5 PWM Duty Ratio

Figures 10.32a,b show the freewheeling durations and angles for various PWM duty ratios, taking commutation modes 1 and 2 into consideration.

The figures show that for mode 1, the freewheeling angle increases with the duty ratio; for mode 2, the freewheeling angle decreases with the duty ratio. The reasons can be explained as follows. First, the relation of the current descent rates can be concluded as

$$\eta_{m1}^{off} > \eta_{m1}^{on} = \eta_{m2}^{on} > \eta_{m2}^{off} \tag{10.69}$$

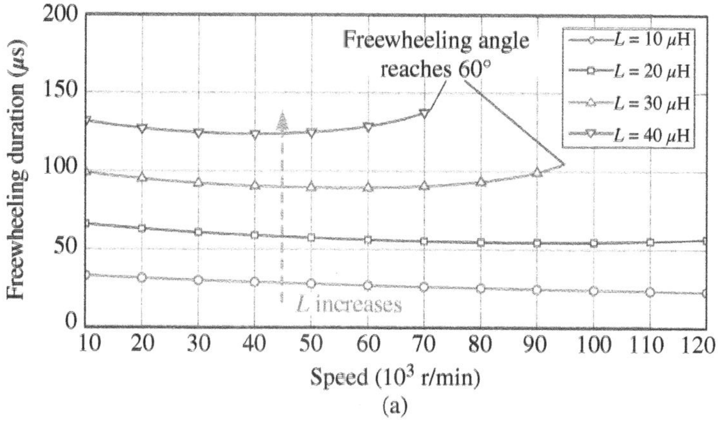

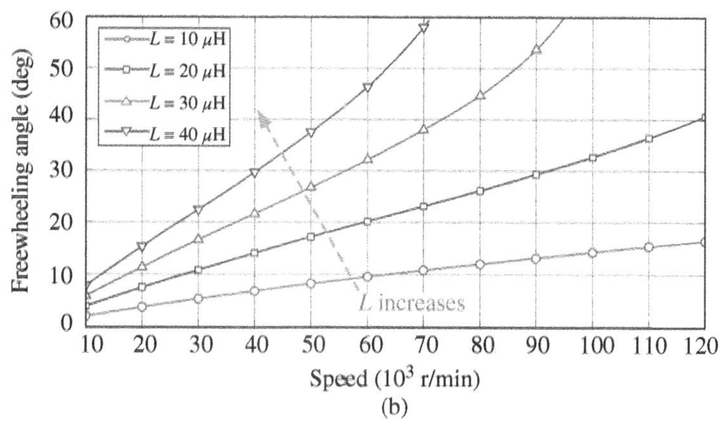

Figure 10.30 Freewheeling durations and angles for various inductances. (R = 20 mΩ, ψ_m = 1 mWb, I_0 = 30 A, V_{dc} = 25 V, D = 1.0). (a) Freewheeling durations. (b) Freewheeling angles. (*Source:* Adapted from [19].)

It shows that for mode 1, the current descent rate of the chop-off interval is higher than that of the chop-on interval. As a result, a higher duty ratio will lead to a reduced chop-off interval, a lower current descent rate, and a larger freewheeling duration (angle). On the contrary, for mode 2, the current descent rate of the chop-off interval is lower than that of the chop-on interval. With similar logic, as the duty ratio rises, the freewheeling duration (angle) decreases. When the duty ratio reaches 1.0, the freewheeling angles of these two modes coincide with each other.

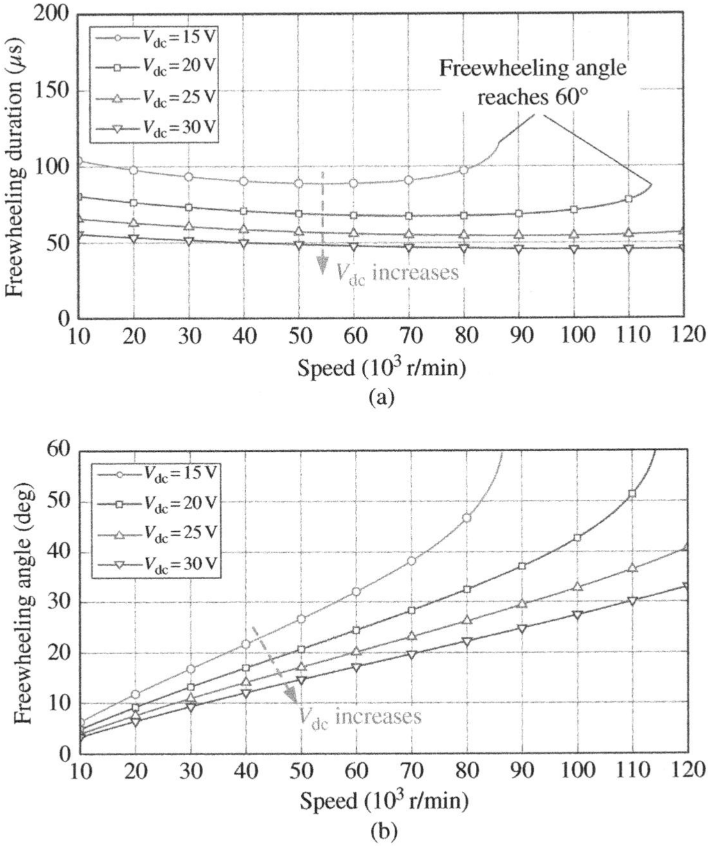

Figure 10.31 Freewheeling durations and angles for various dc-link voltages. (R = 20 mΩ, L = 20 μH, ψ_m = 1 mWb, I_0 = 30 A, D = 1.0). (a) Freewheeling durations. (b) Freewheeling angles. (*Source:* Adapted from [19].)

10.6 Machine Design Effect

This section describes electromagnetic design considerations [3] for high-speed motors equipped with surface-mounted PMs and for sensorless operation based on the detection of the zero-crossing of the back-EMF waveform.

For a given space envelope, there are many possible motor designs that can meet the same torque/speed specification, albeit exhibiting variations in achievable efficiency. However, many would be unsuitable for operation using a "sensorless" drive in which commutation events are controlled by detecting the zero-crossing of the induced back-EMF waveforms. As mentioned in Section 10.5, due to the

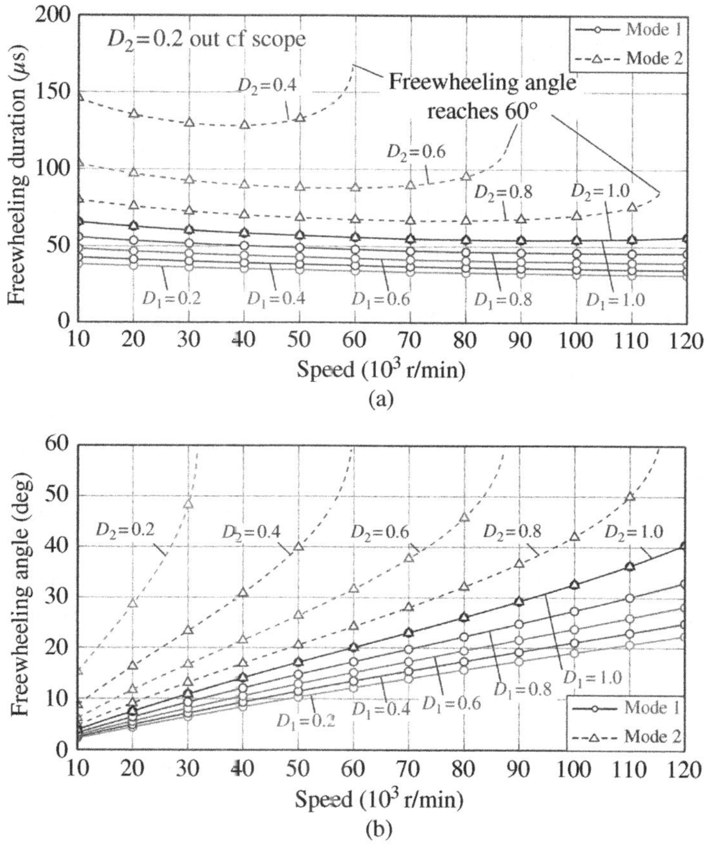

Figure 10.32 Freewheeling durations and angles for various PWM duty ratios, where D_1 and D_2 denote the duty ratios for modes 1 and 2, respectively; (R = 20 mΩ, L = 20 μH, ψ_m = 1 mWb, I_0 = 30 A, V_{dc} = 25 V). (a) Freewheeling durations. (b) Freewheeling angles. (*Source:* Adapted from [19].)

effect of the winding inductance and the conducting duration of the freewheel diodes, the phase currents may flow more or less continuously, thereby preventing the detection of the zero-crossing of the EMF. Therefore, in order to facilitate sensorless commutation, a motor must be designed in such a way that it results in a diode conduction angle that is significantly less than 30° (electrical angle). However, it is also important not to compromise on the performance of the motors, and therefore they are designed to have the maximum efficiency in a specified space envelope. Hence, the motor efficiency and the conduction angle of the freewheeling diodes in the power electronic drive are key design considerations for high-speed brushless dc motors for sensorless operation [3].

By way of example, the motors under consideration are equipped with surface-mounted sintered NdFeB magnets (remanence = 1.2 T) and designed to a specification of 190 W at 120 000 r/min to fit in a total space envelope of 50 mm × 30 mm and employ conventional 0.35 mm silicon steel laminations. When designing a high-speed PM BLDC motor, there are a number of considerations that have to be taken into account in terms of its electrical, magnetic, and mechanical design [22]. Previous work [23] has shown that at high speeds optimal rotor diameter exists for maximum efficiency. However, there is a mechanical limit both in terms of the resonance modes of the rotor [24] and the readily available bearing technologies, and this necessitated fixing the magnet outer diameter at 17 mm for the specific motor designs.

Within a fixed space envelope, the stator flux density can have a significant effect on the geometry of the motor as well as affect the efficiency. However, the primary concern is with reducing the diode conduction angle (freewheeling angle) while maintaining maximal efficiency.

Figures 10.33 and 10.34 show the effect of the stator flux density, the rotor outer diameter, and the motor aspect ratio on the diode conduction angle. When the airgap length and rotor outer diameter are fixed to 4 and 17 mm, respectively, varying the stator flux density, and therefore, the required active length, has a much more

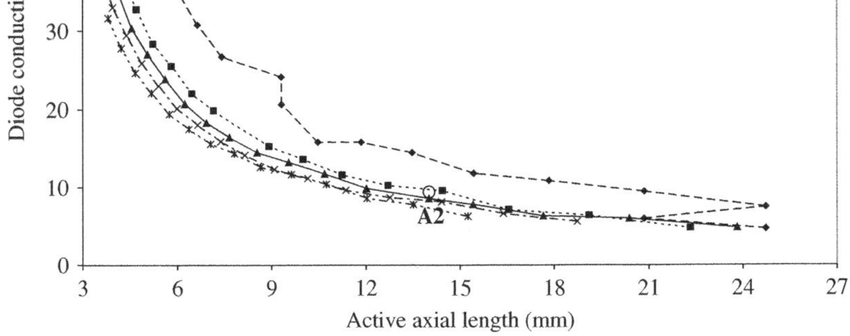

Figure 10.33 Influence of active length and rotor diameter on diode conduction angle. (*Source:* [3]/IOS Press.)

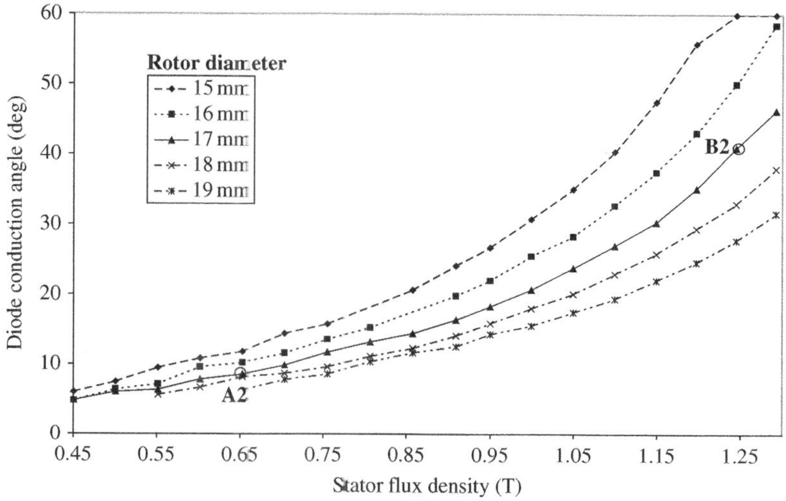

Figure 10.34 Influence of stator flux density and rotor diameter on diode conduction angle. (*Source:* [3]/IOS Press.)

significant effect on the motor efficiency and the diode conduction angle, as can be seen in Figure 10.35. Highlighted on these figures are motors A and B which are selected for prototyping. Motor A has an active length of 14 mm, a stator flux density of 0.65 T, a low number of turns per phase (74), and thus a low winding inductance (0.72 mH), which results in a diode conduction angle of ~9°; while motor B has an active length of just 3.6 mm and a stator flux density of 1.25 T, a high number of turns per phase (220), and thus a high winding inductance (2.172 mH), which results in a diode conduction angle of ~42°. Cross sections of both motors are shown in Figure 10.36 and the main features are summarized in Table 10.7.

It can clearly be seen that motor designs with a relatively low stator flux density and a high stator core length-to-diameter ratio generally use more iron and less copper, which results in fewer turns per phase, and consequently low winding inductances and a low diode conduction angle. Both motors are tested at around the rated speed to highlight the difference in their diode conduction angles. Figure 10.37a shows the current and voltage waveforms for motor A, which is designed specifically for high-speed sensorless operation, while Figure 10.37b shows the corresponding waveforms for motor B, using a Hall sensor-based drive as it is unsuitable for sensorless operation due to its high diode conduction angle and non-existence of zero-crossing of the back-EMF waveform.

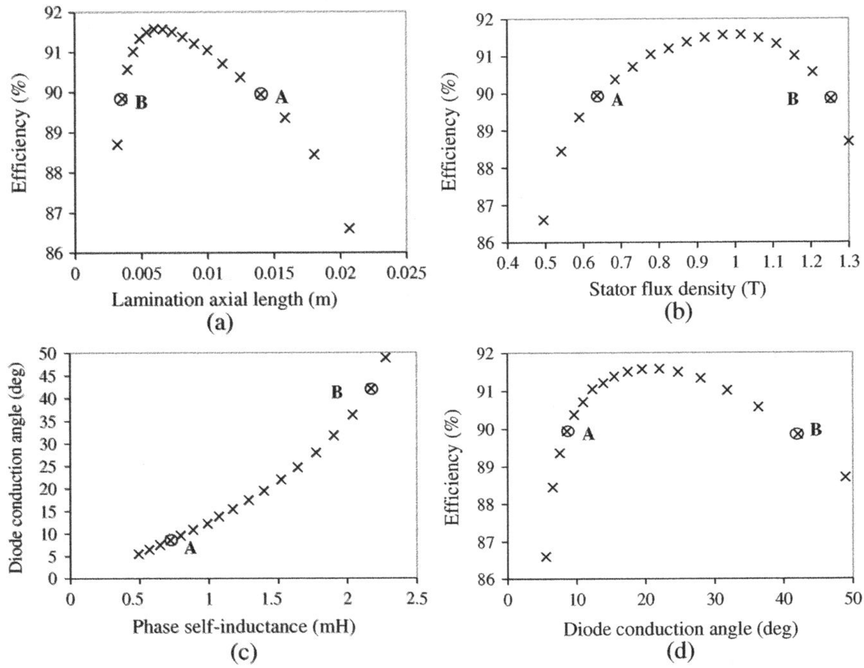

Figure 10.35 Influence of design parameters on motor efficiency, self-inductance, and diode conduction angle. (a) Efficiency against lamination axial length. (b) Efficiency against stator flux density. (c) Diode conduction angle against phase self-inductance. (d) Efficiency against diode conduction angle. (*Source:* [3]/IOS Press.)

Table 10.7 Comparison of 120 000 r/min, 190 W PM BLDC designs.

	Motor A	Motor B		Motor A	Motor B
Supply voltage/ current, V/A	200/1.1	200/1.1	No. of turns per phase	74	220
Outer diameter/ length, mm	50/30	50/30	Winding inductance, mH	0.72	2.172
Airgap length, mm	4	4	Copper/Stator iron weight, g	18.4/134	59.5/21.3
Rotor magnet outer diameter, mm	17	17	Magnet weight, g	14	3.5
NdFeB magnet thickness, mm	3	3	Copper/Iron loss, W	4.6/16.5	13/8.3
Flux density in stator iron, T	0.65	1.25	Efficiency, %	89.9	89.8
Active axial length, mm	14	3.6	Diode conduction angle, ° elec.	8.6	42.1

(*Source:* [3]/IOS Press.)

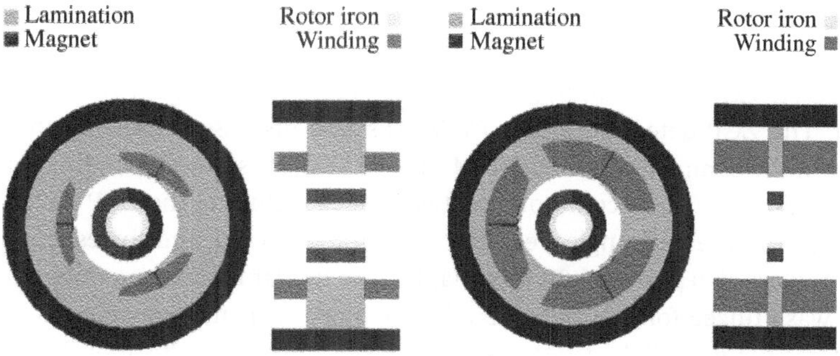

Figure 10.36 Schematic of motors A and B. (a) Cross-sections of motor A. (b) Cross-sections of motor B. (*Source:* [3]/IOS Press.)

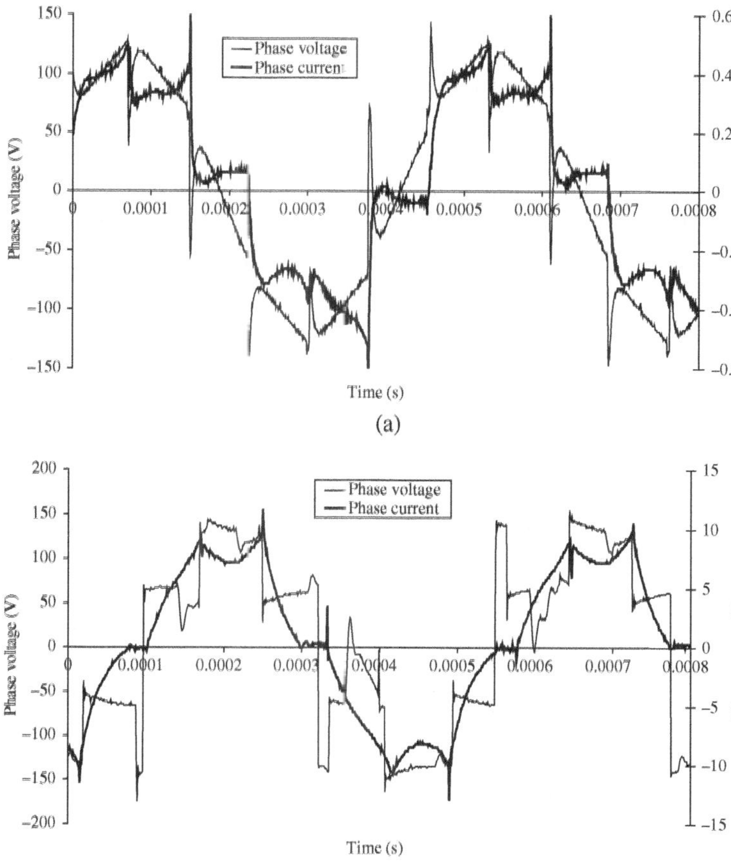

Figure 10.37 Phase voltage and phase current waveforms. (a) Motor A under sensorless operation at 132 kr/min. (b) Motor B operated from Hall sensors at 127 kr/min. (*Source:* [3]/IOS Press.)

10.7 Summary

In this chapter, ZCP detection-based sensorless control of BLDC drives was introduced. The ZCP detection under PWM was also discussed, showing that back-EMF can be sampled at either PWM-on or PWM-off states depending on operating conditions. Then, common practical issues including the ZCP deviation and large freewheeling angle were described and analyzed, and corresponding solutions were also provided. In this chapter, the fundamental component of back-EMF was utilized for ZCP detection. In Chapter 11, an alternative method is introduced, i.e. the ZCP detection of the third harmonic back-EMF of BLDC drives.

References

1 L. Wang, Z. Q. Zhu, H. Bin, and L. Gong, "A commutation error compensation strategy for high-speed brushless dc drive based on adaline filter," *IEEE Trans. Ind. Electron.*, vol. 68, no. 5, pp. 3728–3738, May 2021.

2 L. Wang, Z. Q. Zhu, H. Bin, and L. Gong, "A commutation optimization strategy for high-speed brushless dc drives with voltage source inverter," *IEEE Trans. Ind. Appl.*, vol. 58, no. 4, pp. 4722–4732, Jul. 2022.

3 Z. Q. Zhu, J. D. Ede, and D. Howe, "Design criteria for brushless dc motors for high-speed sensorless operation," *Int. J. Appl. Electromagn. Mech.*, vol. 15, pp. 79–87, 2001/2002.

4 J. Shao, D. Nolan, M. Teissier, and D. Swanson, "A novel microcontroller-based sensorless brushless dc (BLDC) motor drive for automotive fuel pumps," *IEEE Trans. Ind. Appl.*, vol. 39, no. 6, pp. 1734–1740, Nov. 2003.

5 J. Shao, "An improved microcontroller-based sensorless brushless dc (BLDC) motor drive for automotive applications," *IEEE Trans. Ind. Appl.*, vol. 42, no. 5, pp. 1216–1221, Sep. 2006.

6 H.-G. Yee, C.-S. Hong, J.-Y. Yoo, H.-G. Jang, Y.-D. Bae, and Y.-S. Park, "Sensorless drive for interior permanent magnet brushless dc motors," in *1997 IEEE Int. Elect. Mach. Drives Conf. Rec.*, p. TD1/3.1-TD1/3.3, May 1997.

7 Y.-S. Lai and Y.-K. Lin, "A unified approach to zero-crossing point detection of back EMF for brushless dc motor drives without current and hall sensors," *IEEE Trans. Power Electron.*, vol. 26, no. 6, pp. 1704–1713, Jun. 2011.

8 S. Chen, G. Liu, and S. Zheng, "Sensorless control of BLDCM drive for a high-speed maglev blower using low-pass filter," *IEEE Trans. Power Electron.*, vol. 32, no. 11, pp. 8845–8856, Nov. 2017.

9 Q. Jiang, C. Bi, and R. Huang, "A new phase-delay-free method to detect back EMF zero-crossing points for sensorless control of spindle motors," *IEEE Trans. Magn.*, vol. 41, no. 7, pp. 2287–2294, Jul. 2005.

10 T.-W. Chun, Q.-V. Tran, H.-H. Lee, and H.-G. Kim, "Sensorless control of BLDC motor drive for an automotive fuel pump using a hysteresis comparator," *IEEE Trans. Power Electron.*, vol. 29, no. 3, pp. 1382–1391, Mar. 2014.

11 G.-J. Su and J. W. McKeever, "Low-cost sensorless control of brushless dc motors with improved speed range," *IEEE Trans. Power Electron.*, vol. 19, no. 2, pp. 296–302, Mar. 2004.

12 G. Haines and N. Ertugrul, "Wide speed range sensorless operation of brushless PM motor using flux linkage increment," *IEEE Trans. Ind. Electron.*, vol. 63, no. 7, pp. 4052–4060, Jul. 2016.

13 S. Ogasawara and H. Akagi, "An approach to position sensorless drive for brushless dc motors," *IEEE Trans. Ind. Appl.*, vol. 27, no. 5, pp. 928–933, Sep. 1991.

14 N. Matsui, "Sensorless PM brushless dc motor drives," *IEEE Trans. Ind. Electron.*, vol. 43, no. 2, pp. 300–308, Apr. 1996.

15 J. X. Shen, Z. Q. Zhu, and D. Howe, "Sensorless flux-weakening control of permanent-magnet brushless machines using third harmonic back EMF," *IEEE Trans. Ind. Appl.*, vol. 40, no. 6, pp. 1629–1636, 2004.

16 H. Li, S. Zheng, and H. Ren, "Self-correction of commutation point for high-speed sensorless BLDC motor with low inductance and nonideal back EMF," *IEEE Trans. Power Electron.*, vol. 32, no. 1, pp. 642–651, Jan. 2017.

17 X. Zhou, X. Chen, F. Zeng, and J. Tang, "Fast commutation instant shift correction method for sensorless coreless BLDC motor based on terminal voltage information," *IEEE Trans. Power Electron.*, vol. 32, no. 12, pp. 9460–9472, Dec. 2017.

18 L. Yang, Z. Q. Zhu, B. Shuang, and H. Bin, "Adaptive threshold correction strategy for sensorless high-speed brushless dc drives considering zero-crossing-point deviation," *IEEE Trans. Ind. Electron.*, vol. 67, no. 7, pp. 5246–5257, Jul. 2020.

19 L. Yang, Z. Q. Zhu, H. Bin, Z. Zhang, and L. Gong, "Safe operation area of zero-crossing detection-based sensorless high-speed BLDC motor drives," *IEEE Trans. Ind. Appl.*, vol. 56, no. 6, pp. 6456–6466, 2020.

20 Y. Xu, Y. Wei, B. Wang, and J. Zou, "A novel inverter topology for brushless dc motor drive to shorten commutation time," *IEEE Trans. Ind. Electron.*, vol. 63, no. 2, pp. 796–807, Feb. 2016.

21 T. Shi, Y. Cao, G. Jiang, X. Li, and C. Xia, "A torque control strategy for torque ripple reduction of brushless dc motor with nonideal back electromotive force," *IEEE Trans. Ind. Electron.*, vol. 64, no. 6, pp. 4423–4433, Jun. 2017.

22 J. D. Ede, Z. Q. Zhu, and D. Howe, "Design considerations for high-speed PM brushless dc motors," *Proc. Int. Conf. Power Electron. Mach. Drives*, 2004, pp. 686–690.

23 Y. Pang, Z. Q. Zhu, and D. Howe, "Analytical determination of optimal split ratio for permanent magnet brushless motors," *IEE Proc. Elect. Power Appl.*, vol. 153, no. 1, pp. 7–13, 2006.

24 J. D. Ede, Z. Q. Zhu, and D. Howe, "Rotor resonances of high-speed permanent magnet brushless motors," *IEEE Trans. Ind. Appl.*, vol. 38, no. 6, pp. 1542–1548, 2002.

11

Sensorless Control Based on Third Harmonic Back-EMF

11.1 Introduction

Third harmonic back-EMF in the stator phase voltages can be significant due to heavy saturation and/or the character of the designed PM excited flux-linkage, which purposely contains the third harmonic [1]. Third harmonic back-EMF-based sensorless control techniques have been employed for BLDC drive operation [2]–[8] and BLAC drive operation [3], [9], [10]. Compared to the zero-crossing detection method of the fundamental back-EMF, the third harmonic back-EMF-based sensorless control has several advantages:

- Low cost
- Less influenced by PWM noises
- Wider speed operation range
- Not distorted by freewheeling diode conduction
- Applicable to both BLDC and BLAC drives

In this chapter, third harmonic back-EMF-based sensorless techniques are presented, including the principles of operation and practical issues.

11.2 Detection and Methods

Figure 11.1 shows a schematic of a three-phase PM brushless drive and a circuit for sensing the third harmonic back-EMF. In the figure, N is the neutral point of the Y-connected motor windings, S is the star-point of a Y-connected resistor network, which consists of three identical resistors R_N, while M is the mid-point of the dc bus, which can be obtained by using either two identical resistors or two identical capacitors. The third harmonic back-EMF can be measured from the voltage

Sensorless Control of Permanent Magnet Synchronous Machine Drives, First Edition.
Zi Qiang Zhu and Xi Meng Wu.

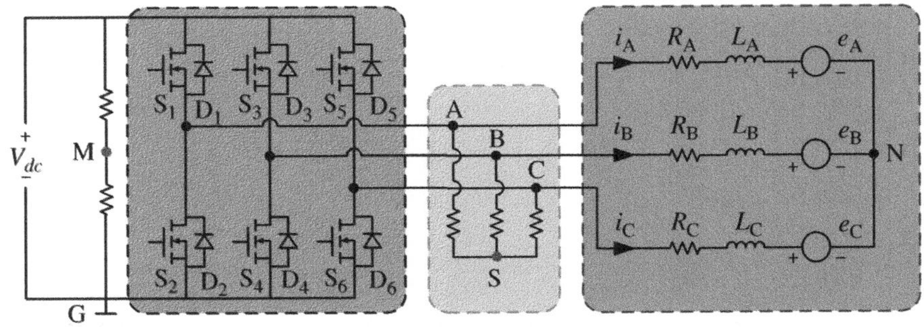

Figure 11.1 Schematic of PM brushless drive and third harmonic back-EMF sensing circuit. (*Source:* Adapted from [3].

difference between points S and N. For the motor without the neutral line, the voltage difference between S and M can be used, i.e. a virtual third harmonic back-EMF.

11.2.1 Third Harmonic Back-EMF

Assuming the inductances of the motor are constant with the rotor position, the motor phase voltages and currents can be expressed as:

$$\begin{cases} v_{AN} = Ri_A + (L-M)pi_A + e_A \\ v_{BN} = Ri_B + (L-M)pi_B + e_B \\ v_{CN} = Ri_C + (L-M)pi_C + e_C \end{cases} \tag{11.1}$$

$$i_A + i_B + i_C = 0 \tag{11.2}$$

Therefore,

$$v_{AN} + v_{BN} + v_{CN} = e_A + e_B + e_C \tag{11.3}$$

Irrespective of the mode of operation, i.e. whether BLDC or BLAC, and with or without PWM, these equations are always applicable [3]. However, the phase back-EMF waveform usually contains the fundamental and odd harmonics. Thus,

$$\begin{cases} e_A = e_{A1} + e_3 + e_{A5} + e_{A7} + e_9 + \cdots \\ \quad = -E_1 \sin\theta_r - E_3 \sin 3\theta_r - E_5 \sin 5\theta_r \\ \quad\quad - E_7 \sin 7\theta_r - E_9 \sin 9\theta_r - \cdots \\ e_B = e_{B1} + e_3 + e_{B5} + e_{B7} + e_9 + \cdots \\ \quad = -E_1 \sin(\theta_r - 2\pi/3) - E_3 \sin 3\theta_r - E_5 \sin 5(\theta_r - 2\pi/3) \\ \quad\quad - E_7 \sin 7(\theta_r - 2\pi/3) - E_9 \sin 9(\theta_r - 2\pi/3) - \cdots \\ e_C = e_{C1} + e_3 + e_{C5} + e_{C7} + e_9 + \cdots \\ \quad = -E_1 \sin(\theta_r + 2\pi/3) - E_3 \sin 3\theta_r - E_5 \sin 5(\theta_r + 2\pi/3) \\ \quad\quad - E_7 \sin 7(\theta_r + 2\pi/3) - E_9 \sin 9(\theta_r + 2\pi/3) - \cdots \end{cases} \tag{11.4}$$

where θ_r is the rotor position. From (11.3) and (11.4), the following equation is derived:

$$v_{AN} + v_{BN} + v_{CN} = 3 \times (e_3 + e_9 + e_{15} + \cdots) \tag{11.5}$$

Further, considering the resistor network, the following relationships exist:

$$\begin{cases} v_{AS} + v_{BS} + v_{CS} = 0 \\ v_{AS} = v_{AN} - v_{SN} \\ v_{BS} = v_{BN} - v_{SN} \\ v_{CS} = v_{CN} - v_{SN} \end{cases} \tag{11.6}$$

Thus, the voltage (v_{SN}) between points S and N is:

$$v_{SN} = \frac{v_{AN} + v_{BN} + v_{CN}}{3} \tag{11.7}$$

From (11.5) and (11.7), v_{SN} is derived as:

$$v_{SN} = e_3 + e_9 + e_{15} + \cdots \tag{11.8}$$

Usually, $e_3 >> (e_9 + e_{15} + \cdots)$. Hence, (11.8) can be approximated to:

$$v_{SN} = e_3 \tag{11.9}$$

Clearly, the voltage v_{SN} represents the third harmonic back-EMF irrespective of the operational mode of the brushless drive, and it is theoretically not distorted by the freewheeling diode conduction [3]. Therefore, v_{SN} is an ideal signal for sensing the third harmonic back-EMF. Hence, if v_{SN} is used for sensorless control, it can overcome the drawbacks of conventional methods by sensing the terminal voltages or detecting the freewheeling diode conducting status [3].

11.2.2 Virtual Third Harmonic Back-EMF

From Figure 11.1, the following voltage equations can be obtained:

$$\begin{cases} v_{AN} = v_{AG} + v_{GM} - v_{SM} + v_{SN} \\ v_{BN} = v_{BG} + v_{GM} - v_{SM} + v_{SN} \\ v_{CN} = v_{CG} + v_{GM} - v_{SM} + v_{SN} \\ v_{GM} = -V_{dc}/2 \end{cases} \tag{11.10}$$

Then, v_{SM} can be calculated as:

$$v_{SM} = \frac{1}{3}(v_{AG} + v_{BG} + v_{CG}) - \frac{1}{2}V_{dc} = \frac{1}{3}(e_A + e_B + e_C) - \frac{1}{2}V_{dc} + v_{NG} \tag{11.11}$$

In the BLAC mode, the terminal voltages (v_{AG}, v_{BG}, and v_{CG}) are either 0 or V_{dc}. Therefore, the voltage v_{SM} comprises a series of pulses, the amplitude being either $V_{dc}/2$ or $V_{dc}/6$, as shown in Figure 11.2. Clearly, v_{SM} cannot be used for the sensorless control of a BLAC drive.

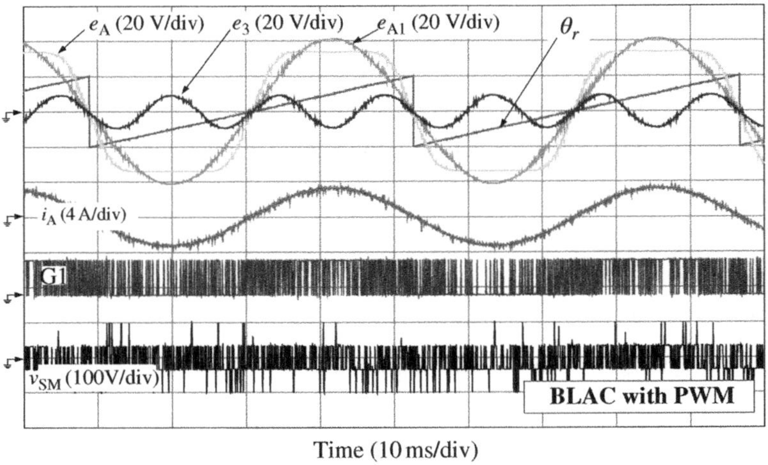

Figure 11.2 Measured v_{SM} under BLAC operation at 460 r/min. (*Source:* Adapted from [3].

Equation (11.11) describes the voltage signal v_{SM} in a general situation. For the six-step commutation method, an electric drive cycle can be divided into six sectors. In each sector, two phases are conducting and the third one is floating. The phases connected to the positive and negative dc-link rails are defined as phases H and L, respectively. The third phase that is floating can be defined as phase O. Then, the voltage v_{NG} can be calculated as:

$$v_{NG} = \frac{1}{2}(v_{HG} + v_{LG}) - \frac{1}{2}(e_H + e_L) = \frac{1}{2}V_{dc} - \frac{1}{2}(e_H + e_L) \tag{11.12}$$

In each sector, the current flows from phase H to phase L. Hence, the current in three phases can be calculated as:

$$i_L = -i_H, \quad i_O = 0 \tag{11.13}$$

By substituting (11.12) and (11.13) into (11.11), the voltage v_{SM} can be obtained as:

$$v_{SM} = \frac{1}{6}(2e_O - e_H - e_L) \tag{11.14}$$

For typical sinusoidal and trapezoidal back-EMFs, it yields:

$$\begin{cases} e_H + e_L = -e_O & \text{sinusoidal EMF} \\ e_H + e_L = 0 & \text{trapezoidal EMF} \end{cases} \tag{11.15}$$

Table 11.1 Summary of third harmonic back-EMF detection methods.

	Third harmonic	Virtual third harmonic
BLDC without PWM	Applicable	Applicable
BLDC with PWM	Applicable	Applicable
BLAC with PWM	Applicable	Inapplicable
Operation mode	Independent	Dependent
Freewheeling diode conduction	Independent	Dependent
Back-EMF component	Third harmonic component	Fundamental component
Neutral point access	Required	Not required

Then, $(2e_O - e_H - e_L)/6$ in (11.19) can be simplified as:

$$\frac{1}{6}(2e_O - e_H - e_L) = \begin{cases} \frac{1}{2}\epsilon_O & \text{sinusoidal EMF} \\ \frac{1}{3}\epsilon_O & \text{trapezoidal EMF} \end{cases} \quad (11.16)$$

Thus, for both sinusoidal and trapezoidal back-EMF-type motors, the voltage v_{SM} contains the floating phase back-EMF signal, which can be used for sensorless control. It is worth noting that v_{SM} does not actually contain the information of the third harmonic back-EMF. Instead, it contains the fundamental component of the floating phase back-EMF [3] and is distorted by PWM and the freewheeling diode conduction as described in Chapter 10.

Based on the analysis in this section, these two introduced detection methods are summarized and compared in Table 11.1.

11.3 BLDC Operation

11.3.1 Without PWM

For third harmonic back-EMF and virtual third harmonic back-EMF-based methods, experimental examples of BLDC operation without PWM are shown in Figure 11.3. By detecting the zero-crossing points, the six particular commutation points essential for BLDC operation, which are $\pi/6$, $\pi/2$, $5\pi/6$, $7\pi/6$, $3\pi/2$, $11\pi/6$, can be detected as shown in Figure 11.4. These discrete rotor positions can be directly employed for BLDC drives. Moreover, in order to improve the steady-state performance, the third harmonic flux-linkage ψ_{SN}, which is the integration of the

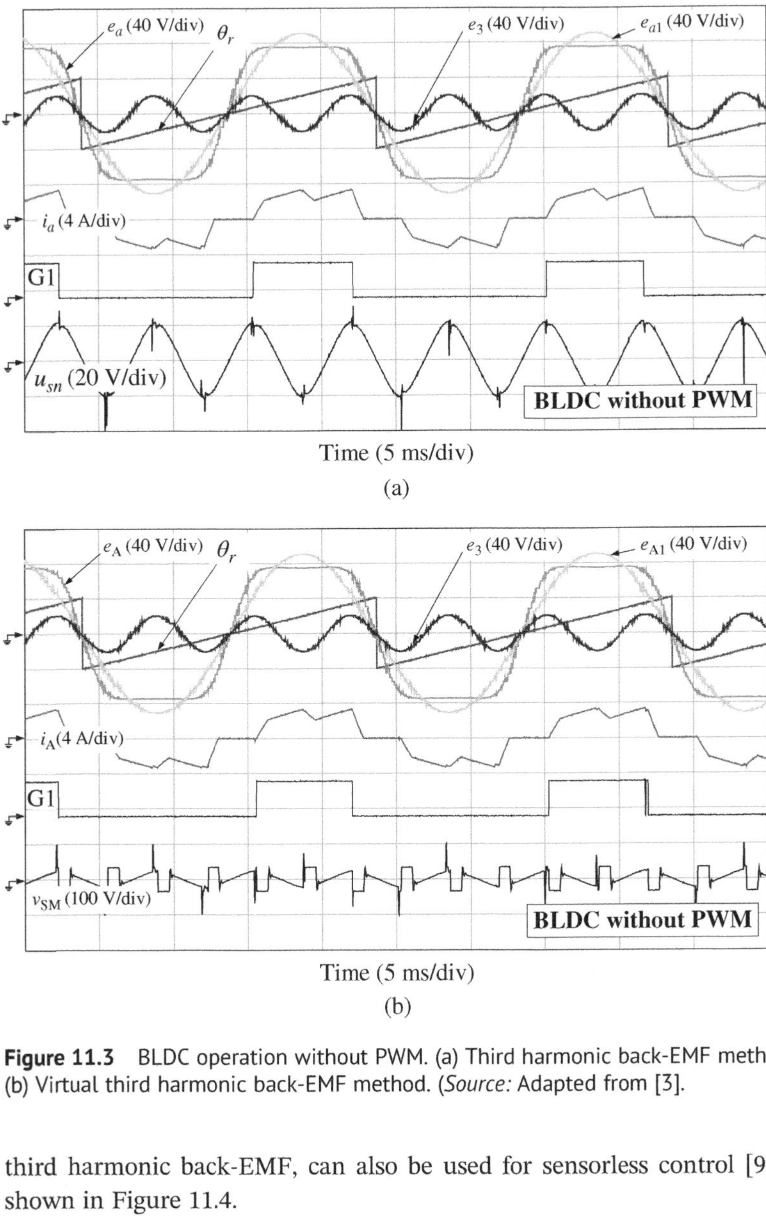

Figure 11.3 BLDC operation without PWM. (a) Third harmonic back-EMF method. (b) Virtual third harmonic back-EMF method. (*Source:* Adapted from [3].

third harmonic back-EMF, can also be used for sensorless control [9], [10] as shown in Figure 11.4.

11.3.2 With PWM

PWM is normally employed in BLDC drives to regulate the control of speed and current. However, it will introduce high-frequency noises in the measurement,

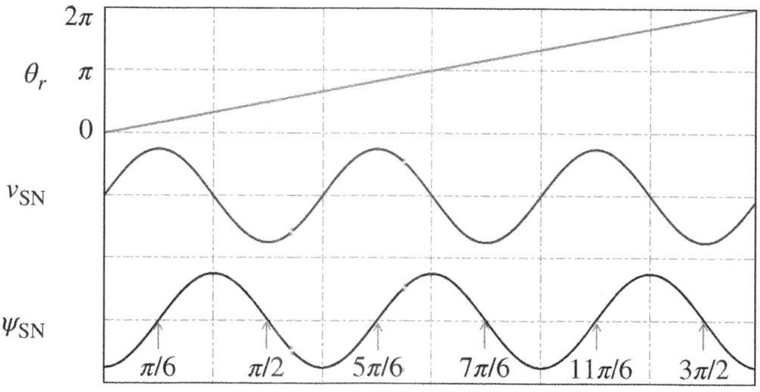

Figure 11.4 Relationship of v_{SN}, its integration ψ_{SN}, and rotor position. (*Source:* [9]/IEEE.)

which increases the difficulty of ZCP detection. Experimental examples of BLDC operation with PWM are shown in Figure 11.5.

11.3.2.1 Low-Pass Filter

To remove the high-frequency noises, low-pass filters (LPFs) or band-pass filters (BPFs) are used in most existing literature [3], [5]–[7]. The usage of filters will cause serious phase lag to the commutation and consequently deteriorate the control performance. Therefore, the corresponding phase compensation delay should be employed.

11.3.2.2 Envelope Construction

The envelope construction method is found to offer an alternative solution [11]. With the help of the analog-to-digital converter (ADC), the back-EMF signal can be reconstructed and then the ZCP can be detected in algorithm. The filters are not necessary, and thereby the resulting phase lag can be avoided. In this section, the scheme of envelope construction of the virtual third harmonic [8] is presented as shown in Figure 11.6.

As introduced in Chapter 10, four common PWM patterns can be used for BLDC drives. In this section, taking the H-PWM-L-ON-based PWM scheme, for instance, the high-side power devices are controlled by PWM chopping signals and low-side power devices keep on or off in each sector.

According to Section 11.2.2, the terminal voltages v_{HG} and v_{LG} can be described as:

$$
\begin{aligned}
v_{HG} &= \Lambda \cdot V_{dc} = \begin{cases} V_{dc} & \text{PWM}-\text{on} \\ 0 & \text{PWM}-\text{off} \end{cases} \\
v_{LG} &= 0
\end{aligned}
\tag{11.17}
$$

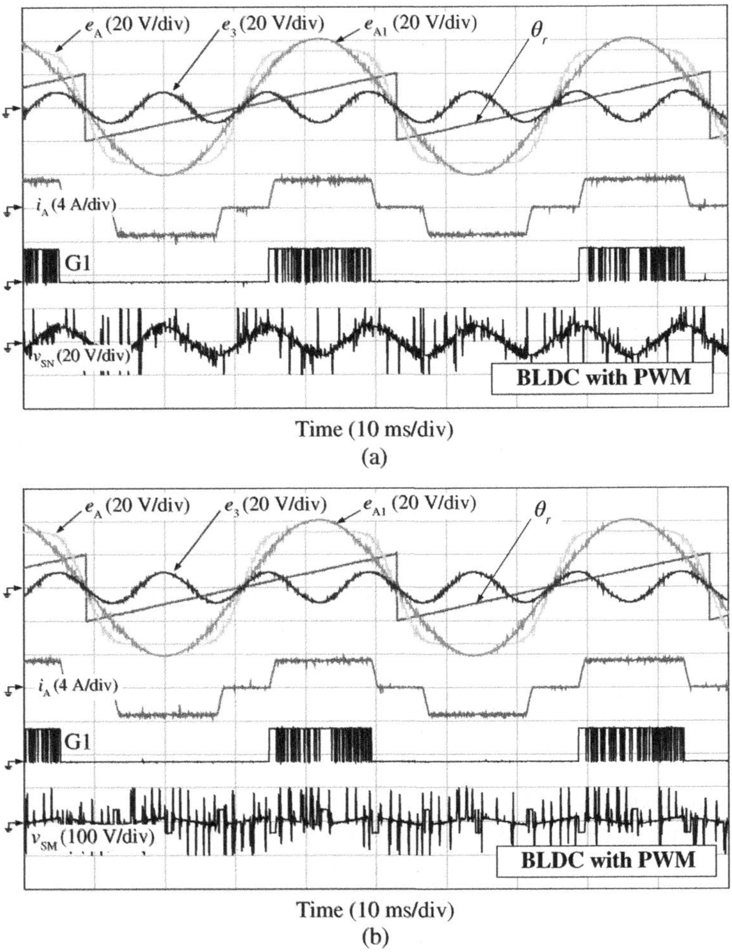

Figure 11.5 BLDC operation with PWM. (a) Third harmonic back-EMF method. (b) Virtual third harmonic back-EMF method. (*Source:* Adapted from [3].)

where Λ is the PWM switching function as:

$$\Lambda = \begin{cases} 1 & \text{PWM} - \text{on} \\ 0 & \text{PWM} - \text{off} \end{cases} \tag{11.18}$$

By substituting (11.40) and (11.41) into (11.39), the voltage v_{SM} can be obtained as:

$$v_{SM} = \underbrace{\frac{1}{6}(2e_O - e_H - e_L)}_{\text{Back}-\text{EMF}} + \underbrace{\frac{1}{2}(\Lambda - 1)V_{dc}}_{\text{PWM fluctuation}} \tag{11.19}$$

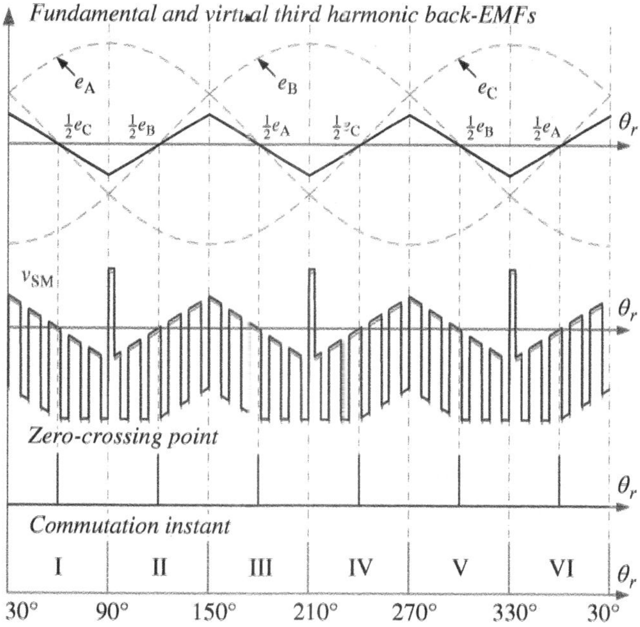

Figure 11.6 ZCPs of virtual third harmonic and commutation positions. (*Source:* [8]/IEEE.)

It can be seen from (11.19) that v_{SM} consists of two parts, i.e. the back-EMF and the voltage fluctuation caused by PWM. Only the first part $(2e_O - e_H - e_L)/6$ contains the rotor position information. As shown in Figure 11.6, although the voltage signal v_{SM} is influenced by high-frequency PWM fluctuation, the back-EMF can be reflected on its upper envelope. The commutation instant can be estimated from the ZCP of this envelope signal. This can be realized by sampling the voltage v_{SM} during the PWM-on period. By substituting $\Lambda = 1$ into (11.19), considering a sinusoidal back-EMF, the upper envelope $\hat{v}_{SM}$ can be expressed as:

$$\hat{v}_{SM} = v_{SM}|_{\Lambda=1} = \frac{1}{2}e_O \tag{11.20}$$

Here, the hat ^ denotes the upper envelope. After obtaining the upper envelope $\hat{v}_{SM}$, the ZCP can be detected with

$$\begin{cases} \hat{v}_{SM}[n-1] \le 0 \\ \hat{v}_{SM}[n] \ge 0 \end{cases} \text{or} \begin{cases} \hat{v}_{SM}[n-1] \ge 0 \\ \hat{v}_{SM}[n] \le 0 \end{cases} \tag{11.21}$$

where $\hat{v}_{SM}[n-1]$ and $\hat{v}_{SM}[n]$ denote the previous and present samplings, respectively. Considering that the excitation voltage term contains PWM switching function Λ, then after obtaining the envelope $\hat{v}_{SM}$, the ZCP can be detected in algorithm without the influence of PWM fluctuation.

Over-Sampling for High-Speed Operation

Figure 11.7 shows the diagram of ZCP detection of the virtual third harmonic.

Since the ADC works in a discrete way, the sampling delay is inevitable in ZCP detection, especially in the high-speed region. It is reasonable to increase the sampling rate to minimize the sampling delay [8]. However, the available sampling duration is limited by the PWM-on period. Thus, to fully utilize the available sampling duration, the number of sampling points per PWM wave N_s is dynamically adjusted according to the PWM duty ratio [8]. As shown in Figure 11.7, assuming the PWM-on duration is T_{on} and the minimum sampling interval is T_s, the number of sampling points N_s can be expressed as:

$$N_s = \text{floor}\left(\frac{T_{on}}{T_s}\right) \tag{11.22}$$

where floor(x) denotes the greatest integer less than or equal to x. As the motor speed increases, there is a higher demand for the ZCP detection precision. Fortunately, the duty ratio also becomes larger with the motor speed. Thus, it provides longer PWM-on duration T_{on} to enable more sampling points. In this way, the number of sampling points N_s is dynamically adjusted according to the PWM duty ratio. The precision of ZCP detection can be guaranteed within a wide speed range.

Experiment Results

Figure 11.8 shows the dynamic adjustment of sampling points, in which the voltage signal v_{SM} and its sampling result $\hat{v}_{SM}$ are given. From the theoretical analysis,

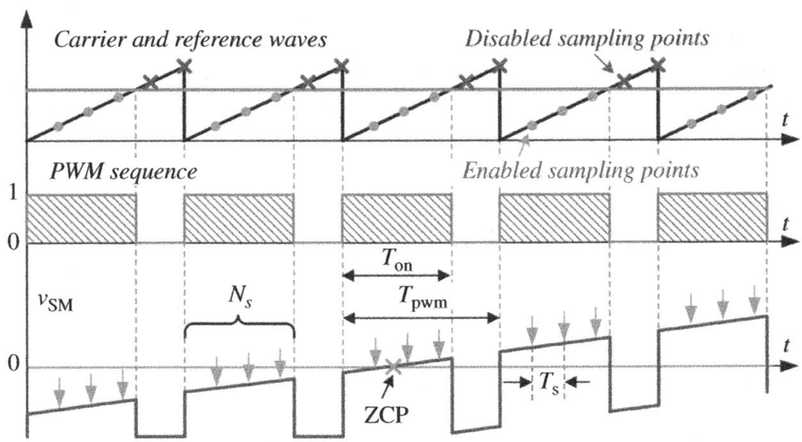

Figure 11.7 Schematic diagram of dynamic over-sampling technique on virtual third harmonic back-EMF. (*Source:* [8]/IEEE.)

the voltage v_{SM} is influenced by the high-frequency PWM fluctuation. The virtual third harmonic signal can be reflected on its upper envelope. By arranging the sampling points on the PWM-on time, this back-EMF envelope can be constructed as $\hat{v}_{SM}$. Also, in the right magnified subplot, some details including the sampling instants are presented. The PWM period is 25 μs and the minimum sampling interval is 5 μs. The number of sampling points N_s is adjusted according to the PWM duty ratio as listed in Table 11.2.

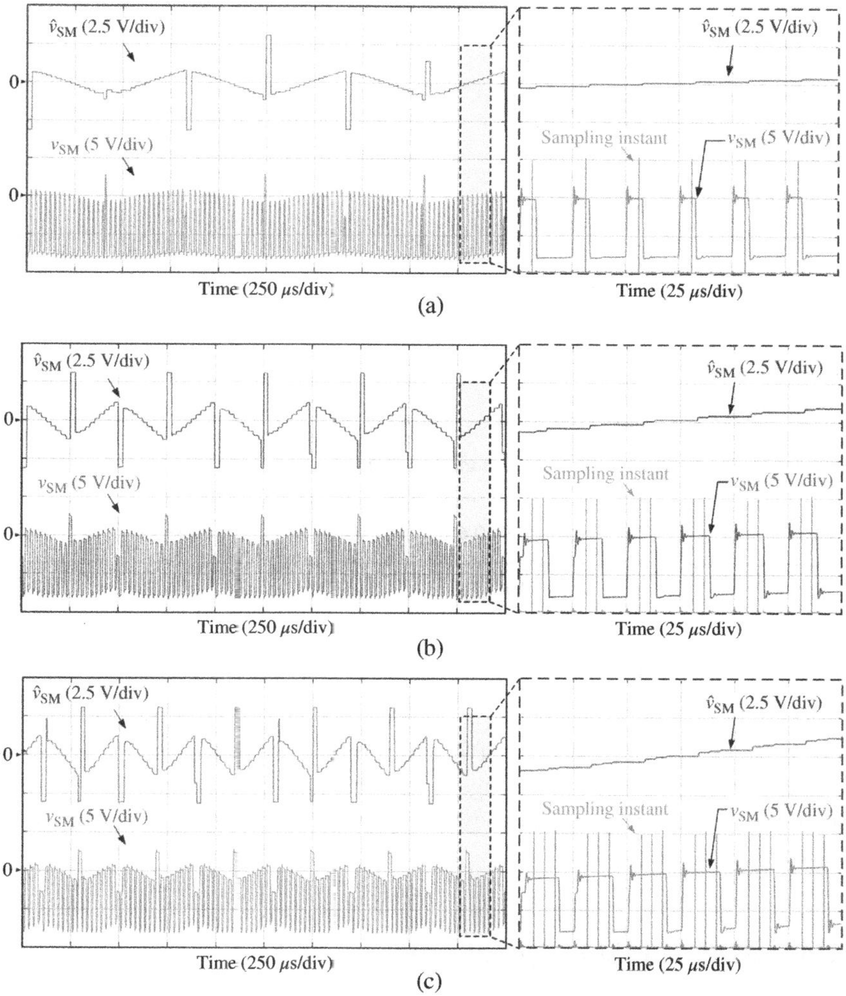

Figure 11.8 Voltage signal v_{SM} and sampling result $\hat{v}_{SM}$. (a) N_s = 1, (b) N_s = 2, (c) N_s = 3, (d) N_s = 4, (e) N_s = 5. (*Source:* [8]/IEEE.)

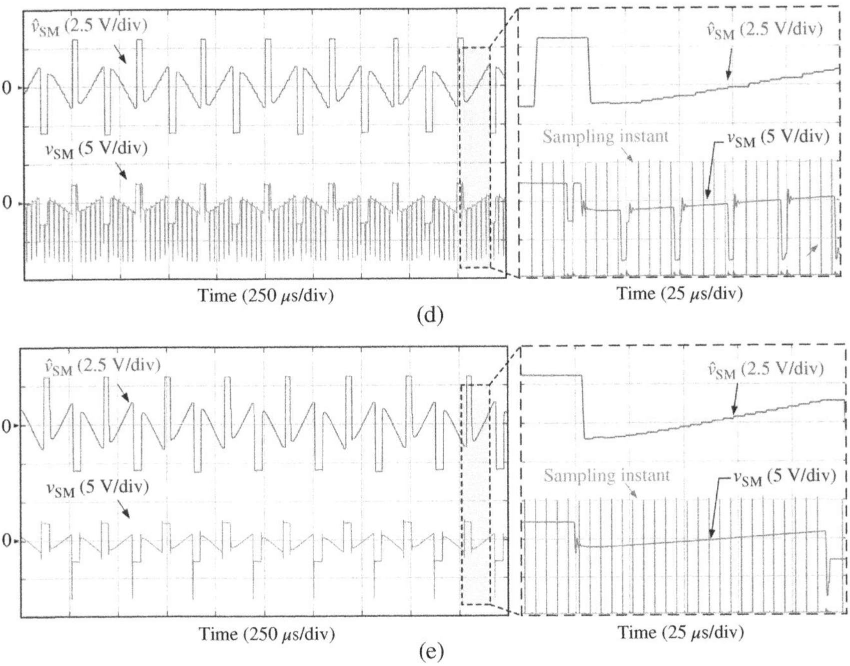

Figure 11.8 (Continued)

Table 11.2 Dynamic adjustment of sampling points.

Duty ratio	**0–0.4**	**0.4–0.6**	**0.6–0.8**	**0.8–1.0**	**1.0**
T_{on} (μs)	0–10	10–15	15–20	20–25	25
N_s	1	2	3	4	5

(*Source:* Adapted from [8].)

As the motor speed increases, the PWM duty ratio becomes higher. There will be a longer PWM-on duration to arrange more sampling points. Therefore, within a wide range of PWM duty ratio, the dynamic over-sampling method is effective to construct the envelope signal $\hat{v}_{SM}$.

11.4 BLAC (PMSM) Operations

As illustrated in Section 11.2, only the third harmonic back-EMF-based method can be used for sensorless control of BLAC drives, while the virtual third harmonic back-EMF-based method cannot be applied in BLAC drives since there is no position information in the measured voltage signal. Therefore, based on the third harmonic back-EMF method, three techniques utilizing third harmonic back-EMF for BLAC [3], [9] are described in this section.

11.4.1 Rotor Position Estimator Based on Integration

A high-resolution estimated rotor position $\hat{\theta}_r$ can be calculated by simple integration [3] as:

$$\hat{\theta}_r = \theta_0 + \int_0^t \hat{\omega}_{ra} dt \tag{11.23}$$

where $\hat{\omega}_{ra}$ is the average estimated rotor speed, which can be calculated by

$$\hat{\omega}_{ra} = \pi/3t_d \tag{11.24}$$

where t_d is the time interval of the previous half cycle between two zero-crossing points.

The calculated $\hat{\omega}_{ra}$ in (11.24) is actually the estimated average speed between the last two zero-crossings rather than the instantaneous speed during the rotor position estimation. The rotor position will be accurate during the steady state since the rotor speed would ideally not change. Furthermore, on each zero-crossing point, the estimated rotor position, which contains some errors due to inaccurate speed estimation, will be forcedly set as the related θ_0. Hence, the rotor position estimation error will not accumulate.

However, during variable speed operation, the estimated average speed $\hat{\omega}_{ra}$ is not fast enough to reflect the speed change. If the rotor speed is high enough, and the speed variation is not too much, the error accumulation will not be great, and will be reset to zero when zero-crossing is detected. However, with low speed, the sharp change of the estimated rotor position will be significant and will degrade the operating performance such as by current distortion and torque pulsation. Employing the phase-locked loop (PLL)-based observer (illustrated in Chapter 2) can improve the speed estimation performance, whereas the insufficient discrete rotor positions from zero-crossings will limit the performance [9].

11.4.2 Rotor Position Estimator Based on Zero-Crossing Correction

To avoid the sharp change on zero-crossings with the estimator based on integration, the position estimator based on zero-crossing correction [9] shown in Figure 11.9 is implemented.

Similarly, the related rotor position θ_0 of each zero-crossing is located and applied as a reference. Then, on each zero-crossing, the rotor position difference between the estimated rotor position and the related reference θ_0 can be calculated and are equivalent to the estimation error between the estimated and actual rotor positions. Thus, a speed correction ω_{cor} can be obtained by the following PI controller as:

$$\omega_{cor} = -\left(K_p + \frac{K_i}{s}\right) \cdot [\Delta\theta \cdot \text{sign}(\hat{\omega}_r)] \tag{11.25}$$

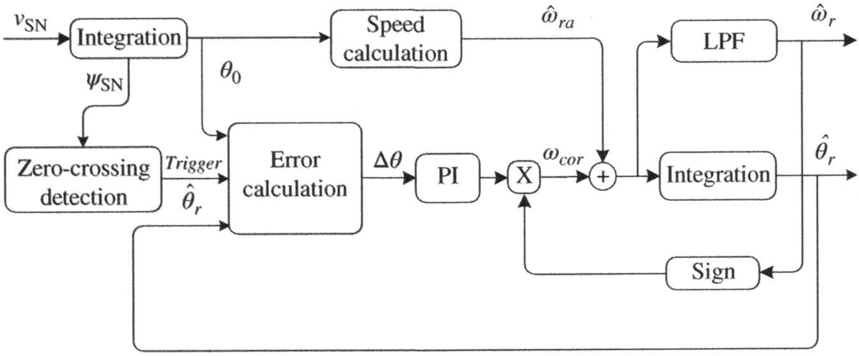

Figure 11.9 Zero-crossing based position estimator. (*Source:* [9]/IEEE.)

where the sign function sign$(\hat{\omega}_r)$ reflects the rotor rotation direction and K_p and K_i represent the proportional and integral gains, respectively.

In (11.25), the speed correction ω_{cor} corresponds to the position estimation error, i.e. the rotor position difference between the estimated and related reference θ_0 on each zero-crossing. Then, the average rotor speed $\hat{\omega}_{ra}$ calculated from the zero-crossings of ψ_{SN} will be appropriately corrected, and the estimated rotor speed $\hat{\omega}_r$ is converged to the actual value. Therefore, a high-resolution rotor position can be estimated by simple integration.

Similar to the rotor position estimator based on integration, as long as the rotor speed is not too low and does not vary too significantly, the error accumulation will not be great and will be reset to zero when zero-crossing is detected. However, for low-speed applications with dynamic performance requirements, the speed calculation error cannot be neglected. Powered by the estimator, speed error during the expected speed change could be compensated for. However, ω_{cor} will only refresh on each zero-crossing. Consequently, the proportional gain K_p and the integral gain K_i in the PI controller cannot be too large in order to avoid divergence. Hence, under the situation of random speed change with low speed, the PI controller cannot follow the speed variation due to insufficient resolution of only six accurate rotor position references. Thus, the dynamic performance will still not be good enough [9].

11.4.3 Rotor Position Estimator Based on Continuous Signal

The calculated $\hat{\omega}_{ra}$ is the estimated average speed between the last two zero-crossings of ψ_{SN}. The estimated rotor position based on that average speed will be accurate during the steady-state operation but inaccurate during variable-speed operation. Therefore, to improve the dynamic performance, speed compensation is required. If the estimation error in the calculated average speed can be compensated for properly, the estimated rotor position based on the compensated speed will be of a high accuracy even under dynamic conditions.

Considering that the third harmonic back-EMF is much bigger than any other triplen harmonics, then the measured v_{SN} can be assumed to contain third harmonic back-EMF only. Therefore, its integration ψ_{SN} will be a continuous sinusoidal signal with constant amplitude. Consequently, the continuous sinusoidal signal of ψ_{SN} can be used as a reference to compensate for the estimation error of the calculated average speed and to further extract the high-resolution rotor position at each sampling time for BLAC sensorless control operation [9].

The block diagram of an improved rotor position estimator based on a continuous signal of ψ_{SN} is shown in Figure 11.10, where $\hat{\psi}_{SN}$ is the virtual third harmonic flux-linkage calculated from the estimated rotor position.

If the estimated rotor position and speed are the same as the actual values, ψ_{SN} and $\hat{\psi}_{SN}$ should be in-phase with each other. In contrast, they will be out of phase when there is an estimation error between the estimated and actual values. The phase difference between the two signals could be used to estimate the rotor position.

ψ_{SN} can be defined as:

$$\psi_{SN} = -A_{mp} \cos(3 \times \theta_r) \tag{11.26}$$

where A_{mp} is the amplitude of ψ_{SN}, which could be obtained from the absolute value of ψ_{SN} when its differential is close to zero.

To estimate the rotor position, the related orthogonal signal of ψ_{SN} also should be calculated as an auxiliary signal based on v_{SN}. However, since the amplitude of v_{SN} varies with the rotor speed, it should be unified as A_{mp} considering the feedback estimated rotor speed $\hat{\omega}_r$ as:

$$v_{SN\_unified} = \frac{K_p}{\hat{\omega}_r} E_3 \sin(3 \times \theta_r) = A_{mp} \sin(3 \times \theta_r) \tag{11.27}$$

where K_p is the unified ratio of amplitudes of the third harmonic flux-linkage and back-EMF. The measured ψ_{SN} and $v_{SN\_unified}$ are shown in Figure 11.11.

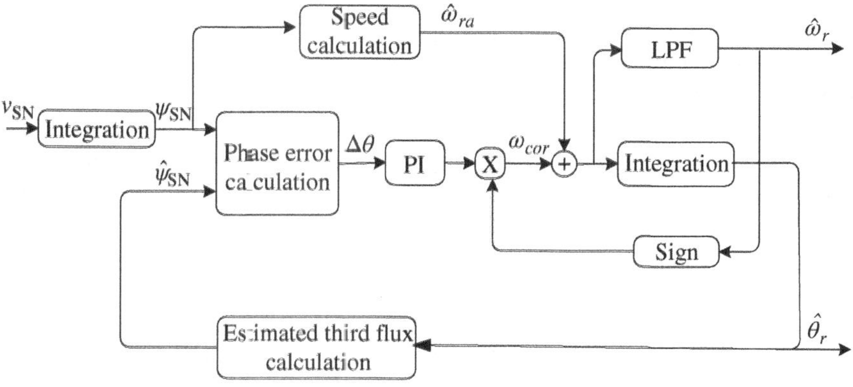

Figure 11.10 Improved rotor position estimator based on speed compensation. (*Source:* [9]/IEEE.)

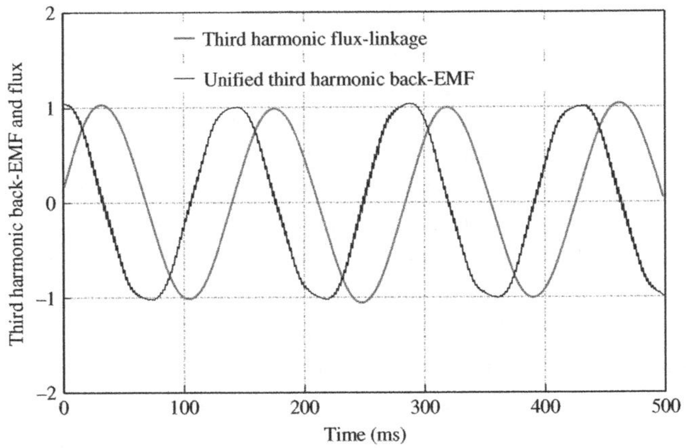

Figure 11.11 Measured ψ_{SN} and $v_{SN\_unified}$. (*Source:* [9]/IEEE.)

Consequently, $\Delta\theta$ between the actual and estimated third harmonic flux-linkages can be derived as:

$$\Delta\theta = -A_{mp} \arcsin\left[\psi_{SN} \times \sin\left(3 \times \hat{\theta}_r\right) + v_{SN\_unified} \times \cos\left(3 \times \hat{\theta}_r\right)\right] \tag{11.28}$$

The phase difference $\Delta\theta$ is also equivalent to three times the rotor position estimation error between the estimated and actual rotor positions. Then, the speed correction ω_{cor} that corresponds to the speed estimation error can be derived by the following PI controller as a loop filter as the same as the one in (11.25) from the phase difference $\Delta\theta$. Then, the rotor speed derived from zero-crossings or by PLL both including the intrinsic speed errors is appropriately corrected, and the estimated rotor speed $\hat{\omega}_{ra}$ including the intrinsic speed error can also be corrected, while the estimated rotor speed $\hat{\omega}_r$ will converge to the actual value. Consequently, the high-resolution rotor position can be estimated by simple integration with a high accuracy from the compensated speed. Even though the source signals may contain certain noise and high-order harmonic components such as ninth, fifteenth, etc., they can be eliminated by the estimator and will have little effect on the accuracy of rotor speed estimation.

However, one cycle of back-EMF contains three identical cycles of third harmonic back-EMF, and the estimated rotor position may align to the other two cycles of third harmonics rather than the correct one. Therefore, the phase A back-EMF should be considered to avoid the misalignment. On the zero-crossing point of phase A back-EMF calculated by a rough flux observer, the estimated rotor position should be located between $-\pi/3$ and $\pi/3$. If the rotor position is out of range, then based on the rotor position estimation error, a rotor position bias that is $-2\pi/3$ or $2\pi/3$ needs to be added to the estimated rotor position. Hence, the accurate rotor position can be derived.

11.4.4 Experiment Results

Several experiments have been carried out to validate the rotor position estimators based on the third harmonic back-EMF for a BLAC drive. The overall control scheme is shown in Figure 11.12. The experiment is based on SPMSM-V (Appendix B). The sensing resistor network is connected with three 20 kΩ power resistors.

The estimators based on the integration, zero-crossing correction, and continuous signal of ψ_{SN} are implemented at the same steady state. Then, the estimated and actual rotor positions and the corresponding errors are shown in Figure 11.13. The position estimator based on a continuous signal has a better steady-state performance with a lower phase error than the other two methods.

The dynamic performance test is carried out when the rotor speed is 10 r/min with the load current changing from 1 to 3 A. Figure 11.14 shows the estimated rotor position and error. It is clearly shown that during load transient, both rotor speeds calculated from zero crossings and by PLL-based speed observer have significant errors due to the insufficiency of six accurate rotor positions. However, the errors can be remarkably compensated for by the estimator based on the continuous signal, and then, the performance of rotor position estimation is significantly upgraded.

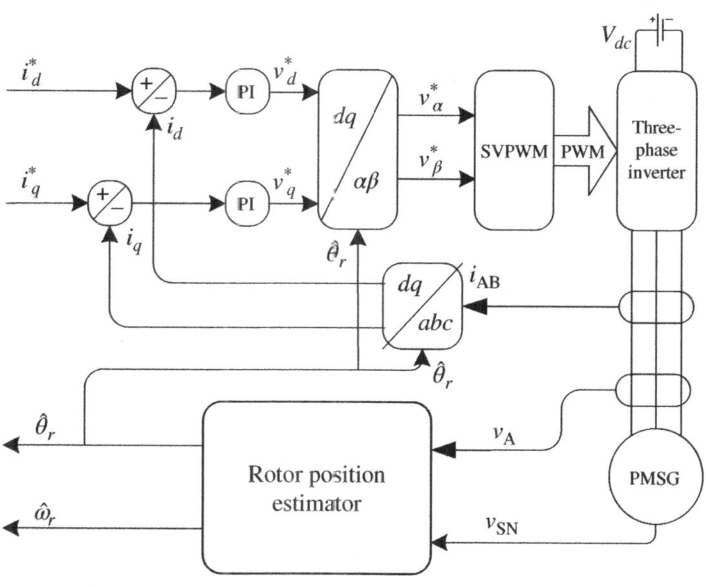

Figure 11.12 Overall control scheme for rotor position estimator based on third harmonic back-EMF. (*Source:* [9]/IEEE.)

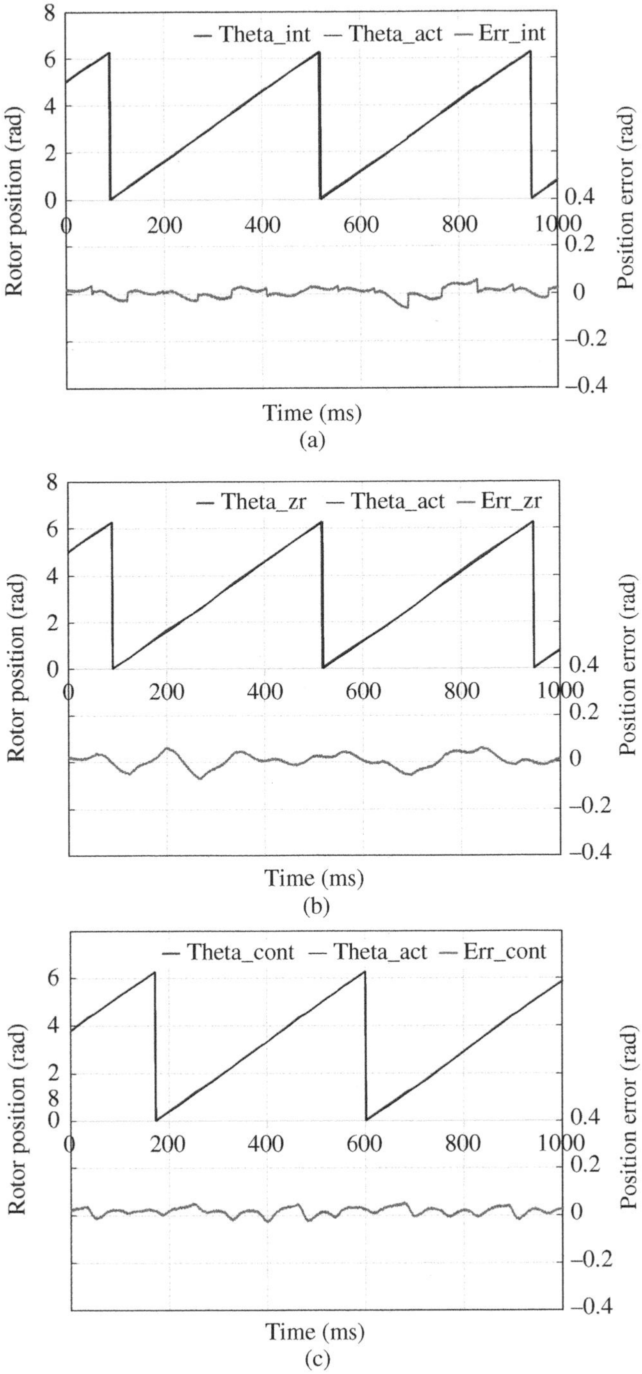

Figure 11.13 Rotor positions estimation in steady state. (a) Estimator based on integration (*Theta_act*, Actual rotor position; *Err_int*, Position estimation error by integration). (b) Estimator based on zero-crossing correction (*Theta_zr*, Estimated rotor position; *Theta_act*, Actual rotor position; *Err_zr*, Position estimation error). (c) Estimator based on continuous signal (*Theta_cont*, Estimated rotor position; *Theta_act*, Actual rotor position; *Err_cont*, Position estimation error). (*Source:* Adapted from [9].)

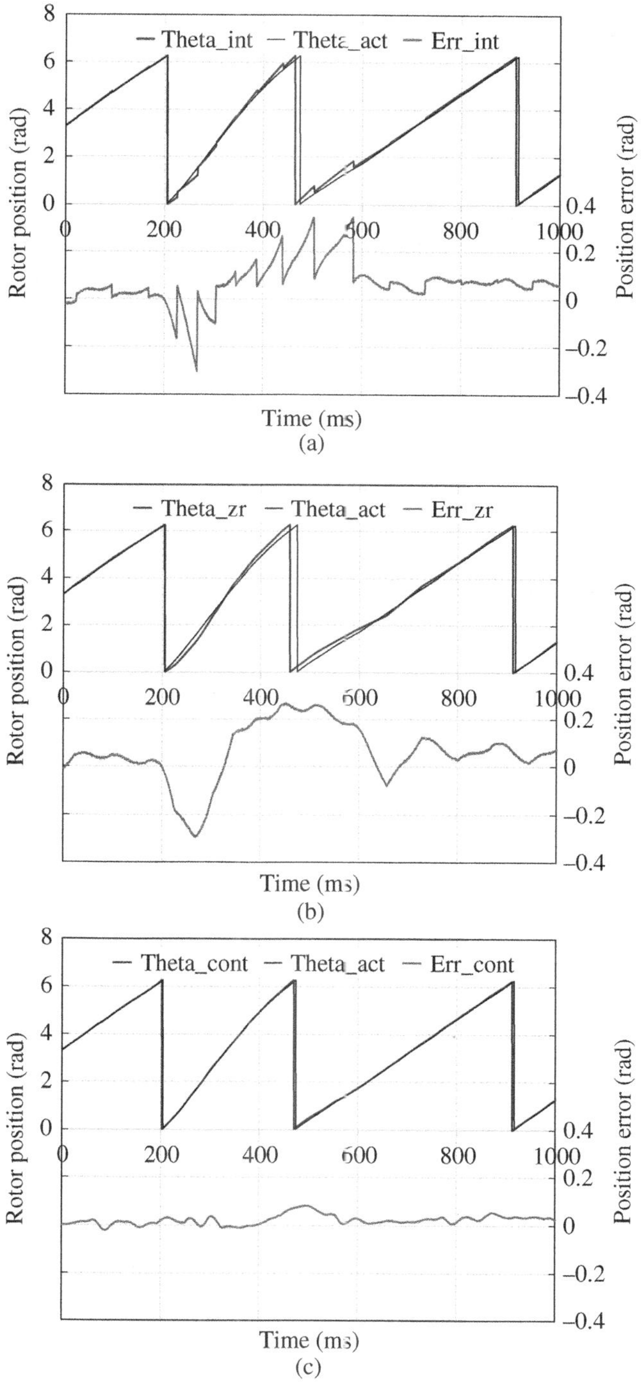

Figure 11.14 Rotor position estimation under step load. (a) Estimator based on integration (*Theta_act*, Actual rotor position; *Err_int*, Position estimation error by integration). (b) Estimator based on zero-crossing correction (*Theta_zr*, Estimated rotor position; *Theta_act*, Actual rotor position; *Err_zr*, Position estimation error). (c) Estimator based on continuous signal (*Theta_cont*, Estimated rotor position; *Theta_act*, Actual rotor position; *Err_cont*, Position estimation error). (*Source:* Adapted from [9].)

11.5 Position Estimation of DTP-PMSMs

11.5.1 Position Estimation Based on Third Harmonic Flux-Linkage

The method introduced in Section 11.4.3 can also be extended to dual three-phase PMSMs [10]. As introduced in Chapter 7, for dual three-phase PMSMs, the neutral points of two sets of windings are usually isolated as shown in Figure 11.15. Hence, there is no electrical connection between the two sets of windings.

For the dual three-phase PMSM, the measured voltages between S1 and N1, and between S2 and N2, could represent the two sets of the third harmonic back-EMFs of the two sets of windings as:

$$\begin{cases} v_{S1N1} = e_{3\_set1} + e_{9\_set1} + e_{15\_set1} + \cdots \approx e_{3\_set1} \\ v_{S2N2} = e_{3\_set2} + e_{9\_set2} + e_{15\_set2} + \cdots \approx e_{3\_set2} \end{cases} \tag{11.29}$$

For most dual three-phase PMSMs, the two sets of windings have a spatial $\pi/6$ elec-rad shift, Hence, e_{3_set1} and e_{3_set2} have the same phase shift under the fundamental reference frame as:

$$\begin{cases} e_{3\_set1} = E_3 \sin(3\theta_r) \\ e_{3\_set2} = E_3 \sin(3 \times (\theta_r - \pi/6)) = -E_3 \cos(3\theta_r) \end{cases} \tag{11.30}$$

For a single three-phase PMSM, the third harmonic flux-linkage ψ_{SN} that results from the integration of the third harmonic back-EMF is usually applied for the

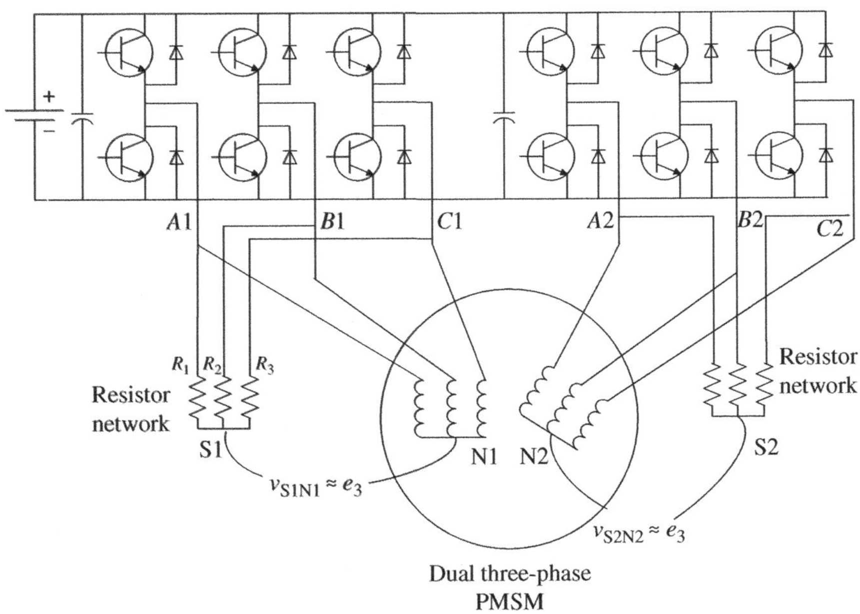

Figure 11.15 Measurement of third harmonic back-EMF in dual three-phase PMSM. (*Source:* [10]/IEEE.)

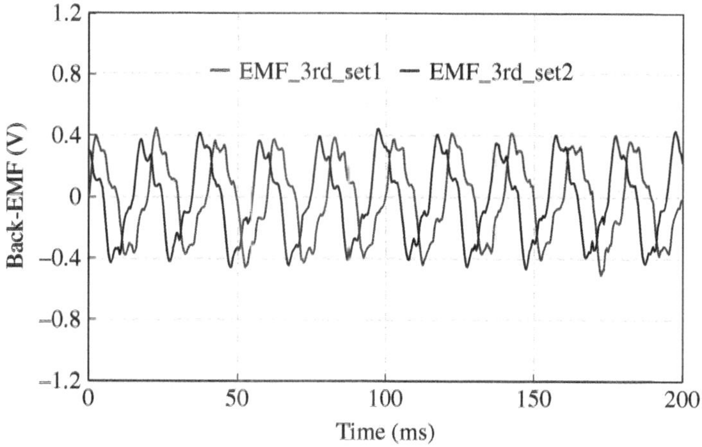

Figure 11.16 v_{SN} of two sets of windings at 200 r/min. (*Source:* [10]/IEEE.)

rotor position estimation due to its better SNR, and it is the same for dual three-phase PMSM. Hence, two sets of ψ_{SN} derived by low cutoff frequency BPF orthogonal to each other under the third harmonic reference frame, i.e.

$$\begin{cases} \psi_{3\_set1} = -\psi_{m3}\cos(3\theta_r) \\ \psi_{3\_set2} = -\psi_{m3}\cos(3\times(\theta_r-\pi/6)) = -\psi_{m3}\cos(3\theta_r-\pi/2) = -\psi_{m3}\sin(3\theta_r) \end{cases} \tag{11.31}$$

The steady-state tests are carried out when the rotor speed is 200 r/min with 1 A of q-axis current. It can be seen that the measured v_{SN} of two sets of windings at 200 r/min shown in Figure 11.16 contain excessive amounts of high-order triplen harmonic noise, while ψ_{SN} of the two sets of windings derived by BPF shown in Figure 11.17 are smoother and have better signal quality. However, the implementation of BPF for getting the third harmonic flux-linkage may lead to unexpected phase delay and poor frequency response, which will deteriorate the performance of the conventional rotor position estimation under dynamic operation.

11.5.2 Position Estimation Using Third Harmonic Back-EMF

From (11.30), it can be found that e_{3_set1} and e_{3_set2} have a shift of $\pi/6$ rad under the fundamental reference frame, which is equivalent to a shift of $\pi/2$ rad under the third harmonic reference frame, i.e.

$$\begin{cases} e_{3\_set1} = E_3\cdot\sin(3\theta_r) \\ e_{3\_set2} = E_3\cdot\sin(3\times(\theta_r-\pi/6)) = -E_3\cdot\cos(3\theta_r) \end{cases} \tag{11.32}$$

As introduced in Chapter 2, by applying the simplified EKF position observer, the contained triplen ninth, fifteenth, etc. harmonics could be eliminated.

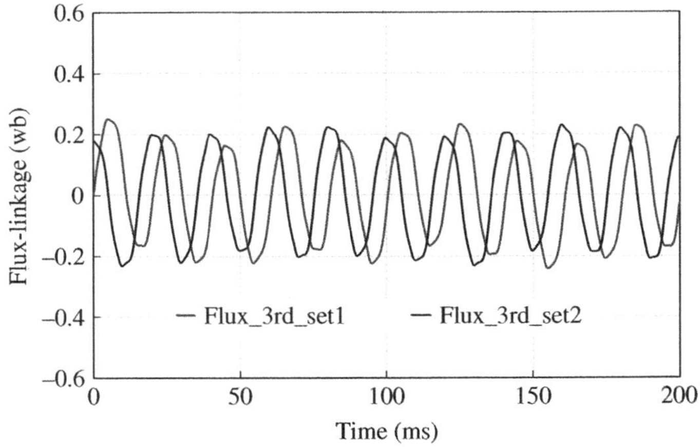

Figure 11.17 ψ_{SN} of two sets of windings at 200 r/min. (*Source:* [10]/IEEE.)

Therefore, with the simplified EKF, it is possible to directly use the third harmonic back-EMFs from two sets of windings, which are also orthogonal to each other as shown in (11.30). Since the BPF filter is eliminated, the phase delay can be minimized and the frequency response in the rotor position estimation can also be enhanced. The block diagram of the simplified EKF for DTP-PMSM is shown in Figure 11.18.

11.5.3 Experiment Results

The overall control scheme of the dual three-phase PMSM is shown in Figure 11.19. The experiment is based on SPMSM-V (Appendix B).

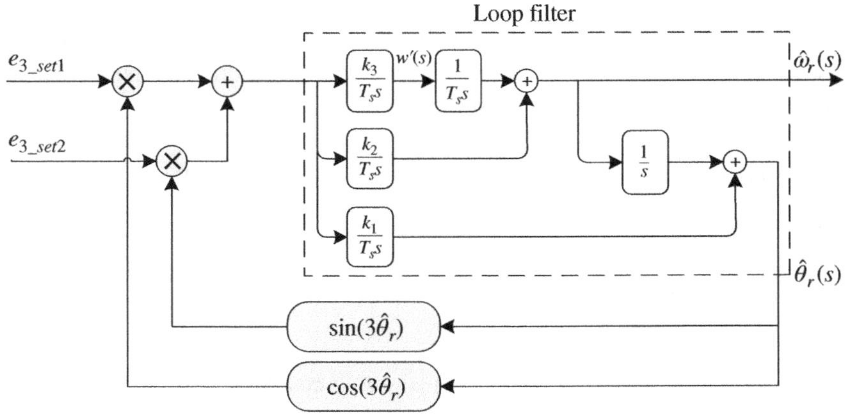

Figure 11.18 Simplified EKF-based rotor position estimator. (*Source:* Adapted from [10].)

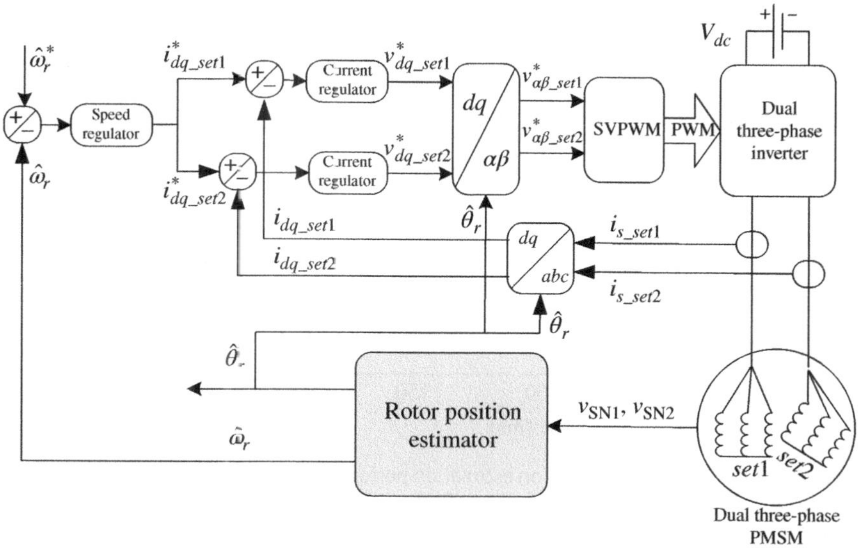

Figure 11.19 Overall control scheme of dual three-phase PMSM. (*Source:* [10]/IEEE.)

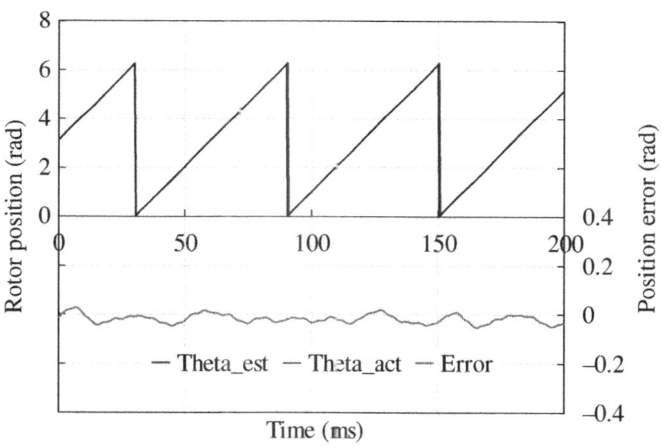

Figure 11.20 Steady-state position estimation performance based on third harmonic flux-linkage (*Theta_est*, Estimated position; *Theta_act*, Actual position; *Error*, Position estimation error). (*Source:* [10]/IEEE.)

Comparing these results in Figures 11.20 and 11.21, it can be found that the accuracy of the two methods is nearly the same. By applying the simplified EKF-based rotor position estimator, the higher order harmonic back-EMFs such as ninth, fifteenth, and so on are eliminated and do not affect the rotor position estimation as shown in Figure 11.21.

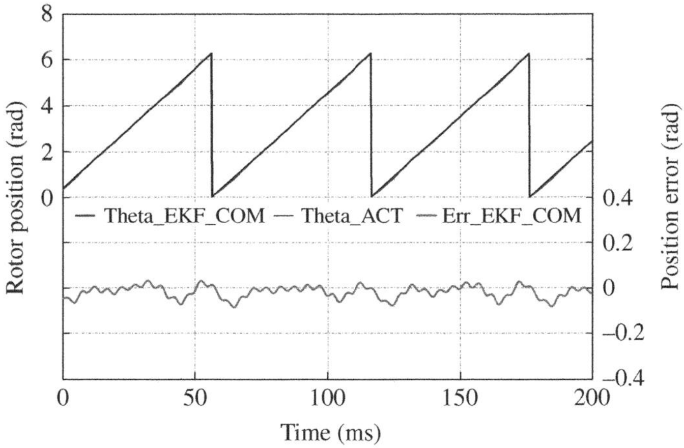

Figure 11.21 Steady-state position estimation performance based on third harmonic back-EMF (*Theta_EKF_COM*, Estimated position; *Theta_act*, Actual position; *Error_EKF_COM*, Position estimation error). (*Source:* [10]/IEEE.)

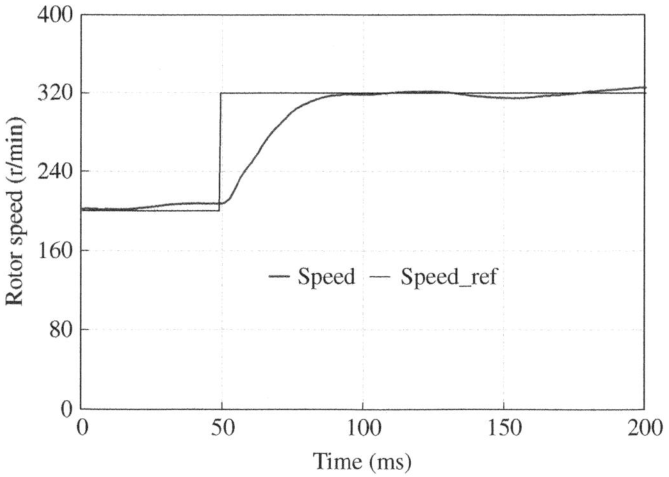

Figure 11.22 Dynamic test situation (*Speed_ref*, Speed reference; *Speed*, Actual rotor speed). (*Source:* [10]/IEEE.)

The dynamic performance tests are carried out when the rotor initial speed is 200 r/min with 1 A q-axis current, and the speed reference changes from 200 to 320 r/min and then the rotor speed response is shown in Figure 11.22. From Figures 11.23 and 11.24, it could be found that the dynamic performance is greatly enhanced by directly using third harmonic back-EMFs.

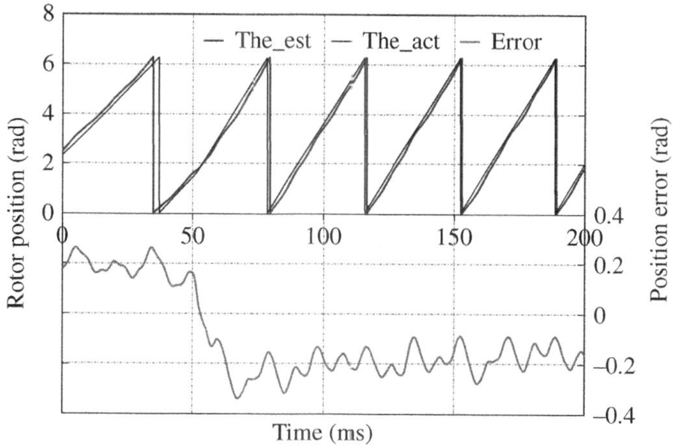

Figure 11.23 Dynamic position estimation performance based on third harmonic flux-linkage (*Theta_est*, Estimated position; *Theta_act*, Actual position; *Error*, Position estimation error). (*Source:* [10]/IEEE.)

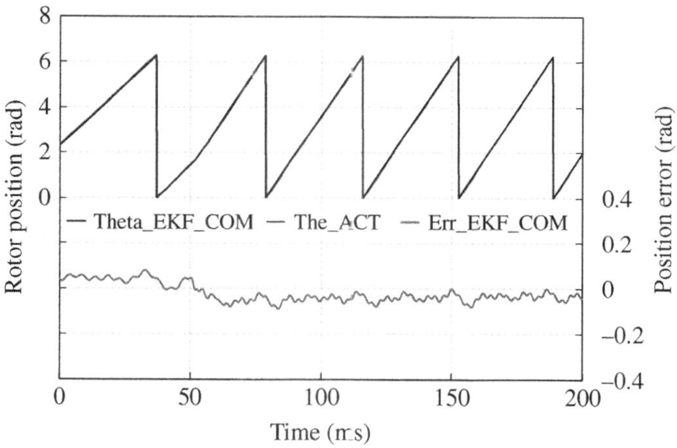

Figure 11.24 Dynamic position estimation performance based on third harmonic back-EMF (*Theta_EKF_COM*, Estimated position; *Theta_act*, Actual position; *Error_EKF_COM*, Position estimation error). (*Source:* [10]/IEEE.)

11.6 Issues with Third Harmonic Back-EMF Detection

This section presents some practical issues in the utility of third harmonic back-EMF in sensorless control of PM machines [12], [13]. Third harmonic back-EMF-based sensorless control is inapplicable and/or has large errors in the following cases:

- The winding neutral point is not accessible.
- The third harmonic component of the excitation field is zero or very small.
- The third harmonic winding factor is zero or very small.
- The winding inductances are not constant due to rotor saliency or other factors.
- The three phases are not symmetrical.

11.6.1 Requirement of Neutral Line

As the third harmonic back-EMF is obtained from the voltage between the star point of the resistor network (point S in Figure 11.1) and the winding neutral (point N in Figure 11.1), a neutral line is therefore required. However, sometimes the motor neutral line is not provided, which means the third harmonic signal is not measurable in this case.

11.6.2 Absence of Third Harmonic Back-EMF

The amplitude of the third-harmonic back-EMF (E_{m3}) is:

$$E_{m3} \propto \omega_r \cdot B_3 \cdot k_{w3} \tag{11.33}$$

where ω_r is the motor speed and B_3 is the amplitude of the third harmonic component of the excitation flux density, while k_{w3} is the winding factor for the third harmonic. k_{w3} is calculated as:

$$k_{w3} = k_{p3} \cdot k_{d3} \cdot k_{s3} \tag{11.34}$$

where k_{p3}, k_{d3}, and k_{s3} are the coil pitch factor, distribution factor, and skew factor, respectively. If either B_3, k_{p3}, k_{d3}, or k_{s3} is zero or very small, the third harmonic back-EMF will be zero or too small to sense. Further, (11.34) also shows that third harmonic back-EMF-based sensorless control is also unsuitable for zero or low-speed operation.

In some PM brushless machines, the third harmonic excitation field is very weak. For example, if a diametrically magnetized magnet is used in a two-pole machine, the airgap field distribution is essentially sinusoidal [12]. Similarly, in some machines the pole-arc is optimized to ~$2\pi/3$ elec-rad in order, for example, to reduce the cogging torque, and hence, the third harmonic airgap field is also very small. Machines with non-overlapping windings and a 3 : 2 ratio of slot number to pole number is very popular for BLDC drives. However, the third harmonic winding factor of such machines is zero. Therefore, in such machines, even if the third harmonic component exists in the airgap field and the waveform of the phase back-EMF is non-sinusoidal, the third harmonic back-EMF is zero.

11.6.3 Rotor Saliency

In the derivation of (11.8) and (11.9), it is assumed that the winding inductances are constant. However, in motors with rotor saliency, such as the IPM motors, the winding inductances vary with the rotor position. Therefore, the expression for v_{SN} needs to be re-derived.

The winding inductances can be expressed as:

$$\begin{cases} L_{AA} = \sum_k L_{sk} \cos k\theta_r \\ L_{BB} = \sum_k L_{sk} \cos k(\theta_r - 2\pi/3) \\ L_{CC} = \sum_k L_{sk} \cos k(\theta_r + 2\pi/3) \end{cases}$$
$$\begin{cases} M_{BC} = M_{CB} = \sum_k M_{sk} \cos k\theta_r \\ M_{CA} = M_{AC} = \sum_k M_{sk} \cos k(\theta_r - 2\pi/3) \quad k = 0, 2, 4, 6, 8, 10, \cdots \\ M_{AB} = M_{BA} = \sum_k M_{sk} \cos k(\theta_r + 2\pi/3) \end{cases} \tag{11.35}$$

while the motor voltage equations are:

$$\begin{bmatrix} v_{AN} \\ v_{BN} \\ v_{CN} \end{bmatrix} = R \begin{bmatrix} i_A \\ i_B \\ i_C \end{bmatrix} + p \left\{ \begin{bmatrix} L_{AA} & M_{AB} & M_{AC} \\ M_{BA} & L_{BB} & M_{BC} \\ M_{CA} & M_{CB} & L_{CC} \end{bmatrix} \cdot \begin{bmatrix} i_A \\ i_B \\ i_C \end{bmatrix} \right\} + \begin{bmatrix} e_A \\ e_B \\ e_C \end{bmatrix} \tag{11.36}$$

From (11.2), (11.4), (11.7), (11.35), and (11.36), v_{SN} is derived as:

$$\begin{aligned} v_{SN} = &(e_3 + e_9 + e_{15} + \cdots) \\ &+ \frac{p}{3} \left\{ \sum_j \left[(L_{sj} - M_{sj}) (i_a \cos j\theta_r + i_b \cos j(\theta_r - 2\pi/3) + i_c \cos j(\theta_r + 2\pi/3)) \right] \right\} \\ j = &2, 4, 8, 10, 14, \cdots \end{aligned} \tag{11.37}$$

The main contents of v_{SN} are the zero sequence components of the back-EMF in which the third harmonic component is dominant. However, if rotor saliency exists, $L_{sj} \neq 0$, $M_{sj} \neq 0$, and usually $L_{sj} - M_{sj} \neq 0$. Therefore, as long as currents flow in the motor windings, the second part on the right side of (11.37) can cause significant distortion in the measured v_{SN}.

The winding inductances of the SPM can be regarded as constant, while they vary in an IPM, which also has a third-harmonic back-EMF. Experimentally captured waveforms of v_{SN} for two prototype motors (IPMSM and SPMSM) are given in Figures 11.25 and 11.26, respectively. The working conditions are given in the captions. Clearly, v_{SN} with the IPM is significantly distorted no matter how large the phase current, but it becomes very clean with only the third harmonic

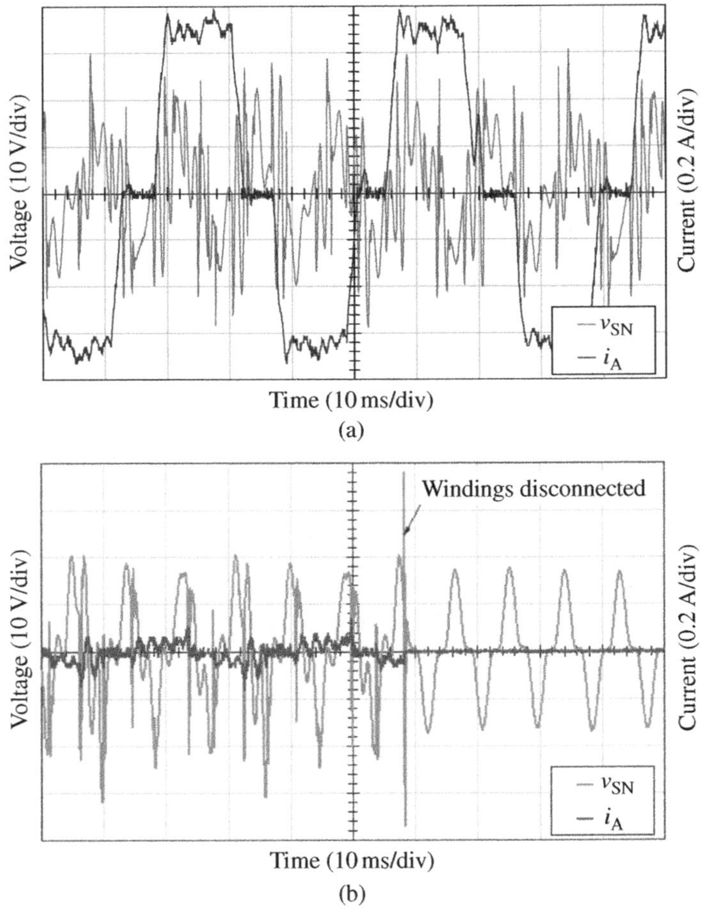

Figure 11.25 Measured voltage (v_{SN}) with an IPM. (a) BLDC operation with PWM (100 Vdc, 0.63 A, 783 r/min). (b) BLDC operation without PWM (100 Vdc, 0.03 A, 1141 r/min). (*Source:* [12]/IEEE.)

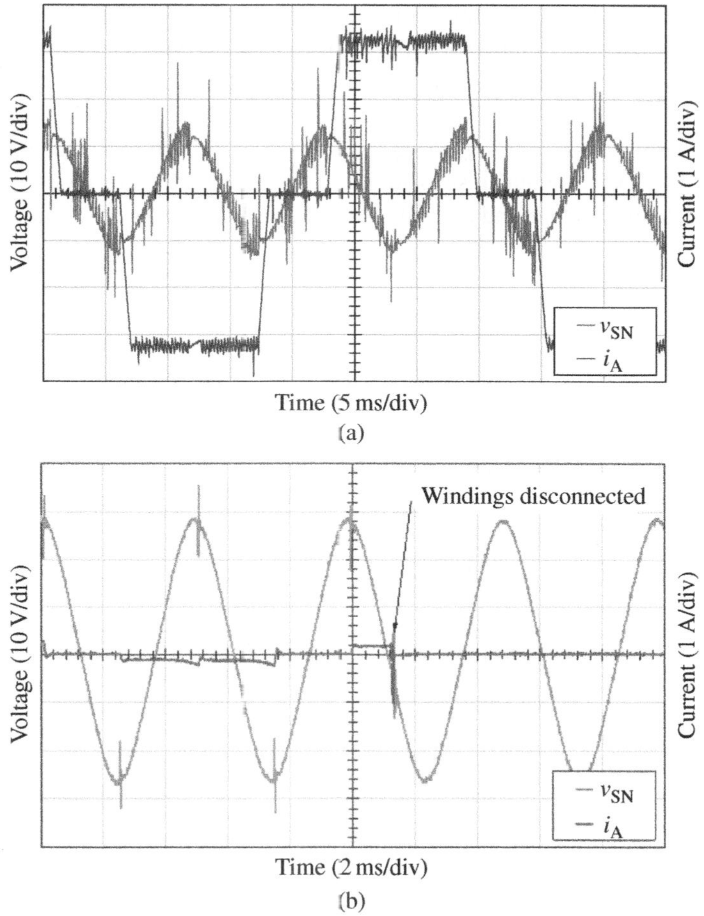

Figure 11.26 Measured voltage (v_{SN}) with a SPM. (a) BLDC operation with PWM (200 Vdc, 1.94 A, 600 r/min). (b) BLDC operation without PWM (200 Vdc, 0.16 A, 1352 r/min). (*Source:* [12]/IEEE.)

back-EMF component remaining when the motor windings are disconnected from the drive. In contrast, v_{SN} with SPM contains a detectable third harmonic back-EMF. In conclusion, if rotor saliency exists, the third harmonic back-EMF-based sensorless control technique is also inapplicable and/or has large errors.

11.6.4 Unbalanced Three Phases

In some surface-mounted PM machines, the three phases are unbalanced in terms of back-EMF amplitude, winding resistance, and inductance [12]. Although the unbalance is not too serious, it can cause a distortion in v_{SN}. Moreover, in practical implementation, v_{SN} usually contains some noises. Hence, v_{SN} is processed with filters, resulting in a modified signal (v_{XN}), which is actually used for the rotor position estimation. Clearly, the distortion due to phase unbalance also occurs in v_{XN}. Furthermore, the distortions in both v_{SN} and v_{XN} are influenced by the motor currents. By way of example, experimental results for the prototype motor are given in Figure 11.27. The dc-link voltage and current values are given in the figure caption.

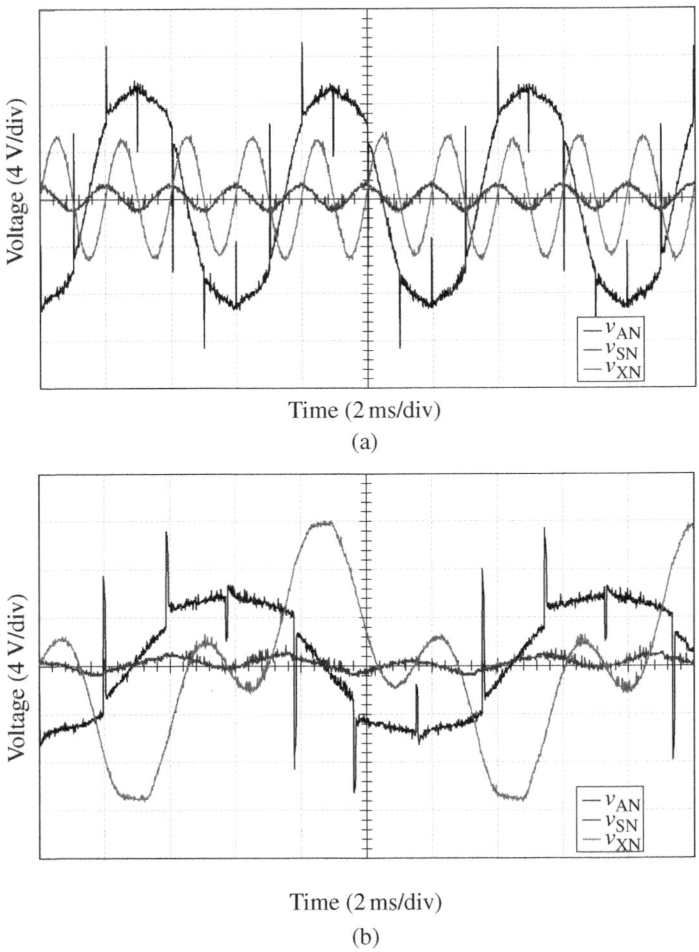

Figure 11.27 Influence of phase unbalance on voltages v_{SN} and v_{XN}. (a) BLDC operation (18.3 Vdc, 0.76 A). (b) BLDC operation (18.3 Vdc, 3.9 A). (*Source:* [12]/IEEE.)

Since v_{SN} is a relatively small signal, its distortion is not very observable. However, the distortion in v_{XN} is very significant if the motor current is large. In conclusion, third harmonic back-EMF-based sensorless control is inapplicable and/or has large errors for motors with unbalance between the three phases.

It will be noted that the motor parameters, such as resistance, inductance, and EMF constant, may vary, for example, due to temperature rise. However, if the three phases are originally balanced, such variations occur simultaneously in all phases and will not cause unbalancing. Furthermore, the third harmonic back-EMF-based sensorless control does not use the exact values of the motor parameters. Instead, it detects the back-EMF zero-crossings. Therefore, the sensorless control is insensitive to motor parameter variations.

11.7 Issues with Virtual Third Harmonic Back-EMF Detection

In sensorless control of BLDC drives, as proved in Section 11.2.2, the virtual third harmonic back-EMF detection-based method is actually utilizing the fundamental component of the back-EMF of the floating phase. The method is similar to the conventional technique of terminal voltage sensing in Chapter 10, which is based on the phase back-EMF. Thus, these two methods have the same issues, viz. (i) they are not suitable for BLAC operation, and (ii) in BLDC operation, the conduction of the freewheeling angle should not be larger than $\pi/6$ rad. These issues will not be explained in this section again. Apart from those issues [8], also shows that parameter asymmetry causes the ZCP deviation and commutation errors with the virtual third harmonic back-EMF detection-based method [8]. This section mainly focuses on this issue, and a solution [8] is provided accordingly.

11.7.1 ZCP Detection Under Asymmetric Parameters

Machine parameter asymmetry is a common problem, especially in high-speed motors that usually have a relatively low number of turns per phase and thus low resistance and inductance. This problem could be caused by cabling, inverter, modular manufacturing, rotor eccentricity, or winding fault. The machine parameter asymmetry can lead to unbalanced ZCP distribution. When parameter asymmetry happens, the ZCP intervals will be repeatedly lengthened and shortened, leading to commutation position errors [8]. Hence, in this section, the position errors caused by parameter asymmetries are analyzed.

For generality, it is assumed that the winding resistances and inductances in three phases are asymmetric. Therefore, the mathematical model of a high-speed BLDC motor in (11.1) can be modified as:

$$\begin{cases} v_{AG} = i_A(R_A + pL_A) + e_A + v_{NG} \\ v_{BG} = i_B(R_B + pL_B) + e_B + v_{NG} \\ v_{CG} = i_C(R_C + pL_C) + e_C + v_{NG} \end{cases} \tag{11.38}$$

where v_{XG} is the terminal voltage, i_X is the current, e_X is the phase back-EMF, v_{NG} is the motor neutral voltage with respect to ground, p is the differential operator, and R_X and L_X are the winding resistance and inductance, respectively. The subscript X denotes phase A, B, or C. Then, v_{SM} can be calculated as:

$$v_{SM} = \frac{1}{3}(i_A Z_A + i_B Z_B + i_C Z_C) + \frac{1}{3}(e_A + e_B + e_C) - \frac{1}{2}V_{dc} + v_{NG} \tag{11.39}$$

where $Z_X = (R_X + pL_X)$ is the phase impedance. According to Section 11.2.2 and (11.39), the voltage v_{NG} can be calculated as:

$$v_{NG} = \frac{1}{2}(v_{HG} + v_{LG}) - \frac{1}{2}(i_H Z_H + i_L Z_L) - \frac{1}{2}(e_H + e_L) \tag{11.40}$$

Accordingly, the current in three phases can be calculated as:

$$i_H = \frac{(v_{HG} - v_{LG}) - (e_H - e_L)}{Z_H + Z_L}, \quad i_L = -i_H, \quad i_O = 0 \tag{11.41}$$

By substituting (11.40) and (11.41) into (11.39), the voltage v_{SM} can be obtained as:

$$v_{SM} = \underbrace{\frac{1}{6}(2e_O - e_H - e_L)}_{\text{Back}-\text{EMF}} + \underbrace{\frac{1}{2}(\Lambda - 1)V_{dc}}_{\text{PWM fluctuation}} - \underbrace{\frac{1}{6}\frac{Z_H - Z_L}{Z_H + Z_L}(\Lambda \cdot V_{dc} - (e_H - e_L))}_{\text{Asymmetric parameters}} \tag{11.42}$$

It can be seen from (11.42) that v_{SM} consists of three parts, i.e. the back-EMF, the voltage fluctuation caused by PWM, and the voltage drift introduced by asymmetric parameters. By substituting $\Lambda = 1$ into (11.42), the upper envelope $\hat{v}_{SM}$ can be expressed as:

$$\hat{v}_{SM} = v_{SM}|_{\Lambda=1} = \frac{1}{2}e_O - \Delta v_{HL} \tag{11.43}$$

where Δv_{HL} is the voltage error caused by asymmetric parameters. Obviously, it is illustrated by (11.43) that the envelope contains the voltage error, which is caused by asymmetric parameters and can be expressed as:

$$\Delta v_{HL} = \frac{1}{6}\frac{Z_H - Z_L}{Z_H + Z_L}(\Lambda \cdot V_{dc} - (e_H - e_L)) \tag{11.44}$$

Consider that the excitation voltage term contains the PWM switching function Λ. The PWM period is much less than the winding electromagnetic time constant L/R. In this situation, within the very short period when the circuit switches from PWM-off ($\Lambda = 0$) to PWM-on ($\Lambda = 1$), the asymmetric impedance term can be simplified as:

$$\frac{Z_{\mathrm{H}} - Z_{\mathrm{L}}}{Z_{\mathrm{H}} + Z_{\mathrm{L}}} \approx \frac{L_{\mathrm{H}} - L_{\mathrm{L}}}{L_{\mathrm{H}} + L_{\mathrm{L}}} \tag{11.45}$$

which means that asymmetric impedances mainly depend on asymmetric inductances. Then, assuming the phase back-EMF amplitude is $\omega_r\psi_m$, where ψ_m is the flux-linkage and ω_r is the electrical speed, the line-line back-EMF $e_{\mathrm{H}} - e_{\mathrm{L}}$ around the ZCP can be expressed as:

$$e_{\mathrm{H}} - e_{\mathrm{L}} = \begin{cases} \sqrt{3}\omega_r\psi_m & \text{sinusoidal EMF} \\ 2\omega_r\psi_m & \text{trapezoidal EMF} \end{cases} \tag{11.46}$$

Finally, letting $V_{\mathrm{RL}} = V_{dc} - (e_{\mathrm{H}} - e_{\mathrm{L}})$ denote the voltage drop on the winding resistance and inductance, (11.44) can be expressed as:

$$\Delta v_{\mathrm{HL}} = \frac{1}{6} \cdot \frac{L_{\mathrm{H}} - L_{\mathrm{L}}}{L_{\mathrm{H}} + L_{\mathrm{L}}} \cdot V_{\mathrm{RL}} \tag{11.47}$$

11.7.2 Commutation Errors Caused by Asymmetric Parameters

As analyzed in the previous section, the machine parameter asymmetry can lead to unbalanced ZCP distribution. When parameter asymmetry happens, the ZCP intervals will be repeatedly lengthened and shortened, leading to commutation position errors. In this section, to illustrate this phenomenon, sectors III and VI are used as an example. In these two sectors, phases B and C are conducting and phase A is floating. In addition, the conducting directions are opposite in the two sectors, which can be described as follows.

In sector III, the current flows from phase B to phase C. Phases B and C are connected to positive and negative dc rails, respectively. As shown in Figure 11.28a, the asymmetric parameters cause a voltage error Δv_{HL} to the envelope, which can be expressed as:

$$\Delta v_{\mathrm{HL}} = \Delta v_{\mathrm{BC}} = \frac{1}{6} \cdot \frac{L_{\mathrm{B}} - L_{\mathrm{C}}}{L_{\mathrm{B}} + L_{\mathrm{C}}} \cdot V_{\mathrm{RL}} \tag{11.48}$$

Oppositely, in sector VI, current flows from phase C to phase B. Phase B connects to the negative dc rail and phase C connects to the positive one. As shown in Figure 11.28b, in this situation, the voltage error can be expressed as:

$$\Delta v_{\mathrm{HL}} = \Delta v_{\mathrm{CB}} = \frac{1}{6} \cdot \frac{L_{\mathrm{C}} - L_{\mathrm{B}}}{L_{\mathrm{C}} + L_{\mathrm{B}}} \cdot V_{\mathrm{RL}} \tag{11.49}$$

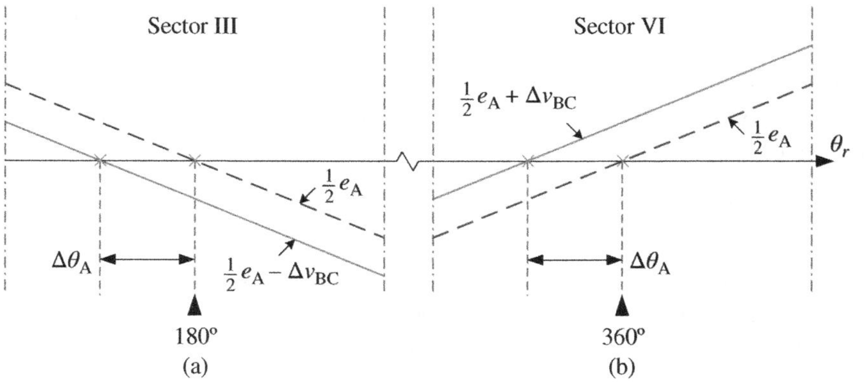

Figure 11.28 Illustration of virtual third harmonic envelopes and ZCP position errors. (a) Sector III. (b) Sector VI. (*Source:* Adapted from [8].)

It can be concluded that $\Delta v_{BC} = -\Delta v_{CB}$. As shown in Figure 11.28, the voltage envelope moves oppositely in sectors III and VI. Similarly, this conclusion can be deduced in all six sectors as:

$$\begin{cases} \Delta v_{AB} + \Delta v_{BA} = 0 \\ \Delta v_{BC} + \Delta v_{CB} = 0 \\ \Delta v_{AC} + \Delta v_{CA} = 0 \end{cases} \tag{11.50}$$

The voltage errors caused by asymmetric parameters will deviate the ZCP of the back-EMF. As shown in Figure 11.28, the back-EMF envelope around ZCP can be regarded as a straight line with slope K_ω. Then, it yields that the ZCP position error is proportional to the voltage error as:

$$[\Delta\theta_C \ \ \Delta\theta_A \ \ \Delta\theta_B] = K_\omega[\Delta v_{AB} \quad \Delta v_{BC} \quad \Delta v_{CA}] \tag{11.51}$$

where

$$K_\omega = \begin{cases} 2/(\omega_r \psi_m) & \text{sinusoidal EMF} \\ \pi/(2\omega_r \psi_m) & \text{trapezoidal EMF} \end{cases} \tag{11.52}$$

The ZCP position errors and the virtual third harmonic envelope are listed in Table 11.3. For the direction of errors, "lag" is defined as positive and "lead" is defined as negative. Figure 11.29 shows the analysis result of ZCP position errors. As it is shown, the ZCP position error increases with the unbalanced inductances. In addition, the error has a nonlinear relationship with the motor speed.

For conventional BLDC sensorless methods, the deviation of ZCP will influence the commutation accuracy. In the ideal case without ZCP position errors, every ZCP interval should be equal to 60° and the next commutation instant is 30° after

Table 11.3 ZCP position errors caused by asymmetric parameters.

Sector	Mode	Envelope	Position errors	Sector	Mode	Envelope	Position errors
I	A + B−	$\frac{1}{2}e_C - \Delta v_{AB}$	$\Delta\theta_C$	IV	B + A−	$\frac{1}{2}e_C + \Delta v_{AB}$	$\Delta\theta_C$
II	A + C−	$\frac{1}{2}e_B + \Delta v_{CA}$	$\Delta\theta_B$	V	C + A−	$\frac{1}{2}e_B - \Delta v_{CA}$	$\Delta\theta_B$
III	B + C−	$\frac{1}{2}e_A - \Delta v_{BC}$	$\Delta\theta_A$	VI	C + B−	$\frac{1}{2}e_A + \Delta v_{BC}$	$\Delta\theta_A$

(*Source:* Adapted from [8].)

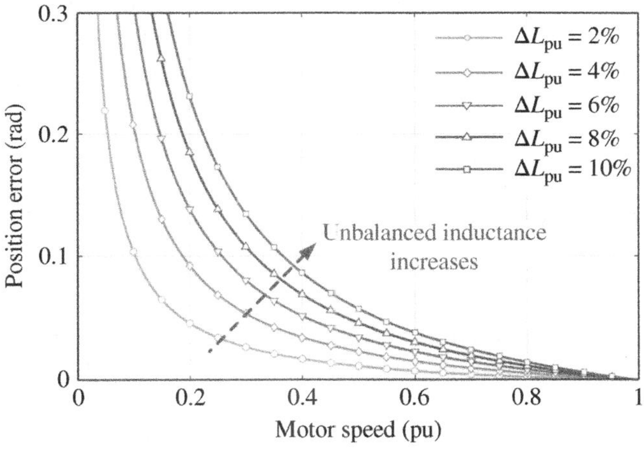

Figure 11.29 Analysis result of ZCP position errors caused by asymmetric machine parameters, where $\Delta L_{pu} = (L_X - L_Y)/(L_X + L_Y)$ (BLDC-III, Appendix B). (*Source:* [8]/IEEE.)

the ZCP. Thus, in conventional methods, the commutation instant is usually calculated with the present ZCP interval as:

$$\varphi[n] = \frac{\pi}{6} = \frac{1}{2}\hat{\theta}[n] \tag{11.53}$$

where $\hat{\theta}[n]$ is the nth measured ZCP interval and $\varphi[n]$ is the nth delay angle from the ZCP to the next commutation instant. However, if the error exists, ZCPs will be detected at inaccurate positions. In addition, the ZCP interval will be repeatedly lengthened and shortened, which means the determination of a 30° delay angle in (11.53) is also incorrect. That is how the asymmetric parameters undermine the conventional BLDC sensorless method.

11.7.3 Phase Compensation of Commutation Errors

In order to suppress the influence of asymmetric parameters on a sensorless drive, a phase compensation is presented in this section. Figure 11.30 shows the overall sensorless control diagram.

First, the signal $\hat{v}_{SM}$ is constructed by the over-sampling approach discussed in Section 11.3.2. Then, the time interval between two adjacent ZCPs is measured by the high-resolution CPU timer, since the ZCP time interval is dependent on the motor speed. The CPU timer starts at the previous ZCP and stops at the present ZCP. In this way, the present ZCP time interval is measured and stored as $t_\theta[n]$ (the previous ZCP time interval is stored as $t_\theta[n-1]$). Finally, the position errors can be calculated and compensated with (11.62), which will be discussed later in this chapter.

Figure 11.31 illustrates the relation between the theoretical ideal and the measured ZCPs. The theoretical ideal ZCPs always locate in the middle of each 60° sector; and measured ZCPs could deviate due to asymmetric impedances.

Then, the ZCP position errors can be modeled as:

$$\begin{aligned}\hat{\theta}_{12}\left(\hat{\theta}_{45}\right) &= \theta_{12}(\theta_{45}) + \Delta\theta_{\mathrm{B}} - \Delta\theta_{\mathrm{C}}\\ \hat{\theta}_{23}\left(\hat{\theta}_{56}\right) &= \theta_{23}(\theta_{56}) + \Delta\theta_{\mathrm{A}} - \Delta\theta_{\mathrm{B}}\\ \hat{\theta}_{34}\left(\hat{\theta}_{61}\right) &= \theta_{34}(\theta_{61}) + \Delta\theta_{\mathrm{C}} - \Delta\theta_{\mathrm{A}}\end{aligned} \tag{11.54}$$

where θ_{xy} represents the ZCP interval from sector x to sector y, and the hat ^ represents measured results. The term in the bracket indicates they have the same

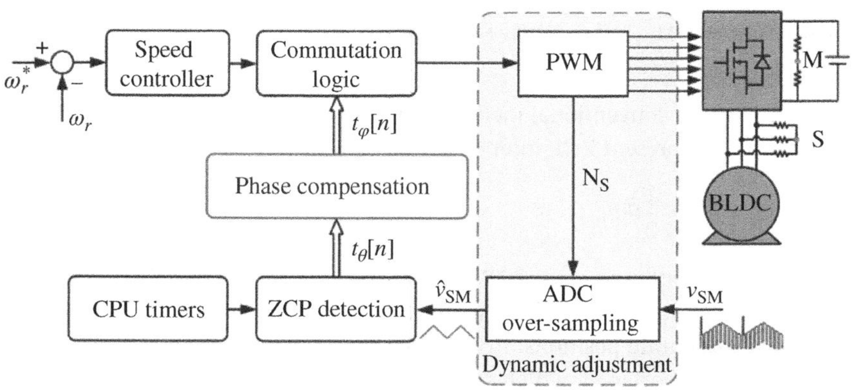

Figure 11.30 Schematic diagram of sensorless control system with commutation phase compensation. (*Source:* [8]/IEEE.)

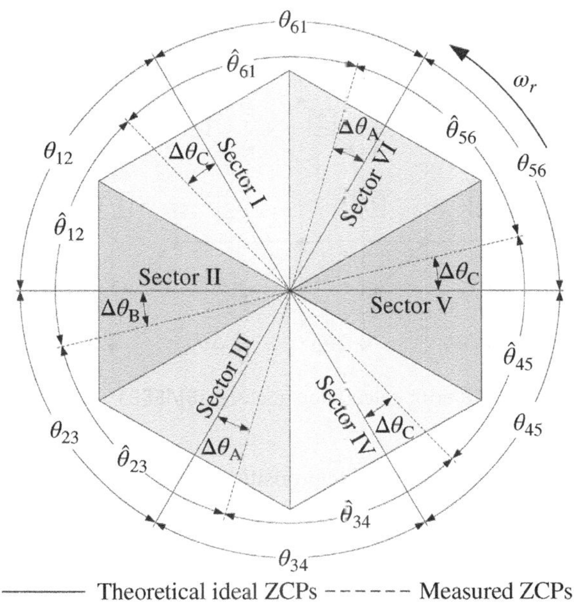

Figure 11.31 Relationship between theoretical ideal and measured ZCPs. (*Source:* [8]/IEEE.).

value, e.g. $\hat{\theta}_{12} = \hat{\theta}_{45}$. It yields that $\theta_{12} \ldots \theta_{61} = \pi/3$. In addition, the sum of three voltage errors can be calculated as:

$$\Delta v_{AB} + \Delta v_{BC} + \Delta v_{CA} = -\frac{1}{6} V_{RL} \cdot \frac{L_A - L_B}{L_A + L_B} \cdot \frac{L_B - L_C}{L_B + L_C} \cdot \frac{L_C - L_A}{L_C + L_A} \tag{11.55}$$

Since every inductance term in (11.55) is very small, their product is approximately equal to zero as:

$$\Delta v_{AB} + \Delta v_{BC} + \Delta v_{CA} \approx 0 \tag{11.56}$$

Then, due to the linear relationship in (11.49), the sum of three ZCP position errors also equals to zero as:

$$\Delta\theta_A + \Delta\theta_B + \Delta\theta_C \approx 0 \tag{11.57}$$

Finally, by solving (11.54) and (11.57), the ZCP position errors can be calculated as:

$$\begin{bmatrix} \Delta\theta_A \\ \Delta\theta_B \\ \Delta\theta_C \end{bmatrix} = \frac{1}{3} \begin{bmatrix} 1 & 2 & 0 \\ 2 & 0 & 1 \\ 0 & 1 & 2 \end{bmatrix} \begin{bmatrix} \hat{\theta}_{12}(\hat{\theta}_{45}) \\ \hat{\theta}_{23}(\hat{\theta}_{56}) \\ \hat{\theta}_{34}(\hat{\theta}_{61}) \end{bmatrix} - \frac{\pi}{3} \begin{bmatrix} 1 \\ 1 \\ 1 \end{bmatrix} \tag{11.58}$$

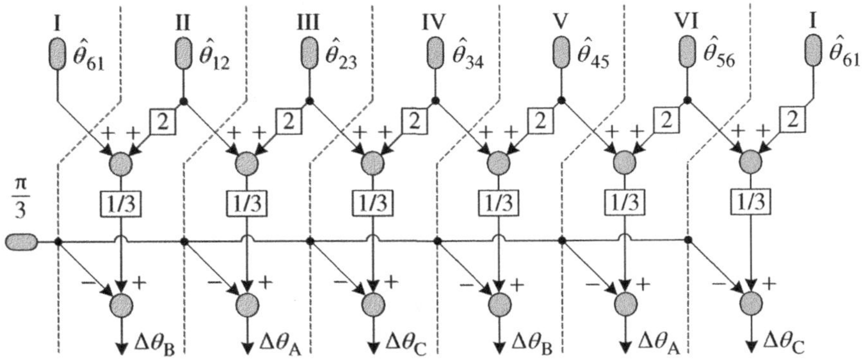

Figure 11.32 Schematic diagram of ZCP error calculation. (*Source:* [8]/IEEE.)

Although the ZCPs deviate from theoretical ideal positions due to asymmetric parameters, according to (11.57), the sum of three consecutive ZCP intervals are identically equal to π as:

$$\hat{\theta}[n] + \hat{\theta}[n-1] + \hat{\theta}[n-2] = \pi \tag{11.59}$$

Figure 11.32 shows the calculation of (11.58), in which it can be derived that the coefficients (2 and 1) in the square matrix of (11.58) always correspond to the present and the previous ZCP intervals, respectively. Thus, (11.58) can be simplified as:

$$\Delta\theta[n] = \frac{1}{3}(2\hat{\theta}[n] + \hat{\theta}[n-1]) - \frac{1}{3}(\hat{\theta}[n] + \hat{\theta}[n-1] + \hat{\theta}[n-2]) = \frac{1}{3}(\hat{\theta}[n] - \hat{\theta}[n-2]) \tag{11.60}$$

where $\Delta\theta[n]$ is the nth calculated ZCP position error as shown in Figure 11.33. Then, considering the ZCP position errors, the delay angle in (11.53) can be modified as:

$$\begin{aligned} \varphi[n] &= \frac{\pi}{6} - \Delta\theta[n] = \frac{1}{6}(\hat{\theta}[n] + \hat{\theta}[n-1] + \hat{\theta}[n-2]) - \frac{1}{3}(\hat{\theta}[n] - \hat{\theta}(n-2)) \\ &= \frac{1}{6}(-\hat{\theta}[n] + \hat{\theta}(n-1) + 3\hat{\theta}(n-2)) \end{aligned} \tag{11.61}$$

In practice, CPU timers are usually employed to measure the ZCP interval and execute the commutation. Finally, (11.61) can be implemented as:

$$t_\varphi[n] = \frac{1}{6}(-t_\theta[n] + t_\theta[n-1] + 3t_\theta[n-2]) \tag{11.62}$$

where $t_\theta[n]$ is the nth ZCP time interval measured by CPU timers and $t_\varphi[n]$ is the nth time delay for commutation instant. In this way, with the last three ZCP time

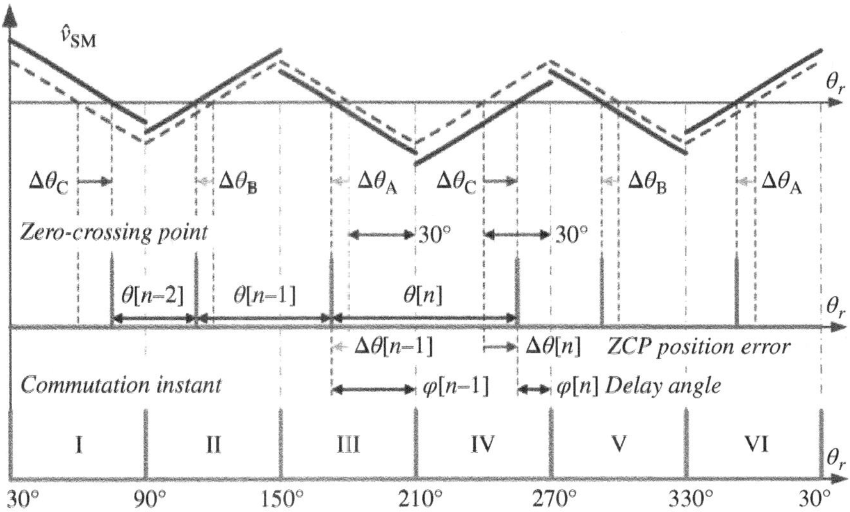

Figure 11.33 Sequence diagram of ZCP interval $\hat{\theta}[n]$, ZCP error $\Delta\theta[n]$, and commutation delay angle $\varphi[n]$. (*Source:* [8]/IEEE.)

intervals obtained from CPU timers, the ZCP errors caused by the parameter asymmetry can be compensated for.

11.7.4 Experiment Results

The compensation algorithm is tested on the BLDC-III (Appendix B). The dc-link voltage is 15 V. The PWM switching frequency is 40 kHz and the minimum ZCP sampling interval is 5 μs. To simulate the asymmetric condition, from Figures 11.34 to 11.36, an extra 5 μH inductor is connected in series to the phase A, which is equivalent to $\Delta L_{pu} = 10.2\%$.

Figures 11.34 and 11.35 compare the results before and after employing the compensation scheme of 20 000 r/min and 60 000 r/min, respectively.

In Figures 11.34 and 11.35, from top to bottom, the voltage v_{SM} with its sampling result $\hat{v}_{SM}$, the commutation interval (measured by the CPU timer), the commutation interrupt signal, and the phase currents are presented. As shown in Figure 11.34a, when the motor operates at 20 000 r/min without compensation, the ZCP interval is repeatedly lengthened and shortened. The commutation interval fluctuates drastically within 45–80°, leading to significant phase current oscillations. It can be seen that the sensorless control performance is deteriorated by the asymmetric inductances. When the compensation is enabled in Figure 11.34b, the ZCP interval can remain at 60° stably. The peak–peak amplitudes of phase currents are reduced from 6.05 to 4.37 A. The comparative results

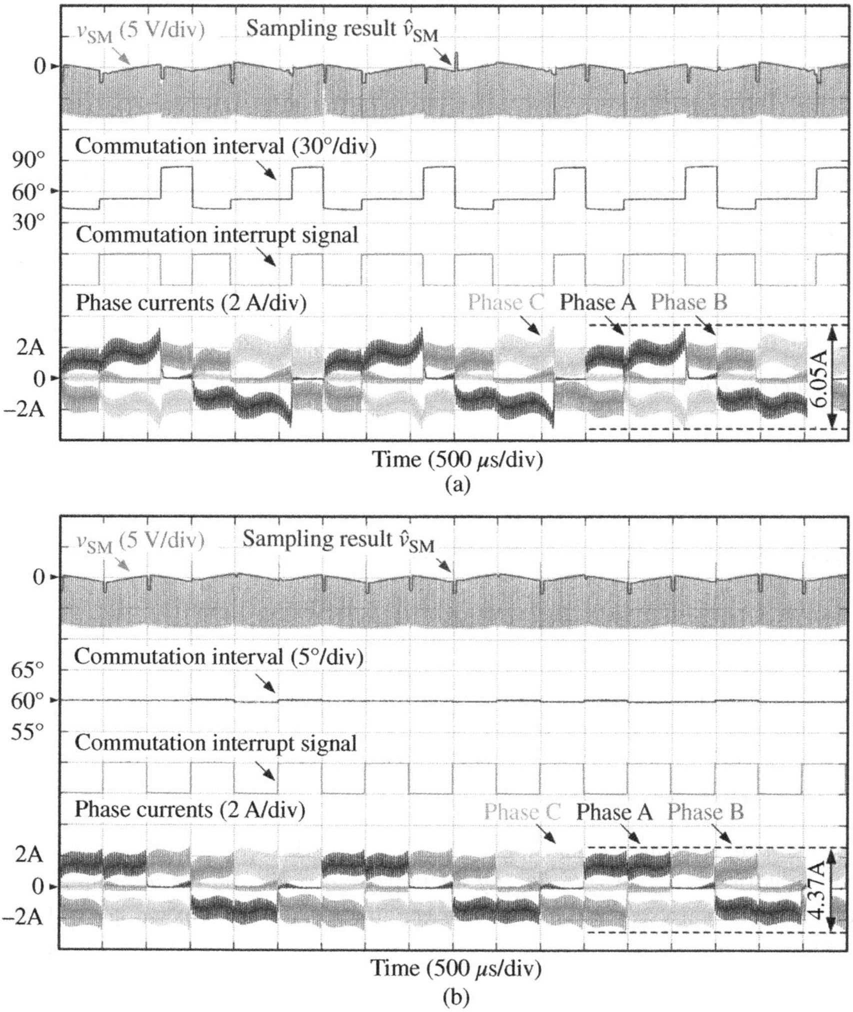

Figure 11.34 Comparative results of steady state at 20 000 r/min. (a) Before compensation. (b) After compensation. (*Source:* [8]/IEEE.)

before and after compensation at 60 000 r/min are shown in Figure 11.35. Likewise, before the compensation, the commutation interval cannot remain at 60°, which makes the phase currents fluctuate drastically. After the compensation, the commutation interval can remain at 60° accurately. The amplitudes of the phase currents are reduced from 24.40 to 23.46 A at 60 000 r/min. Therefore, the method is effective to compensate for the commutation errors caused by asymmetric machine parameters.

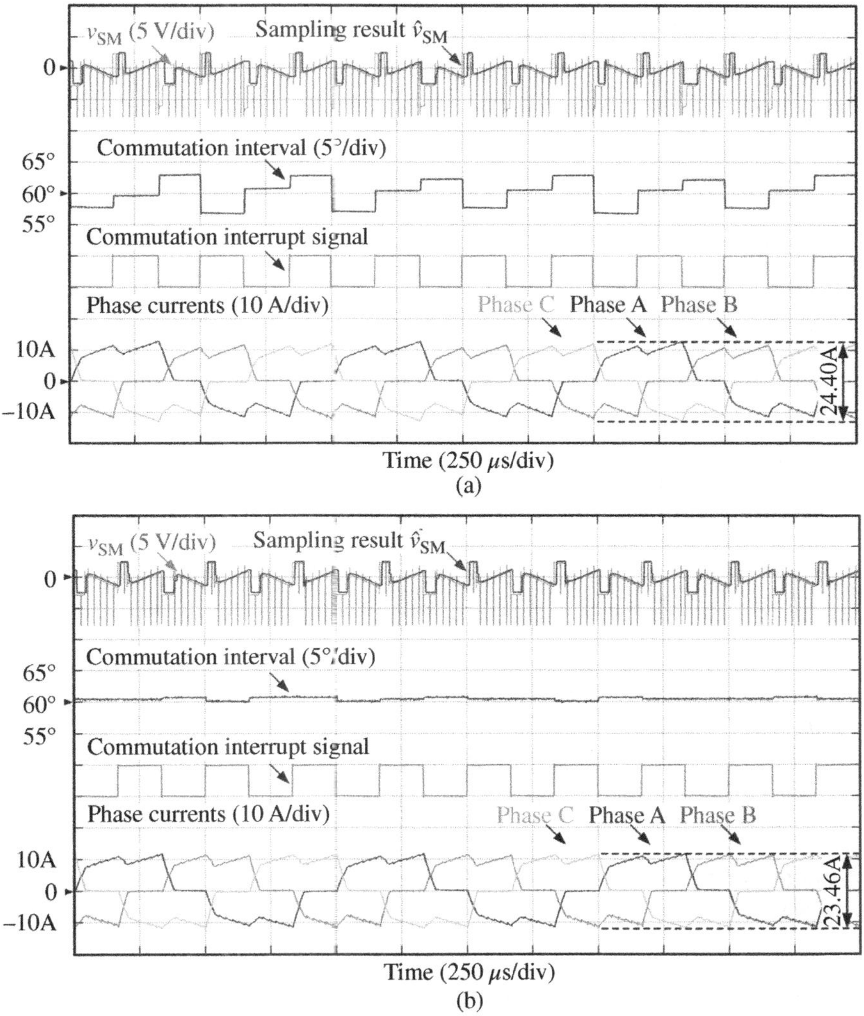

Figure 11.35 Comparative results of steady state at 60 000 r/min. (a) Before compensation. (b) After compensation. (*Source:* [8]/IEEE.)

Figure 11.36 shows the comparative results at transient state with the extra 5 μH inductor. The motor accelerates from 20 000 to 60 000 r/min, and then decelerates back to 20 000 r/min. As shown in Figure 11.36a, before the compensation algorithm is employed, the commutation interval experiences considerable oscillations during acceleration and deceleration processes. Besides, some spikes can be observed on the estimated speed signal, reflecting its poor dynamic performance. On the other hand, when the phase compensation is utilized in Figure 11.36b, the

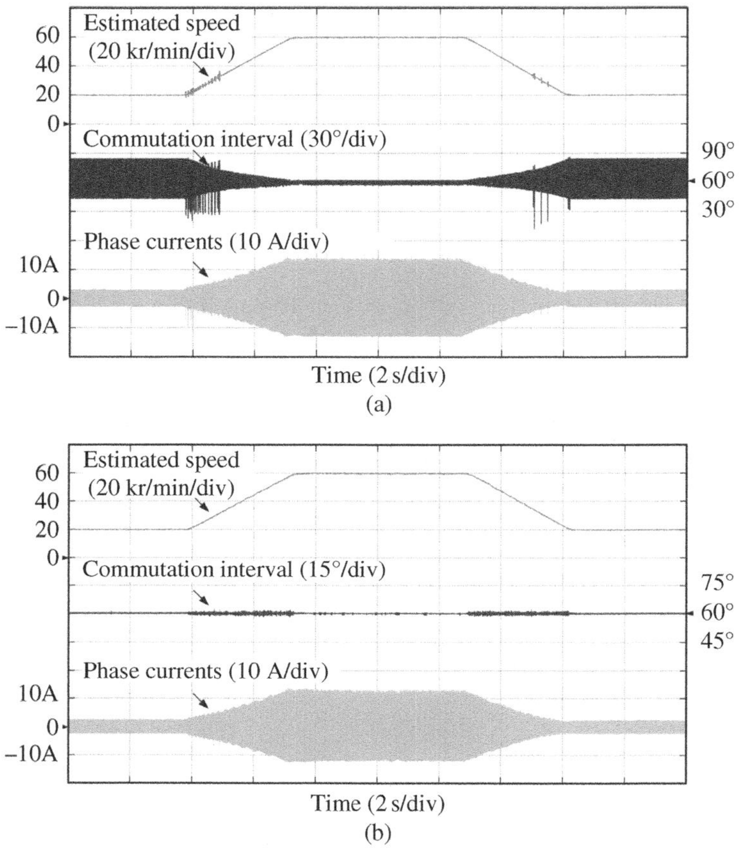

Figure 11.36 Comparative results at acceleration and deceleration processes. (a) Before compensation. (b) After compensation. (*Source:* [8]/IEEE.)

fluctuation range of the commutation interval can maintain at 60° closely, although it amplifies slightly during transient processes. Accordingly, the phase compensation method can improve both the steady and dynamic performances.

11.8 Summary

In this chapter, third harmonic back-EMF detection-based sensorless control techniques for BLDC and BLAC drives were introduced. Two common detection methods including third harmonic back-EMF and virtual third harmonic back-EMF were described. The analysis shows that only the former method mainly contains the third harmonic back-EMF component, while the virtual one utilizes the fundamental component of the back-EMF. Then, these two methods were discussed at different operation modes. The third harmonic back-EMF-based method can be applied to both BLDC and BLAC drives, while the virtual third harmonic back-EMF-based method can only be applied to the BLDC-based method, which is similar to the terminal voltage-based method introduced in Chapter 10. Apart from the single three-phase machines, the third harmonic back-EMF-based method was also applied to a dual three-phase machine, which shows good estimation performance by utilizing the orthogonal property of two third harmonic back-EMF detection signals. Finally, the influences of some common practical issues on third harmonic back-EMF detection were investigated. For the virtual third harmonic back-EMF-based method, the ZCP deviation due to parameter asymmetry was discussed, and a compensation scheme was provided that can effectively suppress the position estimation error.

References

1 J. Moreira and T. A. Lipo, "Modeling saturated AC machines including air gap flux harmonic components," *IEEE Trans. Ind. Appl.*, vol. 28, no. 2, pp. 343–349, 1992.

2 M. Jufer and R. Osseni, "Back EMF indirect detection for self-commutation of synchronous motors," *Proc. 1987 Eur. Conf. Power Electron. Appl.*, EPE 1987, pp. 1125–1129.

3 J. X. Shen, Z. Q. Zhu, and D. Howe, "Sensorless flux-weakening control of permanent-magnet brushless machines using third harmonic back EMF," *IEEE Trans. Ind. Appl.*, vol. 40, no. 6, pp. 1629–1636, 2004.

4 J. X. Shen and S. Iwasaki, "Sensorless control of ultrahigh-speed PM brushless motor using PLL and third harmonic back EMF," *IEEE Trans. Ind. Electron.*, vol. 53, no. 2, pp. 421–428, Apr. 2006.

5 C. Cui, G. Liu, K. Wang, and X. Song, "Sensorless drive for high-speed brushless dc motor based on the virtual neutral voltage," *IEEE Trans. Power Electron.*, vol. 30, no. 6, pp. 3275–3285, Jun. 2015.

6 X. Song, B. Han, S. Zheng, and J. Fang, "High-precision sensorless drive for high-speed BLDC motors based on the virtual third harmonic back-EMF," *IEEE Trans. Power Electron.*, vol. 33, no. 2, pp. 1528–1540, Feb. 2018.

7 X. Song, B. Han, and K. Wang, "Sensorless drive of high-speed BLDC motors based on virtual third-harmonic back EMF and high-precision compensation," *IEEE Trans. Power Electron.*, vol. 34, no. 9, pp. 8787–8796, Sep. 2019.

8 L. Yang, Z. Q. Zhu, H. Bin, Z. Zhang, and L. Gong, "Virtual third harmonic back EMF-based sensorless drive for high-speed BLDC motors considering machine parameter asymmetries," *IEEE Trans. Ind. Appl.*, vol. 57, no. 1, pp. 306–315, 2021.

9 J. M. Liu and Z. Q. Zhu, "Improved sensorless control of permanent-magnet synchronous machine based on third-harmonic back EMF," *IEEE Trans. Ind. Appl.*, vol. 50, no. 3, pp. 1861–1870, May 2014.

10 J. M. Liu and Z. Q. Zhu, "Rotor position estimation for dual-three-phase permanent magnet synchronous machine based on third harmonic back-EMF," in *2015 IEEE Symp. Sensorless Control Elect. Drives (SLED)*, Jun. 2015, pp. 1–8.

11 Y. S. Lai and Y. K. Lin, "A unified approach to zero-crossing point detection of back EMF for brushless dc motor drives without current and hall sensors," *IEEE Trans. Power Electron.*, vol. 26, no. 6, pp. 1704–1713, Jun. 2011.

12 J. X. Shen, Z. Q. Zhu, and D. Howe, "Practical issues in sensorless control of PM brushless machines using third-harmonic back-EMF," in *2006 CES/IEEE 5th Int. Power Electron. Motion Control Conf.*, vol. 2, pp. 1–5, Aug. 2006.

13 Z. Q. Zhu and J. M. Liu, "Influence of stator current and machine saliency on sensorless control performance based on third-harmonic back-EMF," in *7th IET Int. Conf. Power Electron. Mach. Drives (PEMD 2014)*, Manchester, UK, 2014, p. 2.15.01–2.15.01.

12

Application of Modern Control Theories

12.1 Introduction

Based on the state-space model, modern control theories have been studied since the 1960s and have the advantages of robustness, fast dynamics, and the capability of handling nonlinear systems. However, their high computational burden is one of the most critical limits of their real-time application. Fortunately, in recent years, the computation capability of microprocessors has increased significantly, and the modern control theory-based methods have become applicable in real-time calculation. As a result, since the 1990s in the field of sensorless control, many studies have been carried out based on modern control theories [1]–[20].

In this chapter, the applications of the modern control theories to sensorless control will be introduced, which include:

- Model reference adaptive system (MRAS)
- Sliding mode observer (SMO)
- Extended Kalman filter (EKF)
- Model predictive control (MPC)

For the above methods, their basic principles and corresponding sensorless control strategies will be demonstrated with some simulation and test results in this chapter.

12.2 Model Reference Adaptive System

The MRAS technique is widely employed in sensorless control due to its simple concept and low computational burden [1]–[3]. Its basic principle and its three typical sensorless control methods are introduced in this section.

Sensorless Control of Permanent Magnet Synchronous Machine Drives, First Edition.
Zi Qiang Zhu and Xi Meng Wu.

12.2.1 Basic Principle

The diagram of MRAS is shown in Figure 12.1, in which the reference model represents the real machine, while the adjustable model is a fictitious machine based on the fundamental mathematical model. With the same excitation, the difference between the responses from the two models yields an estimation error ε, which is input to the correction controller. Using the correction controller, the adaptive model is adjusted to minimize the estimation error ε. In steady state, ε is very small, which implies that the behavior of the adaptive model is the same as that of the actual machine, and hence, machine states including position information can be accessed from the adaptive mathematical model.

12.2.2 Current Model-Based Observer

In [1], a current model-based method is introduced. The difference between the actual current and the calculated current from the current model is regarded as the position error.

The calculated current and back-EMF in the current model are expressed as:

$$p\begin{bmatrix} i_{d,CM} \\ i_{q,CM} \end{bmatrix} = \frac{1}{L_s}\begin{bmatrix} -R_s & \omega_r L_s \\ -\omega_r L_s & -R_s \end{bmatrix}\begin{bmatrix} \hat{i}_d \\ \hat{i}_q \end{bmatrix} + \frac{1}{L_s}\begin{bmatrix} \hat{v}_d \\ \hat{v}_d \end{bmatrix} - \frac{1}{L_s}\begin{bmatrix} 0 \\ e_c \end{bmatrix} \quad (12.1)$$

$$e_c = \omega_r \psi_m \quad (12.2)$$

where $\hat{v}_d$, $\hat{v}_q$, $\hat{i}_d$, $\hat{i}_q$ are the voltages and currents in the estimated synchronous reference frame, respectively; $i_{d,CM}$,$i_{q,\,CM}$ are the currents of the current model; and e_c is the back-EMF of the current model. Also, ω_r is the electrical rotor speed, R_s is the phase resistance, L_s is the synchronous inductance, ψ_m is the PM flux-linkage, and p is the derivative operator, i.e. $p = d/dt$.

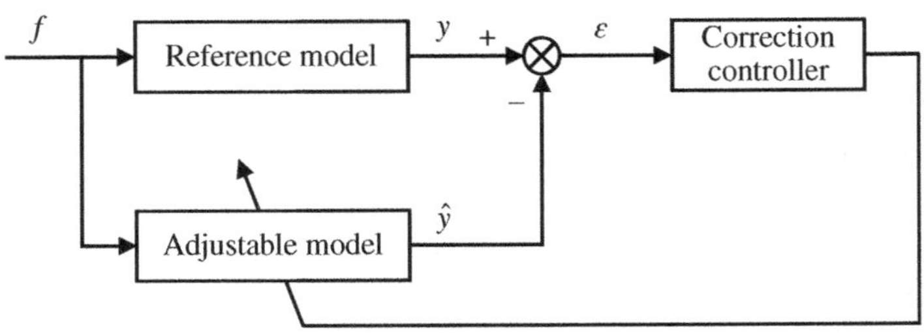

Figure 12.1 Block diagram of MRAS.

The actual machine currents in the estimated reference frame are given by:

$$p\begin{bmatrix}\hat{i}_d\\ \hat{i}_q\end{bmatrix}=\frac{1}{L_s}\begin{bmatrix}-R_s & \omega_r L_s\\ -\omega_r L_s & -R_s\end{bmatrix}\begin{bmatrix}\hat{i}_d\\ \hat{i}_q\end{bmatrix}+\frac{1}{L_s}\begin{bmatrix}\hat{v}_d\\ \hat{v}_d\end{bmatrix}-\frac{1}{L_s}\begin{bmatrix}\hat{E}_d\\ \hat{E}_q\end{bmatrix} \tag{12.3}$$

$$\begin{bmatrix}\hat{E}_d\\ \hat{E}_q\end{bmatrix}=\omega_r\psi_m\begin{bmatrix}-\sin\Delta\theta\\ \cos\Delta\theta\end{bmatrix} \tag{12.4}$$

where $\hat{E}_d$ and $\hat{E}_q$ are the estimated dq-axis back-EMFs, and $\Delta\theta$ is the position error between the real and estimated positions.

Then, (12.1) and (12.3) can be expressed in discrete form as:

$$\begin{bmatrix}\hat{i}_{d,CM}(k)\\ \hat{i}_{q,CM}(k)\end{bmatrix}=\begin{bmatrix}\hat{i}_d(k-1)\\ \hat{i}_q(k-1)\end{bmatrix}+\frac{T_s}{L_s}\left\{\begin{bmatrix}-R_s & \omega_r L_s\\ -\omega_r L_s & -R_s\end{bmatrix}\begin{bmatrix}\hat{i}_d(k-1)\\ \hat{i}_q(k-1)\end{bmatrix}+\begin{bmatrix}\hat{v}_d(k-1)\\ \hat{v}_q(k-1)\end{bmatrix}-\begin{bmatrix}0\\ e_c\end{bmatrix}\right\} \tag{12.5}$$

$$\begin{bmatrix}\hat{i}_d(k)\\ \hat{i}_q(k)\end{bmatrix}=\begin{bmatrix}\hat{i}_d(k-1)\\ \hat{i}_q(k-1)\end{bmatrix}+\frac{T_s}{L_s}\left\{\begin{bmatrix}-R_s & \omega_r L_s\\ -\omega_r L_s & -R_s\end{bmatrix}\begin{bmatrix}\hat{i}_d(k-1)\\ \hat{i}_q(k-1)\end{bmatrix}+\begin{bmatrix}\hat{v}_d(k-1)\\ \hat{v}_q(k-1)\end{bmatrix}-\begin{bmatrix}\hat{E}_d\\ \hat{E}_q\end{bmatrix}\right\} \tag{12.6}$$

where k indicates the sampling instant and T_s is the sampling period.

The difference between (12.5) and (12.6) is used to obtain the estimated rotor position and speed as given by:

$$\begin{bmatrix}\Delta\hat{i}_d\\ \Delta\hat{i}_q\end{bmatrix}=\begin{bmatrix}\hat{i}_d\\ \hat{i}_q\end{bmatrix}-\begin{bmatrix}\hat{i}_{d,CM}\\ \hat{i}_{q,CM}\end{bmatrix}=K_{CM}\begin{bmatrix}\sin\Delta\theta\\ 1-\cos\Delta\theta\end{bmatrix}, K_{CM}=\frac{T_s}{L_s}\omega_r\psi_m \tag{12.7}$$

Clearly, the d-axis current error is proportional to the position error. Based on (12.7), the control scheme of the current model-based method is illustrated in Figure 12.2. For Figure 12.2, the adjustable model is given in (12.5).

The current model-based MRAS observer is tested by simulation. The simulation is based on the prototype SPMSM-I (Appendix B). In the simulation, the machine speed is controlled at 300 r/min. Both the steady and dynamic states are tested. The steady-state position estimation under no load and full load is shown in Figures 12.3 and 12.12, respectively. Moreover, a step load test is shown in Figure 12.13, where a load disturbance is added at 0.5 seconds. It can be seen that a good position estimation performance can be obtained with the current model-based MRAS position observer considering different conditions (Figures 12.4 and 12.5).

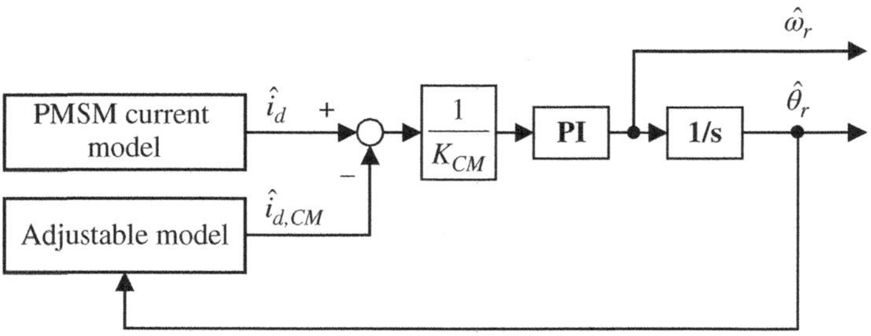

Figure 12.2 Position and speed observer of current model-based method. (*Source:* Adapted from [1].)

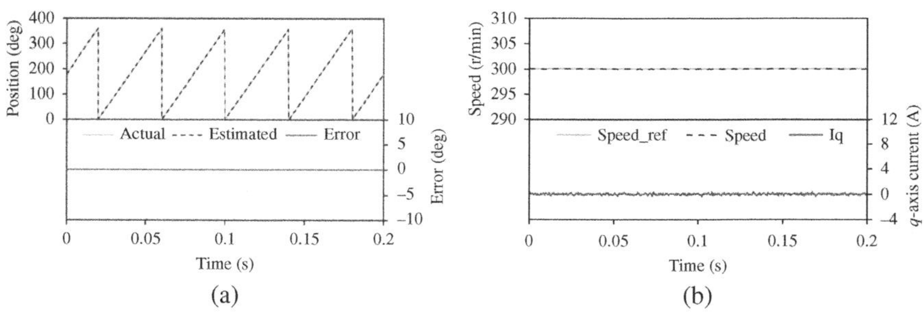

Figure 12.3 Steady-state position estimation with current model observer under no load. (a) Rotor position estimation. (b) Rotor speed and q-axis current.

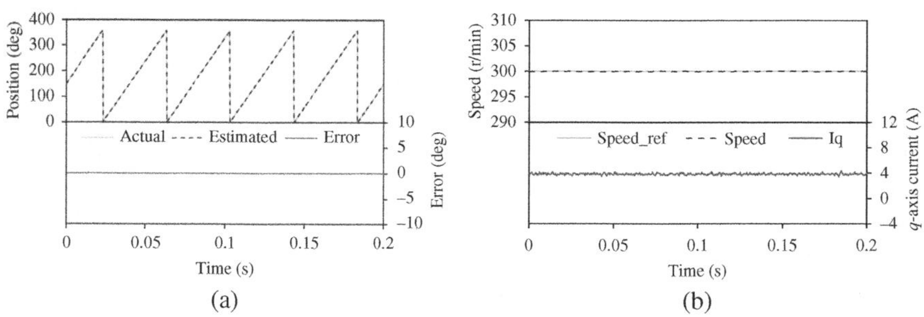

Figure 12.4 Steady-state position estimation with current model observer under full load. (a) Rotor position estimation. (b) Rotor speed and q-axis current.

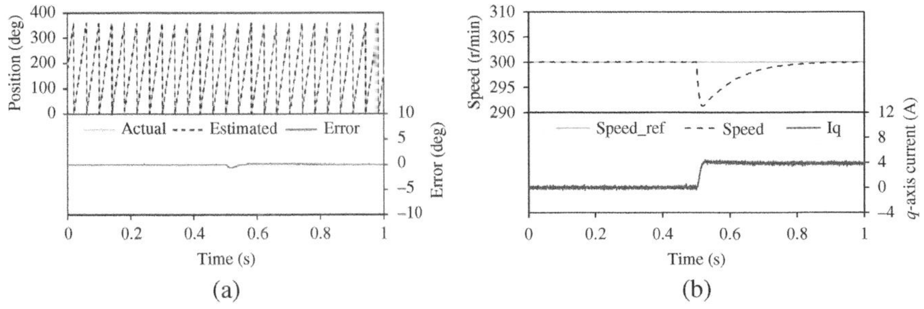

Figure 12.5 Dynamic-state position estimation with current model observer under step load. (a) Rotor position estimation. (b) Rotor speed and q-axis current.

12.2.3 Voltage Model-Based Observer

Apart from the current model, the observer can also be designed based on the voltage model [1]. In this case, the rotor position is estimated by minimizing the error between the reference voltage and the estimated voltage.

For the reference model, the actual voltage applied to the PMSM is expressed as:

$$\begin{bmatrix} \hat{v}_d \\ \hat{v}_q \end{bmatrix} = \begin{bmatrix} R_s + pL_s & -\omega_r L_s \\ \omega_r L_s & R_s + pL_s \end{bmatrix} \begin{bmatrix} \hat{i}_d \\ \hat{i}_q \end{bmatrix} + \omega_r \psi_m \begin{bmatrix} -\sin\Delta\theta \\ \cos\Delta\theta \end{bmatrix} \tag{12.8}$$

Then, for the adjustable model, it is expressed as:

$$\begin{bmatrix} \hat{v}_{d,VM} \\ \hat{v}_{q,VM} \end{bmatrix} = \begin{bmatrix} R_s + pL_s & -\omega_r L_s \\ \omega_r L_s & R_s + pL_s \end{bmatrix} \begin{bmatrix} \hat{i}_d \\ \hat{i}_q \end{bmatrix} + \begin{bmatrix} 0 \\ \omega_r \psi_m \end{bmatrix} \tag{12.9}$$

where $\hat{v}_{d,VM}$ and $\hat{v}_{q,VM}$ are the voltages of the voltage model.

Then, the back-EMF can be estimated from the voltage difference by:

$$\Delta\hat{v}_d = \hat{v}_{d,VM} - \hat{v}_d = K_{VM1}\sin\Delta\theta, K_{VM1} = \omega_r\psi_m \tag{12.10}$$

Based on (12.10), the sensorless control scheme can be developed as shown in Figure 12.6. The adjustable model is given in (12.9).

The voltage model-based MRAS observer is also tested by simulation. The simulation is based on the prototype SPMSM-I (Appendix B). In the test, the machine speed is controlled at 300 r/min. The steady-state position estimation under no load and full load is shown in Figures 12.7 and 12.8, respectively. Moreover, a step load test is shown in Figure 12.9 Clearly, a good position estimation performance can be obtained with the voltage model-based MRAS position observer under different conditions.

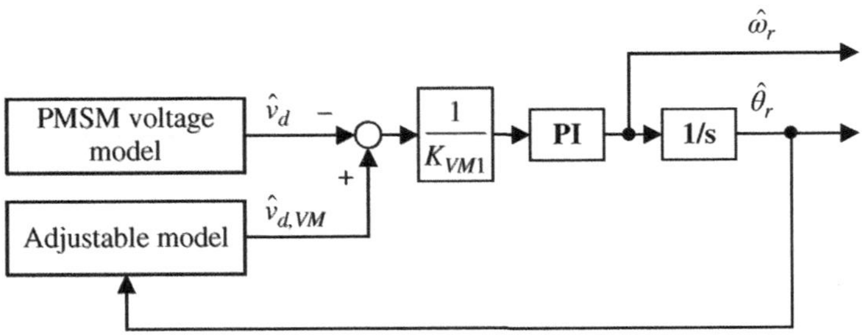

Figure 12.6 Position and speed observer of voltage model-based method. (*Source:* Adapted from [1].)

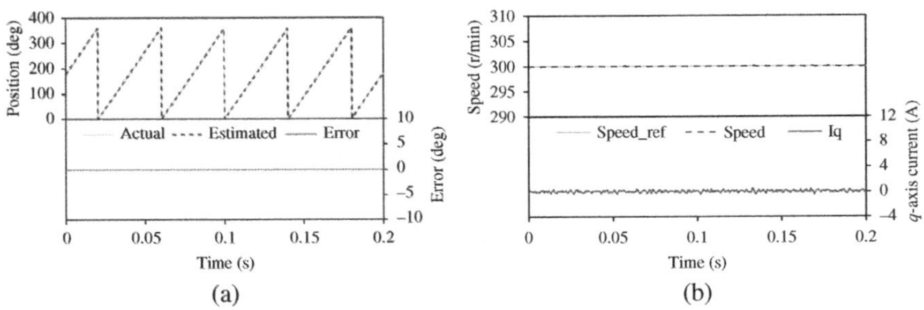

Figure 12.7 Steady-state position estimation with voltage model observer under no load. (a) Rotor position estimation. (b) Rotor speed and q-axis current.

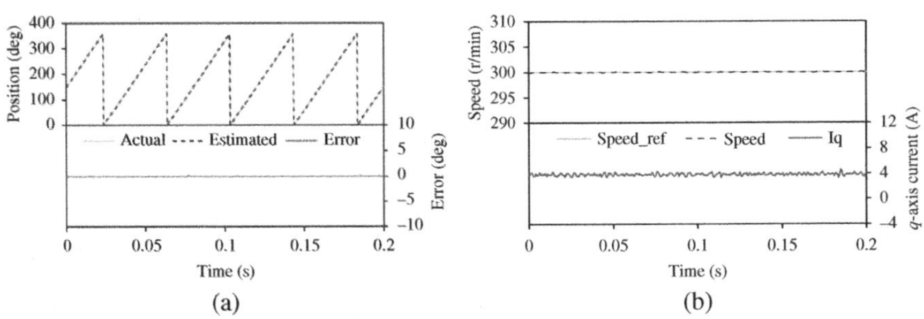

Figure 12.8 Steady-state position estimation with voltage model observer under full load. (a) Rotor position estimation. (b) Rotor speed and q-axis current.

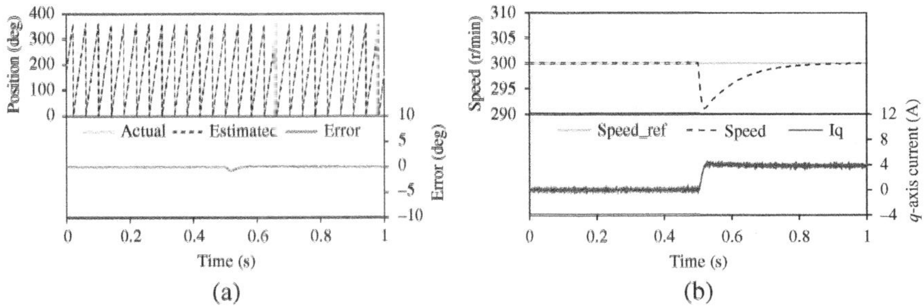

Figure 12.9 Dynamic-state position estimation with voltage model observer under step load. (a) Rotor position estimation. (b) Rotor speed and q-axis current.

12.2.4 Simplified Voltage Model Observer

A simplified voltage model-based MRAS observer is proposed in [2], as shown in Figure 12.10. The output from the d-axis current controller is used as the input estimation error signal.

In Figure 12.10, the feed-forward voltage $\hat{v}_{d\_ff}$ estimated from the adjustable model can be described as:

$$\hat{v}_{d\_ff} = R_s\hat{i}_d - \omega_r L_s \hat{i}_q \tag{12.11}$$

Then, the actual applied voltage in the reference model, i.e. the d-axis voltage command $\hat{v}_d$, can be described as:

$$\hat{v}_d = R_s\hat{i}_d - \omega_r L_s \hat{i}_q - \omega_r \psi_m \sin\Delta\theta \tag{12.12}$$

Based on (12.11) and (12.12), the estimation error between the two models can be expressed as:

$$\Delta\hat{v}_d = K_{VM2}\sin\Delta\theta, K_{VM2} = -\omega_r\psi_m \tag{12.13}$$

Then, with the position observer in Figure 12.10, the estimated rotor position and speed can be obtained.

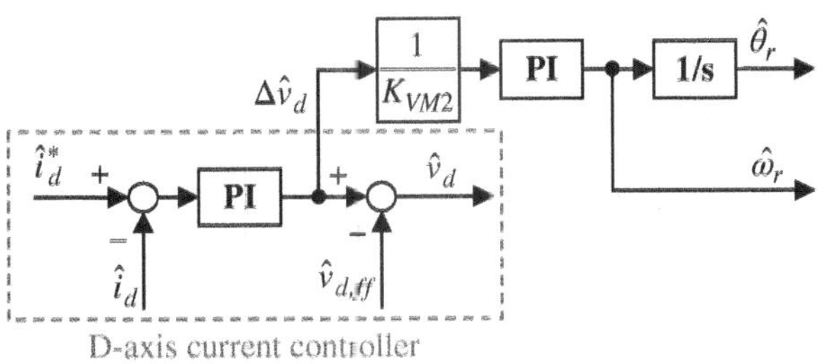

Figure 12.10 MRAS-based sensorless vector control. (*Source:* Adapted from [2].)

Moreover, the simplified voltage model-based MRAS observer is simulated based on the same SPMSM-I. In the simulation, the machine speed is controlled at 300 r/min. Both the steady and dynamic states are simulated and shown in Figures 12.11–12.13. Similarly, a good estimation performance can be obtained with this observer.

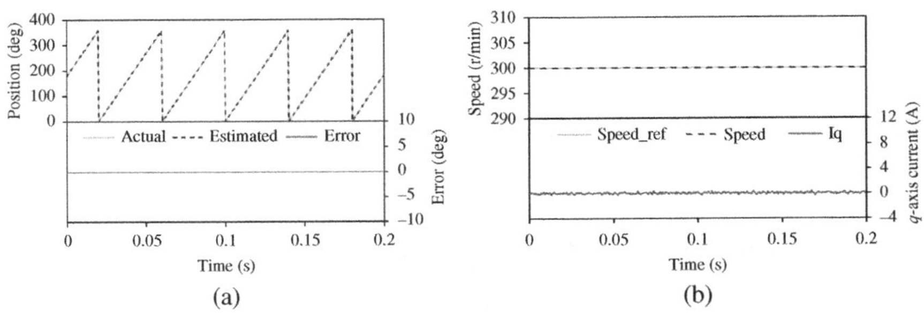

Figure 12.11 Steady-state position estimation with simplified voltage model observer under no load. (a) Rotor position estimation. (b) Rotor speed and q-axis current.

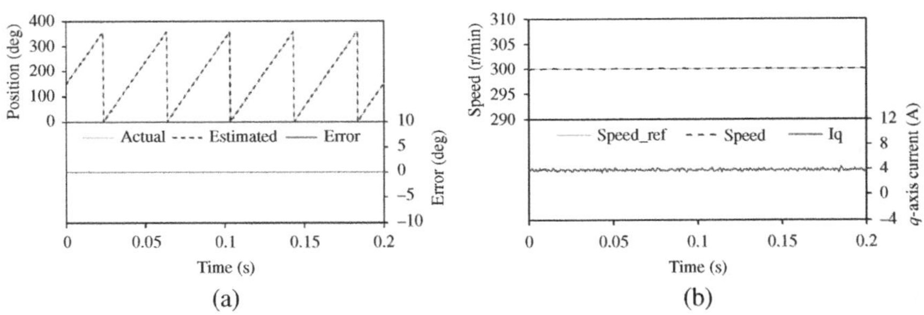

Figure 12.12 Steady-state position estimation with simplified voltage observer under full load. (a) Rotor position estimation. (b) Rotor speed and q-axis current.

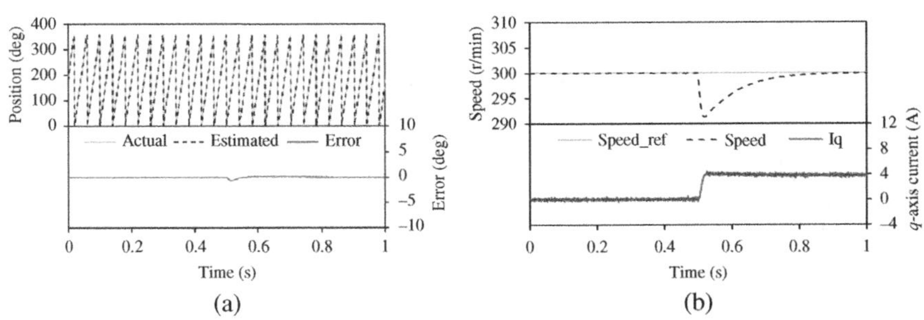

Figure 13.13 Dynamic-state position estimation with simplified voltage observer under step load. (a) Rotor position estimation. (b) Rotor speed and q-axis current.

12.3 Sliding Mode Observer

The SMO is well-known for its robustness, simplicity, and good dynamic performance. Thus, the SMO is widely adopted in sensorless control in middle and high-speed ranges by utilizing the back-EMF [4]–[12]. In this section, the basic principle of sliding model control and a SMO is introduced based on the back-EMF model.

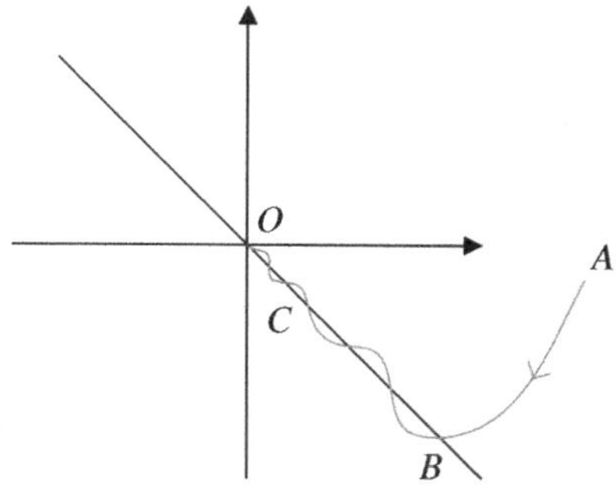

Figure 12.14 Demonstration of the sliding mode control system.

12.3.1 Basic Principle

Sliding mode control is a nonlinear control method. For a typical sliding model control system, there are three modes or stages. The first mode is called the "reaching mode." As shown in Figure 12.14, in path ***AB***, the system states are first driven onto a specified surface, i.e. a sliding surface. Once the sliding surface is reached, they are in the second mode called the "sliding mode." In the sliding mode, i.e. path ***BC***, the trajectories of the states are restrained and kept moving around the sliding surface. Finally, in the steady-state mode, the system reaches its final state, which is the equilibrium point, ***O***.

12.3.2 Conventional SMO

The SMO is designed based on the extended back-EMF model of IPMSMs, as expressed by:

$$\begin{bmatrix} v_\alpha \\ v_\beta \end{bmatrix} = \begin{bmatrix} R_s + pL_d & \omega_r(L_d - L_q) \\ -\omega_r(L_d - L_q) & R_s + pL_d \end{bmatrix} \begin{bmatrix} i_\alpha \\ i_\beta \end{bmatrix} + \begin{bmatrix} e_{ex,\alpha} \\ e_{ex,\beta} \end{bmatrix} \tag{12.14}$$

$$\begin{bmatrix} e_{ex,\alpha} \\ e_{ex,\beta} \end{bmatrix} = E_{ex} \begin{bmatrix} -\sin\theta_r \\ \cos\theta_r \end{bmatrix} \tag{12.15}$$

$$E_{ex} = \omega_r \psi_m + (L_d - L_q)\left(\omega_r i_d - p i_q\right) \tag{12.16}$$

where v_α, v_β, i_α, i_β, and $e_{ex,\alpha}$, $e_{ex,\beta}$ are the stator voltages, currents, and extended back-EMFs, respectively, in the stationary reference frame. L_d and L_q are the dq-axis inductances, and θ_r is the electrical rotor position.

It is worth noting that for a SPMSM, the model in (12.14) can be simplified by applying $(L_d - L_q) = 0$. Then, the state equation of a PMSM can be expressed as follows:

$$p\begin{bmatrix} i_\alpha \\ i_\beta \end{bmatrix} = \frac{1}{L_d}\begin{bmatrix} -R_s & -\omega_r(L_d - L_q) \\ \omega_r(L_d - L_q) & -R_s \end{bmatrix}\begin{bmatrix} i_\alpha \\ i_\beta \end{bmatrix} + \frac{1}{L_d}\begin{bmatrix} v_\alpha \\ v_\beta \end{bmatrix} - \frac{1}{L_d}\begin{bmatrix} e_{ex,\alpha} \\ e_{ex,\beta} \end{bmatrix} \tag{12.17}$$

Then, the SMO can be constructed as:

$$p\begin{bmatrix} \hat{i}_\alpha \\ \hat{i}_\beta \end{bmatrix} = \frac{1}{L_d}\begin{bmatrix} -R_s & -\omega_r(L_d - L_q) \\ \omega_r(L_d - L_q) & -R_s \end{bmatrix}\begin{bmatrix} \hat{i}_\alpha \\ \hat{i}_\beta \end{bmatrix} + \frac{1}{L_d}\begin{bmatrix} v_\alpha \\ v_\beta \end{bmatrix} - \frac{1}{L_d}\begin{bmatrix} z_\alpha \\ z_\beta \end{bmatrix} \tag{12.18}$$

$$\begin{bmatrix} z_\alpha \\ z_\beta \end{bmatrix} = k\begin{bmatrix} \text{sign}(\Delta i_\alpha) \\ \text{sign}(\Delta i_\beta) \end{bmatrix} \tag{12.19}$$

where z is the sliding mode control function and a signum switching function [5] is used in (12.19). $\Delta i_\alpha = \hat{i}_\alpha - i_\alpha$, $\Delta i_\beta = \hat{i}_\beta - i_\beta$, and k is the observer gain. The choice of the gain is a compromise between the operating range of the sliding mode and the resultant ripple in the observed variables. Large gains lead to a wider observer range but may cause excessive ripple in the observed variables, which may lead to the observed variables being unusable for control.

Next, the difference between (12.17) and (12.18) can be obtained as:

$$p\begin{bmatrix} \Delta i_\alpha \\ \Delta i_\beta \end{bmatrix} = \frac{1}{L_d}\begin{bmatrix} -R_s & -\omega_r(L_d - L_q) \\ \omega_r(L_d - L_q) & -R_s \end{bmatrix}\begin{bmatrix} \Delta i_\alpha \\ \Delta i_\beta \end{bmatrix} - \frac{1}{L_d}\begin{bmatrix} z_\alpha \\ z_\beta \end{bmatrix} + \frac{1}{L_d}\begin{bmatrix} e_{ex,\alpha} \\ e_{ex,\beta} \end{bmatrix} \tag{12.20}$$

and the sliding mode surface $\boldsymbol{S}$ can be defined as a function of the error between the estimated current and the actual current as:

$$\boldsymbol{S} = \begin{bmatrix} \Delta i_\alpha \\ \Delta i_\beta \end{bmatrix} = \begin{bmatrix} \hat{i}_\alpha - i_\alpha \\ \hat{i}_\beta - i_\beta \end{bmatrix} \tag{12.21}$$

When the sliding mode is reached, the errors between the estimated and the actual currents are zero, i.e. the sliding mode surface becomes $\boldsymbol{S} = 0$. Then, (12.20) becomes:

$$\begin{bmatrix} z_\alpha \\ z_\beta \end{bmatrix} = \begin{bmatrix} e_{ex,\alpha} \\ e_{ex,\beta} \end{bmatrix} \tag{12.22}$$

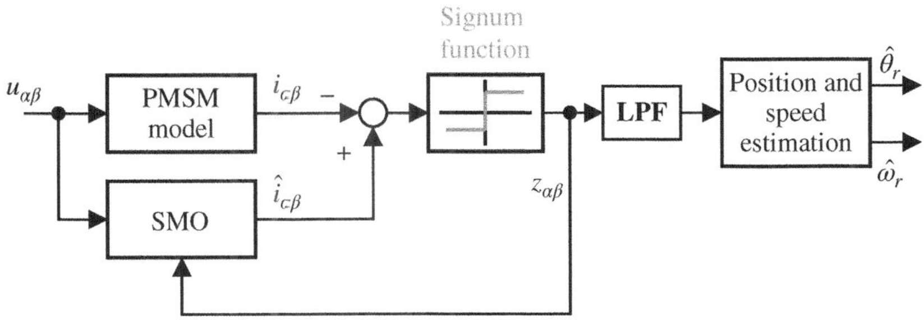

Figure 12.15 Block diagram of conventional SMO.

Clearly, the sliding mode function z contains the rotor position information. Then, the rotor position can be estimated by:

$$\hat{\theta}_r = \arctan\left(-\frac{z_\alpha}{z_\beta}\right) \tag{12.23}$$

In some cases, to avoid the impact from a noisy signal on the position estimation in (12.23), alternatively, a PLL can also be used to derive the position information in the estimated back-EMF. Overall, the block diagram of the SMO can be demonstrated as in Figure 12.15.

12.3.3 Chattering Issue and Solutions

The conventional SMO is based on the signum function, which outputs a discrete signal, and thus, has serious chattering problems. As a result, the high-frequency oscillation components could be introduced in the estimated back-EMFs of the SMO. Simulation results are provided to illustrate the chattering effect. The simulation is based on the prototype SPMSM-I (Appendix B). The rotor speed is 300 r/min. In Figure 12.16, clearly, before filtering, the original estimated back-EMFs are discrete pulses due to the signum function. To solve this issue, a low-pass filter (LPF) is generally used to attenuate the chattering effect. As shown in Figures 12.16 and 12.17, after filtering the estimated back-EMFs are sinusoidal and can be used for position estimation. However, as shown in Figure 12.18a, the implementation of LPF leads to phase delay, causing a position estimation offset. Thus, a compensation is normally required, as shown in Figure 12.18b.

However, introducing an LPF will inevitably affect the dynamic of the position estimation. Thus, in [6], [7], a sigmoid switching function is used to replace the signum function to reduce the chattering effect and thus eliminate the

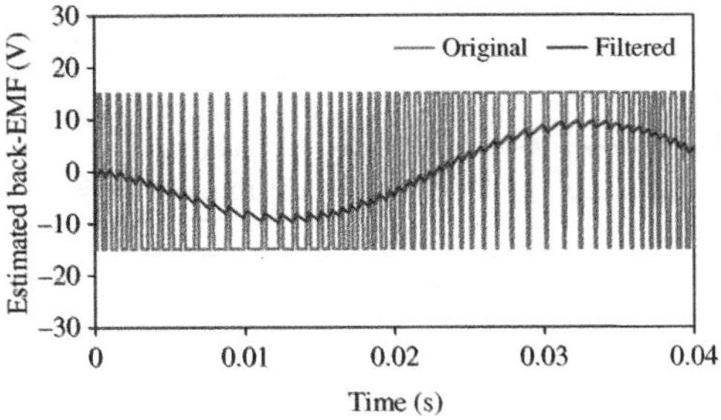

Figure 12.16 Original and filtered estimated back-EMFs.

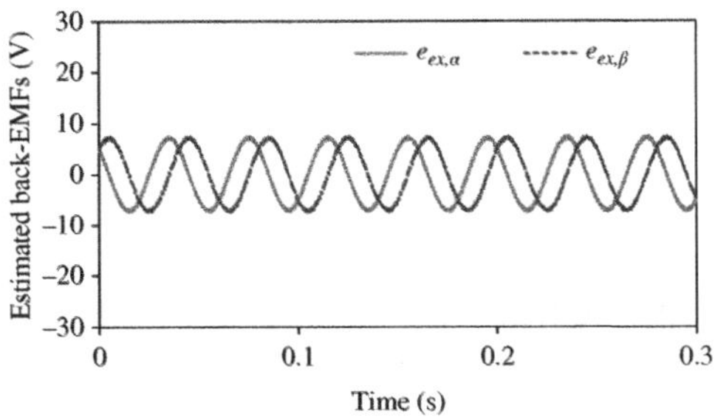

Figure 12.17 Filtered estimated back-EMFs.

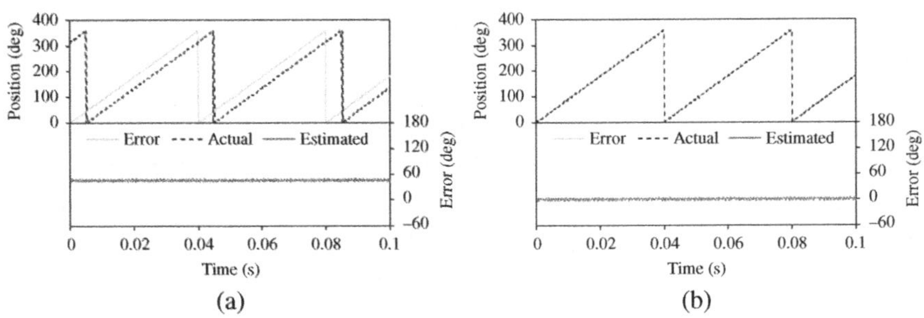

Figure 12.18 SMO position estimation performance with LPF. (a) Without delay compensation. (b) With delay compensation.

requirement of the LPF and the corresponding compensation. The sigmoid function is expressed as:

$$\begin{bmatrix} z_\alpha \\ z_\beta \end{bmatrix} = k \begin{bmatrix} \left(\dfrac{2}{1 + e^{(-a\Delta i_\alpha)}}\right) - 1 \\ \left(\dfrac{2}{1 + e^{(-a\Delta i_\beta)}}\right) - 1 \end{bmatrix} \tag{12.24}$$

where a is a positive constant to regulate the slope of the sigmoid function and k is the gain.

A simulation example with the sigmoid function is shown in Figures 12.19 and 12.20. It can be seen that the chattering effect is greatly suppressed in the original estimated back-EMFs and thus the position estimation can be achieved without the LPF and the corresponding compensations.

Moreover, there are other methods that can also reduce the chattering issues. In [8], [9], an iterative SMO is proposed; by iterating the calculation of SMO for

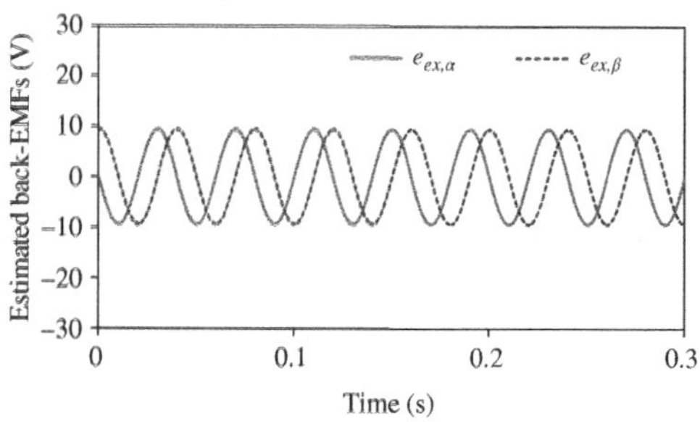

Figure 12.19 Original estimated back-EMFs of SMO with sigmoid function.

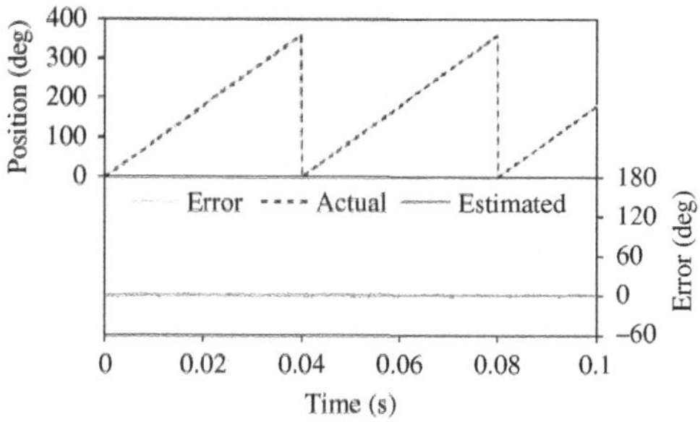

Figure 12.20 Position estimation performance of SMO with sigmoid function.

several times within a current control period, the position estimation error can be reduced with smaller ripples. High-order SMO is another approach to reduce the chattering issues. In [10]–[12], a second-order SMO is used for this purpose. The sliding mode function is designed as:

$$\begin{bmatrix} z_\alpha \\ z_\beta \end{bmatrix} = -k_1 \begin{bmatrix} |\Delta i_\alpha|^{1/2} sign(\Delta i_\alpha) \\ |\Delta i_\beta|^{1/2} sign(\Delta i_\beta) \end{bmatrix} - k_2 \begin{bmatrix} \int sign(\Delta i_\alpha) dt \\ \int sign(\Delta i_\beta) dt \end{bmatrix} \tag{12.25}$$

where k_1 and k_2 are the sliding mode coefficients. Overall, these two methods work effectively but increase the computation burden, which may not be suitable for low-cost applications.

12.4 Extended Kalman Filter

The Kalman filter is an optimal state observation based on least-squares variance estimation for a linear system with Gaussian white noise in the measurements and system states. The EKF is an extension application of Kalman filter in a nonlinear system. It is less influenced by measured noise, and parameter inaccuracy is not as critical as in conventional fundamental estimation methods. The EKF has been implemented in sensorless control for PMSM in the last several decades and has received increasing attention [13]–[20]. In this section, the basic principle of EKF is first illustrated. Both the full-order and reduced-order EKF-based position observers are described with simulation results. Finally, the parameter tuning of the EKF algorithm is discussed.

12.4.1 Basic Principle

In this section, the basic principle of a general EKF algorithm is illustrated [13]. The Kalman filter estimates an unknown state x of a dynamic system and measurements y with random noise:

$$\begin{cases} \dfrac{dx}{dt} = f(x) + g(u) + w \\ y = h(x) + v \end{cases} \tag{12.26}$$

where $f(x)$, $g(x)$, and $h(x)$ are system matrices. The random variables w and v represent the process noise and measurement noise, respectively. They are assumed to be independent, white (zero-mean), and have normal probability distributions:

$$\begin{cases} p(w) \sim N(0, Q) \\ p(v) \sim N(0, R) \end{cases} \tag{12.27}$$

where Q and R are the covariances of process noise and measurement noise, respectively. For a nonlinear system, the EKF algorithm is used and the corresponding Jacobian matrices can be derived as:

$$F(x(t)) = \left.\frac{\partial f(x(t))}{\partial x(t)}\right|_{x=x(t)}, \quad B(t) = \left.\frac{\partial g(u(t))}{\partial u(t)}\right|_{u=u(t)}, \quad H(t) = \left.\frac{\partial h(x(t))}{\partial x(t)}\right|_{x=x(t)} \tag{12.28}$$

In real applications, (12.26) needs to be discretized as:

$$\begin{cases} x(k+1) = f_d(x(k)) + g_d(u(k)) + w(k) \\ y(k) = h_d(x(k)) + v(k) \end{cases} \tag{12.29}$$

$$\begin{cases} f_d(x(k)) = x(k) + f(x(k))T_s \\ g_d(u(k)) = g(u(k))T_s \\ h_d(x(k)) = h(x(k)) \end{cases} \tag{12.30}$$

After defining the system model, the EKF algorithm basically contains two steps to estimate the optimal state $\hat{x}(k)$ and its covariance matrix $P(k)$.

12.4.1.1 Step 1: Prediction

In this step, the prior estimate state $\hat{x}^-(k)$ and its covariance matrix $P^-(k)$ are predicted based on the previous values as shown in (12.31).

$$\begin{cases} \hat{x}^-(k) = f_d(\hat{x}(k-1)) + g_d(u(k-1)) \\ P^-(k) = \Phi(k-1)P(k-1)\Phi(k-1)^T + Q \end{cases} \tag{12.31}$$

where the superscript – indicates the variable is a prior estimate value. Φ is the exponential matrix of the linearized system, which can be derived approximately as:

$$\Phi(k) \cong I + F(k)T_s \tag{12.32}$$

12.4.1.2 Step 2: Innovation

In the innovation step, the prior estimation obtained from the first step is corrected by comparing with the measured variables. The EKF gain K of the error between the measured and the estimated variables is calculated and then applied for the correction. This step can be realized by:

$$\begin{cases} K(k) = \dfrac{P^-(k)H(k)^T}{\left[H(k)P^-(k)H(k)^T + R\right]^{-1}} \\ \hat{x}(k) = \hat{x}^-(k) + K(k)[y(k) - H(k)\hat{x}^-(k)] \\ P(k) = [I - K(k)H(k)]P^-(k) \end{cases} \tag{12.33}$$

12.4.2 PMSM Model Simplification

The mathematical model of a PMSM in the stationary reference frame can be expressed as:

$$\begin{bmatrix} v_\alpha \\ v_\beta \end{bmatrix} = \begin{bmatrix} R_s + pL_d & -\omega_r(L_q - L_d) \\ \omega_r(L_q - L_d) & R_s + pL_d \end{bmatrix} \begin{bmatrix} i_\alpha \\ i_\beta \end{bmatrix} + \begin{bmatrix} e_{ex,\alpha} \\ e_{ex,\beta} \end{bmatrix} \tag{12.34}$$

$$\begin{bmatrix} e_{ex,\alpha} \\ e_{ex,\beta} \end{bmatrix} = E_{ex} \begin{bmatrix} -\sin\theta_r \\ \cos\theta_r \end{bmatrix} \tag{12.35}$$

$$E_{ex} = \omega_r \psi_m + (L_d - L_q)\left(\omega_r i_d - p i_q\right) \tag{12.36}$$

The mechanical equations for a full-order model are:

$$\begin{cases} T_e = J\dfrac{d\omega_r}{dt} + B\omega_r + T_L \\ \omega_r = \dfrac{d\theta_r}{dt} \end{cases} \tag{12.37}$$

where T_L is the load torque, J is the inertia, and B is the viscous damping factor. It is reasonable to assume that the rotor speed remains constant during each control cycle so as to simplify the model. Hence, the mechanical equation has been simplified by assuming that $d\omega_r/dt = 0$ [13].

12.4.3 Full-Order EKF

The EKF can be realized based on a full-order model [14]–[16]. First, the state variables are designed as:

$$\boldsymbol{x} = \left[i_\alpha \;\; i_\beta \;\; \omega_r \;\; \theta_r\right]^T, \;\; \boldsymbol{u} = [v_\alpha \;\; v_\beta]^T, \;\; \boldsymbol{y} = [i_\alpha \;\; i_\beta]^T \tag{12.38}$$

Then, the state-space equations are given by:

$$\begin{cases} x(k+1) = f_d(x(k)) + g_d(u(k)) + w(k) \\ y(k) = h_d(x(k)) + v(k) \end{cases} \tag{12.39}$$

where

$$f_d(x(k)) = x(k) + f(x(k))T_s = \begin{bmatrix} \left(1 - \frac{T_s R_s}{L_s}\right) i_\alpha(k) + \frac{T_s}{L_s}\omega_r \psi_m \sin\theta_r(k) \\ \left(1 - \frac{T_s R_s}{L_s}\right) i_\beta(k) - \frac{T_s}{L_s}\omega_r \psi_m \cos\theta_r(k) \\ \omega_r \\ \theta_r(k) + \omega_r T_s \end{bmatrix} \tag{12.40}$$

$$g_d(u(k)) = \begin{bmatrix} \frac{T_s}{L_s} v_\alpha(k) \\ \frac{T_s}{L_s} v_\beta(k) \\ 0 \\ 0 \end{bmatrix} \tag{12.41}$$

$$h_d(x(k)) = \begin{bmatrix} i_\alpha(k) \\ i_\beta(k) \end{bmatrix} \tag{12.42}$$

To linearize the model, the partial derivative is introduced as:

$$\Phi(k) = I + F(k)T_s = \begin{bmatrix} 1 - \frac{T_s R_s}{L_s} & 0 & \frac{T_s \psi_m}{L_s}\sin\theta_r & \frac{T_s \omega_r \psi_m}{L_s}\cos\theta_r \\ 0 & 1 - \frac{T_s R_s}{L_s} & -\frac{T_s \psi_m}{L_s}\cos\theta_r & \frac{T_s \omega_r \psi_m}{L_s}\sin\theta_r \\ 0 & 0 & 1 & 0 \\ 0 & 0 & T_s & 1 \end{bmatrix} \tag{12.43}$$

$$H(k) = \begin{bmatrix} 1 & 0 & 0 & 0 \\ 0 & 1 & 0 & 0 \end{bmatrix} \tag{12.44}$$

Then, the rotor position and speed can be estimated by substituting (12.43) into (12.31) and (12.33). It is worth noting that the estimated speed ω_r can converge to two points [15], the real value and the opposite one, i.e. $-\omega_r$. It can be corrected by comparing with the derivative of position [15].

The full-order EKF is implemented in the simulation based on the prototype SPMSM-I (Appendix B). In Figure 12.21, the machine operates under no load and a 1 Nm load disturbance is added at 0.5 seconds. Then, in Figure 12.22, the

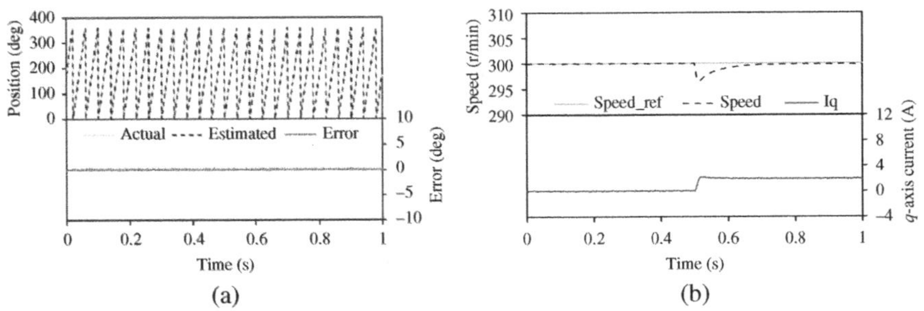

Figure 12.21 Estimation performance of full-order EKF with step load. (a) Rotor position estimation. (b) Rotor speed and q-axis current.

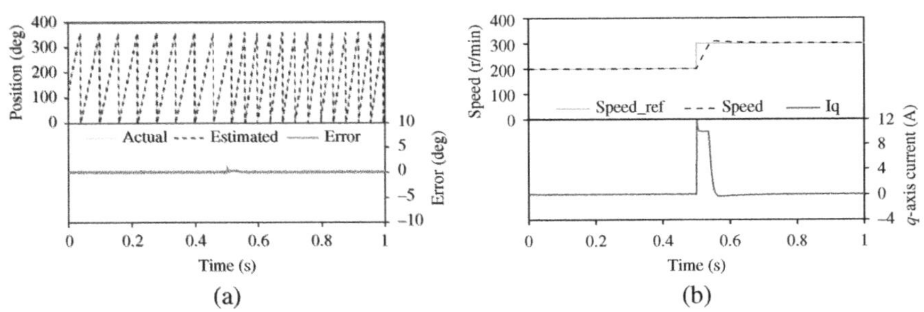

Figure 12.22 Estimation performance of full-order EKF with step speed. (a) Rotor position estimation. (b) Rotor speed and q-axis current.

rotor speed changes from 200 to 300 r/min at 0.5 seconds. From the results, the full-order EKF provides a satisfactory estimation performance under different conditions.

12.4.4 Reduced-Order EKF

If the full states of a system are required to estimate in accordance with measurements, a full-order Kalman filter has to be employed. Obviously, high dimension matrix operations for the Kalman filter recursive algorithms, in particular for the inversion of a matrix of order 4, will be a heavy computational burden for the controller, in terms of both the time and memory requirements. In addition, the tuning of an EKF that involves more state variables will be more difficult and prone to error. However, since some estimated states can be deduced from system measurements and can be used directly without being included in the state variables to be estimated, the order of the state-space can be reduced. Based on this principle, a reduced-order EKF algorithm [13, 17] is used to reduce the complexity of the model and the dimension of the state-space, and hence, reduce the computational burden in a real-time system.

Instead of current, the back-EMFs and rotor speed are chosen as state variables in the reduced-order EKF [17]. The rearranged system model is given as follows:

$$\begin{cases} x(k+1) = f_d(x(k)) + w(k) \\ y(k) = h_d(x(k)) + v(k) \end{cases} \tag{12.45}$$

where

$$x_k = [e_\alpha(k) \quad e_\beta(k) \quad \omega_r(k)]^T \tag{12.46}$$

$$y_k = \begin{bmatrix} i_\alpha(k+1) - \left(1 - \frac{T_s R_s}{L_s}\right) i_\alpha(k) - \frac{T_s}{L_s} v_\alpha(k) \\ i_\beta(k+1) - \left(1 - \frac{T_s R_s}{L_s}\right) i_\beta(k) - \frac{T_s}{L_s} v_\beta(k) \end{bmatrix} \tag{12.47}$$

$$f_d(x(k)) = x(k) + f(x(k))T_s = \begin{bmatrix} e_\alpha(k) - \omega_r T_s e_\beta(k) \\ \omega_r T_s e_\alpha(k) + e_\beta(k) \\ \omega_r(k) \end{bmatrix} \tag{12.48}$$

$$h_d(x(k)) = \begin{bmatrix} -\frac{T_s}{L_s} e_\alpha(k) \\ -\frac{T_s}{L_s} e_\beta(k) \end{bmatrix} \tag{12.49}$$

To linearize the model, the partial derivative is introduced and the discrete state models are given by:

$$\Phi(k) = I + F(k)T_s = \begin{bmatrix} 1 & -\omega_r T_s & -e_\beta T_s \\ \omega_r T_s & 1 & e_\alpha T_s \\ 0 & 0 & 1 \end{bmatrix} \tag{12.50}$$

$$H(k) = \begin{bmatrix} -\frac{T_s}{L_s} & 0 & 0 \\ 0 & -\frac{T_s}{L_s} & 0 \end{bmatrix} \tag{12.51}$$

Then, back-EMF can be estimated based on (12.31)–(12.33). After obtaining back-EMFs, the rotor position can be derived by:

$$\hat{\theta}_r = \arctan^{-1}\left(\frac{-\hat{e}_\alpha}{\hat{e}_\beta}\right) \tag{12.52}$$

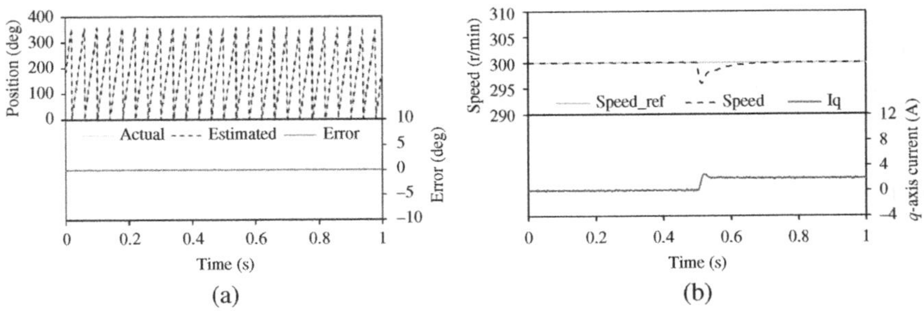

Figure 12.23 Estimation performance of reduced-order EKF with step load. (a) Rotor position estimation. (b) Rotor speed and q-axis current.

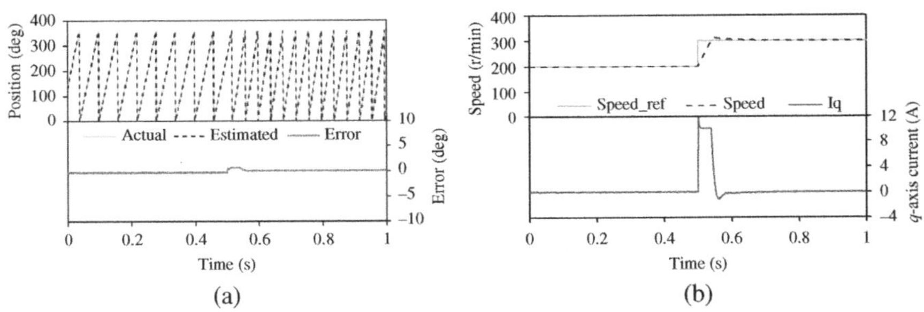

Figure 12.24 Estimation performance of reduced-order EKF with step speed. (a) Rotor position estimation. (b) Rotor speed and q-axis current.

Similarly, the reduced-order EKF is implemented in the simulation based on the prototype SPMSM-I (Appendix B). For the step-load test shown in Figure 12.23, the machine operates under no load and a 1 Nm load disturbance is added at 0.5 seconds. For the step-speed test shown in Figure 12.24, the rotor speed changes from 200 to 300 r/min at 0.5 seconds. Overall, the reduced-order EKF provides a good estimation performance. Compared to the results of the full-order EKF shown in Figures 12.21 and 12.22, the full-order EKF shows a slightly better dynamic performance but at the cost of a higher computational burden.

12.4.5 Parameter Tuning

A critical step of EKF implementation is the parameter tuning [14], [16], [18]–[20]. As introduced in Section 12.4.1, there are three matrices that need to be tuned for EKF implementation including the initial value P_0, the covariance matrices of the measurement noise R, and process noise Q.

According to [14], the initial value P_0 indicates the variances or mean-squared errors in the knowledge of the initial conditions. Different P_0 will influence the magnitude of the transient, but the transient duration and the steady-state conditions remain the same.

Based on (12.53), the calculation of EKF gain K involves the matrices R and Q. A large gain K means a fast dynamic of the EKF. Both R and Q can affect the dynamic- and steady- state performance of the EKF [14].

$$\begin{aligned} P^{-}(k) &= \Phi(k)P(k-1)\Phi(k)^{T} + Q \\ K(k) &= \frac{P^{-}(k)H(k)^{T}}{H(k)P^{-}(k)H(k)^{T} + R} \end{aligned} \tag{12.53}$$

The matrices R and Q work as two weightings for the measurement and model estimation. For the matrix R, a smaller value indicates there is less noise in the measurement and the measurement value is more reliable. Moreover, based on (12.53), a smaller R results in a larger K, and thus a faster EKF dynamic. On the other hand, the matrix Q accounts for the model uncertainty. A smaller Q indicates that there is less noise and a lower uncertainty level of the model. Therefore, the estimation from the model is more reliable. Decreasing Q will reduce the EKF gain K, resulting in a slower EKF dynamic.

For these matrices, it is a common practice to use the diagonal matrices [18], and then normally they are tuned in a trial-and-error manner [16]. In order to generalize the parameters of EKF for different drive systems, the machine parameters and the EKF algorithm can be normalized before tuning [18], which simplifies the tuning process. There are also some algorithms developed in the literature to assist the tuning process. In [19], a real-coded genetic algorithm is applied to optimize the covariance matrices of EKF. In [20] the covariance matrices are determined through the differential evolution algorithm (DEA) and multi-objective DEA.

12.5 Model Predictive Control

Recently, with the development of processors, MPC has started to be employed in power electronics and machine drives [21–23]. For MPC algorithms, the basic idea is to select the optimum control input by evaluating the prediction of the system future behavior based on the system model. MPC has several advantages over conventional PI controllers, including fast dynamic, improved control capability of nonlinear and constraints in the system, etc. Therefore, MPC is widely exploited to replace the conventional PI current controller for machine drives [23].

Typically, MPC can be divided into two categories: continuous-control-set MPC (CCS-MPC) and finite-control-set MPC (FCS-MPC). CCS-MPC outputs a continuous voltage reference and uses a PWM module to generate the desired output voltage of the power converter. On the other hand, the FCS-MPC considers the discrete nature of the power converter and does not require an external modulator.

MPC with a finite control set has gained more popularity in machine drives compared to the CCS-MPC, since CCS-MPC requires more computation effort to handle the nonlinear optimization problems [22]. In FCS-MPC, the modulator is removed by considering the discrete nature of the inverter, which enhances the dynamic performance and simplifies the control strategy. However, due to a lack of the PWM modulator, it is difficult to inject a high-frequency voltage signal. Thus, conventional HFSI sensorless control methods become incompatible with FCS-MPC. Hence, in this section, a simple HFSI sensorless control method of FCS-MPC [23] is presented that is developed based on an FCS-MPC using the deadbeat (DB) solution. With this method, conventional HFSI sensorless control can be simply integrated with FCS-MPC.

12.5.1 Predictive Current Control

In this section, the traditional FCS model predictive current control (FCS-MPCC) is described. The overall block diagram of a traditional FCS-MPCC is shown in Figure 12.25.

The voltage equation of PMSM in discrete domain using the Euler approximation is expressed as:

$$\begin{aligned} v_d(k) &= R_s i_d(k) + \frac{L_d}{T_s}\left[i_d(k+1) - i_d(k)\right] - \omega_r L_q i_q(k) \\ v_q(k) &= R_s i_q(k) + \frac{L_q}{T_s}\left[i_q(k+1) - i_q(k)\right] + \omega_r L_d i_d(k) + \omega_r \psi_m \end{aligned} \tag{12.54}$$

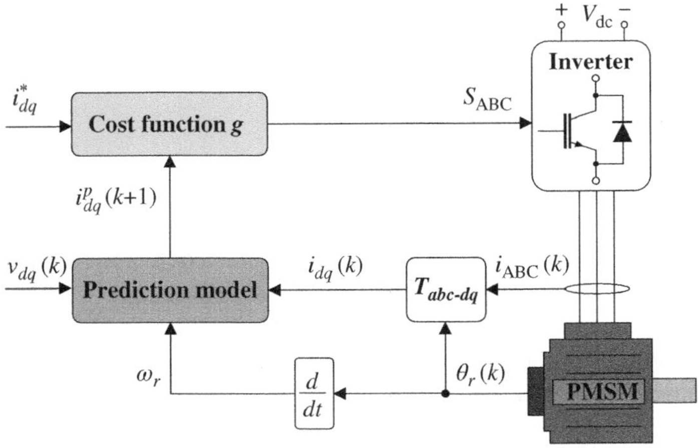

Figure 12.25 Block diagram of conventional FCS-MPCC. (*Source:* [23]/IEEE.)

where v_d, v_q, i_d, i_q are the voltages and currents in the dq-axes, respectively. T_s is the sampling period, L_d and L_q are the dq-axis inductances, θ_r is the electrical rotor position, ω_r is the electrical rotor speed, ψ_m is the PM flux-linkage, and k is the current sampling instant.

Based on (12.54), the current can be predicted by (12.55), where i_d^p and i_q^p are the predicted currents in the dq-axes.

$$\begin{aligned} i_d^p(k+1) &= i_d(k) + \frac{T_s}{L_d}\left[v_d(k) - R_s i_d(k) + \omega_r L_q i_q(k)\right] \\ i_q^p(k+1) &= i_q(k) + \frac{T_s}{L_q}\left[v_q(k) - R_s i_d(k) - \omega_r L_d i_d(k) - \omega_r \psi_m\right] \end{aligned} \tag{12.55}$$

As detailed in Chapter 1, for VSI, which is normally used in the machine drives, there are eight voltage vectors in total, i.e. $\mathbf{v}_0$, ... $\mathbf{v}_7$. These eight voltage vectors are evaluated in (12.55) to calculate the current in the next sampling step. Then, a cost function is designed as (12.56) to select the optimal voltage vector that minimizes the error between the predicted and reference currents, i.e. i_{dq}^p and i_{dq}^*.

$$g = \left|i_d^* - i_d^p(k+1)\right| + \left|i_q^* - i_q^p(k+1)\right| \tag{12.56}$$

Since the PWM module is not included in the FCS-MPCC, it is difficult to inject a high-frequency voltage signal to realize the zero- and low-speed sensorless control.

12.5.2 Predictive Current Control with Deadbeat Solution

The FCS-MPCC can be modified based on the principle of deadbeat control. Based on the deadbeat control, the current reference is transformed into a voltage reference solution that is given by:

$$\begin{aligned} v_d^*(k) &= R_s i_d(k) + \frac{L_d}{T_s}\left[i_d^*(k) - i_d(k)\right] - \omega_r L_q i_q(k) \\ v_q^*(k) &= R_s i_q(k) + \frac{L_q}{T_s}\left[i_q^*(k) - i_q(k)\right] + \omega_r L_d i_d(k) + \omega_r \psi_m \end{aligned} \tag{12.57}$$

According to [24], considering the digital implementation, a one-step delay compensation is applied. Correspondingly, (12.57) is modified as (12.58), where the current is predicted one step ahead based on (12.55).

$$\begin{cases} v_d^*(k) = R_s i_d(k+1) + \frac{L_d}{T_s}\left[i_d^*(k) - i_d(k+1)\right] - \omega_r L_q i_q(k+1) \\ v_q^*(k) = R_s i_q(k+1) + \frac{L_q}{T_s}\left[i_q^*(k) - i_q(k+1)\right] + \omega_r L_d i_d(k+1) + \omega_r \psi_m \end{cases} \tag{12.58}$$

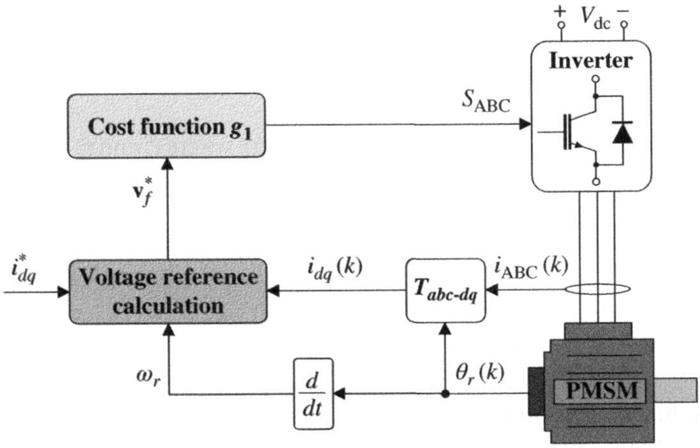

Figure 12.26 Block diagram of FCS-MPCC with DB solution. (*Source:* [23]/IEEE.)

Then, the optimization problem becomes how to select the candidate vector that is the closest to the voltage reference solution. The corresponding cost function is represented by (12.59).

$$g_1 = \left|\mathbf{v}_i - \mathbf{v}_f^*\right| \tag{12.59}$$

where $\mathbf{v}_i$ is the candidate voltage vector and $\mathbf{v}_f^*$ is the fundamental voltage reference solution. The block diagram of the deadbeat solution-based FCS-MPCC algorithm is shown in Figure 12.26. The voltage reference calculation is given in (12.58).

Based on the cost function (12.59), a high-frequency signal injection sensorless control is developed and demonstrated in the next section.

12.5.3 HFSI Sensorless Control

12.5.3.1 High-Frequency Voltage Signal Injection

As shown in the previous section, the cost function is modified to evaluate the voltage reference tracking error. Accordingly, a high-frequency voltage signal reference can be simply added to the cost function as given in (12.60).

$$g_2 = \left|\mathbf{v}_i - \left(\mathbf{v}_f^* + \mathbf{v}_{hf}^*\right)\right| \tag{12.60}$$

where $\mathbf{v}_{hf}^*$ is the high-frequency voltage signal reference. It can be selected as a pulsating or rotating signal. In this section, a pulsating signal is selected due to its simpler implementation. Hence, the high-frequency voltage signal is defined as:

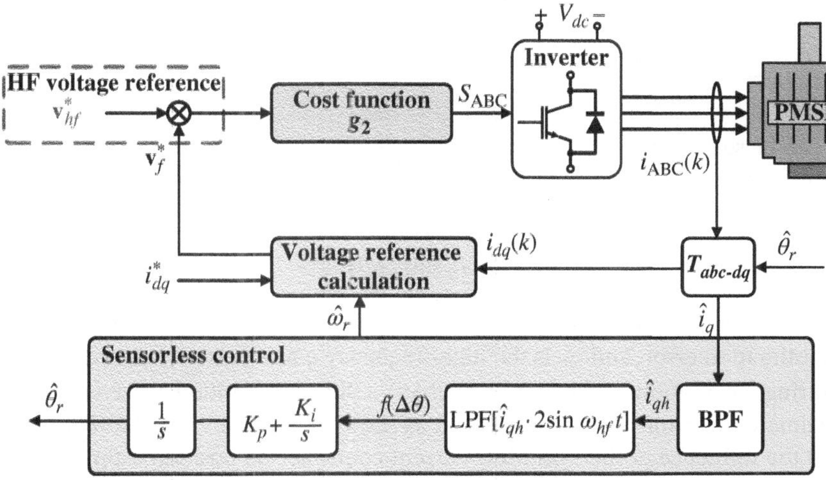

Figure 12.27 Block diagram of HFSI sensorless control for FCS-MPCC with DB solution. (*Source:* [23]/IEEE.)

$$\mathbf{v}_{hf}^* = \begin{cases} V_h \cos\omega_h t \\ 0 \end{cases} \tag{12.61}$$

where V_h and ω_h are the amplitude and frequency of the injected high-frequency voltage signal, and t is the time.

In (12.60), the voltage vector candidates $\mathbf{v}_i$ are compared to the total voltage reference including the fundamental and high-frequency voltage references in one control period. The voltage vector that can minimize the error will be selected. In this way, the output voltage will contain a fundamental part and a high-frequency part. Clearly, with the cost function in (12.60), the high-frequency voltage signal can be injected into the stator winding without modulator, and the HFSI sensorless control can be simply integrated with FCS-MPCC. The overall block diagram of this method is shown in Figure 12.27. The voltage reference calculation is given in (12.58).

12.5.3.2 Signal Demodulation

After injection, a band-pass filter (BPF) can be used to obtain the injected HF component in the q-axis current, which contains the information of the rotor position. Then, the high-frequency q-axis current response is expressed as:

$$\hat{i}_{qh} = I_n \sin(2\Delta\theta)\sin(\omega_h t) \tag{12.62}$$

where

$$I_n = \frac{V_h\left(L_{qh} - L_{dh}\right)}{2\omega_h L_{dh} L_{qh}} \tag{12.63}$$

It can be seen that the high-frequency q-axis current contains the rotor position information and the injected frequency component. Then, the obtained HF q-axis current response is demodulated and expressed as:

$$f(\Delta\theta) = \text{LPF}\left[\hat{i}_{qh} \cdot 2\sin\omega_h t\right] = I_n \sin(2\Delta\theta) \tag{12.64}$$

where L_{dh} and L_{qh} are the incremental inductances in the dq-axes, $\Delta\theta$ is the position estimation error, and $\hat{i}_{qh}$ is the high-frequency q-axis current response. After acquiring the position-dependent signal $f(\Delta\theta)$, a PLL-type position observer is used to estimate the rotor position and speed as shown in Figure 12.27. Moreover, to avoid the influence of the FCS-MPC current controller on position estimation, a LPF is applied to remove the high-frequency component in the current feedback for fundamental current control.

12.5.3.3 Experiment Results

In this section, experimental results for the prototype IPMSM-I (Appendix B) are provided to illustrate the effectiveness of the above method. The position estimation performance at zero and low speed is examined. In the test, a high-frequency pulsating voltage signal is injected into the d-axis to estimate the rotor position. The amplitude and frequency of the injected high-frequency voltage signal are selected as 30 V and 500 Hz to guarantee a reliable estimation.

First, the steady-state estimation performance is shown in Figures 12.28 and 12.29. The speed of the IPMSM is controlled at 60 r/min. The d-axis current is controlled at 0 A. The no load condition is considered in Figure 12.28 and the full load condition is considered in Figure 12.29. It can be seen that good estimation performance is obtained under different load conditions.

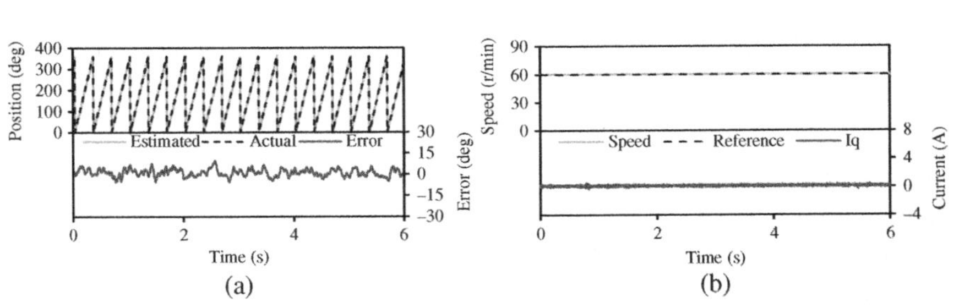

Figure 12.28 Steady-state position estimation under no load. (a) Rotor position. (b) Rotor speed and q-axis current. (*Source:* [23]/IEEE.)

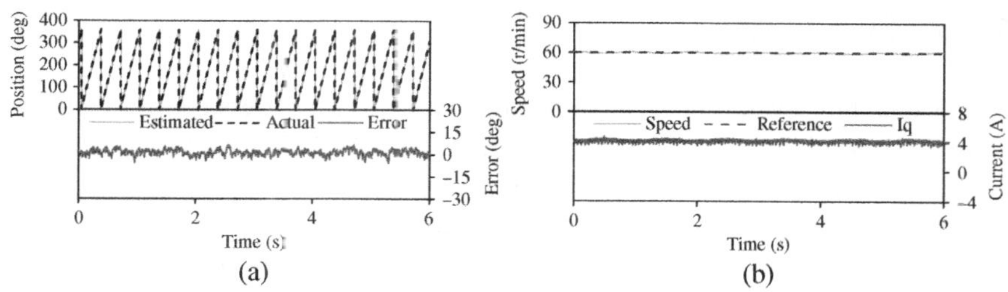

Figure 12.29 Steady-state position estimation under full load. (a) Rotor position. (b) Rotor speed and q-axis current. (*Source:* [23]/IEEE.)

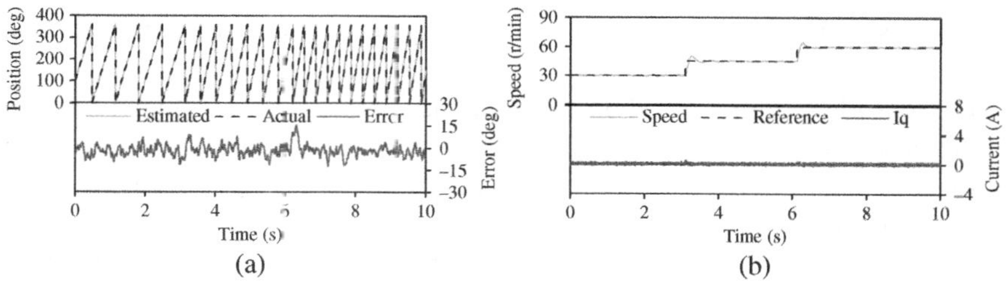

Figure 12.30 Dynamic-state position estimation under step speed. (a) Rotor position. (b) Rotor speed and q-axis current. (*Source:* [23]/IEEE.)

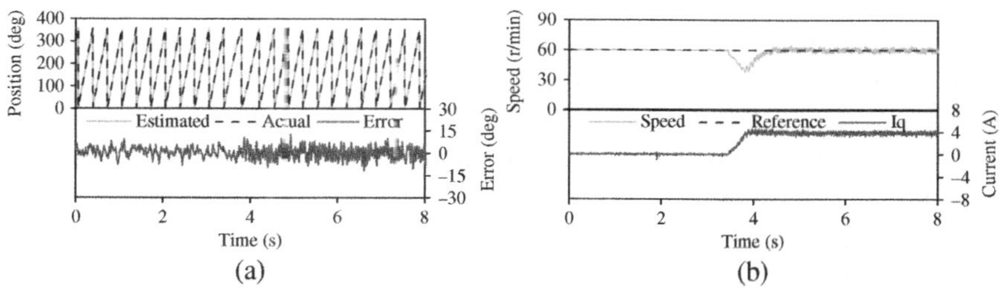

Figure 12.31 Dynamic-state position estimation under step load. (a) Rotor position. (b) Rotor speed and q-axis current. (*Source:* [23]/IEEE.)

Second, the dynamic-state estimation performance is shown in Figures 12.30 and 12.31. In Figure 12.30, the rotor speed is changed to 30–45–60 r/min. A satisfying performance can be observed. In Figure 12.31, the q-axis current is changed from no load to full load (4 A), and a good position tracking performance is maintained.

12.6 Summary

In this chapter, several modern control theory-based sensorless control methods were introduced including MRAS, SMO, EKF, and MPC. For MRAS-based methods, three different adjustable models were introduced that can all work effectively. For SMO-based sensorless control, the common chattering issue was discussed and some solutions were provided. The chattering issue can be effectively suppressed by manipulating the switching functions. For the EKF, a full-order and a reduced-order EKF were described in detail and the parameter tuning process was discussed as well. The MRAS, SMO, and EKF-based methods introduced in this chapter are more suitable for middle and high-speed operations since they are based on the back-EMF, which is proportional to the rotor speed. At zero and low speed, their performance will be degraded due to the unobservable back-EMF.

The sensorless control based on MPC was also discussed. It was found that the HFSI sensorless control cannot be directly applied due to the lack of PWM module in the MPC algorithm. Hence, a simple strategy is presented by utilizing the deadbeat solution to inject the high-frequency voltage signal; thus, the HFSI sensorless control becomes compatible with MPC for the zero- and low-speed sensorless control.

References

1 N. Matsui, "Sensorless PM brushless dc motor drives," *IEEE Trans. Ind. Electron.*, vol. 43, no. 2, pp. 300–308, Apr. 1996.

2 B.-H. Bae, S.-K. Sul, J.-H. Kwon, and J.-S. Byeon, "Implementation of sensorless vector control for super-high-speed PMSM of turbo-compressor," *IEEE Trans. Ind. Appl.*, vol. 39, no. 3, pp. 811–818, May 2003.

3 S. M. Gadoue, D. Giaouris, and J. W. Finch, "MRAS sensorless vector control of an induction motor using new sliding-mode and fuzzy-logic adaptation mechanisms," *IEEE Trans. Energy Convers.*, vol. 25, no. 2, pp. 394–402, Jun. 2010.

4 T. Furuhashi, S. Sangwongwanich, and S. Okuma, "A position-and-velocity sensorless control for brushless dc motors using an adaptive sliding mode observer," *IEEE Trans. Ind. Electron.*, vol. 39, no. 2, pp. 89–95, Apr. 1992.

5 S. Chi, Z. Zhang, and L. Xu, "Sliding-mode sensorless control of direct-drive PM synchronous motors for washing machine applications," *IEEE Trans. Ind. Appl.*, vol. 45, no. 2, pp. 582–590, Mar. 2009.

6 K. Paponpen and M. Konghirun, "An improved sliding mode observer for speed sensorless vector control drive of PMSM," in *2006 CES/IEEE 5th Int. Power Electron. Motion Control Conf.*, vol. 2, pp. 1–5, Aug. 2006.

7 H. Kim, J. Son, and J. Lee, “A high-speed sliding-mode observer for the sensorless speed control of a PMSM,” *IEEE Trans. Ind. Electron.*, vol. 58, no. 9, pp. 4069–4077, Sep. 2011.

8 K.-L. Kang, J.-M. Kim, K.-B. Hwang, and K.-H. Kim, “Sensorless control of PMSM in high speed range with iterative sliding mode observer,” in *19th Annu. IEEE Appl. Power Electron. Conf. Expo., 2004. APEC '04.*, Feb. 2004, vol. 2, pp. 1111–1116.

9 H. Lee and J. Lee, “Design of iterative sliding mode observer for sensorless PMSM control,” *IEEE Trans. Control Syst. Tech.*, vol. 21, no. 4, pp. 1394–1399, Jul. 2013.

10 S. Di Gennaro, J. Rivera, and B. Castillo-Toledo, “Super-twisting sensorless control of permanent magnet synchronous motors,” in *49th IEEE Conf. Decis. Control (CDC)*, pp. 4018–4023, Dec. 2010.

11 L. Zhao, J. Huang, H. Liu, B. Li, and W. Kong, “Second-order sliding-mode observer with online parameter identification for sensorless induction motor drives,” *IEEE Trans. Ind. Electron.*, vol. 61, no. 10, pp. 5280–5289, Oct. 2014.

12 D. Liang, J. Li, and R. Qu, “Sensorless control of permanent magnet synchronous machine based on second-order sliding-mode observer with online resistance estimation,” *IEEE Trans. Ind. Appl.*, vol. 53, no. 4, pp. 3672–3682, Jul. 2017.

13 Y. F. Shi, Z. Q. Zhu, and D. Howe, “Improved sensorless operation of interior PM brushless AC motor drive with reduced-order EKF,” in *3rd IET Int. Conf. Power Electron. Mach. Drives (PEMD 2006)*, Dublin, Ireland, 2006, vol. 2006, pp. 336–340.

14 R. Dhaouadi, N. Mohan, and L. Norum, “Design and implementation of an extended Kalman filter for the state estimation of a permanent magnet synchronous motor,” *IEEE Trans. Power Electron.*, vol. 6, no. 3, pp. 491–497, Jul. 1991.

15 S. Bolognani, L. Tubiana, and M. Zigliotto, “EKF-based sensorless IPM synchronous motor drive for flux-weakening applications,” *IEEE Trans. Ind. Appl.*, vol. 39, no. 3, pp. 768–775, May 2003.

16 P. Niedermayr, L. Alberti, S. Bolognani, and R. Abl, “Implementation and experimental validation of ultrahigh-speed PMSM sensorless control by means of extended Kalman filter,” *IEEE J. Emerg. Sel. Topics Power Electron.*, vol. 10, no. 3, pp. 3337–3344, Jun. 2022.

17 Y.-H. Kim and Y.-S. Kook, “High performance IPMSM drives without rotational position sensors using reduced-order EKF,” *IEEE Trans. Energy Convers.*, vol. 14, no. 4, pp. 868–873, Dec. 1999.

18 S. Bolognani, L. Tubiana, and M. Zigliotto, “Extended Kalman filter tuning in sensorless PMSM drives,” *IEEE Trans. Ind. Appl.*, vol. 39, no. 6, pp. 1741–1747, Nov. 2003.

19 K. L. Shi, T. F. Chan, Y. K. Wong, and S. L. Ho, “Speed estimation of an induction motor drive using an optimized extended Kalman filter,” *IEEE Trans. Ind. Electron.*, vol. 49, no. 1, pp. 124–133, Feb. 2002.

20 E. Zerdali and M. Barut, “The comparisons of optimized extended Kalman filters for speed-sensorless control of induction motors,” *IEEE Trans. Ind. Electron.*, vol. 64, no. 6, pp. 4340–4351, Jun. 2017.

21 J. Rodriguez *et al.*, "Predictive current control of a voltage source inverter," *IEEE Trans. Ind. Electron.*, vol. 54, no. 1, pp. 495–503, Feb. 2007.

22 M. Preindl and S. Bolognani, "Comparison of direct and PWM model predictive control for power electronic and drive systems," in *Proc. 28th Annu. IEEE Appl. Power Electron. Conf. Expo.*, Long Beach, CA, USA, 2013, pp. 2526–2533.

23 X. Wu, Z. Q. Zhu, and N. M. A. Freire, "High frequency signal injection sensorless control of finite-control-set model predictive control with deadbeat solution," *IEEE Trans. Ind. Appl.*, vol. 58, no. 3, pp. 3685–3695, May 2022.

24 P. Cortes, J. Rodriguez, C. Silva, and A. Flores, "Delay compensation in model predictive current control of a three-phase inverter," *IEEE Trans. Ind. Electron.*, vol. 59, no. 2, pp. 1323–1325, Feb. 2012.

Appendix A

Speed Estimation

In this book, various position estimation techniques of sensorless control have been described. For PM machine drives, to achieve a closed-loop speed control, the speed should also be estimated. Hence, in this section, the speed estimation of sensorless control is described. Typically, there are three methods to estimate the rotor speed, as follows:

a) Speed estimation associated with the estimated rotor position [1]–[5]. Speed can be obtained by the differential of the estimated rotor position as shown in Figure A.1a or obtained by a controller (e.g. PI), and its integration is the rotor position as shown in Figure A.1b.
b) Speed estimation based on the machine model by calculating the ratio between PM flux-linkage and back-EMF [6], [8], Figure A.1c.
c) Hybrid speed estimation [7], [8] combining methods (a) and (b), Figures A.1d,e.

It is worth noting that for the fundamental model-based sensorless control methods, all of these three methods can be used. However, for saliency tracking-based sensorless control methods, the accuracy of the machine model is poor at zero and low speed. Thus, only the first type of speed estimation method is normally used.

For the first type of method [1]–[5], the estimated speeds in Figures A.1a,b cannot be directly used in practice since the estimated speed contains errors and noises that could deteriorate the overall control system. A common method is to apply a low-pass filter (LPF) on the estimated speed to obtain the average speed. This method gives the correct value when the motor operates at the steady state. However, its dynamic response is generally not fast enough. Therefore, the second type of method [6], shown in Figure A.1c, is used to determine the speed from the machine model. However, while this method of speed estimation has a fast

Sensorless Control of Permanent Magnet Synchronous Machine Drives, First Edition.
Zi Qiang Zhu and Xi Meng Wu.

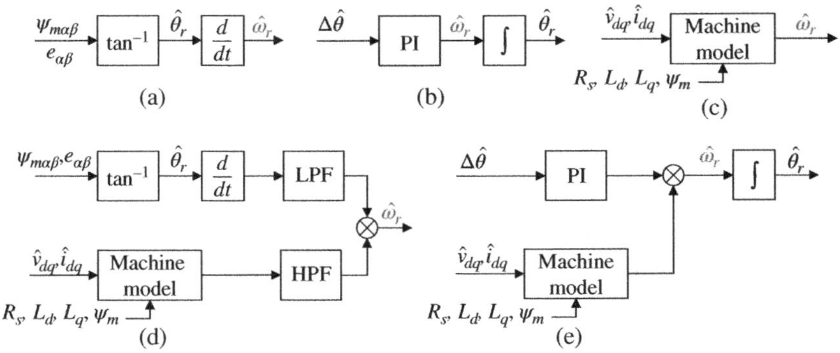

Figure A.1 Block diagram of speed estimation methods. (a) Differential of rotor position [1], [2]. (b) Output of controller [3]–[5]. (c) Calculation from model [6]. (d) Hybrid speed estimation method I [7]. (e) Hybrid speed estimation method II [8].

response, it requires the machine model parameters, which may vary due to variations in temperature and saturation and cause further errors in the estimated speed. Therefore, the hybrid speed estimation methods [7], [8] in Figures A.1d,e can be used, which have the synergies of the foregoing methods.

A.1 Speed Estimation Associated with Rotor Position

First, for speed estimation by the differential of the rotor position as shown in Figure A.1a, the estimated rotor speed $\hat{\omega}_r$ is given by:

$$\hat{\omega}_r = \frac{d\hat{\theta}_r}{dt} \approx \frac{\hat{\theta}_r(k) - \hat{\theta}_r(k-1)}{T_s} \tag{A.1}$$

where $\hat{\theta}_r(k-1)$ and $\hat{\theta}_r(k)$ are the estimated electrical rotor positions at the start and end of the kth time interval T_s, respectively. In general, errors may exist in $\hat{\theta}_r(k-1)$ and $\hat{\theta}_r(k)$, then,

$$\begin{aligned}\hat{\omega}_r &= \frac{[\theta_r(k) - \Delta\theta(k)] - [\theta_r(k-1) - \Delta\theta(k-1)]}{T_s} \\ &= \frac{[\theta_r(k) - \theta_r(k-1)] - [\Delta\theta(k) - \Delta\theta(k-1)]}{T_s} \\ &= \omega_r - \frac{[\Delta\theta(k) - \Delta\theta(k-1)]}{T_s}\end{aligned} \tag{A.2}$$

Since T_s is very short, $\Delta\theta$ may be large and cause a significant error in the estimated speed.

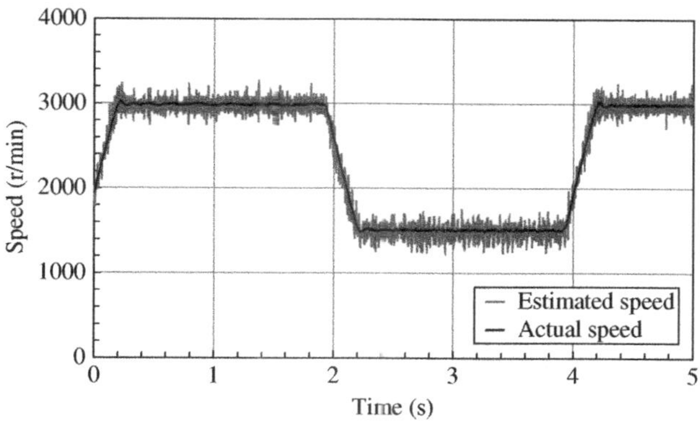

Figure A.2 Estimated speed from the differential of the estimated rotor position [7].

Second, for the speed estimation method shown in Figure A.1b, the estimated position error is the input of a PI controller and the estimated speed is the output. In general, the estimated position error may contain noises. To maintain a satisfying dynamic performance of the position estimation, the bandwidth of the controller should be sufficiently high, which will inevitably introduce high-frequency noises in the estimated speed. The phenomenon of estimated speed is close to the speed estimation by the differential of the rotor position. Therefore, the discussion in this section will be mainly based on speed estimation from the differential of the estimated rotor position.

Based on the prototype SPMSM-VI (Appendix B), an experiment result is shown in Figure A.2 to demonstrate the limited performance of the speed estimation methods in Figure A.1. The speed reference is changed from 1500 to 3000 r/min every two seconds. In addition, the rotor position is estimated by the flux-linkage method as described in Chapter 2, and the speed is estimated from the differential of the estimated rotor position. Figure A.2 shows that the estimated speed contains significant ripples and would be inappropriate for speed feedback in a sensorless drive system.

One simple solution is to employ a LPF with the time constant T_c on the estimated speed $\hat{\omega}_r$, to obtain the filtered average speed, $\hat{\omega}_d$.

$$\hat{\omega}_d = \text{LPF}(\hat{\omega}_r) = \frac{1}{T_c s + 1}\hat{\omega}_r \tag{A.3}$$

An experimental example is shown in Figure A.3 by applying the LPF on the estimated speed by the differential of the rotor position. As shown in Figure A.3, $\hat{\omega}_d$ is in close agreement with the actual speed under steady-state operating conditions. However, it is also observed in the lower section of Figure A.3 that there is a noticeable time delay in the estimated speed.

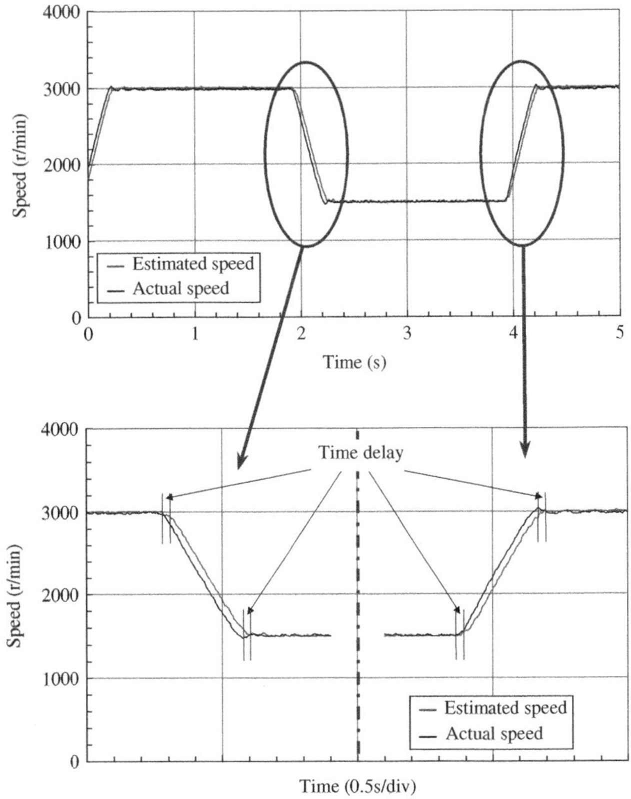

Figure A.3 Average speed estimation [7].

A.2 Speed Estimation from Machine Model

Another way of estimating speed is based on the machine model [6]. The speed can be calculated by the ratio between the PM flux-linkage and the back-EMF. Taking the non-salient PMSM as an example, the q-axis voltage is expressed by:

$$\hat{v}_q = R\hat{i}_q + L_s \cdot p\hat{i}_q + \hat{\omega}_r L_s \hat{i}_d + \hat{\omega}_r \psi_m \tag{A.4}$$

The speed can be estimated from:

$$\hat{\omega}_r = \frac{\hat{v}_q - R\hat{i}_q - L_s \cdot p\hat{i}_q}{\psi_m + L_s\hat{i}_d} \tag{A.5}$$

The main merit of this method is its fast dynamic. However, it has two significant drawbacks. First, the parameters R_s, L_s, and ψ_m are sensitive to variations in temperature and saturation. Second, the speed estimation still involves a differential operation. The differential operation may cause a significant error in the estimated speed. This is evident in Figure A.4a, which shows the d- and q-axis currents (i_d and i_q), calculated from the measured phase currents, when the motor speed is changed from 1500 to 3000 r/min.

Since the estimation of the speed from the voltage model is inherently inaccurate, a further variant, i.e. eliminating the differential operation in (A.5), has been investigated to simplify the calculation. While this increases the potential for error in the estimated speed it ensures a fast response. Moreover, since $L_s\hat{i}_d$ is usually much smaller than ψ_m, especially when the d-axis current reference is set to zero in the vector control, (A.5) may be simplified to:

$$\hat{\omega}_v \approx \frac{\hat{v}_q - R_s\hat{i}_q}{\psi_m} \tag{A.6}$$

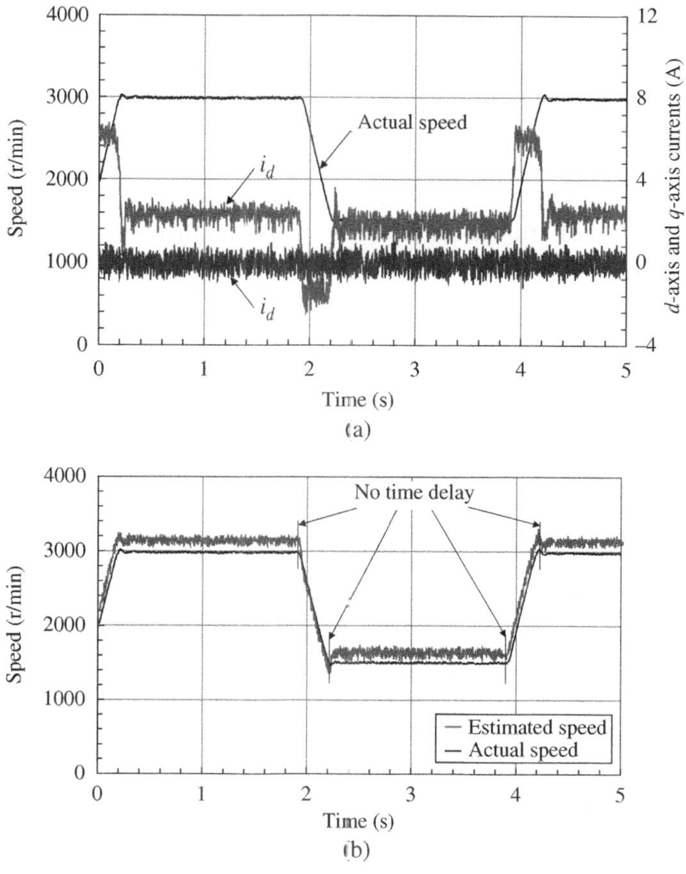

Figure A.4 Estimation of speed from EMF and excitation flux [7]. (a) d-axis and q-axis currents, rich in noise. (b) Comparison of estimated and actual speed.

The motor speed is again varied from 1500 to 3000 r/min every two seconds, and the estimated speed is compared with the actual speed, as shown in Figure A.4b. It can be seen that the response of the estimated speed change is fast, which is the most important feature of the method, although the accuracy of the estimation is not good.

A.3 Hybrid Speed Estimation Method

From the results presented in Figures A.3 and A.4, the following conclusions can be drawn: (i) a phase difference exists between the estimated average speed, $\hat{\omega}_d$, derived from the differential of the estimated rotor position and the actual speed; (ii) a magnitude error exists between the actual speed and the estimated speed, $\hat{\omega}_v$, derived from the voltage model. Thus, neither method is entirely suitable for speed feedback. However, the two methods can be combined to improve the estimation of the speed [7], [8]. In this section, the hybrid method in [7] is introduced as an example.

The following improved estimation of rotor speed, $\hat{\omega}_{hybrid}$, is constructed as:

$$\hat{\omega}_{hybrid} = \hat{\omega}_d \frac{1}{T_c s + 1} + \hat{\omega}_v \frac{T_c s}{T_c s + 1} \tag{A.7}$$

If $s \to 0$, then $\hat{\omega}_{hybrid} \to \hat{\omega}_d$. Thus, under steady-state operation, $\hat{\omega}_d$ plays the dominant role in the speed estimation. However, if $s \to \infty$, then $\hat{\omega}_{hybrid} \to \hat{\omega}_v$. Hence, under transient conditions, $\hat{\omega}_v$ plays the dominant role. (A.7) can be re-expressed as:

$$\hat{\omega}_{hybrid} = \hat{\omega}_d + (\hat{\omega}_v + \hat{\omega}_d)\frac{T_c s}{T_c s + 1} = \hat{\omega}_d + \Delta\omega \frac{T_c s}{T_c s + 1} = \hat{\omega}_d + \omega_{comp} \tag{A.8}$$

Thus, the hybrid speed estimation method compensates for the average speed $\hat{\omega}_d$ by adding ω_{comp}, where ω_{comp} is the output of the high-pass filtering of $\Delta\omega = \hat{\omega}_v - \hat{\omega}_d$. Based on (A.8), it is desirable that the speed compensation ω_{comp} is zero under steady-state conditions, and approaches $\Delta\omega$ under transient conditions. To achieve this, an appropriate value for the time constant T_c of the HPF in (A.8) must be selected. Since it is difficult to calculate T_c analytically, it can be determined experimentally simply by observing the resultant system performance.

Four speed estimation methods are compared by experiments and shown in Figure A.5. The speed reference is varied between 1500 and 3000 r/min every two seconds. As can be seen in Figure A.5a, when the speed feedback is derived from the differential of the estimated rotor position, the speed exhibits a significant high-frequency oscillation as well as a low frequency ripple, while when the estimated average speed is employed as feedback, see Figure A.5b, there are

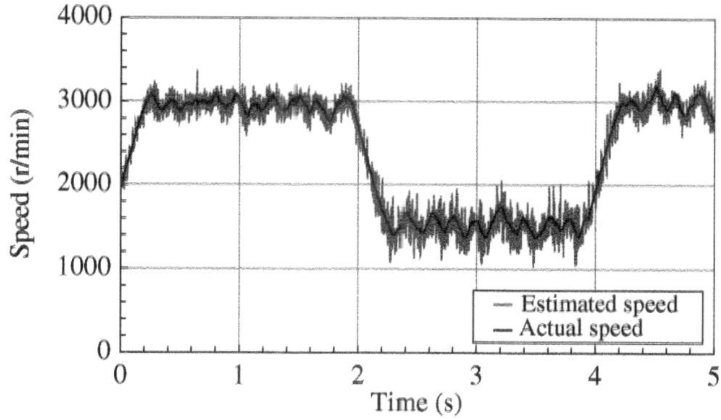

(a)

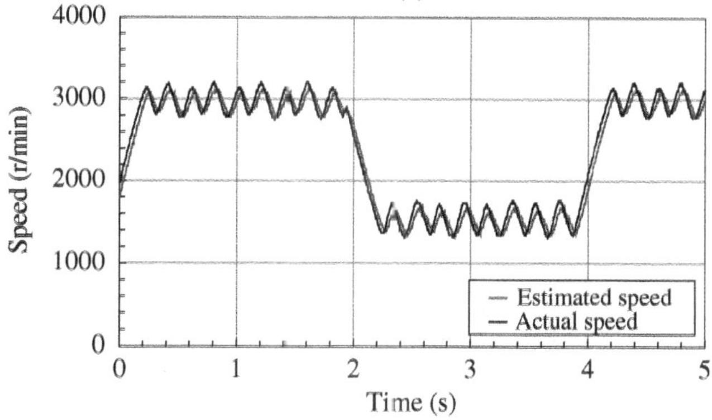

(b)

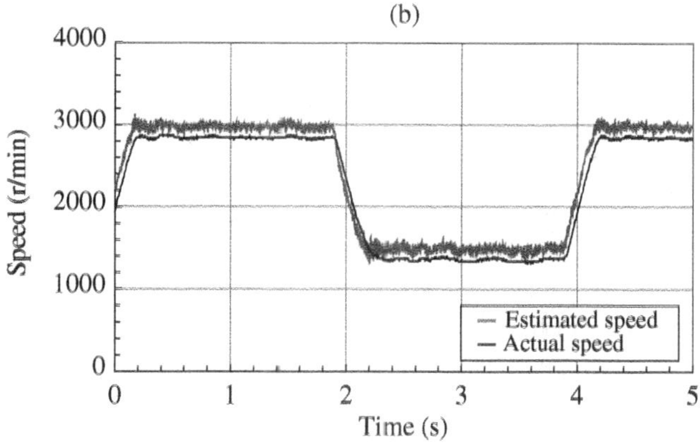

(c)

Figure A.5 Speed control performance with different speed feedback methods [7]. (a) Speed feedback derived from the differential of estimated rotor position. (b) Speed feedback using differential of estimated rotor position after filtering. (c) Speed feedback derived from machine model. (d) Speed feedback using hybrid method of speed estimation.

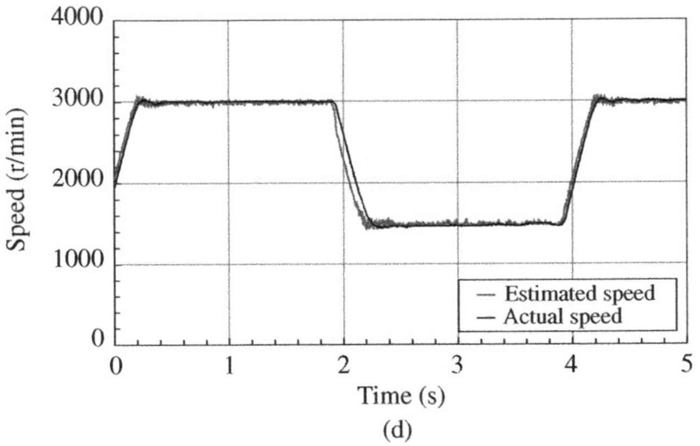

Figure A.5 (Continued)

observable phase errors. When the speed feedback is derived from the voltage model, although the dynamic response is good, a speed error as well as a small speed ripple exist during steady-state operation, as shown in Figure A.5c. However, when the hybrid method of speed estimation is employed as speed feedback, the steady-state speed error is very small, and the dynamic response is good as well, as shown in Figure A.5d.

References

1. R. Wu and G. R. Slemon, "A permanent magnet motor drive without a shaft sensor," *IEEE Trans. Ind. Appl.*, vol. 27, no. 5, pp. 1005–1011, Sep. 1991.
2. S. Ogasawara and H. Akagi, "Implementation and position control performance of a position-sensorless IPM motor drive system based on magnetic saliency," *IEEE Trans. Ind. Appl.*, vol. 34, no. 4, pp. 806–812, Jul. 1998.
3. S. Morimoto, K. Kawamoto, M. Sanada, and Y. Takeda, "Sensorless control strategy for salient-pole PMSM based on extended EMF in rotating reference frame," *IEEE Trans. Ind. Appl.*, vol. 38, no. 4, pp. 1054–1061, Jul. 2002.
4. M. J. Corley and R. D. Lorenz, "Rotor position and velocity estimation for a salient-pole permanent magnet synchronous machine at standstill and high speeds," *IEEE Trans. Ind. Appl.*, vol. 34, no. 4, pp. 784–789, Jul. 1998.
5. J.-H. Jang, S.-K. Sul, J.-I. Ha, K. Ide, and M. Sawamura, "Sensorless drive of surface-mounted permanent-magnet motor by high-frequency signal injection based on magnetic saliency," *IEEE Trans. Ind. Appl.*, vol. 39, no. 4, pp. 1031–1039, Jul. 2003.

6 H. Watanabe, S. Miyazaki, and T. Fujii, "Improved variable speed sensorless servo system by disturbance observer," *IECON '90: 16th Annu. Conf. IEEE Ind. Electron. Soc.*, pp. 40–45, vol. 1, Nov. 1990.

7 J. X. Shen, Z. Q. Zhu, and D. Howe, "Improved speed estimation in sensorless PM brushless ac drives," *IEEE Trans. Ind. Appl.*, vol. 38, no. 4, pp. 1072–1080, Jul. 2002.

8 N. Matsui, "Sensorless PM brushless dc motor drives," *IEEE Trans. Ind. Electron.*, vol. 43, no. 2, pp. 300–308, Apr. 1996.

Appendix B

Specifications of Prototype Machines and Test Rigs

This appendix describes the specifications of the test rigs and prototype machines used in this book, including BLAC and BLDC drives.

B.1 PM BLAC Drives

Most experiments of sensorless control for PM BLAC drives in this book are based on a dSPACE1006 control system. An example of the overall experimental system is shown in Figure B.1. It can be seen that the experimental platform consists of a dc power supply, a two-level voltage source inverter, a dSPACE control system, a prototype PMSM, and a load machine. For the prototype PMSM, three-phase stator currents and the dc-link voltage are measured by Hall-effect sensors. The rotor position and speed of the prototype machine are measured by an incremental encoder mounted on the shaft. The load machine can be controlled in a constant torque mode so that a desired load can be given.

Furthermore, the specifications of the prototype PMSMs that are used in this book are given in Table B.1 for SPMSMs and Table B.2 for IPMSMs.

B.2 PM BLDC Drives

The experimental platform for PM BLDC drives is shown in Figure B.2. The platform contains a digital signal processor (DSP) chip (TMS320F28335), a high-speed BLDC motor, a high-speed motor drive, and a dc power supply. Three Hall-effect

Sensorless Control of Permanent Magnet Synchronous Machine Drives, First Edition.
Zi Qiang Zhu and Xi Meng Wu.

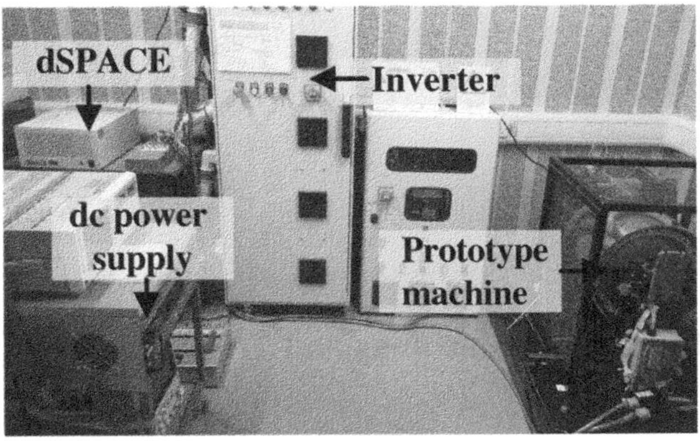

(a)

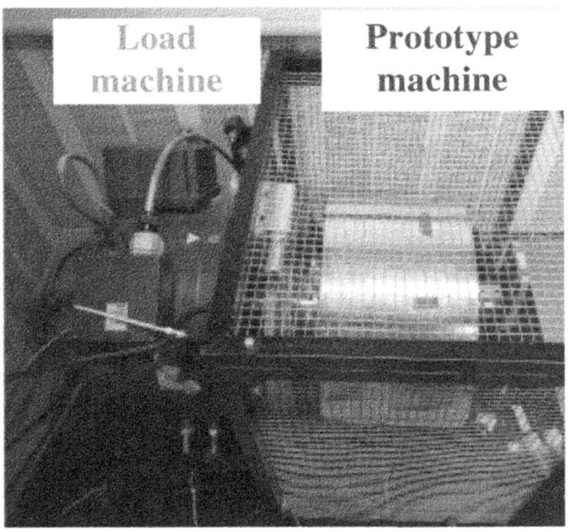

(b)

Figure B.1 Experimental platform. (a) Overall. (b) Machines.

sensors are installed on the stator of the motor to detect the rotor position. The two-phase currents, the dc-link current, and all the voltage signals are sampled for the sensorless BLDC drives.

Moreover, the specifications of the prototype BLDC machines that are used in this book are given in Table B.3.

The parameters of BLDC-I listed in Table B.3 include average values designated by a superscript star symbol *, as the motor has asymmetric parameters. Its three-phase parameters can be found in Table B.4.

Table B.1 Specifications of prototype SPMSM machines.

	SPMSM-I	SPMSM-II	SPMSM-III	SPMSM-IV	SPMSM-V	SPMSM-VI
DC-link voltage (V)	36	70	600	600	600	100
Switching frequency (kHz)	10	10	2.5	2.5	2.5	10
Sampling frequency (kHz)	10	10	2.5	2.5	2.5	10
Rated speed (r/min)	400	3000	170	170	170	3000
Rated current (A, peak)	4	2	4	4	8.2	4.3
Number of pole pairs	5	1	16	16	14	1
Phase resistance (ohm)	1.1	0.9	3.9	3.76	1.75	0.466
d-axis inductance (mH)	2.142	4.2	19.21	17	13.04	4.5
q-axis inductance (mH)	2.142	4.2	19.21	17	13.04	4.5
PM flux-linkage (Wb)	0.0734	0.093	1.03	0.9	1.1315	0.093

Table B.2 Specifications of prototype IPMSM machines.

	IPMSM-I	IPMSM-II
DC-link voltage (V)	158	36
Switching frequency (kHz)	10	10
Sampling frequency (kHz)	10	10
Rated speed (r/min)	1000	400
Rated current (A, peak)	4	10
Number of pole pairs	3	5
Phase resistance (ohm)	6	0.32
d-axis inductance (mH)	40	1.6
q-axis inductance (mH)	60	2.68
PM flux-linkage (Wb)	0.23	0.0707

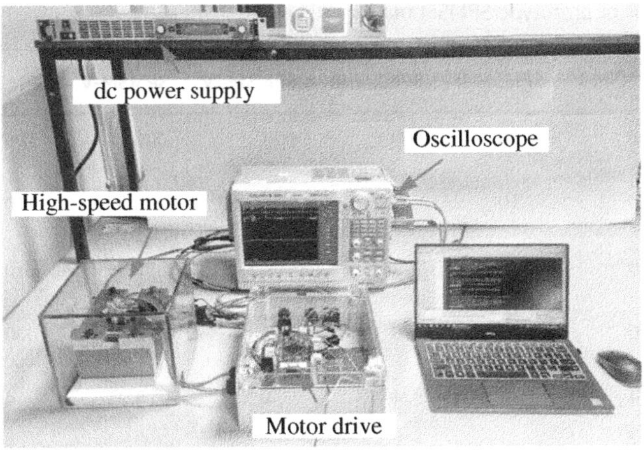

Figure B.2 BLDC drive platform.

Table B.3 Specifications of prototype PM BLDC machines.

	BLDC-I	BLDC-II	BLDC-III
DC-link voltage (V)	15	25	15
Switching frequency (kHz)	40	40	40
Sampling frequency (kHz)	40	40	40
Rated speed (r/min)	54 000	90 000	60 000
Rated current (A, peak)	14	30	30
Number of pole pairs	1	1	1
Phase resistance (Ω)	0.0397*	0.021	0.021
d-axis inductance (mH)	0.032*	0.025	0.025
q-axis inductance (mH)	0.032*	0.025	0.025
PM flux-linkage (Wb)	0.00127	0.00098	0.00098

Table B.4 Three-phase parameters of BLDC-I.

Parameter	Phase A	Phase B	Phase C
Phase resistance (Ω)	0.0381	0.0375	0.0435
Self-inductance (mH)	0.0275	0.0272	0.0286
Mutual inductance (mH)	0.04		

Index

Sensorless Control of Permanent Magnet Synchronous Machine Drives, First Edition.
Zi Qiang Zhu and Xi Meng Wu.

e

f

h

S

Books in the IEEE Press Series on Control Systems Theory and Applications

Series Editor: Maria Domenica Di Benedetto, University of l'Aquila, Italy

The series publishes monographs, edited volumes, and textbooks which are geared for control scientists and engineers, as well as those working in various areas of applied mathematics such as optimization, game theory, and operations.

1. *Autonomous Road Vehicle Path Planning and Tracking Control*
   Levent Güvenç, Bilin Aksun-Güvenç, Sheng Zhu, and Sükrü Yaren Gelbal
2. *Embedded Control for Mobile Robotic Applications*
   Leena Vachhani, Pranjal Vyas, and Arunkumar G. K.
3. *Merging Optimization and Control in Power Systems: Physical and Cyber Restrictions in Distributed Frequency Control and Beyond*
   Feng Liu, Zhaojian Wang, Changhong Zhao, and Peng Yang
4. *Model-Based Reinforcement Learning: From Data to Continuous Actions with a Python-based Toolbox*
   Milad Farsi and Jun Liu
5. *Disturbance Observer for Advanced Motion Control with MATLAB/Simulink*
   Akira Shimada
6. *Control over Communication Networks: Modeling, Analysis, and Design of Networked Control Systems and Multi-Agent Systems over Imperfect Communication Channels*
   Jianying Zheng, Liang Xu, Qinglei Hu, and Lihua Xie
7. *Advanced Control of Power Converters: Techniques and Matlab/Simulink Implementation*
   Hasan Komurcugil, Sertac Bayhan, Ramon Guzman, Mariusz Malinowski, Haitham Abu-Rub
8. *Sensorless Control of Permanent Magnet Synchronous Machine Drives*
   Zi Qiang Zhu and Xi Meng Wu

Printed and bound by CPI Group (UK) Ltd, Croydon, CR0 4YY
17/09/2026
14964309-0006